Computational and Experimental Methods in Structures – Vol. 1

BUCKLING AND POSTBUCKLING STRUCTURES

Experimental, Analytical and Numerical Studies

Computational and Experimental Methods in Structures

Series Editor: Ferri M. H. Aliabadi *(Imperial College London, UK)*

Vol. 1 Buckling and Postbuckling Structures: Experimental, Analytical and Numerical Studies
edited by B. G. Falzon and M. H. Aliabadi
(Imperial College London, UK)

Vol. 2 Advances in Multiphysics Simulation and Experimental Testing of MEMS
edited by A. Frangi, C. Cercignani (Politecnico di Milano, Italy), S. Mukherjee (Cornell University, USA) and N. Aluru (University of Illinois at Urbana Champaign, USA)

Computational and Experimental Methods in Structures – Vol. 1

BUCKLING AND POSTBUCKLING STRUCTURES

Experimental, Analytical and Numerical Studies

edited by

B G Falzon & **M H Aliabadi**

Imperial College London, UK

ICP

Imperial College Press

Published by

Imperial College Press
57 Shelton Street
Covent Garden
London WC2H 9HE

Distributed by

World Scientific Publishing Co. Pte. Ltd.
5 Toh Tuck Link, Singapore 596224
USA office: 27 Warren Street, Suite 401-402, Hackensack, NJ 07601
UK office: 57 Shelton Street, Covent Garden, London WC2H 9HE

British Library Cataloguing-in-Publication Data
A catalogue record for this book is available from the British Library.

Computational and Experimental Methods in Structures — Vol. 1
BUCKLING AND POSTBUCKLING STRUCTURES
Experimental, Analytical and Numerical Studies

ISBN-13 978-1-86094-794-0
ISBN-10 1-86094-794-8

Printed in Singapore by B & JO Enterprise

PREFACE

The use of slender or thin-walled construction in aeronautical, space, civil, maritime and offshore structures, necessitates the need to consider their stability under compressive loads. A number of case histories exist which attribute structural failure to buckling. These include the collapse of silos, bridges still under construction, oil platforms and structural failure of aircraft operating beyond their design ultimate loads. In most cases, the desire to utilise this form of construction arises from the need to minimise weight. This is the primary objective in the development of aircraft structures where a light airframe leads to reduced payload costs which, in turn, minimises environmental impact. The emergence of carbon-fibre composites, with their superior specific weight and stiffness, has greatly contributed to the development of even lighter structures. New generations of aircraft are utilising an increased level of this material in their primary structure and the possibility of developing postbuckling primary composite aerostructures paves the way for the development of very lightweight airframes. Composites also have the added advantage that they do not corrode and have excellent fatigue properties compared to metals.

This book brings together a number of established researchers in the field of structural stability and presents the state-of-the-art. An emphasis is placed on the structural stability of composite structures. Despite their obvious advantages, their anisotropic nature, coupled with a relatively weak through-thickness strength, present numerous challenges to the structural analyst. It will be shown that more sophisticated tools are required to characterise their structural response. The results of a number of experimental programmes are also presented which enhance our understanding of stability phenomena and serve as a valuable database for validating numerical and analytical models.

In the design of plate structures, it has been known for decades that a metallic plate, suitably supported along its edges, may be designed to carry load beyond initial buckling. Further demonstrations followed whereby thin-skinned panels, with stiffeners attached, were shown to be capable of sustaining load beyond initial skin-buckling. In the 1980s a number of papers were published highlighting the same postbuckling load-carrying capability in thin-skinned composite structures with stiffeners co-cured, co-bonded or secondary bonded onto the skin. Most of these experimental studies were conducted under uniaxial loading and a number of numerical studies, using the finite element method (FEM), were undertaken to predict their structural response.

Chapter 1 builds on this experimental database by presenting the results of a new experimental programme where curved stiffened composite panels, which are arguably more representative of wing or fuselage sections, are loaded in compression until failure. Torsion boxes, constructed from two of these stiffened panels, mounted on aluminium side panels, are also tested in compression, torsion and combined loading. This work formed part of a large European Union funded programme. In Chapter 2, experimental results on stiffened cylindrical shells under loading conditions similar to those used on the boxes in Chapter 1 are presented. Indeed, this work also formed a part of the same research programme and together these two chapters present a wealth of experimental results which yield valuable insight into the postbuckling response of thin-skinned composite structures and may be used reliably in the validation of analytical or computational models.

In Chapter 3, the observed phenomenon of 'mode-jumping', where highly postbuckled structures undergo secondary instabilities beyond initial buckling, is discussed and the difficulties that this poses to standard nonlinear finite element solution schemes are highlighted. A numerical methodology is proposed which combines aspects of quasi-static implicit and pseudo-transient schemes. This is shown to be robust and capable of predicting mode-jumping with good accuracy.

The central aim of Chapter 4 is to provide less conservative guidelines for the design of imperfection-sensitive composite shell structures. A similar experimental procedure to Chapter 2 is followed whereby geometric imperfections are measured on a set of composite shells which are subsequently tested in compression and combined compression-torsion loading. These measured imperfections are used to determine a manufacturing-process-specific imperfection signature for the shells.

Chapter 5 also deals with stiffened composite structures. A number of curved stiffened panels, with different lay-up, dimensions and material system to the ones presented in Chapter 1, are tested in uniaxial compression. The challenges presented to numerical analysis support the concerns raised in Chapter 3 and a nonlinear finite element solution strategy, using commercially-available finite element software, is presented. Collectively, the first five chapters attest to the interest in characterising the structural behaviour of thin-walled stiffened structures which predominate most advanced composite airframe construction.

The anisotropy associated with composite materials is hardly ever exploited and is often seen as a hindrance in the design of composite structures. This is partly due to the added analytical complexity which must be dealt with even at a preliminary design stage. In Chapter 6, buckling and postbuckling formulae are derived for anisotropic plates and shells which demonstrate the structural efficiency that may be gained by utilising this anisotropy. These expressions are particularly useful at a preliminary design stage prior to the use of numerical methods for detailed analysis and design. The use of genetic algorithms and a finite-strip buckling analysis program to optimise the buckling load of stiffened composite panels is presented in Chapter 7.

Tubes and pipelines find widespread use in engineering applications across scales ranging from millimetres to kilometres. Their susceptibility to local buckling, when subjected to bending or external pressure loads, is demonstrated through the non-linear finite element method in Chapter 8. Both metallic and composite tubes and pipelines are considered and the problem of delamination in composite tubes under external pressure is discussed at some length.

Chapter 9 presents a mixed variational finite element formulation for the modelling of imperfection-sensitive spherical caps subjected to an external pressure load. The results obtained from the numerical modelling are validated using a comprehensive experimental programme where initial geometric imperfections on thin membrane polymer caps were measured prior to loading.

The use of sandwich construction, where metallic (or, indeed, composite) face sheets are bonded onto a softer core, are being increasingly utilised in the design of advanced lightweight structures. Chapter 10 deals with the complex global and local instabilities which may arise in such structures. Instabilities in these types of structures are also investigated in Chapter 13.

In Chapter 11, boundary element method (BEM) formulations for the analysis of the buckling and postbuckling of plates and shells are explained. This provides an alternative numerical approach to the more established traditional one of using the finite element method. A special feature of the BEM presented, is its ability to model nonlinear problems and still retain its boundary only modelling philosophy.

Chapter 12 presents a progressive failure analysis of compression-loaded composite structures where a novel approach for accounting for fibre microbuckling, a highly localised instability which leads to structural failure, is presented. Highly localised instabilities in cellular materials, at the micro and meso levels, are modelled in Chapter 13 using detailed finite element analysis. This chapter shows how instabilities at the micro level propagate to create instabilities at higher levels leading to structural failure at much larger length scales.

In developing this book, we have endeavoured to ensure that enough introductory material was included in each chapter to make it accessible to a wider readership. Engineering graduates new to this field can gain an overview of our current understanding of structural stability and the strategies being adopted to model this behaviour. Researchers, academics and practicing engineers who are familiar with these issues will find it useful as a comprehensive reference text encompassing the latest developments in the field.

We are eternally grateful to all contributors who have displayed great enthusiasm for this book. Their prompt response to our numerous e-mails and phone calls over the past couple of years, the willingness of some to meet us and discuss their chapter contributions whenever they were passing through London, or indeed, when either of us were crossing their paths in other parts of the world is deeply appreciated.

B. G. Falzon
M. H. Aliabadi

CONTENTS

CHAPTER 1

EXPERIMENTAL STUDIES OF STIFFENED COMPOSITE PANELS UNDER AXIAL COMPRESSION, TORSION AND COMBINED LOADING

H. Abramovich

Technion-Israel Institute of Technology, Israel
E-mail: haim@aerodyne.technion.ac.il

Experimental results on the behavior of nine single panels and of four torsion boxes, each comprising of two stringer-stiffened cylindrical graphite-epoxy composite panels are presented, these were tested under axial compression, torsion and combined loading. The buckling and postbuckling behavior of these single panels and torsion boxes demonstrated consistent results. Prior to performing the buckling tests, the initial geometric imperfections of the panels and boxes were scanned and recorded. The tests were complemented by finite element calculations, which were performed for each panel and box. These detailed calculations have also assisted in identifying critical regions of the boxes and the boxes were reinforced accordingly to avoid premature failure. The investigation on the single panels revealed a good correlation between the predictions of the finite element codes and the experimental results. The tests indicated that the torsion carrying capacity of the boxes is laminate lay-up dependent; axial compression results were in very good agreement with previous tests performed with single identical panels; and that both the panels and the boxes have a very high postbuckling carrying capacity.

1.1 Introduction

It is well known[1] that stiffened panels can have considerable postbuckling reserve strength, enabling them to carry loads significantly in excess of their initial buckling load. If appropriately designed, their load carrying capacity will even appreciably exceed that corresponding to an equivalent weight unstiffened shell (i.e. a shell of identical radius and thicker skin and which is also more sensitive to geometrical imperfections). In these shells, initial buckling of the panel in a local mode takes place, i.e. skin buckling between stiffeners, and not in an overall mode, i.e., an Euler or wide column mode.

The design of aerospace structures places great emphasis on minimizing weight and reducing lifecycle costs. An optimum (minimum mass) design approach based on initial buckling, stress or strain, and stiffness constraints, typically yields an idealized structural configuration characterized by almost equal critical loads for local and overall buckling. This, of course, results in little postbuckling strength capacity and susceptibility to premature failure. However, the optimum design approach can be modified to produce lower weight designs for a given loading by requiring the initial local buckling to occur considerably below the design limit load and allowing for the response characteristics known to exist in postbuckled panels,[2] i.e. capability to carry loads higher than their initial buckling load. To meet the requirements of low weight, advanced lightweight laminated composite elements are increasingly being introduced into new designs of modern aerospace structures for enhancing their structural efficiency and performance. In recognition of the numerous advantages that composites offer, there is a steady growth in replacement of metallic components by composite ones in marine structures, ground transportation, robotics, sports and other fields of engineering.

Loading of single curved panels in buckling tests, which represent a non-symmetric structure, poses a tough problem, particularly when they represent a segment of a structure, e.g. fuselage, and conclusions from the tests have to be drawn for the full structure. Therefore, though being much more expensive in testing, it is more appropriate to test symmetric closed type structures, which consist of two or more identical panels, thus avoiding non-symmetric loading introduction difficulties. This approach has also been adopted in the present test program described in this chapter.

Many theoretical and experimental studies have been performed on buckling and postbuckling behavior of flat stiffened composite panels.[3-5] Recently, a wide body of description and detailed data on buckling and postbuckling tests has been compiled.[13]

Experimental and theoretical investigations on buckling and postbuckling behavior of composite-stiffened-curved panels and shells are quite scarce (all of which are concerned with aerospace applications) and are barely documented in the open literature.[6-30] Among these are the analytical and experimental studies carried out at the Aircraft Division of the Northrop Corporation, Hawthorne, California, by Agarwal;[8] at the NASA Langley Research Center by Knight and Starnes;[6] the joint programs of NASA Langley Research Center, Lockheed Engineering and Sciences Company and Boeing Commercial Airplane group, the Douglas Aircraft Company in California and the ALENIA Company in Italy which were reported by McGowan et al.[9] and by Bucci and Mercuria;[12] the studies conducted by Israel Aircraft Industries (IAI) together with the Aerospace Structures Laboratory, Technion, Israel and the recent studies performed within the framework of the POSICOSS consortium funded by the 5th 'European Union Framework Program' initiative program.[22-30]

In light of the above considerations, it has been suggested that permitting postbuckling below ultimate load of fuselage structures, i.e. alleviation of design constraints, may provide a means for meeting the objectives for the design of next generation aircraft, where the demand is a reduction of weight without prejudice to cost and structural life.

This approach has been undertaken in the present experimental study (Improved **PO**stbuckling **SI**mulation for the Design of Fiber **CO**mposite **S**tiffened Fuselage **S**tructures - POSICOSS project) as a part of an ongoing effort for the design of low-cost and lightweight airborne structures initiated by the 5th European Union Framework Program. It aimed at supporting the development of improved, fast and reliable procedures for analysis and simulation of postbuckling behavior of fiber composite stiffened panels of future generation fuselage structures and their design.

The present chapter presents the results of tests on nine laminated composite 'stringer-stiffened' curved panels under axial compression and four torsion boxes under various combinations of axial and shears loads (through torsion), the local buckling of their skins; their behavior in postbuckling under combined loading and their collapse under torsion. These tests have been conducted within the framework of the POSICOSS consortium. The tests aimed at demonstrating the safe operation of postbuckled composite cylindrical stiffened panels, as well as providing part of a database for the development of "fast tools" for the reliable design of these types of structures.

1.2 Testing of stiffened composite panels under axial compression

Within the framework of the POSICOSS effort, Israel Aircraft Industries (IAI) designed and manufactured nine Hexcel IM7 (12K)/8552(33%) graphite-epoxy 'blade-stiffened' composite panels, using a co-curing process. The nominal radius of each panel was R=938 mm and its total length L=720 mm (which included two end supports of height 30 mm, each). The stringer lay-up was $(\pm 45^0, 0^0{}_2)_{3S}$ for the blade. Each layer had a nominal thickness of 0.125 mm. Panels PSC-1, PSC-2, PSC-4, had five blade stringers with a height of 20 mm, panels PSC-3, PSC-5 and PSC-6 had five blade stringers with a height of 15 mm and panels PSC-7, PSC-8 and PSC-9 had six blade stringers with a height of 20 mm[20-21] (see Table 1.1). The guidelines for the design were based on the following requirements: the first buckling load of the skin of the panel was to coincide with the design limit load of the structure and the first failure in buckling of the stringers was to comply with the ultimate load requirements. This yielded two well defined buckling points, where the ultimate load of the panel was at least 1.5 times its first buckling load (local buckling mode, between the stringers).

Table 1.1. Geometrical and material properties: panels PSC-1 – PSC-9.

Properties	PSC-1,PSC-2, PSC-4	PSC-5,3PSC-PSC-6	PSC-7,PSC-8, PSC-9
Total panel length [mm]	720	720	720
Free panel length [mm]	660	660	660
Radius [mm]	938	938	938
Arc length [mm]	680	680	680
Num. of stringers	5	5	6
Stringer distance [mm]	136	136	113
Laminate lay-up of skin	$[0°,\pm45°,90°]_s$	$[0°,\pm45°,90°]_s$	$[0°,\pm45°,90°]_s$
Laminate lay-up of stringer	$[\pm45°,0°_2]_{3s}$	$[\pm45°,0°_2]_{3s}$	$[\pm45°,0°_2]_{3s}$
Ply thickness [mm]	0.125	0.125	0.125
Type of stringer	blade	blade	blade
Stringer height [mm]	20	15	20
Stringer feet width [mm]	60	60	60
E_{11} [N/mm^2]	147300	147300	147300
E_{22} [N/mm^2]	11800	11800	11800
G_{12} [N/mm^2]	6000	6000	6000
ν_{12}	0.3	0.3	0.3

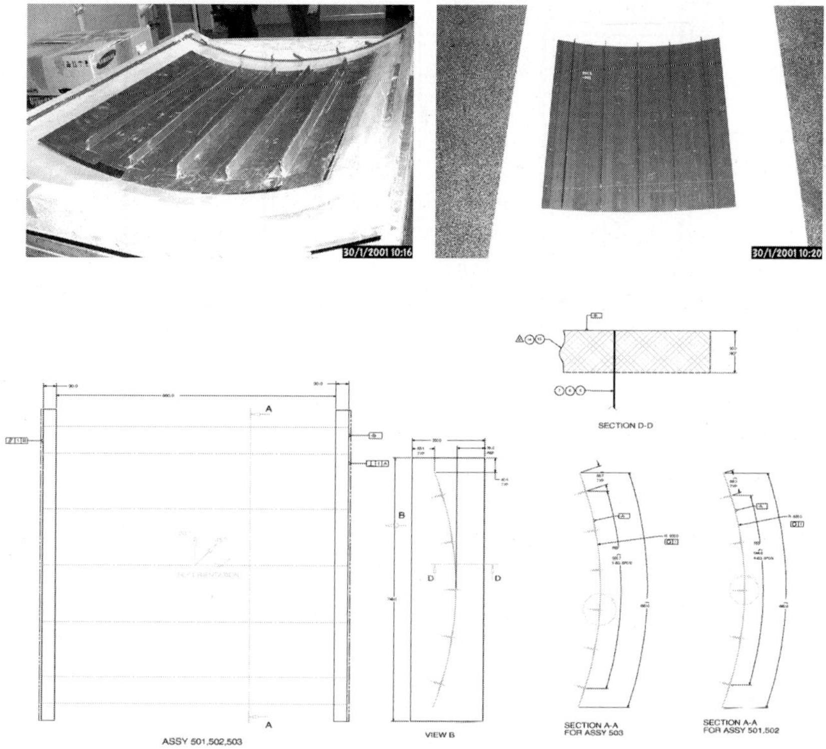

c.

Fig. 1.1. A stiffened cylindrical composite panel: a. on its tooling, b. after being removed, c. various details of the panels.

Figure 1.1 depicts the panels before and after being removed from their tooling and their geometrical dimensions. Figure 1.2 presents the distribution of strain gages and various linear variable differential transducers (LVDTs) on panel PSC-2. The relative large number of strain gages (66) bonded back-to-back were aimed to measure the load distribution across the panel skin as well as in the stringers while the LVDTs were used to measure the panel end-shortening and lateral deflections. Similar distributions of strain gages and LVDTs were used for all the tested panels.

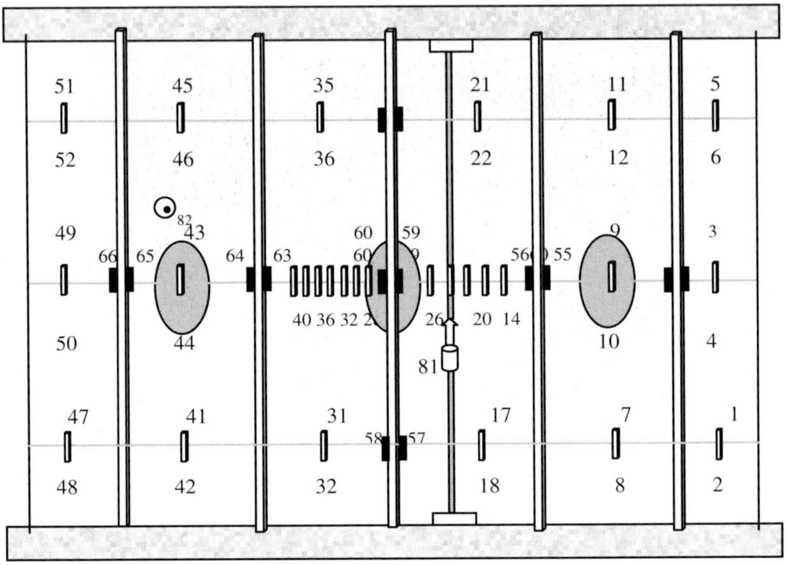

Fig. 1.2. Panel PSC-2: The locations of the strain gages (▯ = strain gage). The number below is for a strain gage in front of the panel and the number above is for a strain gage on the back of the panel, #81 LVDT axial shortening, #82 LVDT lateral displacement.

Before being tested, each panel was measured to find its initial geometric imperfections. Figure 1.3 shows the LVDT probe used for the measurement of the initial geometric imperfections. The probe was moved circumferentially at 12 predetermined heights of the panel and the data was recorded using an analog-to-digital (A/D) card and a PC. Typical raw data (panel PSC-2) is shown in Fig. 1.4a. To eliminate rigid body motions, a special code was applied yielding the reduced data used to generate the map of initial geometric imperfections for a given panel (in this case panel PSC-2: see Fig. 1.4b).

Each panel had two end pieces made of hardened polymer at its curved boundaries simulating clamped boundary conditions. The boundary conditions at the panel's straight longitudinal boundaries were assumed to be simply supported. This was achieved using two fixture plates with a V groove (see Figs. 1.5a, 1.5b). A 500kN MTS loading, machine was used to apply the axial compression forces. To visualize the lateral deflections of the panel during its axial compression loading a Moiré grid was used (see Fig. 1.6).

Fig. 1.3. The LVDT probe used for measurement of the initial geometric imperfections.

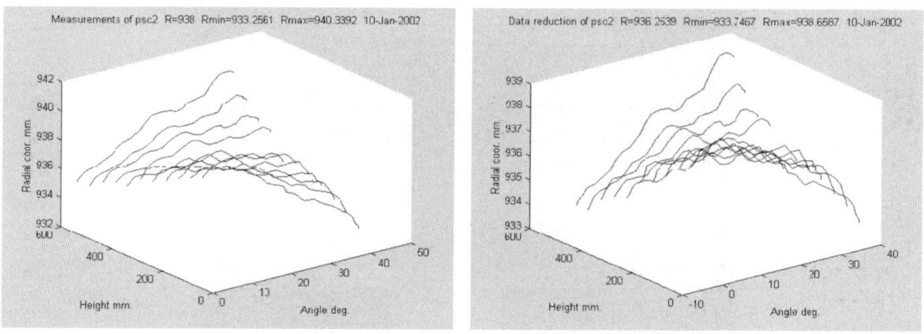

a. b.

Fig. 1.4. Initial geometric imperfections of panel PSC-2: a. raw data, b. reduced data (eliminating rigid body motions).

A typical test included the following steps:

a. Loading the panel until the first buckling of the skin. Releasing the load.
b. Loading and unloading the panel above the first buckling load several times.
c. Loading the panel until collapse.

Strain gages, end-shortening and lateral readings as a function of the axial compression loading were recorded using a PC at each of the above steps, accompanied by video recording and photographs.

a. b.

Fig. 1.5. Panel PSC-1in the testing rig, including the longitudinal fixture plates: a. front view, b. back view.

Fig. 1.6. Panel PSC-8 in the 500 kN MTS loading machine. Note the visualization of the buckling waves obtained with the Moiré method (axial loading= 270 kN).

1.3 Testing of stiffened composite panels under torsion and combined torsion and axial compression

Within the framework of the POSICOSS effort, Israel Aircraft Industries (IAI) has continued conceptual work and designed and manufactured 12 Hexcel IM7 (12K)/8552(33%) graphite-epoxy 'stringer-stiffened' composite panels, using a co-curing process. The nominal radius of each panel was R=938 mm and its total length L=720 mm (which included two end supports what a height of 30 mm, each). The nominal test length was Ln=680 mm and the panel arc-length was L_{al}=680 mm. The skin lay-up was quasi-isotropic $(0°,\pm45°,90°)_S$. Each layer had a nominal thickness of 0.125 mm. Eight of these panels were used to form 4 torsion boxes, with co-cured stringers. Each box consisted of two curved stringer-stiffened curved panels that were connected together by two flat non-stiffened aluminum side plates.

Two of the boxes comprised of panels with blade type stringers (see Fig. 1.7a), one box had short flange "J" type stringers (see Fig. 1.7b) and the 4th box had long flange "J" type stringers (see Fig. 1.7c). The dimensions and properties of the panels are given in Table 1.2.

Each panel was stiffened by five stringers, except for the panel with the long flange "J" type stringers, which had only four stringers. Since each pair of panels was tested as part of a closed box, it was assumed that by applying a torque to the box, the two panels were subjected to

Table 1.2. Data used for load calculations of torsion boxes BOX 1-4.

	Stringer type		
	Blade	Short flange "J"	Long flange "J"
Specimens	BOX1, BOX2	BOX3	BOX4
Total panel length [mm]	720	720	720
Free panel length [mm]	660	660	660
Radius [mm]	938	938	938
Arc length [mm]	680	680	680
Num. of stringers	5	5	4
Stringer distance [mm]	136	136	174
Laminate lay-up of skin	$[0°,\pm45°,90°]_s$	$[0°,\pm45°,90°]_s$	$[0°,\pm45°,90°]_s$
Laminate lay-up of stringer	$[\pm45°,0°_2]_{3s}$	$[\pm45°,0°]_{3s}$	$[\pm45°,0°_2]_{3s}$
Ply thickness [mm]	0.125	0.125	0.125
Type of stringer	blade	J	J
Stringer height [mm]	20	20.5	20.5
Stringer feet width [mm]	60	60	60
Stringer flange width [mm]	---	10	20
E_{11} [N/mm^2]	147300	147300	147300
E_{22} [N/mm^2]	11800	11800	11800
G_{12} [N/mm^2]	6000	6000	6000
ν_{12}	0.3	0.3	0.3

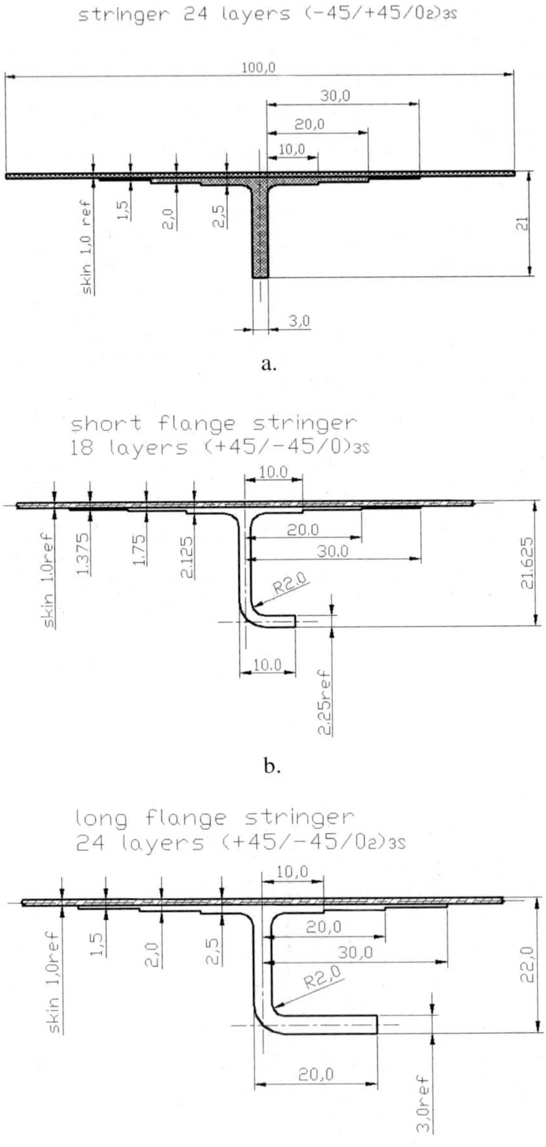

Fig. 1.7. Dimensions, geometry and lay-ups of panels: a. BOX 1 and BOX 2, b. narrow flange stringer, BOX 3, c. wide flange stringer, BOX 4.

"identical" shear. Various tests were performed on each box, before reaching its collapse under torsion, including combined axial compression and torsion loadings. For each box, the "first" buckling load was observed in the skin, for axial compression only, torsion only and combinations of axial compression and torsion, while the collapse test was performed under torsion only.

Fig. 1.8. BOX 1 in the loading equipment in the laboratory at the Politecnico di Milano.

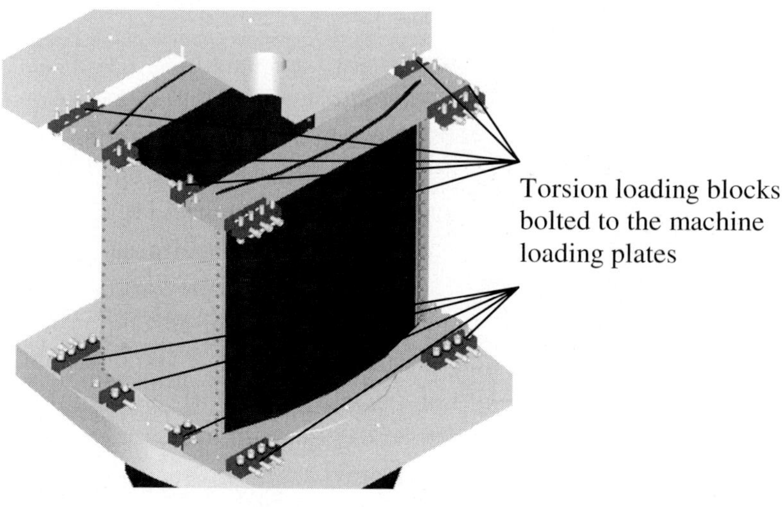

Fig. 1.9. Schematic drawing of the torsion box, indicating the way torsion is transferred into curved stringer stiffened panels.

The four torsion boxes were tested at the structures laboratory at the Politecnico di Milano[25] employing their controlled position and loading equipment (see Fig. 1.8 and Fig. 2.3).

The way the shear stresses were introduced into the torsion boxes is schematically depicted in Fig. 1.9. Four aluminum blocks were bolted to the upper and lower heavy loading plates of the loading machine. The blocks were closely tightened against the aluminum loading end-pieces of the boxes. Thus, when rotating the lower loading plate of the machine, the loading blocks bolted to it reacted against the lower box end-pieces and thus introduced torsion into the lower end of the box. At the same time the aluminum loading blocks, which were bolted to the fixed upper machine loading plate, reacted against the upper box end pieces and prevented the upper end of the box from rotating. Thus, the box was exposed to a couple that introduced a torque into it and consequently uniform shear stresses into the curved panels.

Prior to performing the buckling tests, a comprehensive FE failure detailed analysis of the torsion boxes was conducted, indicating high local stresses at the corners of each panel. Those corners were subsequently reinforced locally, outside and inside, by bonded aluminum patches (see Fig. 1.10).

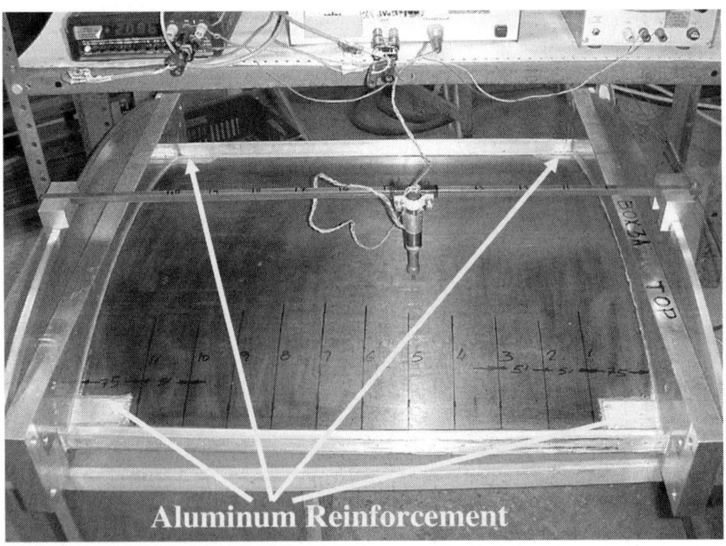

Fig. 1.10. Panel A of BOX 3 in the set up for measurement of initial geometric imperfections. Note the four aluminum reinforcements at the corners of the panel.

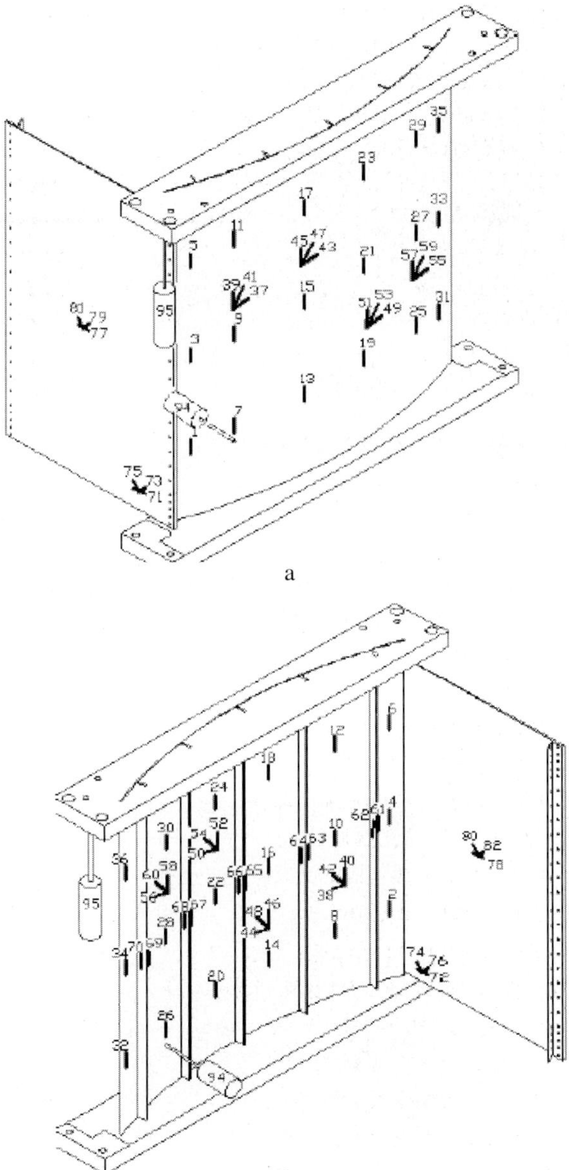

a

b

Fig. 1.11. Typical strain gages and LVDT locations: a. external view, b. internal view.

Fig. 1.12a. A panel with side aluminum plate and bonded strain gages.

Fig. 1.12b. Panel A of BOX 1 with bonded strain gages prior to being assembled.

During the tests, the values of axial displacement, rotation, axial compression load and torque of the box were measured by LVDTs and a load cell. In parallel, the strains at designated points were monitored by eighty strain gages that were bonded back-to-back, both on the skin and on the stringers of each panel. The strain gages map and the LVDT locations are shown in Fig. 1.11, while Fig. 1.12a depicts a photo of a panel with the bonded strain gages and the side aluminum panel while Fig. 1.12b shows panel A of BOX 1 with bonded strain gages prior to being assembled.

1.4 Experimental results – Axial compression

The nine PSC panels manufactured by IAI and equipped with end-pieces at their curved boundaries were transferred to the Technion and the test program was initiated and performed at the Aerospace Structures Laboratory, (ASL), Faculty of Aerospace Engineering, Technion, I.I.T., Israel.

All the specimens were equipped with strain gages to monitor the strains on the skin and the stringers of the panels. Prior to testing, the outer non-stiffened curved surface of the panels were scanned to determine their initial geometric imperfections.[18-22] Furthermore, the panels were excited and their first natural frequencies were measured and compared with calculated values[18-22] (see also Table 1.1). Then the panels were placed in a 500 kN MTS loading machine, loaded in axial compression and responses of the gages as well as of the end-shortening of the panels were monitored and recorded up to comprehensive failure of the panel. A typical axial end-shortening of a panel[20] can be seen in Fig. 1.13. As already mentioned in Section 1.2, in order to monitor the first local buckling and the development of the buckles as a function of axial loading, the Moiré technique was employed, yielding consistent results (see for example Figs.1.14a, 1.14b). Finally, each panel was tested to determine its collapse load (see for example Figs. 1.15a, 1.15b).

The collapse was characterized by a loud noise for those panels stiffened by high (20mm) stringers, while the panels with 15 mm stringers, collapsed with only a soft noise. In both cases, breakage of fibers, local delaminations, and tearing of the skin were observed. Some typical results of the strain distribution along the panel, as a function of the axial load, are given in Figs. 1.16a-c.

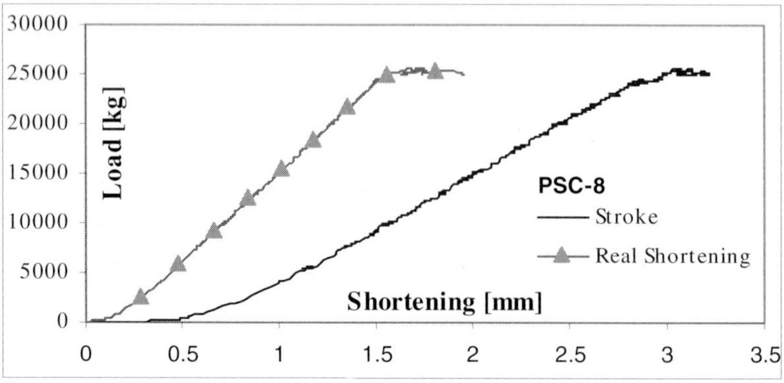

Fig. 1.13. Axial shortening vs. axial loading, panel PSC-8.

a. b.

Fig. 1.14. Panel PSC-4 at various postbuckling axial loads: a. at 198 kN ,b. at 210 kN.

a. b.

Fig. 1.15. Panel PSC-1 after collapse at 212.7 kN: a. front ,b. back.

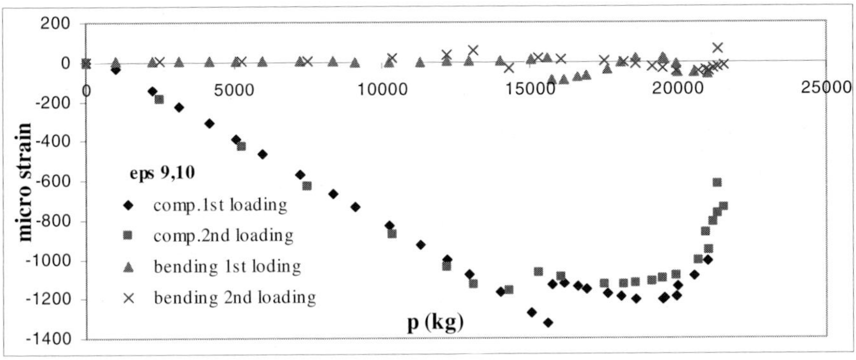

a.

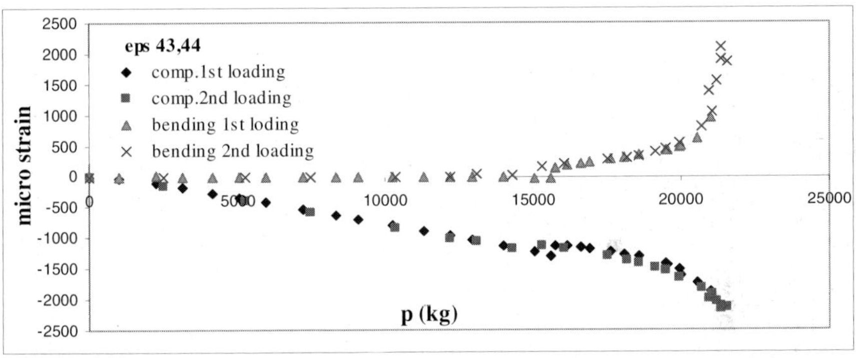

b.

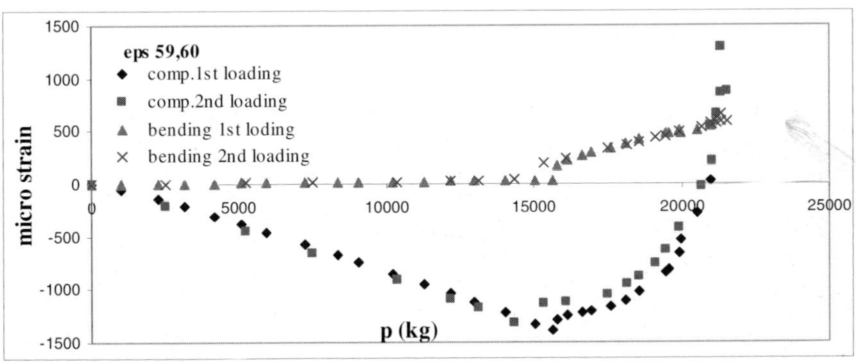

c.

Fig. 1.16. Panel PSC-2: strain distribution along the panel as a function of the axial load a. S.G.# 9,10, b. S.G.# 43,44, c. S.G.# 59,60.

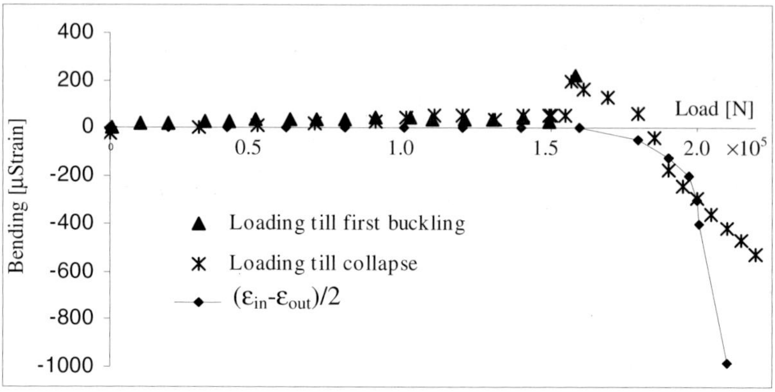
Bending [µStrain]
400
200
0
-200
-400
-600
-800
-1000
Load [N]
0.5
1.0
1.5
2.0 ×10^5
Loading till first buckling
Loading till collapse
(εin-εout)/2

a.

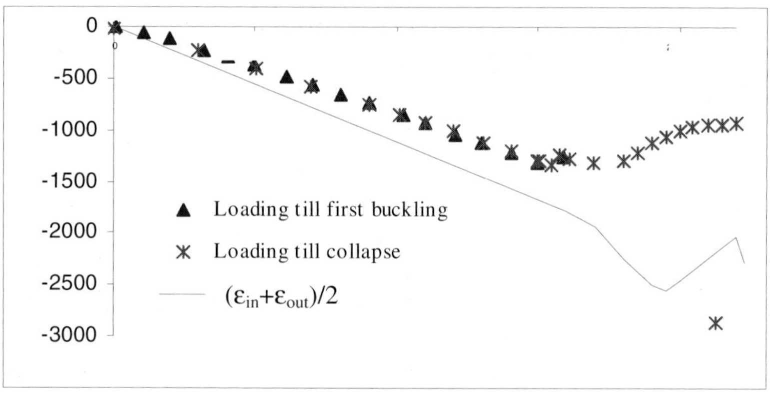
0
-500
-1000
-1500
-2000
-2500
-3000
Loading till first buckling
Loading till collapse
(εin+εout)/2

b.

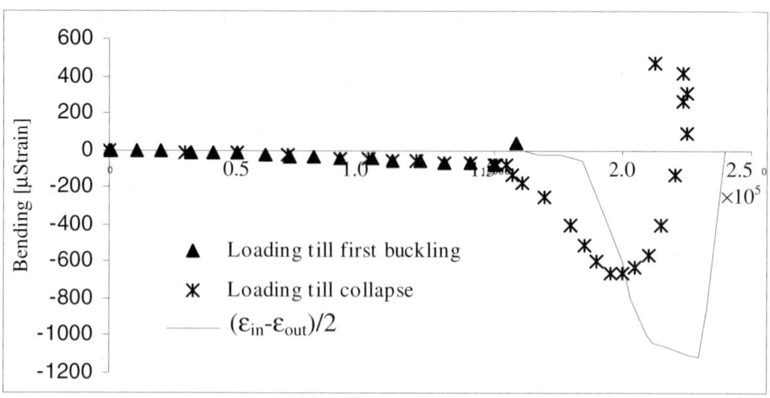
Bending [µStrain]
600
400
200
0
-200
-400
-600
-800
-1000
-1200
0.5
1.0
2.0
2.5
×10^5
Loading till first buckling
Loading till collapse
(εin-εout)/2

c.

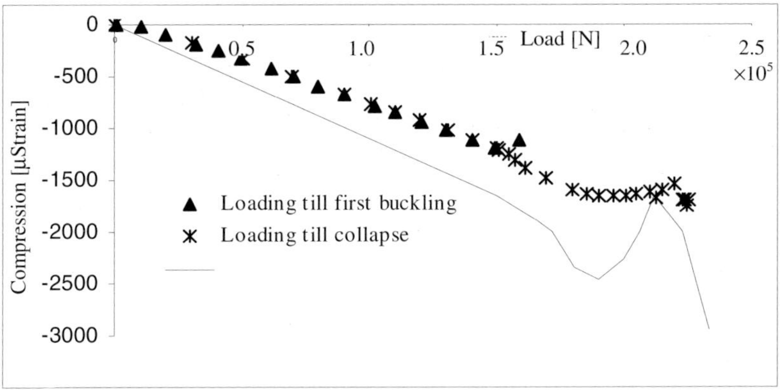

d.

Fig. 1.17. Panel PSC-2 -A comparison between the ABAQUS (Refs. 31,32 finite element predictions of the strains at typical locations on the panel surface, as compared to the experimental results: a. bending strains # 3,4, b. compression strains # 3,4, c. bending strains # 49,50, d. compression strains # 49,50.

A comparison between the ABAQUS[31,32] finite element code predictions of the strains, at typical locations on the panel surface, and the experimental results is presented in Figs. 1.17a-d. Good correlation was found though the experimental strains were lower than the predicted ones.

The test results are given in Table 1.3. Good correlation with the predicted buckling behavior of the panels was achieved. However the predicted local buckling loads underestimated the test results, while the predicted collapse loads were in good correlation with the test results.

Although the exact value of the first buckling load of a curved stiffened stringer panel has only a limited importance, it is worthwhile to investigate the possible reasons for the consistent higher loads obtained in the tests as compared with the finite element predictions.

Table 1.4 presents a comparison between the finite element predictions, using the MSC NASTRAN code and different experimental methods. When industry uses material data, they use what is called "B-Basis" properties, which are reduced properties. It turns out that after we checked experimentally the material properties of the laminated composite structure of the panels we found that the real properties where higher by a factor of 1.15 than the ones used in the finite element

Table 1.3. Test results on nine stiffened composite cylindrical panels – comparison with MSC-NASTRAN[33] predictions.

Type of panel			Prediction (MSC NASTRAN)			Experiment		
Specimen	No. of blades	Blade height [mm]	1st buckling [kN]	Collapse buckling [kN]	1st natural frequency [Hz]	1st buckling [kN]	Collapse buckling [kN]	1st natural frequency [Hz]
PSC-1	5	20	92	235	205	131.0	212.7	-
PSC-2	5	20	92	235	205	150.0	227.0	193
PSC-4	5	20	92	235	205	158.5	229.2	-
PSC-3	5	15	89	128	188	136.0	162.0	-
PSC-5	5	15	89	128	188	113.0	152.6	185
PSC-6	5	15	89	128	188	126.0	140.0	-
PSC-7	6	20	139	250	246	228.5	280.0	-
PSC-8	6	20	139	250	246	240.0	270.0	212
PSC-9	6	20	139	250	246	244.0	280.0	-

Table 1.4. Definition of the first buckling load (in kN) using different methods-comparison of finite element predictions vs. experimental results.

	Finite Element MSC NASTRAN		Experiment		
Specimen	B-base Material Properties	Real Material properties	Moiré	Bending vs. Compression Strains	Lateral Deflection
PSC-1	92.0	105.8	131.0	80.0	90.0
PSC-2	92.0	105.8	150.0	82.0	90.0
PSC-4	92.0	105.8	158.5	86.7	100.0
PSC-3	89.0	102.5	136.0	80.2	113.0
PSC-5	89.0	102.5	113.0	70.0	113.0
PSC-6	89.0	102.5	126.0	70.0	80.0
PSC-7	139.0	159.9	228.5	155.0	180.0
PSC-8	139.0	159.9	240.0	121.0	100.0
PSC-9	139.0	159.9	244.0	120.0	140

calculations. The experimental results as listed in Table 1.3 were based on the Moiré method, namely the first appearance of a buckle, which normally is accompanied by a noise. A widely used method for experimental determination of buckling loads is based on strain gages readings, and plotting of a curve of bending strains vs. axial compression strains. At the vicinity of the buckling load the bending strains tend to increase rapidly. The point when the curve changes its slope is defined as the buckling load. Another experimental method is by plotting the lateral deflection of the curved panel as a function of the axial compression load. The lateral deflection is similar to the bending strain in the previous method while the axial compression load "plays the role" of the compression strain. The buckling load is defined as the point where the deflection vs. axial compression load changes its slope, a similar criterions as used previously.

Comparing the second column in Table 1.4 with the results presented in columns 4-5 of the same table, reveals quite a new trend , namely the predicted buckling loads are consistently higher then the experimental ones, which is normally found when comparing experimental results with the numerical ones.

1.5 Experimental results – Torsion and combined loading

Tables 1.5-1.8 summarize the experimentally observed buckling loads of the four torsion boxes. One should note that the buckling loads presented in these tables are associated with the first appearance of a local single buckle.

The experimentally observed and numerically predicted interaction curves for the "first" buckling loads of BOX 1-BOX 4 (skin buckling) are presented in Figs. 1.18a-d. One should note the asymmetry of the numerical interaction curve calculated using the MSC NASTRAN finite element code that even exists for the blade type stringer stiffened panels of BOX 1 and BOX 2 (Figs. 1.18a, 1.18b). This asymmetry was barely experienced during the tests on these torsion boxes. This behavior was anticipated for the panels with non-symmetric "J" type stiffeners and indeed it was more noticeable in the tests with BOX 3, with the smaller "J" type stiffeners (Fig. 1.18c) and quite pronounced in the tests of BOX 4 with large "J" type stringers (Fig. 1.18d).

Table 1.5. Experimental Buckling Loads – BOX 1.

Test No.	Initial P const.(kN)	Initial Tconst. (kNm)	Panel A Pcr(kN) or Tcr (kNm)	Panel B Pcr(kN) or Tcr (kNm)	Remarks
1	0	0	P_{cr}=240.6	P_{cr}=268	Pure axial compression test
2	11	0		T_{cr}=12.5 ($P_{CONST.}$=10.5)	CCW* "pure" torsion test
3	11	0		T_{cr}=13.0 ($P_{CONST.}$=13.3)	CW** "pure" torsion test
4	0	9 (CCW)	P_{cr}=54.5 ($T_{CONST.}$=9)	P_{cr}=34 ($T_{CONST.}$=9)	Constant torsion, buckling under axial compression
5	0	6 (CCW)	-	P_{cr}=85 ($T_{CONST.}$=6)	Constant torsion, buckling under axial compression
6	0	3 (CCW)	-	P_{cr}=190 ($T_{CONST.}$=3)	Constant torsion, buckling under axial compression
7	85	0	-	T_{cr}=5.2 ($P_{CONST.}$=85)	Constant axial compression, buckling under torsion (CCW)
8	34	0	-	T_{cr}=7.6 ($P_{CONST.}$=34)	Constant axial compression, buckling under torsion (CCW)
9	12	0	T_{cr}= 47.4 ($P_{CONST.}$=18.5)	T_{cr}=47.4 ($P_{CONST.}$= 18.5)	Collapse under torsion (CCW) with a small axial compression
*CCW –Counter clockwise; **CW - Clockwise					

The discrepancy between the numerical results and the experimental ones may be explained as follows. The numerical results were generated using the common industrial FE code, MSC NASTRAN[33], which was shown to predict higher buckling values (in the range of 5-10%) as compared with other non-linear codes, like ABAQUS.[31,32] Another important issue is the fact that the experimental buckling values were associated with the appearance of a local single buckle, at a relatively low combination of loading, and the appearance of a fully developed buckled surface (as numerically predicted) was usually experienced at a higher load (usually 30% higher). One should notice that the graphs in Figs. 1.18a-d, present the initiation of the buckling (at the lower load).

Table 1.6. Experimental Buckling Loads – BOX 2.

Test No.	Initial P const. (kN)	Initial Tconst. (kNm)	Panel A Pcr(kN) or Tcr (kNm)	Panel B Pcr(kN) or Tcr (kNm)	Remarks
1	0	0	P_{cr}=231.0	-	Pure axial compression test
2	15	0	-	T_{cr}=14.0 ($P_{CONST.}$=16.2)	CCW "pure" torsion test
3	15	0	-	T_{cr}= 18.0 ($P_{CONST.}$=18.0)	CW "pure" torsion test
4	60	0	T_{cr} =12.0 ($P_{CONST.}$=60)	T_{cr} =12.0 ($P_{CONST.}$=60)	Constant axial compression, buckling under torsion (CCW)
5	120	0	T_{cr} =7.4 ($P_{CONST.}$=120)	T_{cr} =7.4 ($P_{CONST.}$=120)	Constant axial compression, buckling under torsion (CCW)
6	180	0	T_{cr} =4.0 ($P_{CONST.}$=180)	-	Constant axial compression, buckling under torsion (CCW)
7	180	0	T_{cr}= 48.0 ($P_{CONST.}$=182)	T_{cr}= 48.0 ($P_{CONST.}$= 182)	Collapse under torsion (CCW) with a large axial compression

For all of the four torsion boxes, the comparison of the numerical with the experimental results for "pure" axial compression and "pure" torsion was good, whereas the experimental results for combined loading were consistently lower as compared with the numerical ones. A plausible explanation might be that the repeating buckling procedure used in the present test series, to produce the experimental interaction curve, might have induced residual stresses which influenced the skin buckling load capacity yielding a lower buckling load.

In any case, the "first buckling load" of the tested torsion boxes was associated with the buckling of the skin only, which had no great significance when taking into account the high postbuckling carrying capacity of the box. Therefore, the interaction curves in Figs. 1.18a-d are presented to exhibit the "first skin buckling" behavior of a torsion box under combined torsion –axial compression loading.

Table 1.7. Experimental Buckling Loads – BOX 3.

Test No.	Initial P const. (kN)	Initial Tconst. (kNm)	Panel A Pcr(kN) or Tcr (kNm)	Panel B Pcr(kN) or Tcr (kNm)	Remarks
1	15	0	T_{cr}=12.8 ($P_{CONST.}$=13.5)	-	CCW "pure" torsion test
2	15	0	T_{cr}=15.0 ($P_{CONST.}$=16.2)	T_{cr}= 20.0 ($P_{CONST.}$=16.3)	CW "pure" torsion test
3	0	0	P_{cr}=150.0	P_{cr}=200	Pure axial compression test
4	0	4 (CCW)	P_{cr}=87.0 ($T_{CONST.}$=4)	-	Constant torsion, buckling under axial compression
5	0	7.5 (CCW)	P_{cr}=65.0 ($T_{CONST.}$=7.5)	P_{cr}=68.0 ($T_{CONST.}$=7.5)	Constant torsion, buckling under axial compression
6	0	11 (CCW)	P_{cr}=50.6 ($T_{CONST.}$=7.5)	P_{cr}=50.6 ($T_{CONST.}$=7.5)	Constant torsion, buckling under axial compression
7	16	0	T_{cr}=51.2 ($P_{CONST.}$=21.3)	T_{cr}= 51.2 ($P_{CONST.}$= 21.3)	Collapse under torsion (CCW) with a small axial compression

Typical measurements of the strain gages are given in Fig. 1.19. It presents the results experienced by three pairs of strain gages (for reference see Fig. 1.11) of panel A, BOX 1: gages 7 and 8 on the skin closer to the lower edge of the panel, gages 23 and 24 on the skin near the upper end of the panel and gages 63 and 64 located at the middle of the forth blade type stringer. The results (see Table 1.7) are presented for counter clockwise torsion only (Test 1), constant axial compression only (Test 3) and constant counter clockwise torsion combined with increasing axial compression (Test 6) in Figs 1.19a-c, respectively. For comparison, the strain gages readings of the counterpart panel B, BOX-1, are also presented in Figs. 1.19d-f under the same combinations of loading. Note that to make a correct comparison, the readings of strain gages 25-26, 17-18, 67-68 of panel B were compared with their

Table 1.8. Experimental Buckling Loads – BOX 4.

Test No.	Initial $P_{const.}$ (kN)	Initial T_{const} (kNm)	Panel A P_{cr}(kN) or T_{cr} (kNm)	Panel B P_{cr}(kN) or T_{cr} (kNm)	Remarks
1	0	0	P_{cr}=115.0	P_{cr}=115.0	Pure axial compression test
2	10	0	T_{cr}= 8.0 ($P_{CONST.}$=10.0)	T_{cr}= 8.0 ($P_{CONST.}$=10.0)	CCW "pure" torsion test
3	10	0	T_{cr}= 22.0 ($P_{CONST.}$=10.0)	T_{cr}= 16.0 ($P_{CONST.}$=10.0)	CW "pure" torsion test
4	0	4 (CCW)	P_{cr}=70.0 ($T_{CONST.}$=4)	P_{cr}=50.0 ($T_{CONST.}$=4)	Constant torsion, buckling under axial compression
5	0	2 (CCW)	P_{cr}=80.0 ($T_{CONST.}$=2)	P_{cr}=74.0 ($T_{CONST.}$=2)	Constant torsion, buckling under axial compression
6	0	6 (CCW)	P_{cr}=50.0 ($T_{CONST.}$=6)	P_{cr}=60.0 ($T_{CONST.}$=6)	Constant torsion, buckling under axial compression
7	60	0	T_{cr} =5.4 ($P_{CONST.}$=60)	T_{cr} =3.0 ($P_{CONST.}$=60)	Constant axial compression, buckling under torsion (CCW)
8	0	8 (CW)	P_{cr}=130.0 ($T_{CONST.}$=8)	P_{cr}=121.0 ($T_{CONST.}$=8)	Constant torsion, buckling under axial compression
9	0	12 (CW)	P_{cr}=126.0 ($T_{CONST.}$=12)	P_{cr}=70.0 ($T_{CONST.}$=12)	Constant torsion, buckling under axial compression
10	0	4 (CW)	P_{cr}=120.0 ($T_{CONST.}$=4)	P_{cr}=130.0 ($T_{CONST.}$=4)	Constant torsion, buckling under axial compression
11	10	0	T_{cr}= 69.0 ($P_{CONST.}$= 15)	T_{cr}= 69.0 ($P_{CONST.}$= 15)	Collapse under torsion (CCW) with a small axial compression

counterpart strain gages 7-8, 23-24 and 63-64 (see Fig. 1.11), respectively. As one can see, the readings of the strain gages on both panels, A and B of BOX 1 seem very similar, thus indicating that the introduction of the loads into both panels was equal and balanced.

Moiré fringes, as shown in Figs. 1.20a-r, were used to detect and identify the buckling and postbuckling patterns of each panel. Figures 1.20a-r present the behavior and the associated changes in deformation patterns that were observed with increase in load. Figure 1.21 presents the typical collapse mode that was observed for panel B of Box 3. Readings of the twist angle vs. the applied torsion that was observed in the collapse test of BOX 1 are presented in Fig. 1.22.

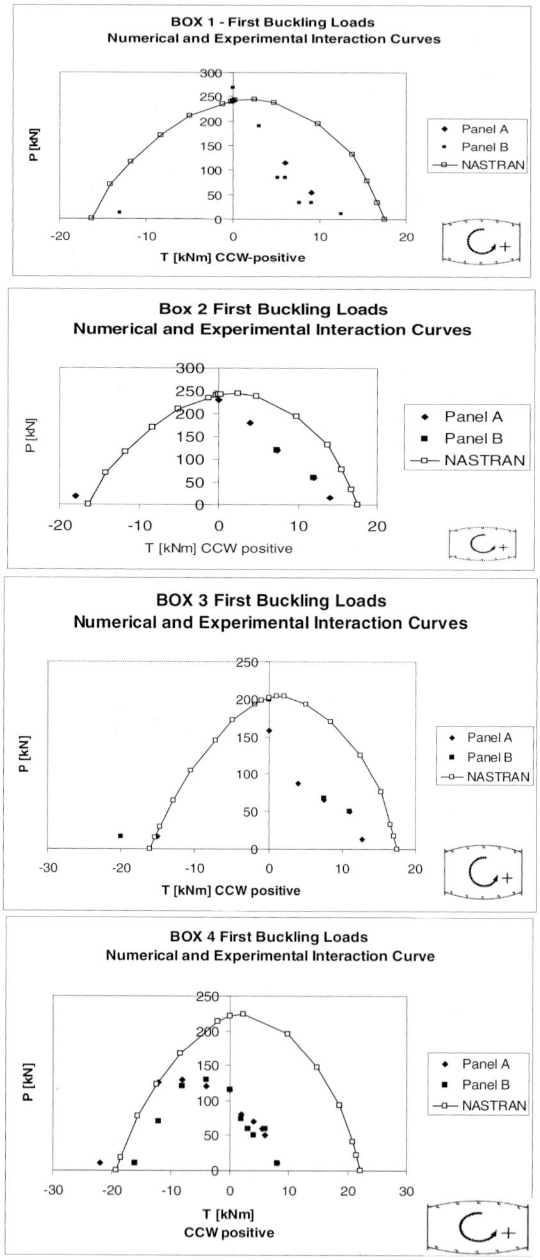

Fig. 1.18. Typical experimental and numerical interaction curves: BOX 1-4.

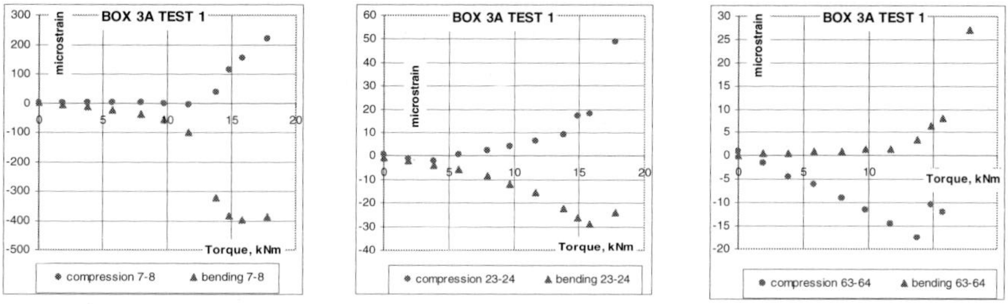
BOX 3A TEST 1
microstrain
Torque, kNm
compression 7-8
bending 7-8
BOX 3A TEST 1
microstrain
Torque, kNm
compression 23-24
bending 23-24
BOX 3A TEST 1
microstrain
Torque, kNm
compression 63-64
bending 63-64

a.

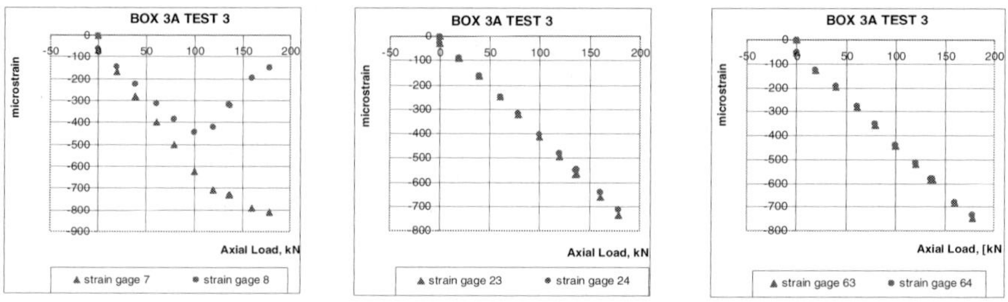
BOX 3A TEST 3
microstrain
Axial Load, kN
strain gage 7
strain gage 8
BOX 3A TEST 3
microstrain
Axial Load, kN
strain gage 23
strain gage 24
BOX 3A TEST 3
microstrain
Axial Load, [kN
strain gage 63
strain gage 64

b.

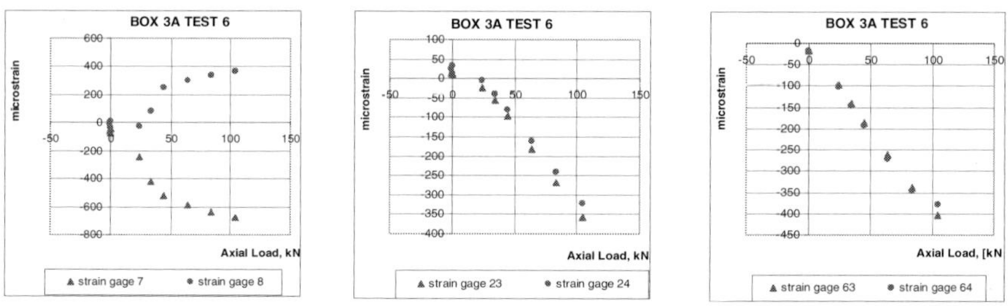
BOX 3A TEST 6
microstrain
Axial Load, kN
strain gage 7
strain gage 8
BOX 3A TEST 6
microstrain
Axial Load, kN
strain gage 23
strain gage 24
BOX 3A TEST 6
microstrain
Axial Load, [kN
strain gage 63
strain gage 64

c.

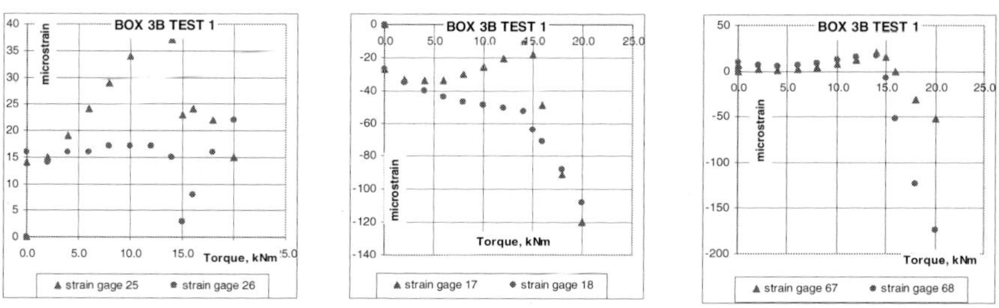
BOX 3B TEST 1
microstrain
Torque, kNm
strain gage 25
strain gage 26
BOX 3B TEST 1
microstrain
Torque, kNm
strain gage 17
strain gage 18
BOX 3B TEST 1
microstrain
Torque, kNm
strain gage 67
strain gage 68

d.

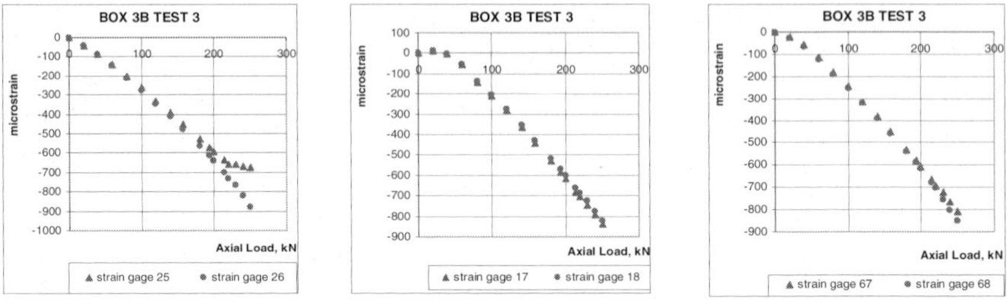

e.

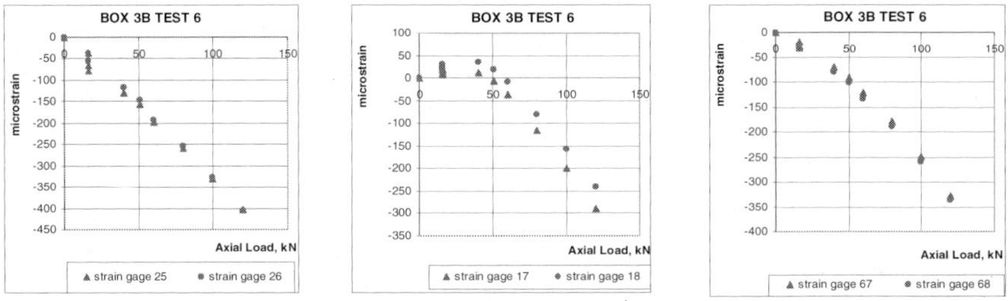

f.

Fig. 1.19. Typical strain gages measurements. a. Test 1, b. Test 3, c. Test 6– panel A BOX 3; d. Test 1, e. Test 3, f. Test 6– panel B BOX 3.

The experimental results were compared with predictions obtained by the finite element analyses performed with the ABAQUS code. Figure 1.23 shows a typical result of the torque as a function of the circumferential displacement (representing the twist angle), that was calculated by the ABAQUS code. The mode shapes at various critical points were also calculated. The numerical "first" skin buckling torque for BOX 1 and BOX 2 was evaluated at 13.2 kN·m. It was in good agreement with the experimental ones, measured at 12.5 kN·m and 14 kN·m, respectively (see Tables 1.5-1.6). Collapse for this type of box was numerically obtained at 50 kN·m, which is comparable to the experimental value of 47.4 kN·m for BOX 1(see Table 1.5).

To increase the accuracy of the numerical predictions, the refined F.E. model of Fig. 1.24 was used. This model was employed for the numerical predictions of Figs. 1.18a-d. It simulated the connections and the stiffening of the side flat aluminum panels as can be seen in Fig. 1.25.

a.

b.

c.

d.

e.

f.

AXIAL
15.7
KN
TORQ. 28.0
KNm

j.

21.2.04
AXIAL
20,0 KN
BOX3A
TEST_7
TORQ.
50.0
KNm

h.

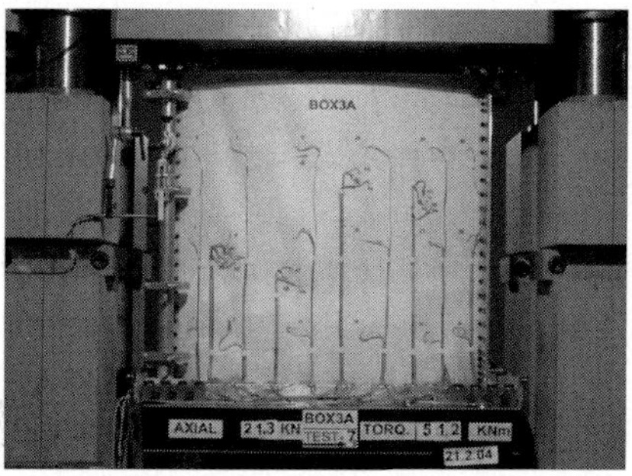
BOX3A
AXIAL
21.3 KN
BOX3A
TEST_7
TORQ.
51.2
KNm
21.2.04

i.

AXIAL
13.3
KN
BOX3B
TEST-1
TORQ.
16.0
KNm

g.

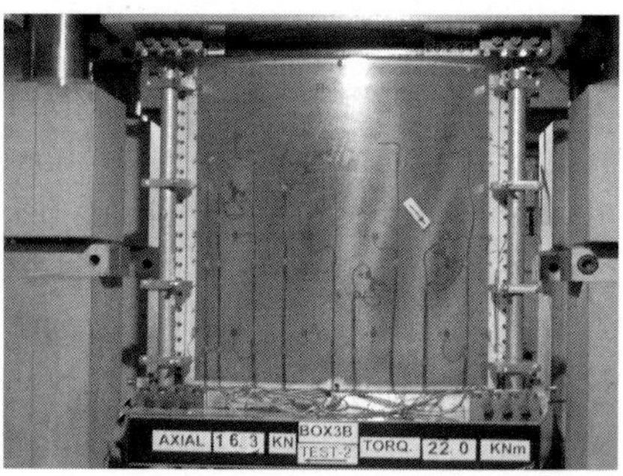
AXIAL
16.3
KN
BOX3B
TEST-2
TORQ.
22.0
KNm

k.

20.2.04
AXIAL 250.0 KN
BOX3B
TEST-3
TORQ.
0.0
KNm

l.

m.

n.

o.

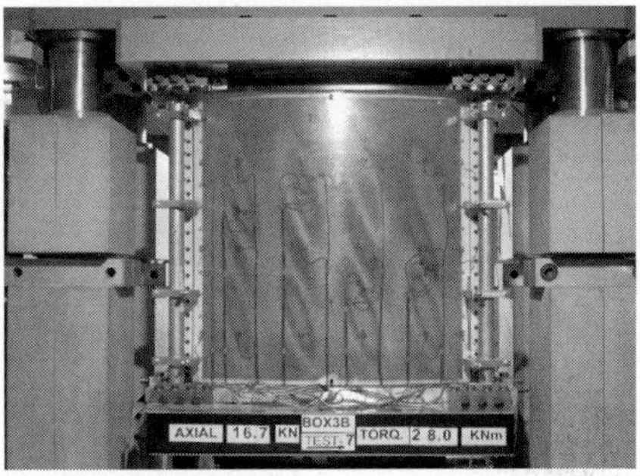

p.

q.

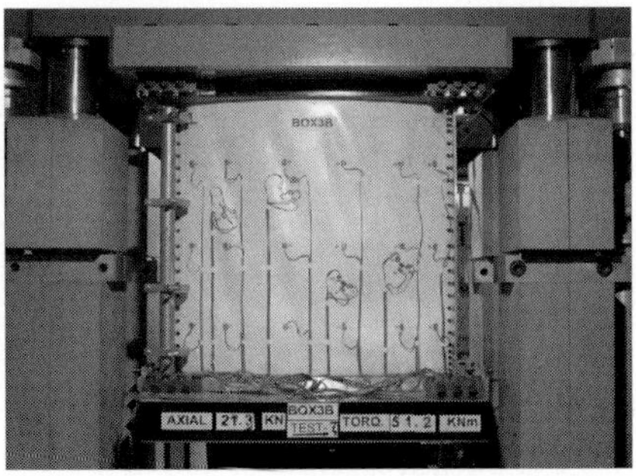

r.

Fig. 1.20. Typical postbuckling patterns of BOX 3 (a-i, Panel A, j-q Panel B):

a. Test 1 : P_{axial}= 13.0 kN ,T= 18.0 kNm,
b. Test 2 : P_{axial}= 16.3 kN ,T= 22.0 kNm,
c. Test 3 : P_{axial}= 250.0 kN ,T= 0.0 kNm,
d. Test 6 : P_{axial}= 100.0 kN ,T= 11.0 kNm,
e. Test 7 : P_{axial}= 17.0 kN ,T= 18.0 kNm,
f. Test 7 : P_{axial}= 16.7 kN ,T= 20.0 kNm,
g. Test 7 : P_{axial}= 16.7 kN ,T= 28.0 kNm,
h. Test 7 : P_{axial}= 20.0 kN ,T= 50.0 kNm,
i. Test 7 : P_{axial}= 21.3 kN ,T= 51.2 kNm,
j. Test 1 : P_{axial}= 13.3 kN ,T= 16.0 kNm,
k. Test 2 : P_{axial}= 16.3 kN ,T= 22.0 kNm,
l. Test 3 : P_{axial}= 250.0 kN ,T= 0.0 kNm,
m. Test 6 : P_{axial}= 120.0 kN ,T= 11.0 kNm,
n. Test 7 : P_{axial}= 17.0 kN ,T= 18.0 kNm,
o. Test 7 : P_{axial}= 16.7 kN ,T= 20.0 kNm,
p. Test 7 : P_{axial}= 16.7 kN ,T= 28.0 kNm,
q. Test 7 : P_{axial}= 20.0 kN ,T= 50.0 kNm,
r. Test 7 : P_{axial}= 21.3 kN ,T= 51.2 kNm.

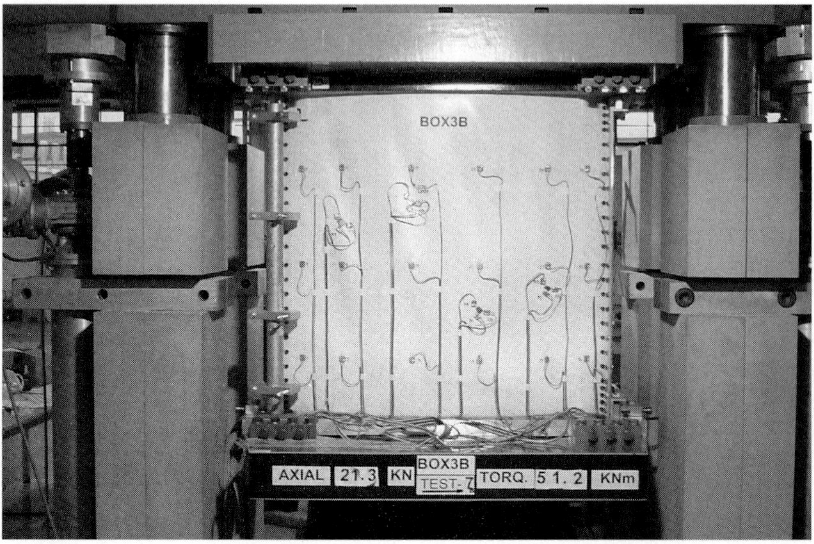

Fig. 1.21. Typical collapse pattern –panel B of BOX 3.

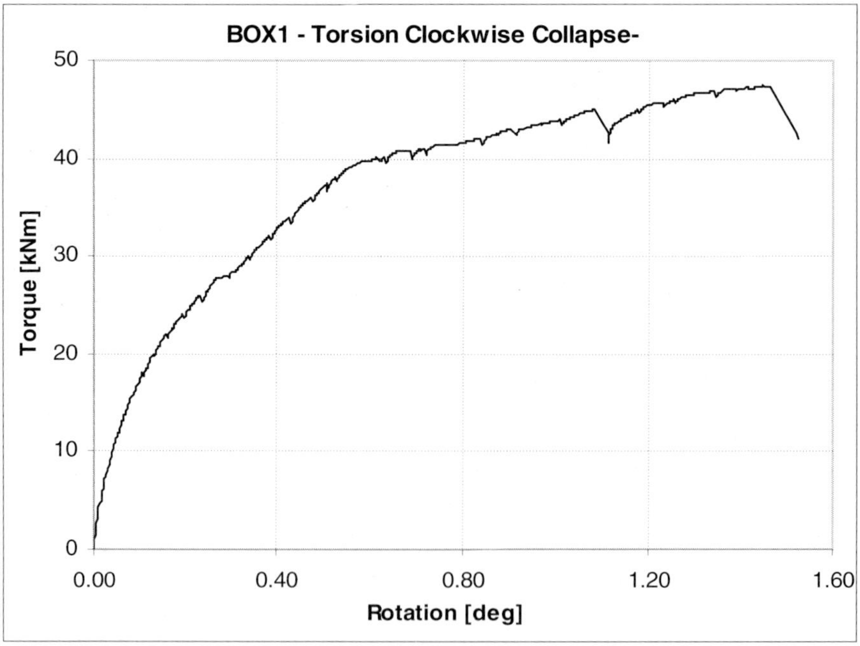

Fig. 1.22. Torque vs. rotation angle – BOX 1.

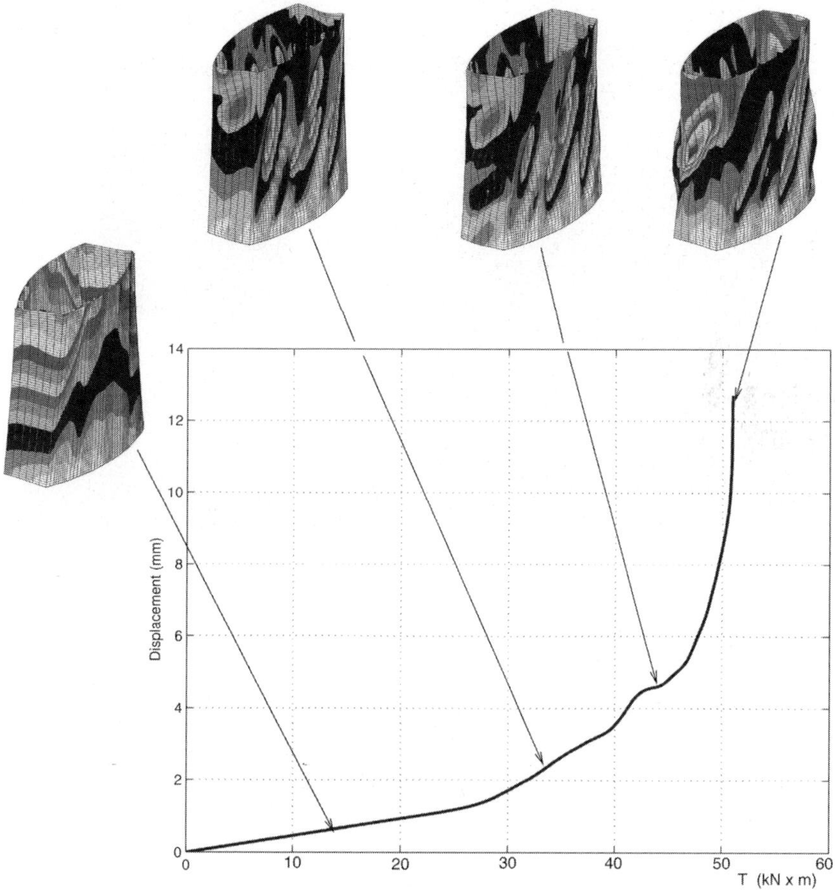

Fig. 1.23. ABAQUS results: torque vs. circumferential displacement and associated mode shapes at various critical points – BOX 1 and BOX 2.

To simulate the connections and the stiffening of the flat side aluminum panels (see Fig. 1.25), four rows of nodes (three rows of elements) at the bottom part of the composite panels were clamped, whereas four rows of nodes at the upper part of the panel (three rows of elements) were connected together to form a rigid body to a reference node at the center of the box (see Fig. 1.24). The axial force and the torsion moment were applied at this reference node. The side flat

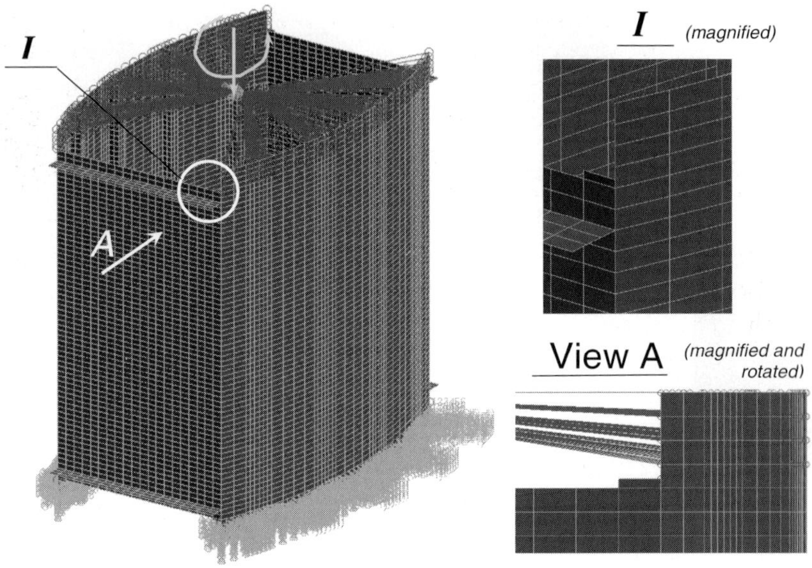

Fig. 1.24. Refined finite element model.

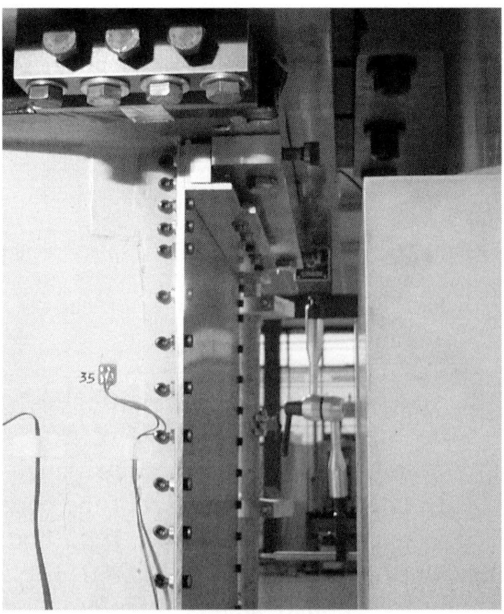

Fig. 1.25. Details of the stiffening and the connections of the side flat aluminum plate.

aluminum panels were 3.3 mm thick and were connected with the composite panels 7 mm beneath the rigid body, so that no direct axial force was introduced into them during loading (see detail I and view A of Fig. 1.24). All of the elements of the model were "CQUAD4" shell type elements. For BOX 1 and BOX 2 11096 elements were used, while for BOX 3 and BOX 4 11816 elements and 10872 elements were employed, respectively. In general, the FE predictions yielded by this model, that were applied in the present test program, were found to be in good correlation with the experimental observed results for all of the loading combinations.

1.6 Conclusions

- The behavior of nine curved stringer-stiffened cylindrical graphite-epoxy composite panels, subjected to axial loading was investigated, both experimentally and numerically, in the postbuckling region up to collapse.
- Deep postbuckling behavior of four torsion boxes, each comprising of two curved 'stringer-stiffened' cylindrical graphite-epoxy composite panels that have been subjected to torsion, axial loading and their combinations, was investigated experimentally and numerically.
- Consistent results of buckling and postbuckling behavior of both the single panels and the torsion boxes was demonstrated.
- The correct comparison between experimental results and numerical predictions should include the real material properties of the laminated composite structure of the panels.
- The first buckling load of a reinforced curved laminated composite panel should be calculated using various methods (bending strain vs. compressive strain, end shortening or lateral deflection vs. axial compression load).
- Detailed finite elements calculations have assisted in identifying critical regions of the boxes. The boxes were reinforced accordingly to avoid their premature failure.
- The tests indicated that: the torsion carrying capacity is laminate lay-up dependent ; axial compression results were in very good agreement with previous tests performed with single identical panels; and the boxes have a very high postbuckling load carrying capacity. Comparisons of the experimentally

experienced "first" skin buckling and collapse torques with those predicted numerically by finite element analyses were found to be in good agreement for "pure" torsion and axial compression whereas the experimental results for combined loading were consistently lower as compared with the numerical ones. A plausible explanation might be that the repeating buckling procedure used in the present test series, to produce the experimental interaction curve, might have induced residual stresses which influenced the skin buckling load capacity yielding a lower buckling load.

- "First" experimentally observed skin buckling is a very local nature. Hence comparisons with analysis might be not conclusive. Since we are dealing with "deep" postbuckling behavior and consequently a very high margin of reserve of load carrying capacity it is suggested to compare the analysis with the fully developed experimentally observed "first" skin buckling to enable fair and conclusive conclusions.
- To obtain reliable experimental results in the deep postbuckling region, the longitudinal straight free boundaries of the curved panel were simply supported using specially designed fixture plates with a V type groove.
- The measured initial geometric imperfections have little influence on the first buckling load of a laminated composite stringer stiffened curved panel. However, those initial imperfections might influence the collapse loads of these panels.
- In general, good correlation was found between the finite element predictions and the experimental results.
- It was shown that to obtain the correct collapse load, one should employ non-linear finite analysis codes, like ABAQUS explicit rather than the MSC NASTRAN program.

1.7 Acknowledgements

This work was partly supported by the European Commission, Competitive and Sustainable Growth Program, Contract No. G4RD-CT-1999-00103, project POSICOSS (http://www.posicoss.de). The information in this paper is provided as is and no guarantee or warranty is given that the information is fit for any particular purpose. The user thereof uses the information at its sole risk and liability.

The author would like to acknowledge the valuable conversations with Prof. T. Weller, the outstanding experimental work performed by Mr. A. Grunwald and his exceptional assistance in setting the tests and dedicated assistance in performing them, the assistance of Mrs. R. Yaffe in the data reduction and monitoring of the test results, the numerical/finite element studies conducted by Dr. P. Pevsner, and Mrs. L. Edery-Azulay for editing the present manuscript, all from the Aerospace Structures Laboratory, Faculty of Aerospace Engineering, Technion, Haifa, Israel and the collaboration with Prof. C. Bisagni and Mr. P. Cordisco, from Politechnico di Milano.

1.8 References

1. J. W. Hutchinson and W. T. Koiter, *App. Mech. Rev.*, 23 : 1353 (1970).
2. M. Lilico, R. Butler, G. W. Hunt, A. Watson, D. Kennedy, and F.W. Williams, *AIAA J* , 40(5) : 996 (2002).
3. Y. Frostig, G. Siton, A. Segal, I. Sheinman and T. Weller, *AIAA J. of Aircraft*, 28(7) : 471 (1991).
4. A. Segal, G. Siton and T. Weller, Proc., *ICCM and ECCM,6th Int. Con. on Comops. Mat. and 2nd Eur. Con. on Compos. Mat.*, F.L. Matthews, N.C.R. Bushnell, J.H. Hodgkinson and J. Morton, eds., Imperial College, London, p.5.69 (1987).
5. J. H. Jr. Starnes, N. F. Knight and M. Rouse, Proc., *AIAA/ ASME/ ASCE/ AHS/ 23rd Struct., Struct. Dyn. and Mat. Con.*, New Orleans, La., p. 464 (1982).
6. N. F. Jr. Knight and J. H. Jr. Starnes, *AIAA J.*, 26(3) : 344 (1988).
7. L. H. Sobel and B.L. Agarwal, *J. of Compos. and Struct.*, 6(3) : 193 (1976).
8. B.L. Agarwal, *Exp. Mec.*, 22 : 231 (1982).
9. D.M. McGowan, R.D. Young,, G.D. Swancon, and W.A. Waters, *5th NASA/DoD Adv. Compos. Tec. Con.*, Seattle, Wash., Paper No. A94-33140 (1994).
10. P. Vestergen and L Knutsson, *Proc. of the 11th Con. of the Int. Council of the Aeronautical Sciences (ICAS)*, J. Singer and R. Staufenbiel, eds., Lisbon, Portugal, p.217 (1978).
11. G. Romeo, *AIAA J.*, 24(11) : 1823 (1986).
12. A. Bucci and U. Mercuria, *Utiliz. of Adv. Compos. in Military Aircraft*, AGARD-785, San Diego, Calif., p.12.1 (1996).
13. J. Singer, J. Arbocz and T. Weller, John Wiley and Sons, Inc., New York (2002).
14. M. M Lei and S. Cheng, S, *J. of App. Mech., Tran. of the ASME*, p. 791 (1969).
15. R. Jr. Johnson, NASA CR-3026 (1978).
16. R.C. Tennyson, D.B. Muggeridge, K.H. Chan and N.S. Khot, TR-72-102, Air Force Flight Dyn. Lab., Wright-Patterson Air Force Base (1972).
17. M. F. Card, NASA TN D-3522 (1966).

18. H. Abramovich, A. Grunwald, P. Pevsner, T. Weller, A. David, G. Ghilai, A. Green, and N. Pekker, *ICCES'04 Con., Theme: Compos. – Exp. and Ana.*, Portugal (2004).
19. H. Abramovich, T. Weller and C. Bisagni, 46th AIAA/ASME/ASCE/AHS/ASC Struct., Struct. Dyn. and Mat. Con., Texas, USA (2005).
20. H. Abramovich, A. Grunwald and T. Weller, TAE Report No. 990 (2003).
21. H. Abramovich, P. Pevsner and T. Weller, TAE Report No. 991 (2003).
22. C. Bisagni and P. Cordisco, *Compos. Struct.*, 60 : 391 (2003).
23. E. Gal, R. Levy, H. Abramovich and P. Pevsner, *Compos. Struct.*, 73(2) : 179 (2006).
24. R. Zimmermann and R. Rolfes, *Compos. Struct.*, 73(2) : 171 (2006).
25. C. Bisagni and P. Cordisco, *Compos. Struct.*, 73(2) : 138 (2006).
26. H. Abramovich, P. Pevsner, T. Weller, N. Pecker and G. Ghilai, *Int. Con. on Buckling and Postbuckling Behavior of Compos. Laminated Shell Struct.*, Eilat, Israel (2004).
27. R. Zimmermann, H. Klein and A. Kling, *Compos. Struct.,* 73(2) : 150 (2006).
28. Abramovich, H., Pevsner, P., Weller ,T. and Bisagni, C., Int. Con. on Buckling and Postbuckling Behavior of Compos. Laminated Shell Struct., Eilat, Israel (2004).
29. R.S. Thomson and M.L. Scott, Int. Con. on Buckling and Postbuckling Behavior of Compos. Laminated Shell Struct. Eilat, Israel (2004).
30. B.G. Falzon and M. Cerini, *Compos. Struct.*, 73(2) : 186(2006).
31. Hibbit, Karlsson and Sorensen. ABAQUS/ Standard User's Manual. Vol. 1, Version 6.1 (2000).
32. Hibbit, Karlsson and Sorensen. ABAQUS/Explicit User's Manual. Vol. 1, Version 6.1 (2000).
33. The MacNeal-Schwendler Corporation. MSC/PATRAN MSC/NASTRAN Preference Guide (2000).

CHAPTER 2

BUCKLING AND POSTBUCKLING TESTS ON STIFFENED COMPOSITE PANELS AND SHELLS

Chiara Bisagni

Department of Aerospace Engineering, Politecnico di Milano
Via La Masa 34, 20156 Milano, Italy
E-mail: chiara.bisagni@polimi.it

First, the test equipment used for various types of buckling experiments in the Department of Aerospace Engineering at the Politecnico di Milano is presented. It can apply axial and torsion loading, separately and in combination, using displacement control. Then, the results of an experimental investigation on stiffened shells and curved panels, made from graphite-epoxy, are presented. The experimental data acquired during the first non-destructive buckling tests and during the final destructive failure tests demonstrate clearly the strength capacity of these structures to work in the postbuckling range allowing further weight savings. The results show that the structures are able to sustain load in the postbuckling range without any damage. On the negative side, the collapse is sudden and destructive. The measured data are also useful for the development and validation of analytical and numerical high-fidelity methods. These validated analysis tools can provide design criteria that are less conservative than the existing ones.

2.1 Introduction

A great effort is underway in the aircraft industry to improve the effectiveness of current aircraft structures. Advanced composite materials are extremely important due to their high strength-to-weight and stiffness-to-weight ratios and are more and more in use. Indeed, the lower weight of advanced composite structures, compared to conventional metallic structures, leads to a substantial weight reduction.

However, there is the possibility for further weight savings with composite structures by allowing structures to operate in the postbuckling range. Unfortunately, prediction of the buckling and postbuckling strength is difficult for these structures, and their damage and collapse prediction is even more complex.

Recent advances at NASA, on the development of aircraft structures and launch vehicles, have indicated that the existing monographs on structural stability need to be updated and expanded.[1-2] For example, the original NASA monograph,[3] that reports recommendations for the design of buckling-resistant cylindrical structures, provides reliable but often overly conservative means of designing shells by using simple, linear analytical models and an empirical correction factor, the "knock-down factor". It contains practically no design information for lightweight high-performance composite shells. The interest in updating the monographs is mainly driven by significant advances in computer technology and computational analysis tools as well as in the experimental methods and techniques.

The improved computational analysis tools have made it possible to introduce more sophisticated analytical and numerical models for the non-linear structural response, allowing the investigation of complex geometries, loading conditions, boundary conditions as well as introducing initial geometric imperfections. The non-linear buckling behaviour of several structures has been numerically studied in recent years. Numerical simulation has proven capable of performing structural optimization of composite panels under buckling, postbuckling and strength constrains,[4-7] and can be used to establish buckling behaviour trends and to perform sensitivity studies on a wide range of parameters to define design recommendations.[8-10]

In any case, the analytical and numerical models need to be validated with test results, before they can be used with full confidence. At the same time, the advancements in experimental methods and techniques provide more carefully controlled experiments and high-fidelity test results.[11] For example, technology is now available to measure accurately the initial geometric imperfections of composite test specimens,[12-16] that can be introduced in the numerical models, to understand the effect of the initial geometric imperfection shape and magnitude.

Experimental data available for composite unstiffened cylindrical shells[17-19] and for composite stiffened cylindrical shells[20-21] are still scarce, especially in the postbuckling range, all the way to failure. There are only few experimental data for composite stiffened curved panels under axial compression and even more scarce are those under combined loading.[22-26] These structures are the most common ones in the aeronautical and aerospace industries, but unfortunately to test these structures under combined loading is very complicated. On the other hand, further weight savings, required in future aerostructural designs, can be reached with composite stiffened structures by moving the operational regime into the postbuckling range, as for metallic structures. For this reason, the postbuckling response must be clearly identified and understood as well as the collapse load and the collapse mode, through experiments.

A selective testing approach, in conjunction with numerical simulation, is particularly important when considering the costs of conducting experiments and the costs of test specimens such as those made of fiber-reinforced composite materials. Consequently, selective experiments can be identified and accurately conducted to establish, together with high-fidelity analysis methods, credible design recommendations and design criteria less conservative than current ones.

Test equipment for performing buckling tests has been developed in the Department of Aerospace Engineering at the Politecnico di Milano, and different structures have been tested in past years to investigate the effect of axial compression and torsion applied separately and in combination.

The results obtained on stiffened shells and curved panels in graphite-epoxy are presented. In particular, the experimental data acquired during the first non-destructive buckling tests and during the final destructive collapse tests are shown in terms of graphs of axial compression load versus displacement or torque versus rotation and of postbuckling deformation evolution.

2.2 Test specimens

Various composite structures were tested in the Department of Aerospace Engineering at the Politecnico di Milano. In particular, stiffened composite shells and panels are described herein. These stiffened cylindrical shells were fabricated by Agusta/Westland.[27-28] Different configurations were designed to be able to operate in the postbuckling range. Consequently, the number of stiffeners and their lay-up were optimized so as to guarantee local skin buckling and a large postbuckling capability before final collapse.

The results of two configurations are presented, with the geometry reported in Table 2.1. They present an internal diameter and a length equal to 700 mm. The actual length of the shell was limited to the central part of the height and was equal to 540 mm. This was due to the necessity of including two tabs at the top and at the bottom of the shells, in order to permit the fixing into the test equipment.

The shells were made from graphite-epoxy. The material was laid up on a mandrel and cured in an autoclave. The ply material properties are reported in Table 2.2, and each ply had a nominal thickness of 0.33 mm.

Table 2.1. Shells characteristics.

		First shell configuration	Second shell configuration
Total shell length [mm]		700	700
Free shell length [mm]		540	540
Shell diameter [mm]		700	700
Stiffeners:	Number	8	8
	L length [mm]	700	700
	L width [mm]	25 x 32	25 x 32
Lay-up:	Skin	[45°/-45°]	[45°/-45°]
	Skin (reinforcements)	[45°/-45°/0°/45°/-45°]	[45°/-45°/0°/45°/-45°]
	Stiffeners	$[0°/90°]_{3S}$	$[0°/90°]_{2S}$
Ring:	Height [mm]	/	40
	Lay-up	/	$[0°/90°]_{2S}$

Table 2.2. Material properties of the shells ply.

	Ply
Young's modulus E_{11} [N/mm^2]	57765
Young's modulus E_{22} [N/mm^2]	53686
Shear modulus G_{12} [N/mm^2]	3065
Poisson's ratio ν_{12}	0.048
Density ρ [kg/mm^3]	1510
Ply thickness [mm]	0.33

The first configuration (Fig. 2.1) was designed to offer a ratio between the collapse load and the first buckling load, under axial compression, of approximately 3. It was characterized by eight L-shaped stiffeners, equally oriented and equally spaced in the circumferential direction. The blade of each stiffener was 25 mm long, while the flange, attached to the skin of the shell, was 32 mm wide. The stiffeners present a rounded corner, with an average radius of 7 mm, due to the manufacturing process. Three reinforcement layers, 40 mm wide and 700 mm long were added on the outer side of the skin corresponding to each stiffener location.

The second configuration (Fig. 2.1) was designed to work under torsion offering a ratio between the collapse torque and the first buckling torque of approximately 3. As the presence of a central reinforcing ring, co-cured on the external surface at the centre of the cylinder, has been demonstrated to be the best method to increase postbuckling range under torsion without influencing buckling torque,[21,29] the second configuration consisted of a ring-stiffened cylindrical shell with longitudinal stiffeners. The reinforcing ring in the central part of the cylinder was 40 mm high, while the stiffeners were equal in number and dimensions to those of the first configuration. Both the stiffeners and the reinforcing ring consisted of eight plies. Even in this case, the stiffeners present a rounded corner and three reinforcement layers were added on the outer side of the skin, corresponding to the stiffener locations.

Fig. 2.1. Stiffened cylindrical shells: first shell configuration (left) and second shell configuration (right).

In both configurations, the stiffeners were bonded to the inner side of the skin and then nine rivets were added to each stiffener for safety to avoid debonding. Also, the reinforcing ring was bonded to the outer side of the skin and then riveted onto the stiffeners.

The two stiffened curved panels were designed and manufactured by Israel Aircraft Industries (IAI)[25] and were first introduced in Section 1.3 as 'BOX 2'. The overall dimensions are re-stated in Table 2.3. These two panels had a nominal radius of 938 mm, an arc-length of 680 mm, and a total length of 720 mm, which included two end supports with a depth of 30 mm each. There were five blade stiffeners in each panel. The blade of each stiffener was 20 mm long, while the flange was 60 mm wide (Fig. 1.7). The panels were made from Hexcel IM7 graphite-epoxy, using a single co-curing process. The ply material properties are reported in Table 2.4, and each ply had a nominal thickness of 0.125 mm.

To test the panels under combined axial compression and shear, they were assembled to form a torsion box[30-31] where the two curved stiffened panels were connected together by two flat non-stiffened side aluminium plates (Fig. 2.2).

Table 2.3. Panels characteristics.

		Curved panel
Total panel length [mm]		720
Free panel length [mm]		660
Radius [mm]		938
Arc length [mm]		680
Stiffeners:	Number	5
	L length [mm]	720
	L width [mm]	20 x 60
Lay-up:	Skin	[0°/45°/-45°/90°]s
	Stiffeners	[45°/-45/0$_2$]$_{3S}$

Table 2.4. Material properties of the panels ply.

	Ply
Young's modulus E_{11} [N/mm^2]	147300
Young's modulus E_{22} [N/mm^2]	11800
Shear modulus G_{12} [N/mm^2]	6000
Poisson's ratio ν_{12}	0.3
Density ρ [kg/mm^3]	1580
Ply thickness [mm]	0.125

Fig. 2.2. Stiffened curved panel with side aluminium plate, and final box consisting of two stiffened curved panels and two side aluminium plates.

2.3 Test equipment

The test equipment, shown in Fig. 2.3, enables buckling tests under axial compression, torsion and combined axial compression and torsion, using displacement control.[9, 15, 19, 27-28, 30-31]

To apply axial compression, the loading platform is pushed by a hydraulic ram against four ball-screw supports placed at the four corners of the platform. At the beginning, the load given by the ram is completely supported by the four screws, which distribute the real applied load on the structure during the test. Indeed, the screws motion is computer-controlled, producing exactly the desired displacement to the loading platform, using four stepping motors through four reduction gears. Thus, the load level, which is transferred smoothly to the structure, depends only on the platform displacement and on the structure elastic response, and does not substantially depend on the load magnitude due to the hydraulic ram acting on the platform.

To apply torsion, the rotation is applied to the bottom of the structure by a torsion lever. The lever motion is computer-controlled, producing the desired displacement of a screw by a stepping motor through a reduction gear, as in the case of axial compression.

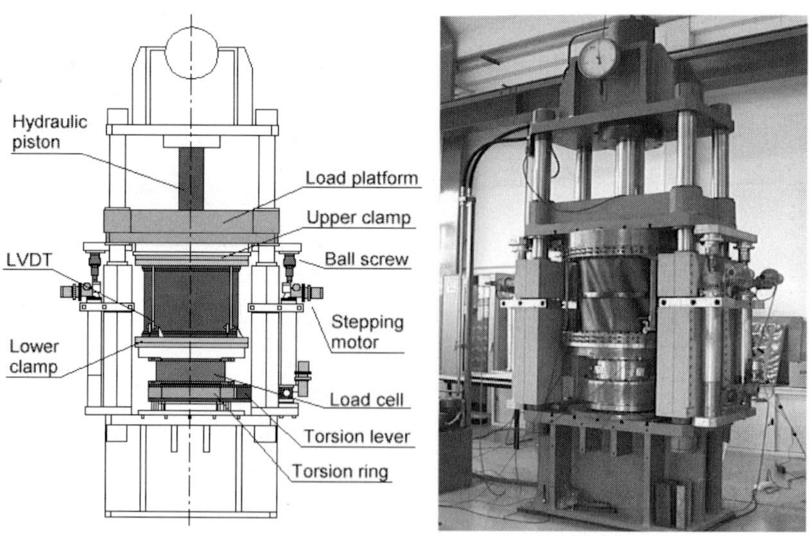

Fig. 2.3. Equipment for the buckling tests.

All the five stepping motors can be computer-controlled separately or simultaneously. In this way it is possible, on one hand, to apply axial compression and torsion in any sequence, in order to perform combined tests, and, on the other hand, to constrain the two ends of the structures to remain parallel during the tests.

Three LVDT transducers give directly the axial displacement of the structure at three equally spaced points, measuring the distance between the inner surface of the upper clamp and of the lower clamp. Three other LVDT transducers measure the tangential displacement of the structure bottom with respect to the top, at the same three points used during axial compression. The load cell, situated under the lower clamp, allows the measurement of both the compression load at three points and the torque.

During the tests of the cylindrical shells, the inner surface was scanned by means of an optical system consisting of five laser displacement sensors (Fig. 2.4). In particular, the specimens' inner surfaces were measured in terms of initial geometric imperfections, prebuckling shape and progressive change of postbuckling deformations. The equipment was placed inside the shell during the tests, avoiding any contact with the structure and consequently without influencing the buckling behaviour. The laser displacement sensors were fixed on a slide, capable of rotating and translating vertically. The combination of the two movements allowed the measuring of the whole inner surface of the shells. These movements were generated by two stepping motors and the vertical position of the slide was determined by an incremental encoder. The laser displacement sensors were placed at a distance of 40 mm from the shell inner surface, and guarantee a measurement range of ± 10 mm with a resolution of 15 μm. In this way, it was possible to measure the geometrical imperfections (some tens of micrometers), the prebuckling shape and the progressive change of postbuckling deformations (about 10-20 mm). The measurements were taken over a uniform grid with steps of 10 mm in the axial direction and 10 mm in the circumferential direction. The time required to scan the whole inner surface took only 4 minutes.

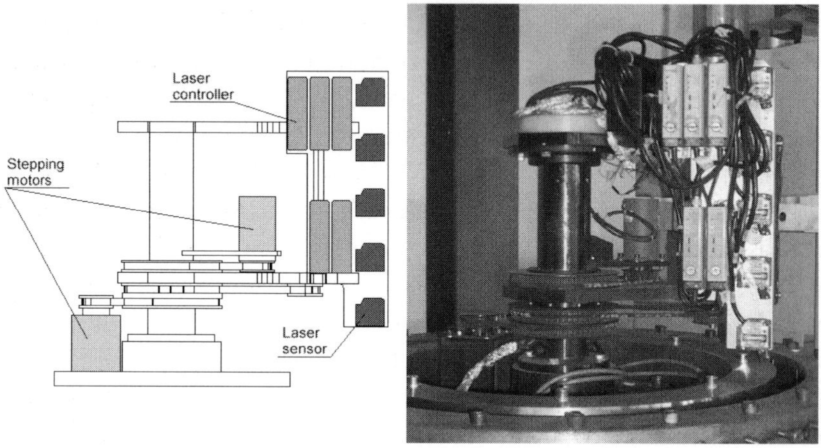

Fig. 2.4. Equipment for the measurement of the imperfections and of the postbuckling deformations of the cylindrical shells.

2.4 Test procedures and measurements

Each structure was first tested under pure axial compression and under pure torsion, separately. Both axial displacement and angular rotation were applied by steps, imposed via software. After each step, the effective axial displacement and the rotation angle were measured by the LVDT transducers, while the compression load and the torsion moment were measured by the loading cell. The graph of axial load versus displacement as well as the graph of torsion versus rotation were recorded in real-time.

Each structure was then tested under combined axial and torsion loading before the final collapse tests. The procedure was different during the combined tests on the cylindrical shells, and on the boxes of curved panels. Also some different test measurements were taken for the shells and the curved panels.

To perform combined tests on the cylindrical shells, three different procedures were investigated. The first procedure involved twisting the shell in one direction (clockwise or counter-clockwise) to a pre-set torsion torque level and then loading axially until buckling. The second one involved axially loading the shell to a pre-set axial load level and then twisting it into one or the opposite direction until buckling. The third procedure consisted of applying fixed steps of axial load and torsion torque with different ratios of axial load and torsion torque levels. During the three procedures, the axial and tangential displacements were recorded using the LVDT transducers throughout the whole test, as well as the axial compression load and the torsion torque using the load cell. All the shells were loaded into the postbuckling field, and then unloaded, always using the displacement control. It was noted that the sequence of application of axial compression and torsion in the combined tests influence neither the buckling load nor the buckling deformation.

Finally, the collapse test was executed under axial compression on the first configuration, and under torsion on the second configuration. The graphs of axial compression versus displacement and of torque versus rotation were recorded in real-time. During the torsion test, in order to perform a completely displacement-controlled test, it was decided to constrain the axial displacement of the upper loading platform and to apply a controlled rotation to the lower one. Since the axial shortening of the shells was constrained, a considerable axial reaction force develops in the tests.

The inner surface of the cylindrical shells was measured 15-20 times during each test using the lasers system. The time required to measure a complete shell surface was limited to 4 minutes. The collapse tests were also digitally recorded by a pair of high resolution cameras, to see the postbuckling deformation evolution and to obtain the failure mode.

To perform combined tests on the stiffened curved panels, the box was loaded under axial compression to a pre-determined value, and then clockwise torque or counter-clockwise torque was applied until local buckling was reached. Even in these cases, the panels were loaded into the postbuckling field, and then the box was unloaded, always using displacement control. The axial and tangential displacements were recorded using LVDT transducers throughout the whole test, as well as the axial compression load and the torsion torque using the load cell.

Then the collapse test was executed. It was a collapse test under combined load. Indeed the axial load was kept constant and equal to 180 kN, and the counter-clockwise torque was applied. In this way, the buckling load was reached, and the test was continued until the final collapse.

The effective values of displacement, rotation, axial load and torsion moment were measured by the LVDTs and the load cell. The graphs of torque versus rotation were recorded in real-time, during the tests.

On each panel, 80 strain gauges were bonded back-to-back, both on the skin and on the stiffeners, to be able to measure the panels' strains and to detect the evolution of buckling deformation. Moiré fringes were placed on the two stiffened panels, during the tests, so to highlight the postbuckling shape. The collapse test of the curved panels was also digitally recorded by the pair of high resolution cameras.

2.5 Results on shells

Each of the stiffened shells was at first tested under pure axial compression and pure torsion, separately, until local buckling. Then the combined tests were performed. Tables 2.5 and 2.6 summarize the experimentally observed buckling loads in different tests for the first shell configuration and the second shell configuration, respectively.

Then the cylindrical shells were tested until collapse. The first shell was tested until collapse under axial compression. Fig. 2.5 reports the evolution of the postbuckling deformation together with the graph of axial compression versus displacement. The dot on the graph highlights when the photograph was taken using the camera system. The corresponding envelope of the internal surface of the shell, measured by the five laser sensors, are reported in Fig. 2.6.

The graph of axial compression versus displacement presents a constant slope in its first part, even after the first local buckling occurs at 92.7 kN. The reason is that the largest part of the load was supported by the stiffeners, which continue working even after the buckling of the skin. This was due to the numbers of the plies and to their orientation. Indeed, while the skin was composed of only two [45°/-45°] plies, the stiffeners were made by superimposing twelve plies $[0°/90°]_{3S}$.

Table 2.5. Buckling loads of the first shell configuration.

Type of test	Axial compression load [kN]	Torque [kNm]
Pure axial compression test	92.7	0
Combined test	66.0	0.7
Combined test	38.1	1.4
Combined test	15.1	2.1
Pure torsion test	0	2.8

Table 2.6. Buckling loads of the second shell configuration.

Type of test	Axial compression load [kN]	Torque [kNm]
Pure axial compression test	172.5	0
Combined test	154.0	3.6
Combined test	119.9	7.2
Combined test	75.8	10.8
Pure torsion test	0	15.2

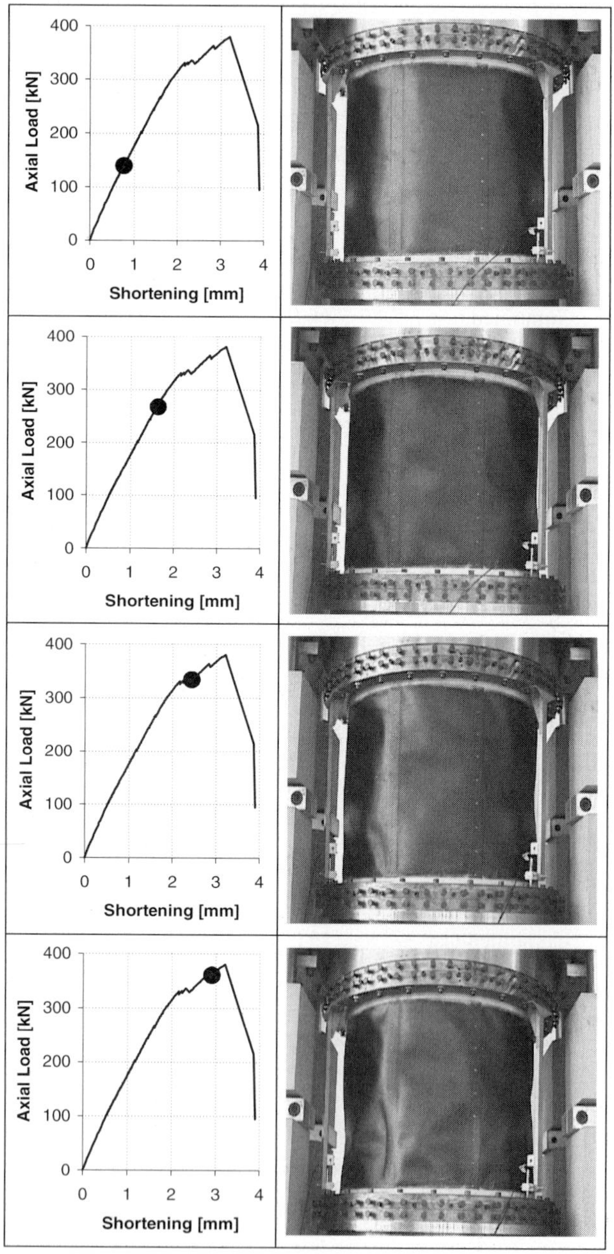

Fig. 2.5. Evolution of the first shell configuration: axial load = 152 kN; axial load = 267 kN; axial load = 338 kN; axial load = 364 kN.

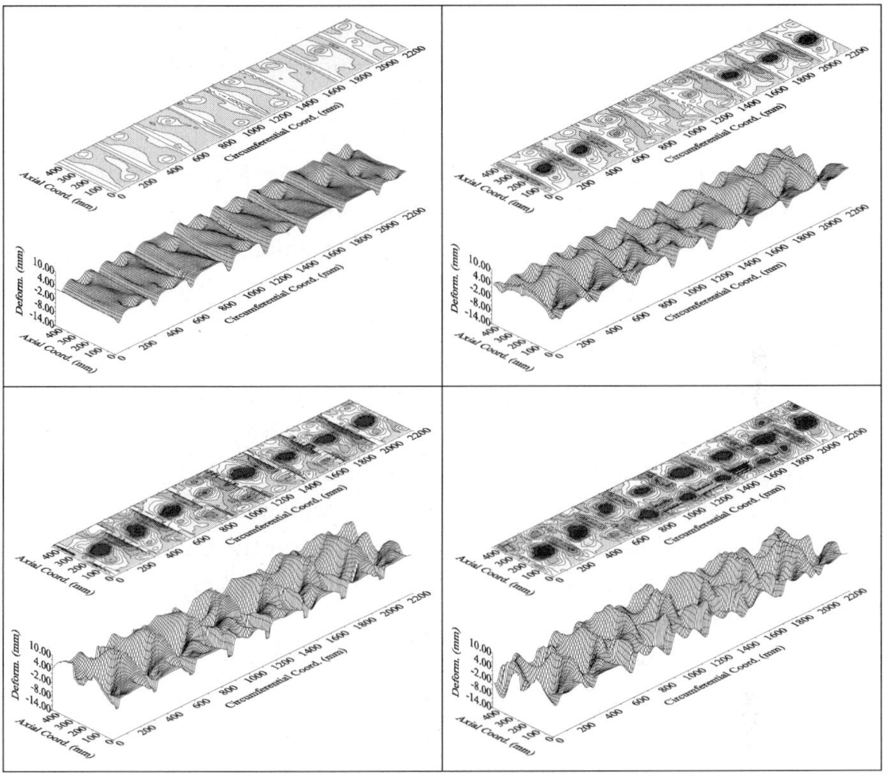

Fig. 2.6. Evolution of the first shell configuration measured by the laser sensors: axial load = 152 kN; axial load = 267 kN; axial load = 338 kN; axial load = 364 kN.

At about 152 kN, the out-of-plane deformation had already spread all over the skin and was characterized by a double series of half-waves in five sectors and by a single bigger half-wave in the remaining three sectors. Increasing the axial displacement, a third series of half-waves appeared, and at 267 kN the waves in five sectors evolved, in the center of the specimen, into a single big one. At this load, the mean displacements normal to the surface reached 8 mm both inward and outward.

In the postbuckling range, the change in load distribution and the large increase in load carried by the stiffeners caused the buckling of these stiffeners. At about 330 kN, a progressive reduction in the stiffness associated with drops in load started, probably due to the buckling of the stiffeners, causing a reduction in the slope of the graph. With a further increase in axial displacement, at 364 kN, the buckling of the stiffeners became more evident and the displacements normal to the surface increased, reaching 12.7 mm inward and 9.3 mm outward.

No failure mechanisms were visible in the postbuckling range. Only mild noises, maybe connected with limited delamination, were heard just before the collapse, that occurred at 388 kN suddenly and very audibly. Collapse was caused by the simultaneous breakage of all the stiffeners. The failure mechanisms start from the stiffeners in the central area of the cylinder height, far from the clamping tabs. It caused the release of energy in the failure area and consequently extensive fracture both in the stiffeners and in the adjacent skin (Fig. 2.7). No damage or failure occurred at the stiffener-skin interfaces. The rivets did not break and no fracture started from the rivet holes. After the collapse, the specimen was not able to sustain further load. The ratio between the collapse load and the first buckling load was equal to about 4.2, showing a substantial postbuckling capability.

Fig. 2.7. Collapse of the first shell configuration.

The second configuration was tested until collapse under torsion. Fig. 2.8. reports the evolution of the shell postbuckling shape together with the graph of torque versus rotation. The dot on the graph highlights when the photograph was taken from the camera system. The corresponding envelope of the internal surface of the shell measured by the five laser sensors are reported in Fig. 2.9.

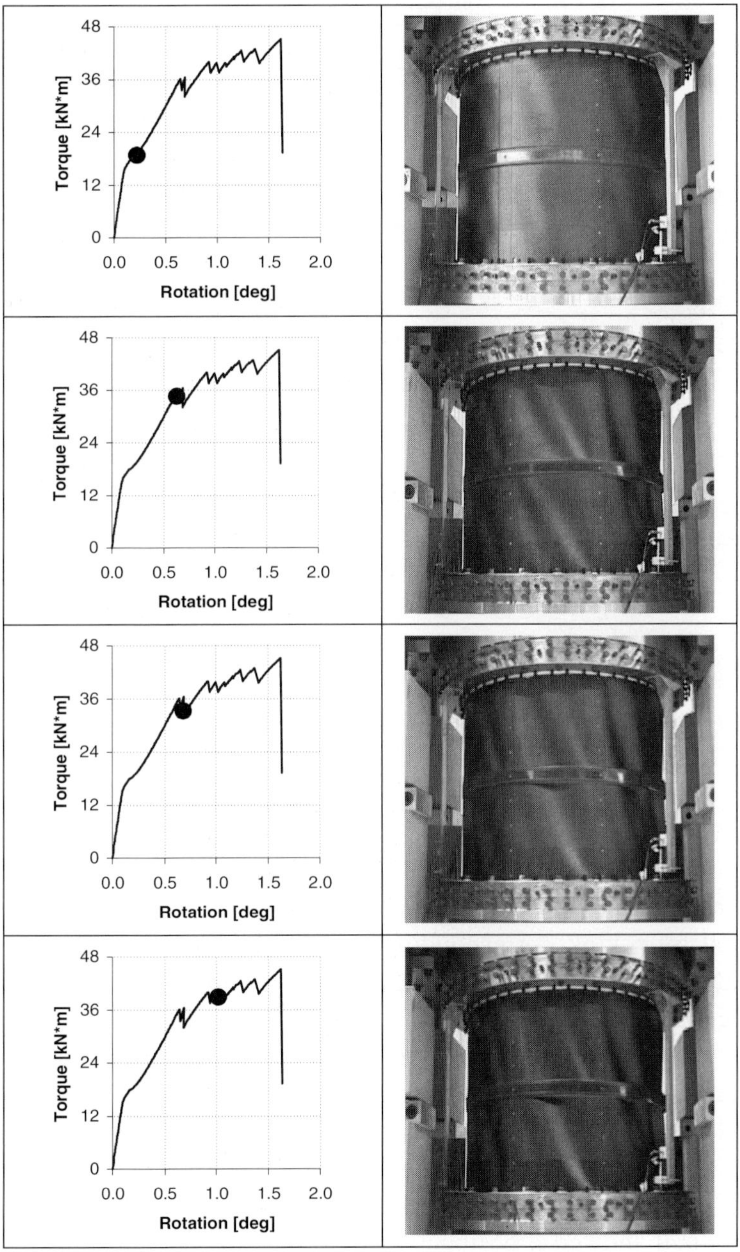

Fig. 2.8. Evolution of the second shell configuration: torque = 19 kNm; torque = 33.4 kNm; torque = 32.2 kNm; torque = 37.5 kNm.

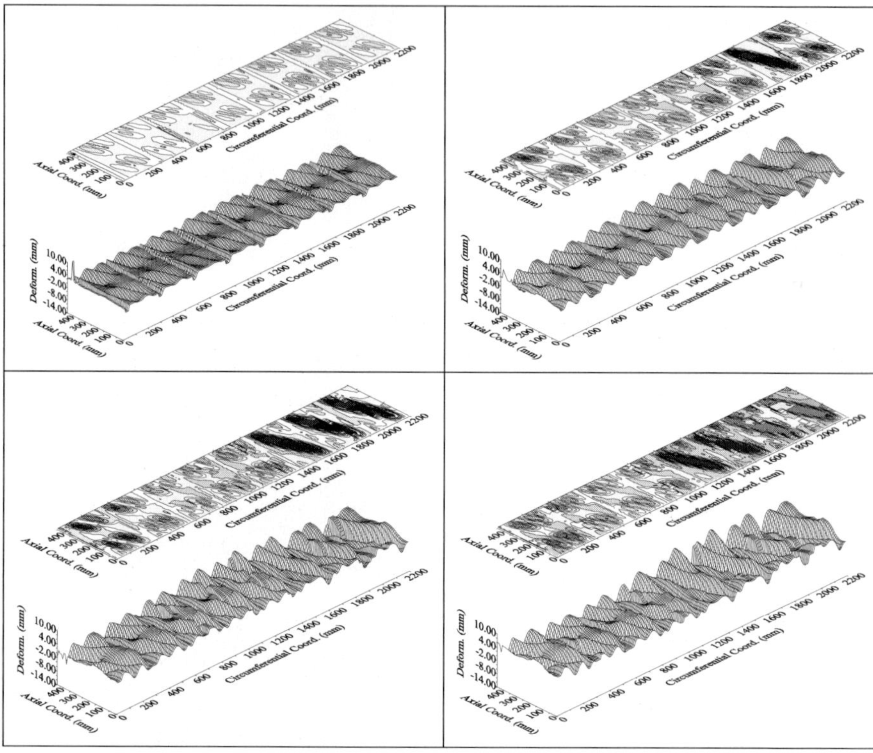

Fig. 2.9. Evolution of the second shell configuration measured by the laser sensors: torque = 19 kNm; torque = 33.4 kNm; torque = 32.2 kNm; torque = 37.5 kNm.

Initially the slope of the graph of torque versus rotation was constant, followed by a kink corresponding to the first local buckling mode, at 15.2 kNm. The stable postbuckling mode was characterized by the presence of two diagonal regular and well-defined half-waves in each sector limited by two stiffeners in the circumferential direction and by the external reinforcing ring in the axial direction.

At 19 kNm all the skin was completely buckled, and the waves were perfectly defined and with a regular shape. The normal displacements were 5.5 mm inward and 3.5 mm outward.

The shell showed a postbuckling residual torsional stiffness equal to about 27 % of the prebuckling one, as it is equal to 43 kNm/degree against a prebuckling one of 158 kNm/degree. Under torsion, indeed, the role played by the axial stiffeners was negligible, so the stiffness of the specimen sensibly decreased after skin buckling. In any case, the central reinforcing ring offered a residual stiffness, and helped the specimen to carry torque even in the postbuckling phase.

Increasing the torque, the out of plane deformations increased, but they did not change their shape until, at about 33.4 kNm, with a rotation angle equal to about 0.7°, the central reinforcing ring debonds from the skin in one of the shell sectors. This was the direct consequence of the way the skin deformed under torsion. Indeed, the postbuckling pattern of a shell without a central reinforcing ring was characterized by ideal diagonal lines from the lower-right to the upper-left corners of each sector. In this way, traction stresses were concentrated along the bay, diagonally, and the fibres of these skin areas were free to shorten significantly and to move internally with respect to the initial shell radius. This was prevented by the central reinforcing ring that, remaining of the same radius, constrained the panel skin to buckle in two different diagonal waves in each of the half shell height. Thus, increasing the torque, significant interlaminar stresses appeared between the panel skin and the reinforcing ring, which produced their separation. Once the central ring was debonded, the skin pattern changed suddenly.

As the torque increased again, the ring debonded in other sectors and the radial displacements reach 14 mm inward and 10 mm outward. At a rotation of about 1.4° the circumferential ring was completely debonded from the skin and remained connected to the skin only by the eight rivets, which fixed it to the stiffeners. Unfortunately, at this rotation the laser scanning system was unable to measure the inner surface any longer as the stiffeners were already significantly buckled and touched the laser sensors.

The structural collapse happened, very audibly and was clearly visible, at a torque of 45.1 kNm and at an imposed rotation of 1.6°. The ratio between the collapse torque and the first buckling torque was approximately 3. The collapse was due to the simultaneous breakage of all the stiffeners. Due to the torque applied, the failure of the stiffeners started near the clamping rings. It caused extensive fracture in the skin and stiffeners, so that, after collapse, the load suddenly dropped significantly. A photograph of a failed shell is shown in Fig. 2.10.

Fig. 2.10. Collapse of the second shell configuration.

2.6 Results on panels

The tests on the stiffened curved panels were performed with the two panels assembled in a closed box. This particular test is also discussed in Chapter 1 as part of a larger test programme in which both the Technion and Politecnico di Milano were involved (Refer to BOX 2 in Section 1.5). At first, the pure axial compression test and the pure torsion test were performed. Then the combined tests were performed. Table 2.7 summarizes the experimentally observed buckling loads of the box in the different tests.

Figure 2.11 presents the behaviour that was observed during the collapse test with a constant axial load of 180 kN and an increase in torque. In particular, the evolution of the panels' postbuckling shape and of the graph of torque versus rotation are reported.

The dot on the graph highlights when the photograph was taken by the camera system. The first three photographs were taken with the Moiré fringes that were used to detect and identify the buckling and postbuckling patterns. The last photo was taken without the Moiré fringes. Indeed the fringes were removed due to the panels' large deformation. The skin buckling torque was measured at 6 kNm, while the collapse at 48 kNm.

Table 2.7. Buckling loads of the curved panels.

Type of test	Axial compression load [kN]	Torque [kNm] Panel A	Torque [kNm] Panel B
Pure axial compression test	231		
Combined test	180	6	6.2
Combined test	120	7.4	7.4
Combined test	60	12	12
Pure torsion test - counterclockwise	15		14
Pure torsion test - clockwise	15		18

Typical measurements of the strain gauges are given in Fig. 2.12. It presents the measurements experienced by two pairs of strain gauges. The first pair shows the responses of the back-to-back gauges 27 and 28 on the skin at middle height of the panel, in pure axial compression. The buckling load is clearly visible where there is the separation of the two gauge measurements. The second pair shows the response of the back-to-back gauges 43 and 44 that belong to a rosette placed on the skin of the panel, during the collapse test. It is possible to see that there are many changes in the behaviour before reaching the collapse at 48 kNm. In all the tests, the readings of the strain gauges on both panels of the box were very similar, thus indicating that distribution of the loads in both panels was equal and balanced. A photo of the panel failure is shown in Fig. 2.13. The failure occured at 43 kNm, but the box was able to sustain torque until 48 kNm.

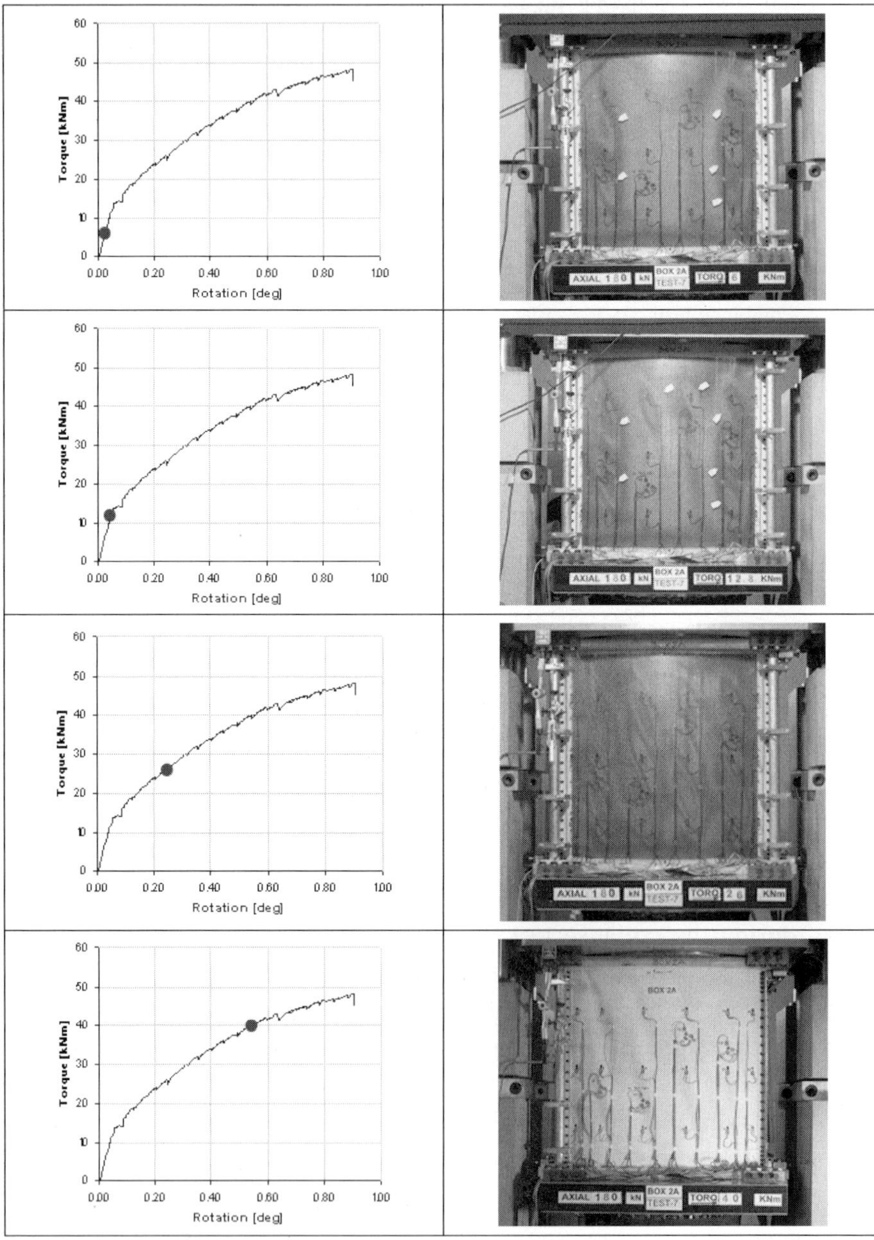

Fig. 2.11. Evolution of the box consisting of two curved panels: torque = 6 kNm; torque = 12.8 kNm; torque = 26 kNm; torque = 40 kNm, with an applied axial compression load = 180 kN.

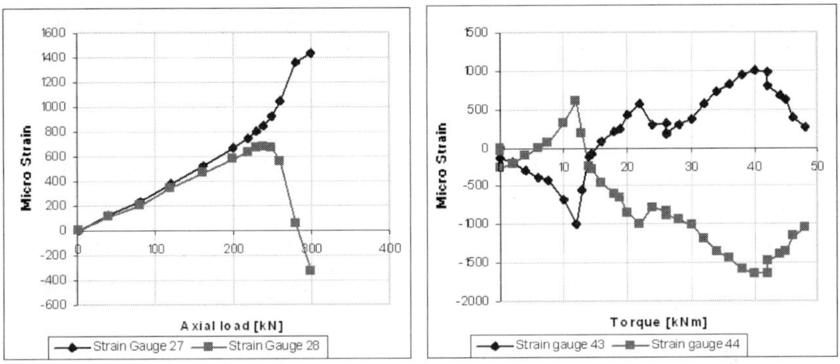

Fig. 2.12. Strain gauges measurements on a curved panel: axial compression test, and collapse test under combined load.

Fig. 2.13. Collapse of a curved panel.

2.7 Conclusions

The test equipment in the Department of Aerospace Engineering at the Politecnico di Milano was utilized to perform buckling tests using a displacement control, applying axial compression and torsion separately and in any sequence combination by means of five stepping motors. During the tests, the diagrams of axial load versus displacement and of torsion versus rotation, measured using a load cell and LVDT transducers, were recorded in real-time. The results of an experimental investigation on buckling and postbuckling behaviour of stiffened cylindrical shells and stiffened panels in graphite-epoxy were presented and discussed.

The cylindrical shells were at first tested under axial compression and torsion applied separately and in combination. Then, a stiffened cylindrical shell was tested until collapse under axial compression, and a ring-stiffened cylindrical shell with stiffeners was tested until collapse under torsion. During the tests, a laser scanning system measured the progressive change in the out of plane deformations of the shells in the postbuckling field.

The stiffened curved panels were tested as part of an assembled closed box, consisting of two stiffened curved panels and two side aluminium plates. The box was at first tested under axial compression and torsion applied separately and in combination. Then the box was tested until collapse under combined loading. In particular, axial load compression equal to 180 kN was kept constant, and the torque was increased until collapse. During the tests, strain gauge measurements were taken, and Moiré fringes were applied.

The data acquired during the first non-destructive tests and during the destructive collapse tests demonstrate that the structures were able to sustain load in the postbuckling field without any damage. Indeed no failure mechanisms were visible in the postbuckling range. On the other hand, the collapse was sudden and destructive. It causes extensive fracture in the skin and in the stiffeners, so that, after the collapse, the load suddenly dropped to approximately zero.

The experimental data show clearly the strength capacity of these structures to work in the postbuckling range, allowing further weight savings, demanded by future aerospace structures. The ratio between the collapse load and the first buckling load was higher then 3 for the cylindrical shells, while it was about 8 for the curved panels assembled in a box. The curved panels presented a very high postbuckling load carrying capacity.

The measured experimental data are also fundamental for the development and validation of analytical and numerical high-fidelity methods, and together with these validated analysis tools, they can contribute to the definition of improved design criteria in the postbuckling field.

2.8 Acknowledgments

The author is grateful to Prof. Vittorio Giavotto for sharing his expertise and for helping with many valuable suggestions, and to Mr. Potito Cordisco for his contribution in the execution of the tests. The author would like also to thank Prof. Haim Abramovich, Prof. Tanchum Weller and Mr. Ariel Grunwald from Technion, Haifa, Israel, with whom the tests on the curved panels were conducted.

This work was supported by the European Commission, Competitive and Sustainable Growth Programme, Contract No. G4RD-CT-1999-00103, project POSICOSS. The information in this paper is provided as is and no guarantee or warranty is given that the information is fit for any particular purpose. The reader thereof uses the information at his or her sole risk and liability.

2.9 References

1. M. P. Nemeth and J. H. Jr. Starnes, *The NASA Monographs on Shell Stability Design Recommendations*, NASA/TP-1998-206290 (1998).
2. J. Arbocz and J. H. Jr. Starnes, *Thin Wall. Struct.*, 40: 729 (2002).
3. Anon., *Buckling of Thin-Walled Circular Cylinders*, NASA SP-8007 (1968).
4. S. Nagendra, D. Jestin, Z. Gürdal, R. T. Haftka and L. T Watson, *Comput. Struct.*, 58: 543 (1996).
5. J. F. M. Wiggenraad, P. Arendsen and J. M. da Silva Pereira, *AIAA-98-1750* (1998).
6. C. Bisagni and L. Lanzi, *Compos. Struct.*, 58: 237 (2002).
7. R. Rikards and J. Auzins, *AIAA-2002-5654* (2002).
8. G. J. Simitses, *Appl. Mech. Rev.*, 39: 1517 (1986).
9. C. Bisagni, *Compos. Part B- Eng.*, 31: 655 (2000).
10. L. Wullschleger and H.-R. Meyer-Piening, *Int. J. Nonlinear Mech.*, 37: 645 (2002).
11. J. Singer, J. Arbocz and T. Weller, *Buckling Experiments – Experimental Methods in Buckling of Thin-Walled Structures, Vol. 2* (John Wiley & Sons Inc. 2002).
12. J. Singer and H. Abramovich, *Thin Wall. Struct.*, 23: 379 (1995).
13. J. Arbocz and J. M. A. M. Hol, *Thin Wall. Struct.,* 23: 131 (1995).
14. M. K. Chryssanthopoulos, V. Giavotto and C. Poggi, *Compos. Manuf.*, 6: 93 (1995).
15. C. Bisagni, *AIAA J.*, 47: 276 (1999).
16. M. W. Hilburger and J. H. Jr. Starnes, *Int. J. Nonlinear Mech.*, 37: 623 (2002).
17. M. K. Chryssanthopoulos, A. Y. Elghazouli and I. E. Esong *Compos. Part B- Eng.*, 30: 335 (1999).

18. H.-R. Meyer-Piening, M. Farshad, B. Geier and R. Zimmermann, *Compos. Struct.*, 53: 427 (2001).
19. C. Bisagni and P. Cordisco, *Compos. Struct.*, 60: 391 (2003).
20. M. Booton and R. C. Tennyson, *AIAA J.*, 17: 278-287 (1979).
21. R. F. Lennon and P. K. Das, *Thin Wall. Struct.*, 38: 229 (2000).
22. B. L. Agarwal, "Postbuckling Behavior of Composite-Stiffened Curved Panels Loaded in Compression", *Exp. Mech.*, 22: 231 (1982).
23. N. F. Jr. Knight and J.H., Jr. Starnes, *AIAA J.*, 26: 344 (1988).
24. D. M. McGowan, R. D. Young, G. D. Swancon and W. A. Waters, *5th NASA/DoD Advanced Composite Technology Conference*, Seattle (WA), Paper No. A94-33140 (1994).
25. H. Abramovich, A. Grunwald, P. Pevsner, T. Weller, A. David, G. Ghilai, A. Green and N. Pekker, *AIAA-2003-1793* (2003).
26. L. Lanzi, "A Numerical and Experimental Investigation on Composite Stiffened Panels into Post-Buckling", *Thin Wall. Struct.*, 42:1645 (2004).
27. C. Bisagni and P. Cordisco, "Testing of Stiffened Composite Cylindrical Shells in the Postbuckling Range until Failure", *AIAA J.*, 42:1806 (2004).
28. C. Bisagni and P. Cordisco, *Compos. Struct.*, 73:138 (2006).
29. P. K. Das, A. Thavalingam and Y. Bai, *Thin Wall. Struct.*, 41:69 (2003).
30. C. Bisagni, P. Cordisco, H. Abramovich and P. Pevsner, "Buckling and Post-Buckling Behavior of Stringer Stiffened Laminated Composite Curved Panels Subjected to Shear", in *Advances in Computational & Experimental Engineering & Sciences*, Forsyth (USA), Tech Science Press, 339-344 (2004).
31. H. Abramovich, T. Weller and C. Bisagni, *AIAA-2005-1933* (2005).

CHAPTER 3

MODE-JUMPING IN POSTBUCKLING STIFFENED COMPOSITE PANELS

Brian G. Falzon

Department of Aeronautics, Imperial College London
South Kensington Campus, London, SW7 2AZ, United Kingdom
E-mail: b.falzon@imperial.ac.uk

The phenomenon of secondary instabilities in stiffened composite panels, loaded in uniaxial compression, is studied through a number of experimental investigations. This behaviour is shown to manifest itself as a 'mode-jump' – a sudden change in the out-of-plane deformation of a buckled panel. Use of high-speed digital speckle photogrammetry is made to gain further insight into an I-stiffened panel's response during the transient phase associated with mode-jumping. At high loading, this sudden dissipation of energy will be shown to be able to cause failure in vulnerable structures. Predicting this structural response reliably, using the finite element method, poses considerable numerical challenges and the shortcomings of current non-linear solution schemes are discussed. A robust and efficient strategy, which utilizes an automated quasi-static/pseudo-transient hybrid scheme, is presented and validated using a number of experimental tests. This approach is shown to be able to predict mode-jumping with good accuracy.

3.1 Introduction

The prospect of allowing composite aerostructures to operate in a postbuckled state, between the design limit and ultimate loads and possibly even below the design limit load, can potentially lead to significantly lighter aerostructures. The study of postbuckling stiffened composite panels, loaded in uniaxial compression, has received much attention in recent years.[1-7] A considerable number of experimental programmes worldwide have established the ability of these structures to support load beyond initial skin-buckling. The European Union's objectives, outlined in a document setting out its vision of Aeronautics in 2020,[8] provide added impetus for this research. One goal is a 50% reduction in fuel consumption and an 80% reduction in nitrous oxide emissions. A significantly lighter structure will have a major impact towards meeting these targets. Further plans to design advanced composite aerostructures, which allow for the onset of damage below the ultimate design load, are also being considered within a number of research institutes in Europe.

The reluctance to fully exploit the postbuckling strength of composite stiffened panels is due to the relatively weak skin-stiffener interface resulting from co-curing, co-bonding or secondary bonding the stiffeners to the skin. The alternative is to use mechanical fasteners but it is recognised that their use in composite structures should be kept to a minimum. The added stress concentrations which may arise around a fastener hole due to slight dimensional inaccuracies are not easily diffused in a composite component, as they are in metallic structures, due to the brittle nature of the material. It is worth noting that metallic aerostructures, constructed of a thin-skin to which stiffeners are fastened using rivets, have been routinely designed with postbuckling strength.

Numerous challenges need to be overcome before a design capability is developed to realise the next generation of advanced lightweight aerostructures. For example, as has already been mentioned, the bonding schemes used to attach the stiffeners to the skin, need to be used with caution, particularly for postbuckling structures. Buckling in stiffened composite aerostructures gives rise to a dramatic increase in the interlaminar stresses at the skin-stiffener interfaces which are only resisted by the relatively weak through-thickness strength of the adhesive or resin. Damage initiation and progression in these vulnerable regions is still difficult to predict and this has lead to conservative composite designs in aerostructures, in some cases imposing a 'no-buckling' principle up to the ultimate design load.

Various experimental studies[1-4] have shown that the postbuckling response of stiffened thin-skinned structures is further complicated by the observed sudden change in buckling mode-shape. This phenomenon is referred to as a mode-jump, mode-switch or a secondary instability. Primary skin-buckling is characterised by a stable bifurcation point which defines the intersection of the primary equilibrium path with a secondary stable path. For secondary instabilities in plate structures, the

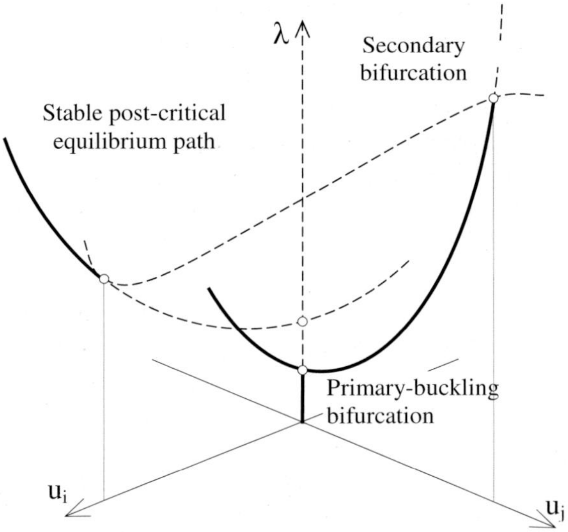

Fig. 3.1. Equilibrium path showing primary and secondary instabilities (solid lines represent stable states and broken lines unstable states).

bifurcation point is the intersection of the current stable equilibrium path with at least one unstable equilibrium path, which can only lead to a stable post-critical branch through a dynamic event. This concept is shown schematically in Fig. 3.1, where u_i and u_j are generalised coordinates representing a deformation mode and λ is a load parameter.

In this chapter the mechanics of postbuckling are briefly discussed. An experimental approach, utilising high-speed non-contact measuring techniques for directly observing transient changes during a secondary instability, is presented. At high loading, this sudden energy-dissipating phenomenon may release enough energy to cause damage at the first occurrence. Previous experimental tests conducted by Falzon *et al.*[2] and Romeo and Frulla[9] have yielded indirect evidence of failure of some test panels by mode-jumping. This chapter will present more substantial evidence of this mode of failure through the results of a recently-tested hat-stiffened panel.

It is imperative that the future design of postbuckling structures will utilise robust computational tools which are capable of predicting such abrupt behaviour. Since this occurs within a predominantly quasi-static loading history, numerical modelling of postbuckling structures has conventionally been performed using implicit continuation methods. It is well established that standard non-linear solution schemes, such as the Newton-Raphson methods, fail when they encounter such behaviour. While the arc-length method,[10-12] was developed to handle highly non-linear behaviour, it may still run into difficulty in the presence of mode-jumping. One possibility of analysing these structures is to use a full dynamic analysis but this is computationally expensive. A more effective strategy for capturing mode-jumping is to use a combined static-dynamic analysis as was demonstrated by Riks *et al.*[13] In this study, the switching from one solution procedure to the other was performed interactively using restarting schemes available in most finite element packages and utilising standard implicit or explicit dynamic schemes.

This chapter also presents an automated quasi-static/pseudo-transient procedure for modelling this phenomenon. The static part of the solution process uses an implicit continuation scheme. A modified explicit dynamic routine, which is more computationally efficient than standard implicit and explicit dynamic schemes, is used for the transient phase. Procedures for determining the location of critical points along an equilibrium path, and switching to a secondary path, are presented.

3.2 Experimental Observations of Mode-Jumping

3.2.1 *Hat-stiffened panel (I)*

A discussion of the hat-stiffened panel, shown in Fig. 3.2(a), serves as a good example for explaining the phenomenon of mode-jumping. This panel was manufactured from unidirectional pre-impregnated carbon-fibre laminae (T300/934) and incorporated two fully integrated hat-stiffeners in a quasi-isotropic skin $[\pm 45/0/90]_s$.[4] It was loaded in uniaxial compression and the central skin bay buckled at a loading of 9.4 kN, in a three half-waves buckling mode, as shown by the Moiré fringe patterns in Fig. 3.2(b). The graph presented in Fig. 3.3 shows the strain gauge readings from a strain gauge pair (1-2) mounted back-to-back at the centre of the panel.

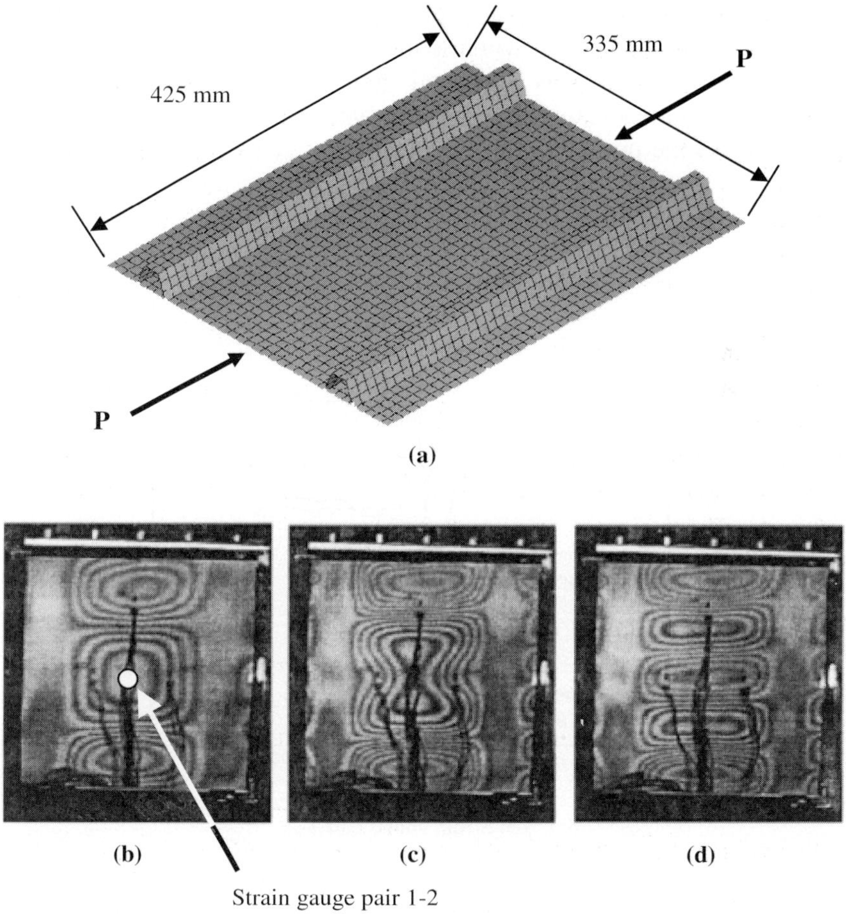

Fig. 3.2. Postbuckling of a hat-stiffened composite panel loaded in uniaxial compression; (a) Panel geometry showing major dimensions, (b) Moiré fringe pattern at P = 40 kN, (c) P = 66 kN (t = -0.04s) and (d) P = 66 kN (t = 0.04s).

Immediately after buckling, the strain gauges recorded an initial increase in bending strain followed by a gradual reduction beyond a loading of 23 kN. At a loading of approximately 40 kN, the bending strain was reduced to zero, indicating that the peak of the central crest had flattened. This phenomenon was described by Koiter[14] in his work on the postbuckling of isotropic plates where the natural half-waves were observed to flatten out when loaded far beyond the buckling point. A deflected form of this flattened half-wave was first postulated by Cox[15]. The rather square-like central half-wave shown in Fig. 3.2(b) corresponds to the point at which the curvature at its centre is reduced to zero.

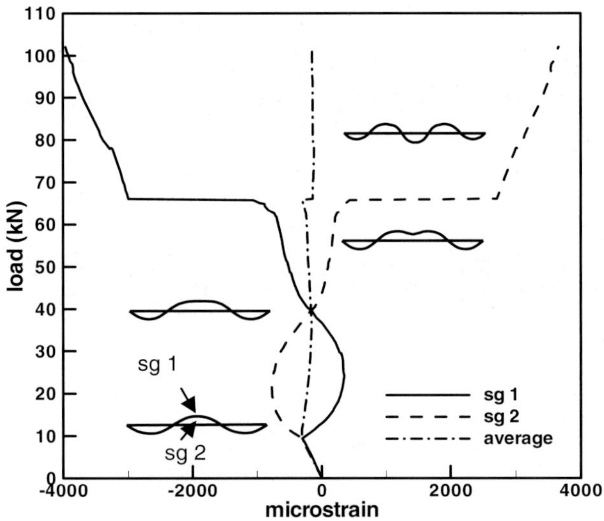

Fig. 3.3. Strain gauge pair 1-2.

At higher loading, the centre region was observed to reverse its bending direction as shown by the strain readings in Fig. 3.3. In Fig. 3.2(c) it is noted that this corresponded to the formation of an ‘hourglass’ Moiré fringe pattern where the central crest seemed to split into two peaks. At a loading of 66 kN there was an abrupt mode-jump to five half-waves as shown in Fig. 3.2(d) and Fig. 3.3. The flattening out of the central crest resulted from the increased membrane stresses in the skin. This constraint to further bending led to the observed local buckling of the central flat region of the crest which subsequently initiated the mode-jump. This explanation was further confirmed by Chai[16] in his study of unilaterally constrained polycarbonate plates under uniaxial compression. In this experimental work, the constraint restricting the out-of-plane deflections was a rigid wall, placed in close proximity to the plate, as opposed to the build-up of membrane stresses. The Moiré fringe patterns obtained from this experiment are reproduced in Fig. 3.4 where it is seen that the similarity with those obtained for the hat-stiffened panel is readily apparent.

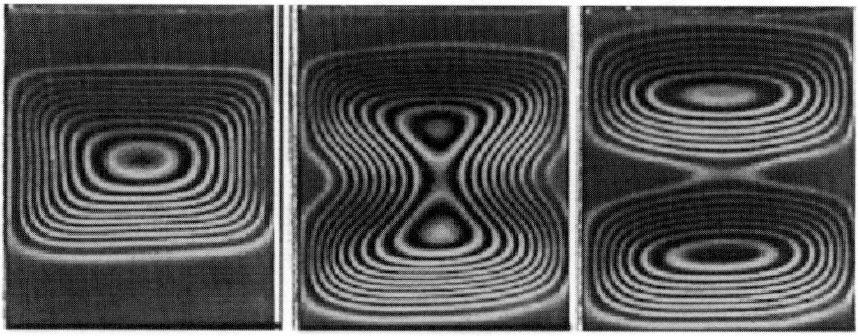

(a) **(b)** **(c)**

Fig. 3.4. Moiré fringe pattern for a unilaterally constrained polycarbonate plate under uniaxial compression; (a) beyond initial buckling, (b) at the onset of secondary instability and (c) splitting into two half-wavelengths. Source: (Chai, 2001).

3.2.2 *I-stiffened panel*

Two I-stiffened panels, designated I-P1 and I-P2, were manufactured from T300/914C unidirectional pre-impregnated composite (material properties are shown in Table 3.1) with a nominal skin-thickness of 2.0 mm. Four equally-spaced stiffeners were secondary-bonded onto the skin using FM300 adhesive film. Panel I-P1 was tested in a custom-built 250 ton hyperstiff testing machine[17] in the Department of Aeronautics at Imperial College London while I-P2 was tested at the DLR Institute of Structural Mechanics in Braunschweig, Germany, using a 1 MN servo-hydraulic testing facility and a newly developed high-speed non-contact full-field measuring system to directly observe mode-jumping. The effective test sections were 790 mm x 604 mm for I-P1 and 686 mm x 604 mm for I-P2. A schematic showing the major dimensions of IP-2 and the lay-up details of the stiffener is given in Fig. 3.5.

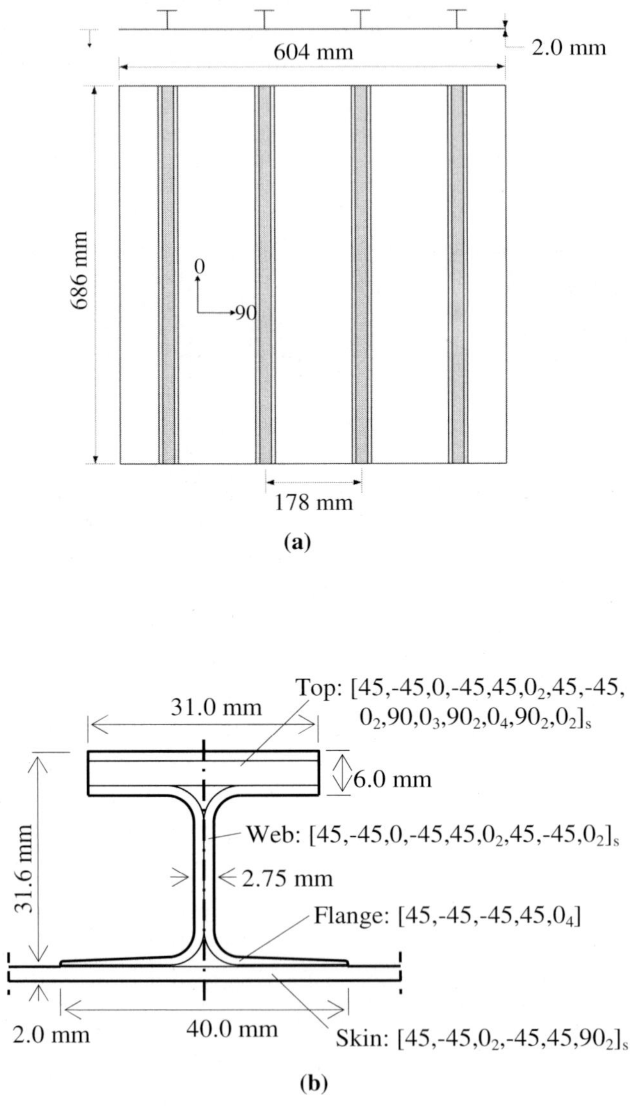

Fig. 3.5. I-stiffened panel, I-P2; (a) major dimensions, (b) details of the stiffener cross-section.

Table 3.1. Material properties for carbon-fibre unidirectional prepreg T300/914C.

Property	Value (RT/Dry)
Longitudinal tensile Young's modulus, E_{1t}	135 GPa
Longitudinal compressive Young's modulus, E_{1c}	120 GPa
Transverse tensile Young's modulus, E_{2t}	9.0 GPa
Transverse compressive Young's modulus, E_{2c}	9.0 GPa
In-plane shear modulus, ν_{12}	0.28
Nominal ply thickness	0.125 mm

In a numerical study conducted by Bushnell *et al.*[18] it was postulated that the possibility of damage during a mode-jump depended on the frequency of occurrence in service, the number and amplitude of stress oscillation resulting from the jump, whether these oscillations involved stress reversal and the energy released in the process. To gain further insight into this phenomenon, the ARAMIS[19] high-speed non-contact full-field measuring system was used. The ARAMIS set-up requires that the surface being measured is painted white and then sprayed with a black speckle pattern to create a distribution of greyscale values. The basic system uses two cameras and processing software to track changes in the relative position of the speckles. Displacement and strain values for a structure under loading are obtained using photogrammetry principles. The high-speed set-up at DLR uses four ARAMIS systems each with a pair of high-speed cameras able to sample at a quoted rate of 1 kHz. In practice, the maximum capture rate that was achieved was 833 Hz. The recording of images between these four systems was staggered, resulting in a maximum sampling speed of 3.33 kHz. Fig. 3.6(a) shown the panel mounted in the servo-hydraulic compression testing machine. Fig. 3.6(b) shows the speckle pattern applied on the panel's outer surface (the stiffeners are mounted on the other side) and Fig. 3.6(c) shows the eight mounted cameras focussed on the panel in the foreground.

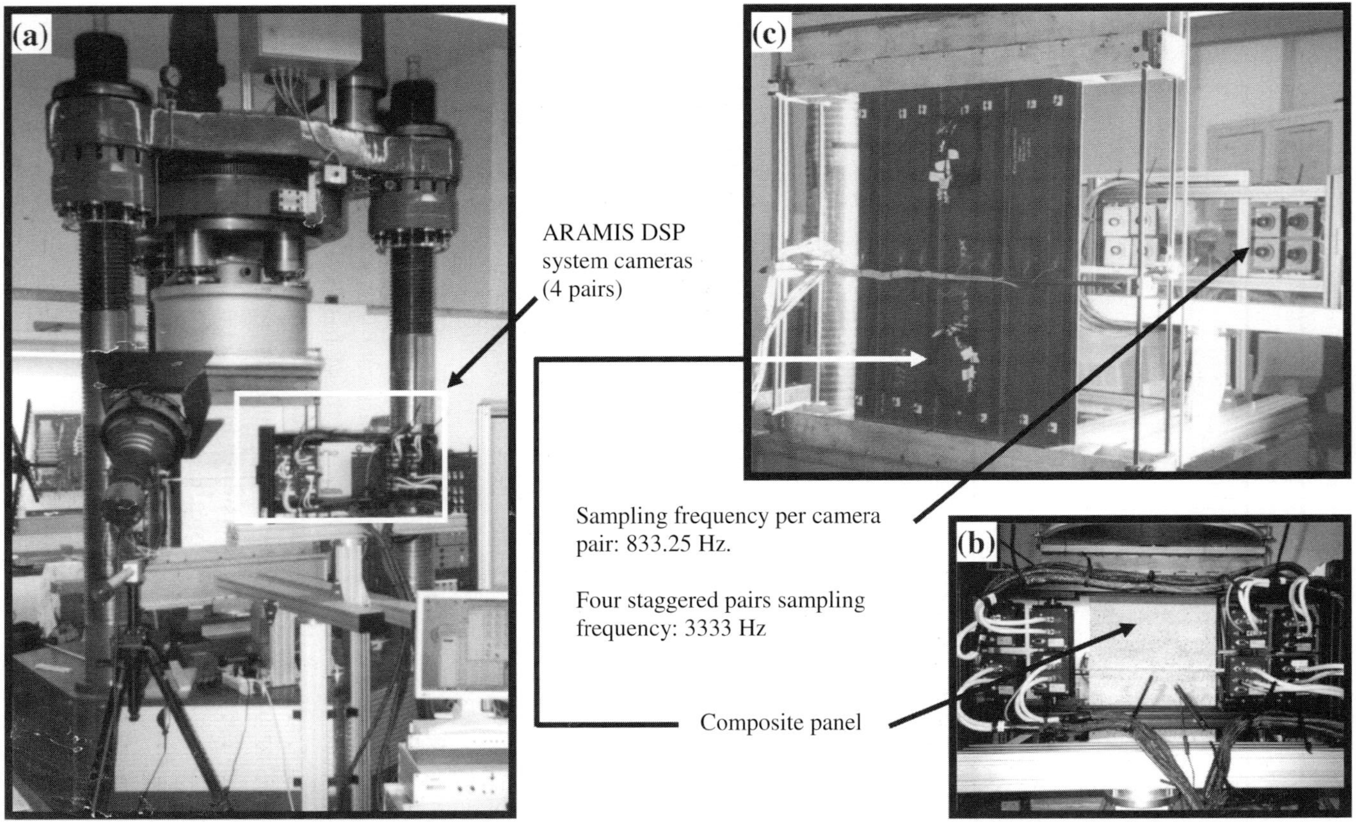

Fig. 3.6. Experimental set-up; (a) panel mounted in compression testing machine, (b) detail of the panel's back face with black speckle paint, (c) view showing stiffener side of panel in the foreground and DSP cameras in the background.

Both panels were loaded in uniaxial compression; IP-1 until failure while the test on IP-2 was terminated at a loading of 387 kN. The aspect ratio of the skin-bays in IP-1 was 4.43 while for IP-2 the aspect ratio was 3.85. Each panel displayed a five half-waves initial skin-buckling pattern. Fig. 3.7(a) shows the Moiré fringe pattern obtained from the test on IP-1 and Fig. 3.7(b) shows a contour plot of panel IP-2 obtained from the DSP system.

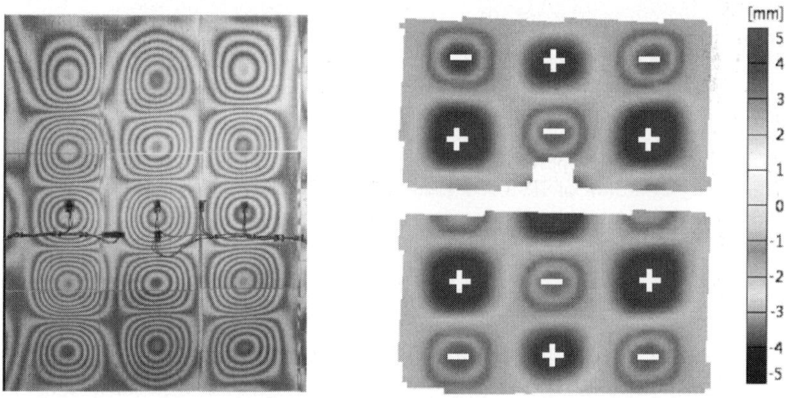

(a) (b)

Fig. 3.7. (a) Moire fringe pattern of IP-1 at a loading of 190 kN, (b) Digital speckle photogrammetry contours of IP-2 at a loading of 349 kN. The (+) and (-) signs on each half wave indicate whether the displacement is into the page (+) or out of the page (-).

Strain gauge wiring mounted across the panel's centre masked the speckle in that region resulting in the missing contour sections. The initial skin buckling load for Panel IP-1 was recorded at 120 kN and for IP-2 at 125 kN. This difference is consistent with the variation in aspect ratio of the respective skin-bays. Both panels displayed a sudden secondary instability to six half-waves but this occurred at different loading in each panel: at 244 kN for IP-1 and 366 kN for IP-2, shown in Fig. 3.8. It is evident that while the different aspect ratios had a small effect on the initial buckling load, the influence on the secondary instability was more pronounced.

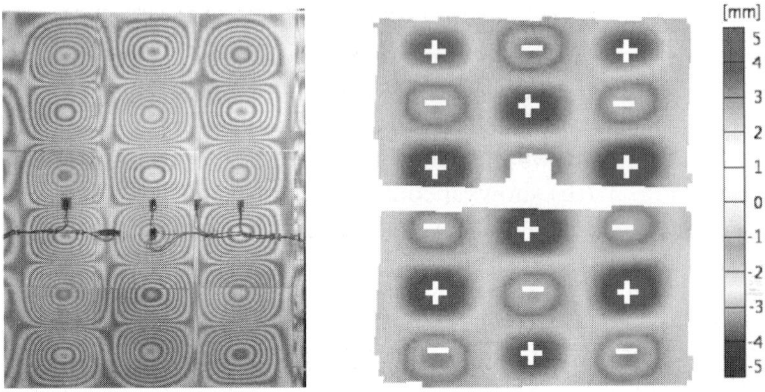

(a) (b)

Fig. 3.8. (a) Moiré fringe pattern of IP-1 at a loading of 245 kN (b) Digital Speckle Photogrammetry contours of IP-2 at a loading of 387 kN. The (+) and (-) signs on each half wave indicate whether the displacement is into the page (+) or out of the page (-).

A major motivation for using the high-speed DSP system was to gain further insight into the transient phenomenon associated with secondary instabilities. Fig. 3.9 shows the out-of-plane displacement recorded in the vicinity of a node-line. The mode-jump is characterized by high-frequency oscillations which are damped over a period of approximately 106 milliseconds. There is a jump in out-of-plane displacements from 0.8 mm to 3.55 mm but with peak displacements reaching 4.93 mm. These short-term peaks are also evident from the strain plot shown in Fig. 3.10 where higher strain peaks than the quasi-static value, achieved after the mode-jump, were recorded. There was a large strain-jump, from approximately 700 microstrain to 2700 microstrain. Such jumps may be enough to cause damage, particularly if this region corresponds to a vulnerable structural feature, such as a skin-stiffener interface, which has been compromised in some way. Local strain peaks in excess of 3000 microstrain were observed during this transient phase.

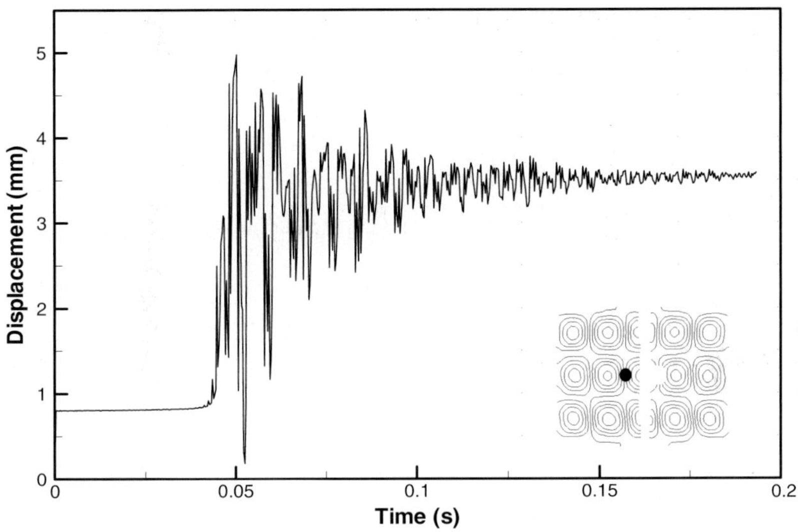

Fig. 3.9. Out-of-plane displacement close to an initial node-line during the mode-jump.

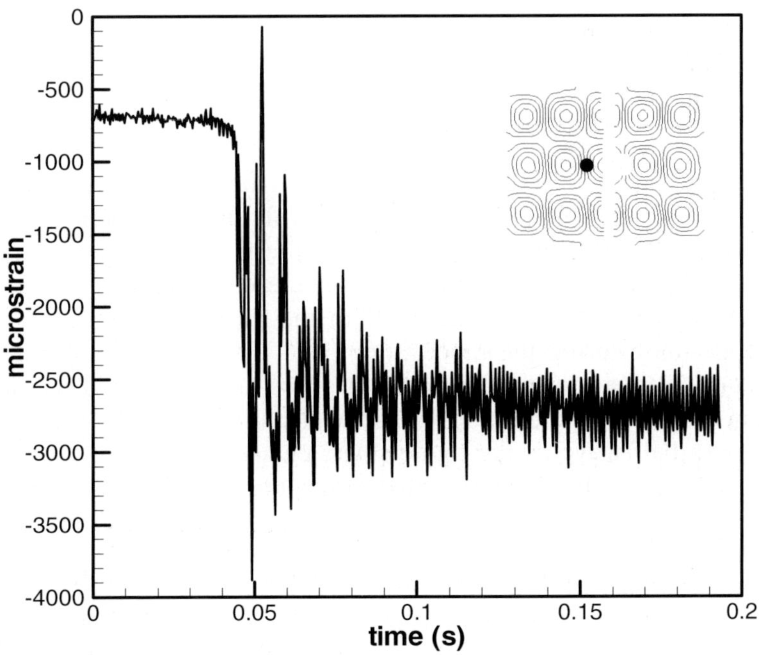

Fig. 3.10. Longitudinal direct strain close to an initial node-line during the mode-jump.

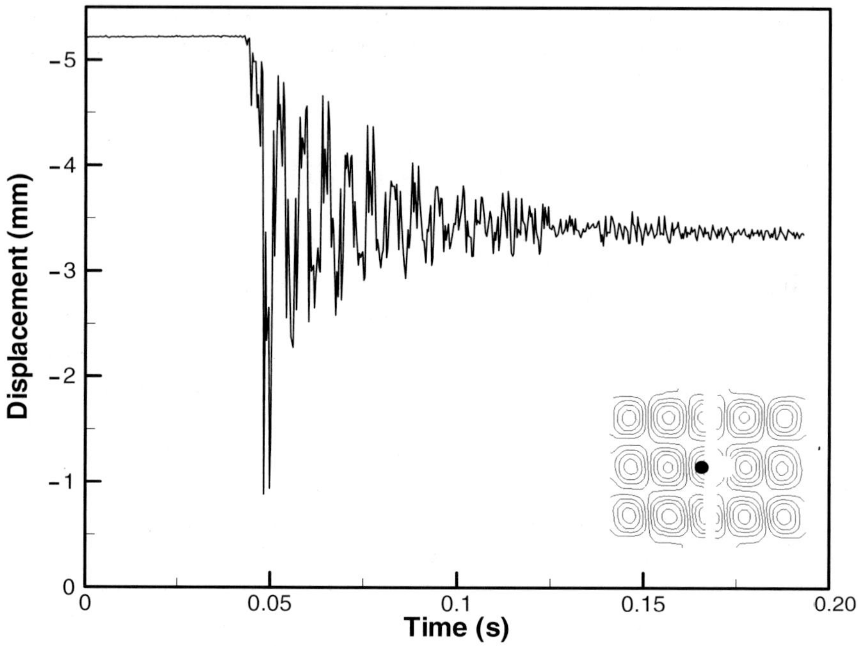

Fig. 3.11. Out-of-plane displacement close to an initial crest during the mode-jump.

A higher level of noise is observed in the strain results. This is attributed to additional computations involving the formation of a deformation gradient tensor which relates the position of the speckle in the undeformed state to their position in the deformed state. Fig. 3.11 shows the out-of-plane displacement in the vicinity of an initial crest (anti-node-line) during the mode jump. The high-frequency oscillations are again observed during the transient phase and it is noted that at this location there is a sudden drop in displacement arising from the creation of an extra half-wave as the mode-shape changes from five to six half-waves and where this location now corresponds to a region closer to a node-line.

It is plausible that sudden jumps in strain may cause damage and a structural analyst will need to be able to reliably predict this phenomenon in the design of postbuckling composite structures. The next section outlines the challenges faced when using standard non-linear finite element solution schemes and presents a novel technique which is robust and computationally efficient.

3.3 Numerical Analysis

3.3.1 *Background*

The high non-linearity associated with secondary instabilities often results in convergence difficulties using standard non-linear solution strategies currently found in most commercial finite element packages. The presence of a bifurcation point along an equilibrium path presents particular problems. For example, some solution algorithms, such as the class of Newton-Raphson methods are blind to bifurcations. The presence of a bifurcation is usually detected by observing the changes in the sign of entries in the diagonal matrix, $\boldsymbol{D}$, of the factored tangential stiffness matrix, $\boldsymbol{K}_t=\boldsymbol{LDL}^T$, where $\boldsymbol{L}$ is a lower triangular matrix. This is equivalent to identifying the presence of a negative eigenvalue of $\boldsymbol{K}_t$ which also implies a negative determinant.[20] When this occurs, the stiffness matrix is no longer positive-definite, that is, for a permissible displacement vector, $\boldsymbol{u}$, the condition $\boldsymbol{u}\ \boldsymbol{K}_t\ \boldsymbol{u}^T > \boldsymbol{0}$ no longer holds. There are no computational difficulties in the factorisation and subsequent forward and back substitutions associated with matrix inversion but the buckling response, where the resulting displacements are orthogonal to the applied loading, is not captured.

The presence of a bifurcation point means that a primary equilibrium path is intersected by a secondary path and a solution strategy must be able to switch from one path to another. Newton-Raphson and other continuation methods are incapable of this. The introduction of imperfections can remove bifurcations and reduces this response to one which is represented by a single equilibrium path, albeit one which may display the presence of limit points. These are points along an equilibrium path where the stiffness is reduced to zero. A number of solution techniques have been developed over the years to traverse these limit points and the interested reader is referred to the monograph by Falzon and Hitchings[21] which discusses a number of these strategies.

3.3.2 *The arc-length method*

The arc-length method[10-12] has emerged as one technique which is being widely adopted in a number of non-linear finite element packages. This method is based on augmenting the system of non-linear equilibrium equations,

$$\boldsymbol{g}(\boldsymbol{u},\lambda) = \boldsymbol{f}(\boldsymbol{u}) - \lambda \boldsymbol{f}_{ext} = 0\,, \tag{3.1}$$

(where $\boldsymbol{g}$ is the vector of residual forces, $\boldsymbol{f}(\boldsymbol{u})$ the vector of internal forces, $\boldsymbol{u}$ is the displacement vector, $\boldsymbol{f}_{ext}$ is a fixed external load vector and λ a load multiplier) with an incremental constraint which is a function of both the incremental displacements and the load increment,

$$\Delta\boldsymbol{u}^T \Delta\boldsymbol{u} + \Delta\lambda^2 \psi^2 \boldsymbol{f}_{ext}^T \boldsymbol{f}_{ext} = \Delta l^2\,. \tag{3.2}$$

$\Delta\boldsymbol{u}$ and $\Delta\lambda$ are the incremental changes in displacement and load respectively, Δl is the arc-length constraint and ψ is a scaling parameter. Riks' formulation[11] iterates along a hyperplane, orthogonal to the tangent produced to length Δl, from a previously converged point along the equilibrium path. Crisfield's method[12] iterates along a hypersphere, where the arc-length constraint is also employed for corrector iterations, and is less likely to fail in the vicinity of sharp limit points. Crisfield has further shown that the loading terms have little influence on the solution process and advocated setting $\psi = 0$. The iterations within each increment are usually still performed using a Newton-Raphson scheme. More specifically, a modified Newton-Raphson scheme is employed whereby the tangential stiffness matrix is held fixed within each increment, as opposed to updating at every iteration.

Solving Eq. (3.1) and Eq. (3.2) simultaneously is not readily amenable to finite element analysis since the augmented stiffness matrix will be neither symmetric nor banded. Using an approach suggested by Batoz and Dhatt[22] this system of equations reduces to a quadratic in $\delta\lambda$ which represents an iterative change in the applied load:

$$a_1 \delta\lambda^2 + a_2 \delta\lambda + a_3 = 0 \tag{3.3}$$

The expressions for the quadratic coefficients are given by Crisfield.[23] This implies two possible solutions for the load increment. The correct choice of $\delta\lambda$, at the predictor stage, is central to this method's effectiveness. One possibility is to match the sign of $\delta\lambda$ with the determinant of the stiffness matrix. In practice this is achieved by looking at terms of the diagonal matrix, $\boldsymbol{D}$, of the factored tangential stiffness matrix. If all terms on the diagonal are positive this implies the stiffness matrix is positive definite and hence a positive sign for the load increment is chosen. If a negative diagonal is encountered, this is equivalent to detecting a negative eigenvalue and implies that a limit point has been passed. Hence a negative load increment is applied. The arc length method is able to trace both the stable and unstable sections of an equilibrium curve although it is still liable to fail in the vicinity of 'sharp' limit points.

If the arc-length routine is used in the presence of a bifurcation point the solution will tend to oscillate. This occurs because if such a point is passed, then a negative entry in $\boldsymbol{D}$ will invoke a negative load increment. This may well bring the solution below the critical load which would imply that $\boldsymbol{K}_t$ is positive-definite and hence a positive load increment is used which, in turn, could push the solution past the bifurcation point and therefore as this process is repeated, oscillation occurs about the bifurcation point.

3.3.3 *Dynamic methods*

The dynamic nature of mode-jumping suggests that the problem could be solved using dynamic analysis. Indeed this approach has been adopted in a number of numerical studies.[7,24] The dynamic equations of motion are to be solved at some time $\tau = t + \Delta t$ where Δt is the time-step (or time increment) and t is the time at the start of the step. Two solution methods are available for solving these equations; an *implicit* solution where the 'exact' equations are solved for $\tau = t + \Delta t$, or an *explicit* solution where the equations are solved approximately.

For the implicit solution the equations of motion are solved at $\tau = t + \Delta t$,

$$ {}^{t+\Delta t}\boldsymbol{M}\,{}^{t+\Delta t}\ddot{\boldsymbol{u}} + {}^{t+\Delta t}\boldsymbol{C}\,{}^{t+\Delta t}\dot{\boldsymbol{u}} + {}^{t+\Delta t}\boldsymbol{K}_t\,{}^{t+\Delta t}\boldsymbol{u} = {}^{t+\Delta t}\lambda \boldsymbol{f}_{ext}\,, \tag{3.4} $$

where $\boldsymbol{u}$ is the displacement vector, $\dot{\boldsymbol{u}}$ the velocity vector and $\ddot{\boldsymbol{u}}$ the acceleration vector. $\boldsymbol{K}_t$ is the stiffness matrix, $\boldsymbol{C}$ the damping matrix and $\boldsymbol{M}$ the mass matrix. Assuming a linear acceleration over the step, it may be shown that[21]

$$^{t+\Delta t}\dot{\boldsymbol{u}} = \frac{1}{2}\Delta t\, ^{t}\ddot{\boldsymbol{u}} + \frac{1}{2}\Delta t\, ^{t+\Delta t}\ddot{\boldsymbol{u}} + ^{t}\dot{\boldsymbol{u}}, \tag{3.5}$$

$$^{t+\Delta t}\boldsymbol{u} = \frac{\Delta t^2}{3}\, ^{t}\ddot{\boldsymbol{u}} + \frac{\Delta t^2}{6}\, ^{t+\Delta t}\ddot{\boldsymbol{u}} + \Delta t\, ^{t}\dot{\boldsymbol{u}} + ^{t}\boldsymbol{u}. \tag{3.6}$$

By substituting Eqs. (3.5) and (3.6) into Eq. (3.4),

$$\begin{gathered}\left(^{t+\Delta t}\boldsymbol{M} + \frac{1}{2}\Delta t\, ^{t+\Delta t}\boldsymbol{C} + \frac{\Delta t^2}{6}\, ^{t+\Delta t}\boldsymbol{K}_t \right) {}^{t+\Delta t}\ddot{\boldsymbol{u}} = \\ ^{t+\Delta t}\lambda \boldsymbol{f}_{ext} - ^{t+\Delta t}\boldsymbol{C}\left(\frac{\Delta t}{2}\, ^{t}\ddot{\boldsymbol{u}} + ^{t}\dot{\boldsymbol{u}} \right) - ^{t+\Delta t}\boldsymbol{K}\left(\frac{\Delta t^2}{6}\, ^{t}\ddot{\boldsymbol{u}} + \Delta t\, ^{t}\dot{\boldsymbol{u}} + ^{t}\boldsymbol{u}. \right)\end{gathered} \tag{3.7}$$

This gives the acceleration $^{t+\Delta t}\ddot{\boldsymbol{u}}$, at the end of the step. Unfortunately, the matrix to be inverted to find the acceleration at the end of the step,

$$\boldsymbol{M} + \frac{1}{2}\Delta t\, ^{t+\Delta t}\boldsymbol{C} + \frac{\Delta t^2}{6}\, ^{t+\Delta t}\boldsymbol{K}_t, \tag{3.8}$$

is a function of the stiffness $^{t+\Delta t}\boldsymbol{K}_t$ and damping $^{t+\Delta t}\boldsymbol{C}$ at the end of the step. These are not known and the equations have to be solved iteratively. Initially it is assumed that $^{t+\Delta t}\boldsymbol{K}_t = {}^{t}\boldsymbol{K}_t$ and $^{t+\Delta t}\boldsymbol{C} = {}^{t}\boldsymbol{C}$. The equations are solved giving an approximation for the displacements at the end of the step. This allows better estimates for $^{t+\Delta t}\boldsymbol{K}_t$ and $^{t+\Delta t}\boldsymbol{C}$ to be found and an iterative solution to be formed for the step. This has to be repeated for each step, making this process inefficient for non-linear dynamics. An exception to this is for linear problems using the Newmark method[20] where large time steps can be used (i.e. the method is unconditionally stable) and the matrix to be inverted is only formed once.

For the explicit solution it is assumed that a small time-step is used. With this assumption, the displacement will change only by a second order over the step and hence the stiffness force, $\boldsymbol{K}_t\boldsymbol{u}$, will not change significantly over the step. The damping is assumed to be small, hence, although the velocity has a first order change over the step, the damping force is small compared to the inertia or stiffness forces. Therefore the initial condition is approximated as,

$$\left(\boldsymbol{M}\,{}^{t+\Delta t}\ddot{\boldsymbol{u}} + {}^{t}\boldsymbol{C}\,{}^{t}\dot{\boldsymbol{u}} + {}^{t}\boldsymbol{K}_t\,{}^{t}\boldsymbol{u}\right) = {}^{t+\Delta t}\lambda \boldsymbol{f}_{ext}, \qquad (3.9)$$

where the superscript on $\boldsymbol{M}$ has been dropped since it is assumed that the mass remains constant. As before, ${}^{t+\Delta t}\dot{\boldsymbol{u}}$ and ${}^{t+\Delta t}\boldsymbol{u}$ are calculated using Eqs. (3.5) and (3.6). This method is very stable for highly non-linear behaviour but it is conditionally stable, i.e., the time-step has to be less than a certain critical value which is of the order of the inverse of the maximum resonant frequency of the finite element model. In practice, explicit codes calculate the critical time step based on the time required for an elastic stress wave to propagate across the shortest dimension of the smallest element, L_{min}, in the mesh, i.e. $\Delta t \le L_{min}/c$, where c is the dilatational wave speed in the material. For an elastic isotropic material c is a function of Young's modulus, E, and density ρ.

The solution is approximate but computationally efficient, especially if the mass matrix is assumed to be diagonal. Since the time step, Δt, is small, many steps are required but each step can be computed quickly. The efficiency is made even greater by recognising that the product ${}^{t}\boldsymbol{K}_t\,{}^{t}\boldsymbol{u}$ can be carried out on an element-by-element basis and the global stiffness matrix need not be formed.

While the time-step size is controlled by the highest resonant frequency, ω_{max}, the duration of the calculation to achieve a static solution, as would be required beyond the mode-jump, is governed by the lowest resonant frequency, ω_{min}. This implies that the number of cycles required to recover a static solution is proportional to the ratio of the highest to lowest resonant frequencies. An artificial mass matrix is often used to reduce the span of the resonant frequencies of the structure. However the use of a diagonal matrix, for computational efficiency, limits what can be done. Selectively scaling the mass of specific elements allows some control over reducing ω_{max}. Another means of speeding up the solution time is to increase the load rate but this needs to be introduced with caution, to ensure that inertial effects are not introduced during the static phase of the actual loading history.

3.3.4 *An automated combined quasi-static/pseudo-transient method*

It has been argued that the computational time required to recover a static solution using explicit dynamic analysis is governed by the use of a diagonal mass matrix and the ratio $\omega_{max}/\omega_{min}$. This ratio may be reduced to unity if we replace $\boldsymbol{M}$ and $\boldsymbol{C}$ by $\boldsymbol{K}_t$,

$$\bar{\boldsymbol{K}}_t\,{}^{t+\Delta t}\ddot{\boldsymbol{u}} + \boldsymbol{K}_t\,{}^{t}\dot{\boldsymbol{u}} + {}^{t}\boldsymbol{K}_t\,{}^{t}\boldsymbol{u} = {}^{t+\Delta t}\lambda \boldsymbol{f}_{ext}\,. \tag{3.10}$$

This scheme was implemented by the author and co-worker[5] where it is noted that a bar is added over the stiffness matrix replacing the mass matrix to indicate that this matrix was only updated periodically to reduce computational effort. One small drawback here is that we no longer have a diagonal form of the matrix to be inverted as the emphasis is shifted to reducing $\omega_{max}/\omega_{min}$. In practice, the solution scheme was further improved by removing second order terms altogether so that a transient scheme was used,

$$\bar{\boldsymbol{K}}_t\,{}^{t+\Delta t}\dot{\boldsymbol{u}} + {}^{t}\boldsymbol{K}_t\,{}^{t}\boldsymbol{u} = {}^{t+\Delta t}\lambda \boldsymbol{f}_{ext}\,. \tag{3.11}$$

By assuming a linear variation in velocity over the time-step, Δt, the following equations were solved incrementally,

$$\begin{aligned} {}^{t+\Delta t}\dot{\boldsymbol{u}} &= \bar{\boldsymbol{K}}_t^{-1}\left({}^{t+\Delta t}\lambda \boldsymbol{f}_{ext} - {}^{t}\boldsymbol{K}_t\,{}^{t}\boldsymbol{u}\right), \\ {}^{t+\Delta t}\boldsymbol{u} &= \left(1 + \frac{\Delta t}{2}\right){}^{t}\dot{\boldsymbol{u}} + \frac{\Delta t}{2}\,{}^{t+\Delta t}\dot{\boldsymbol{u}}\,. \end{aligned} \tag{3.12}$$

This method also allows for a large time-step in comparison to a standard explicit analysis. In fact, it can easily be shown that if $\boldsymbol{K}_t$ is used instead of $\bar{\boldsymbol{K}}_t$ (at the expense of increased computational time when inverting this matrix at every increment) the condition for stability is that Δt should be less than or equal to unity. In practice, because $\bar{\boldsymbol{K}}_t$ was used, Δt was limited to 0.4 in the analysis presented later in this chapter. The ratio of the norms of the out-of-balance force to the applied load $\left\| {}^{t+\Delta t}\lambda \boldsymbol{f}_{ext} - {}^{t}\boldsymbol{K}_t {}^{t}\boldsymbol{u} \right\| / \left\| {}^{t+\Delta t}\lambda \boldsymbol{f}_{ext} \right\|$ was used to check convergence and the change of this was used to adjust Δt, to speed up convergence if the change was small and to reform the current tangent stiffness matrix estimate, $\bar{\boldsymbol{K}}_t$, if the out-of-balance force was increasing significantly.

This pseudo-transient method may be used on its own to solve problems involving mode-jumping.[5] The experimental observations discussed earlier show that for the most part of the loading history, the panel only displays 'mild' non-linear behaviour which is easily captured using standard methods of non-linear numerical analysis. It is only when a critical point is reached (whether a bifurcation or a limit point) that a more robust scheme needs to be employed.

The strategy adopted for this present study was to use either a Newton-Raphson or arc-length algorithm to initiate the analysis and to continue with this scheme as long as the tangential stiffness matrix remained positive definite. This is represented by the bold curve in Fig 3.12(a) where point A along the equilibrium curve represents a stable configuration. For simplicity these curves have been plotted as a function of the load parameter, λ, and two generalized displacements u_i and u_j. When point C is passed, a negative pivot (diagonal term in $\boldsymbol{D}$) is detected and a bracketing procedure[25] is invoked to reduce the distance between points A and B to within a pre-determined arc-length. Critical point, C, was estimated by linear interpolation of the smallest eigenvalues of the tangential stiffness matrices $\boldsymbol{K}_{tA}$ and $\boldsymbol{K}_{tB}$. If $\gamma_A > 0$ is the smallest eigenvalue of $\boldsymbol{K}_{tA}$ and $\gamma_B < 0$ the smallest eigenvalue of $\boldsymbol{K}_{tB}$ the interpolation parameter α,

$$\alpha = \gamma_A / (\gamma_A - \gamma_B), \tag{3.13}$$

takes a value between 0 and 1. The displacement vector at C, $\boldsymbol{u}_C$ and the corresponding load parameter λ_C were therefore approximated as,

$$\begin{aligned} \boldsymbol{u}_C &= \boldsymbol{u}_A + \alpha \Delta \boldsymbol{u}_{B-A} , \\ \lambda_C &= \lambda_A + \alpha \Delta \lambda_{B-A} , \end{aligned} \tag{3.14}$$

where $\Delta \boldsymbol{u}_{B-A} = \boldsymbol{u}_B - \boldsymbol{u}_A$ and $\Delta\lambda_{B-A} = \lambda_B - \lambda_A$. It is worth noting that while eigenvalue extraction for large problems is computationally expensive, it may be shown[26] that λ and the corresponding eigenvector, $\boldsymbol{\varphi}$, may be extracted directly from the $\boldsymbol{L}^T\boldsymbol{D}\boldsymbol{L}$ decomposition of $\boldsymbol{K}_t$,

$$\lambda \approx \frac{(d_{ii})_{min}}{|\boldsymbol{l}_i|^2} \quad , \qquad \varphi \approx \frac{\boldsymbol{l}_i}{|\boldsymbol{l}_i|} , \tag{3.15}$$

where $(d_{ii})_{min}$ is the smallest diagonal entry in $\boldsymbol{D}$ and $\boldsymbol{l}_i$ is the corresponding column of $(\boldsymbol{L}^T)^{-1}$.

The critical eigenvector, φ, at point C, may be determined from the tangential stiffness, $\boldsymbol{K}_{tC}$. At this point the nature of this critical point may be determined by noting the following conditions,

$$\begin{aligned} \varphi^T \boldsymbol{f}_{ext} \neq 0 &\quad \Rightarrow \quad \textit{limit point} , \\ \varphi^T \boldsymbol{f}_{ext} \approx 0 &\quad \Rightarrow \quad \textit{bifurcation point} . \end{aligned} \tag{3.16}$$

As has already been discussed, a standard Newton-Raphson scheme would fail to converge. If imperfections were added to the panel geometry then it is likely that a limit point is encountered but this would depend on the nature of this imperfection. Provided that the arc-length method was implemented for the quasi-static response, there is a high probability of the solution being able to traverse this point. Even so, if the limit point is very sharp this may still lead to convergence difficulties. For this present study, the checks in Eq. (3.16) were not carried out and the solution was switched to the pseudo-transient scheme when a critical point was encountered.

The choice of suitable starting parameters for the pseudo-transient phase is important to ensure that the solution does not 'back-track' along the same equilibrium path. Use is made of the orthogonality of the eigenvector, φ, with respect to the deformation state at the critical point. Point D in Fig. 3.12(b) denotes the starting conditions for the pseudo-transient phase. Point C_1 may be considered as a point above the critical point C without eigenmode injection. Eq. (3.17) gives the initial conditions for the pseudo-transient phase,

$$\begin{aligned} \boldsymbol{u}_D &= \boldsymbol{u}_A + \alpha \Delta \boldsymbol{u}_{B-A} + \mu \varphi \,, \\ \lambda_D &= \lambda_A + \tilde{\alpha} \Delta \lambda_{B-A} \quad \text{where } \tilde{\alpha} = \alpha + \delta\alpha \,, \end{aligned} \tag{3.17}$$

μ is a scaling parameter for the eigenvector and $\delta\alpha$ is a small fraction of α. The transient phase is stepped through pseudo-time until a steady-state response is reached upon which the solution scheme switches back to the quasi-static method.

3.4 Finite Element Modelling

3.4.1 *I-stiffened panel*

The I-stiffened panel, I-P1, discussed in Section 3.2.2, was modelled using an in-house finite element code FE77[27] which incorporated the developments presented in Section 3.3.4. The panel model was developed using quadratic three-dimensional thick-shell composite elements incorporated in the code[28]. This allowed for detailed modelling of the stiffeners' geometric features including the flange taper (refer to Fig. 3.5) and the curvatures of the transition sections between the flanges and the stiffener webs.

Each skin bay was modelled using 46 elements along the length and eight elements along the width. The triangular regions at the foot and top of the web were filled with resin and therefore these regions were modelled using material properties corresponding to the composite resin 914C. No initial imperfections were introduced into the panel to remove bifurcations points since the solution procedure switched to the transient scheme whenever the tangential stiffness matrix was no longer positive definite, indicating the presence of a critical point. Fig. 3.13 shows the location of back-to-back strain gauges and LVDT which were used to validate the numerical model.

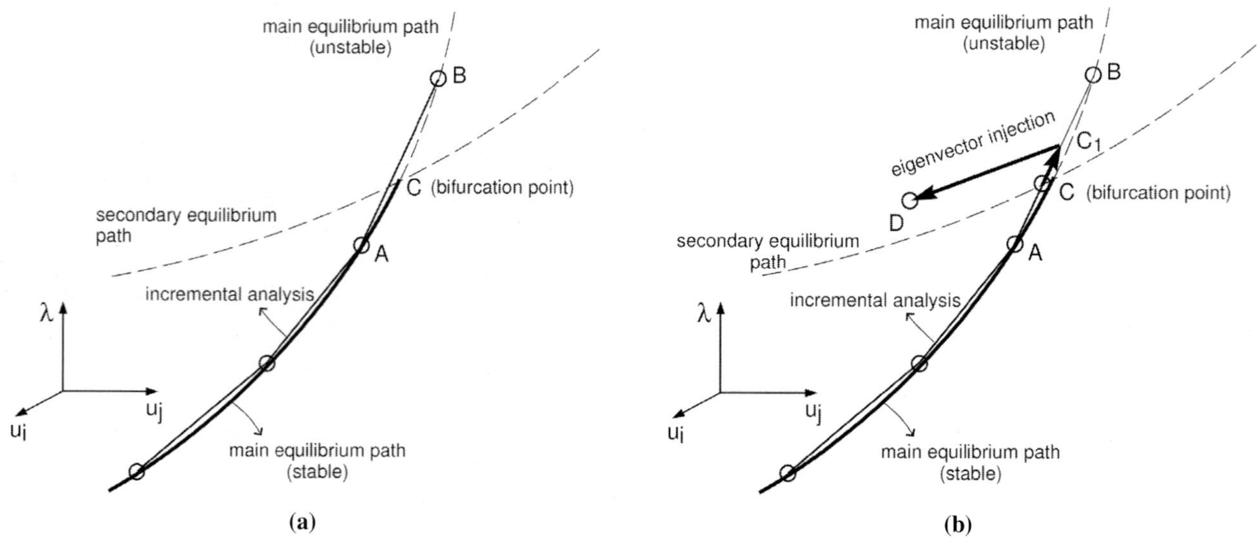

Fig. 3.12. Intersecting equilibrium paths showing; (a) the location of a critical point C and (b) point D which denotes the initial conditions for the pseudo-transient phase.

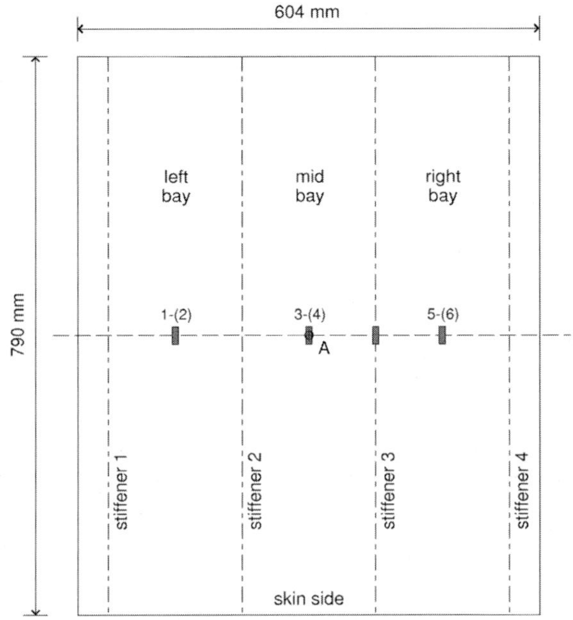

Fig. 3.13. I-stiffened panel, I-P1, showing back-to-back strain-gauges and LVDT location (a bracketed strain-gauge number denotes a strain-gauge mounted on the stiffener side).

Skin buckling into five half-waves occurred at a loading of 110kN which compared favourably with the observed value of 120 kN. Fig. 3.14(a) shows a comparison of a contour plot of out-of-plane displacements with the Moiré fringe pattern, obtained from experiment, just beyond initial buckling. Fig. 3.14(b) shows the gradual development of the five half-wave patterns in the three skin-bays. Between 250 kN and 254 kN the model underwent a mode jump to six half-waves in all skin-bays, consistent with the behaviour observed experimentally and in good quantitative agreement with the finite element solution where this was observed to occur at 244 kN (Fig. 3.14(c)).

Deep in the postbuckling regime the correlation was comparatively poor. The finite element model predicted a jump at a loading of 553 kN, associated with the change in buckled pattern at the outer edges of the panel. This was 13.8% higher than the load at which a mode-jump was observed experimentally, characterized by the outer skin-bays jumping to seven half-waves while the central skin-bay remained at six half-waves until failure. The buckled patterns on the outer skin-bays of the finite element model were observed to undergo a gradual change following the change in deformation of the outer edges. As shown in Fig. 3.14(d), the second half-waves from the bottom edge of the panel are shown to elongate and exhibit an 'hourglass' contour pattern similar to that shown in Fig. 3.2(c). It will be recalled that this deformation was shown to be a precursor to a mode-jump. It is therefore postulated that in the experiment, mode-jumps in the outer edges may have triggered the jump in the outer skin-bays while computationally this influence seemed to have happened gradually where the outer skin-bays did eventually undergo a mode-jump but at a loading of 647 kN. This discrepancy is due to the considerable extent of matrix micro-cracking which occurred under this high compressive loading, reducing the overall stiffness and leading to the observed secondary mode-jump occurring earlier than predicted. This damage was not accounted for in the finite element model. Fig. 3.15 shows the out-of-plane displacement measured by the LVDT. Figs. 3.16-3.18 show the results of back-to-back strain gauges at locations shown in Fig. 3.13. Note that in these figures, the compressive strain is positive. It is seen that good correlation was achieved for most of the loading history before the effects of matrix cracking became significant.

3.4.2 *Hat-stiffened panel (II)*

A hat-stiffened panel with skin-bays of identical lay-up and dimensions to I-P1 was manufactured. The stiffeners were designed to have a similar proportion of 0°, 90° and ±45° plies to that on the I-stiffeners resulting in a second moment of area of the cross-section which was approximately 17% higher.[29] The stiffeners were secondary bonded onto the skin and the manufactured panel, before trimming, is shown in Fig. 3.19(a). Fig. 3.19(b) shows the panel mounted in the 250T compression testing machine.

This panel was loaded in compression and failed prematurely at a loading of 380 kN without a mode-jump being detected. Extensive stiffener debonding was observed. This panel was expected to sustain a load at least equal to that sustained by I-P1 (well over 500kN) before final failure.

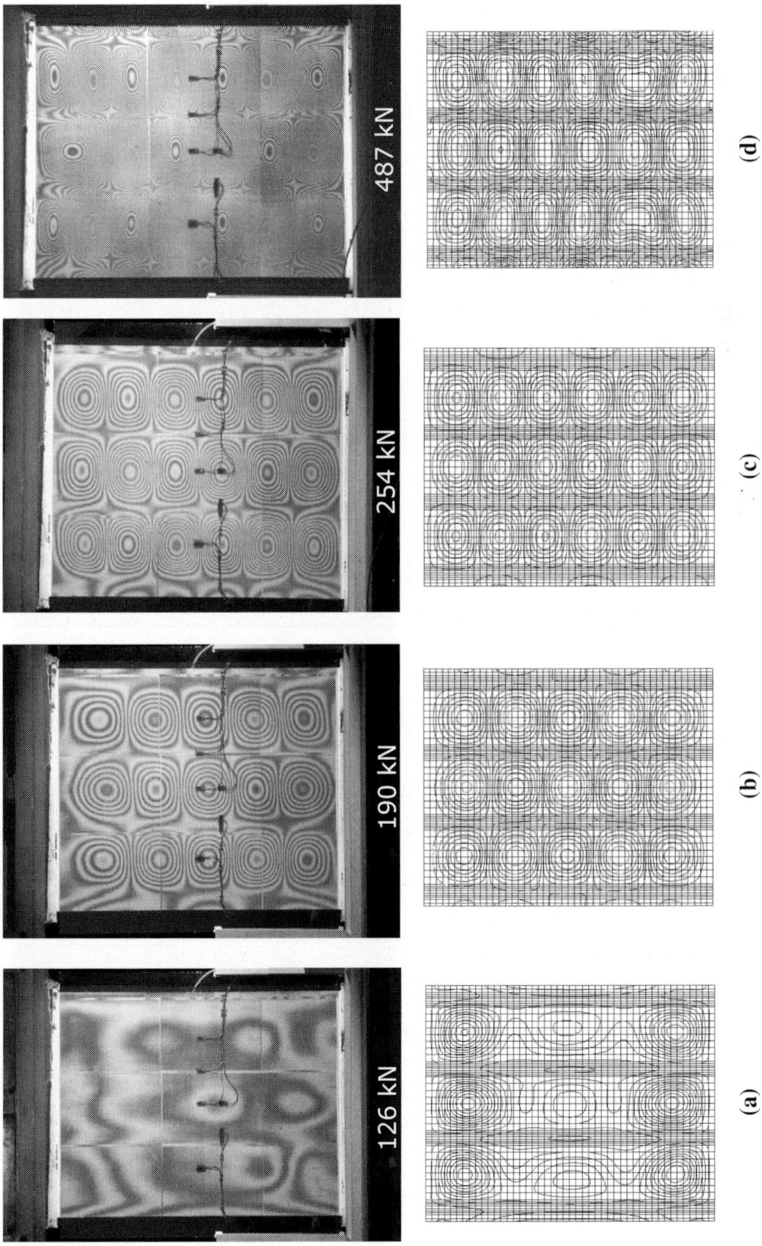

Fig. 3.14. Shadow Moiré fringe patterns at different load levels (top) and corresponding finite element contour plots of out-of-plane displacements (bottom).

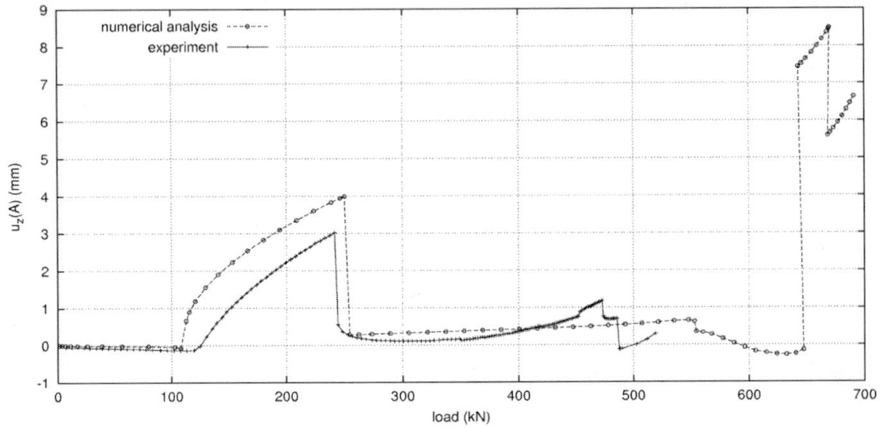

Fig. 3.15. Out-of-plane displacement at the centre of panel I-P1 (point A in Fig. 3.13).

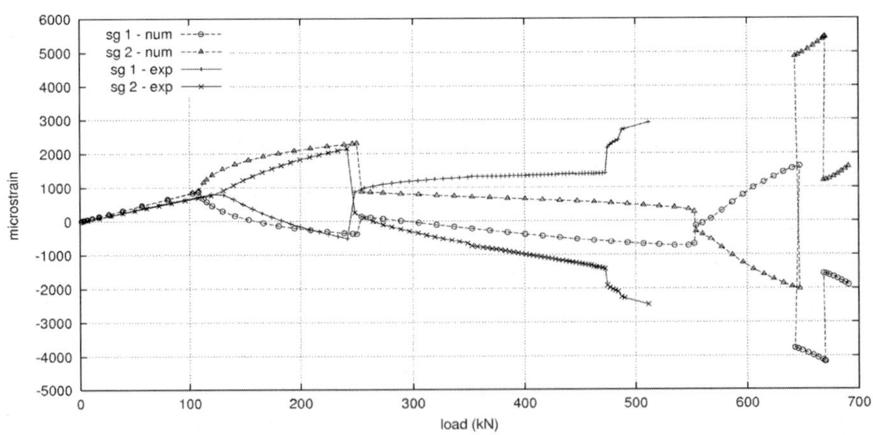

Fig. 3.16. Back-to-back strain gauge pair 1-2 on panel I-P1 (location shown in Fig. 3.13).

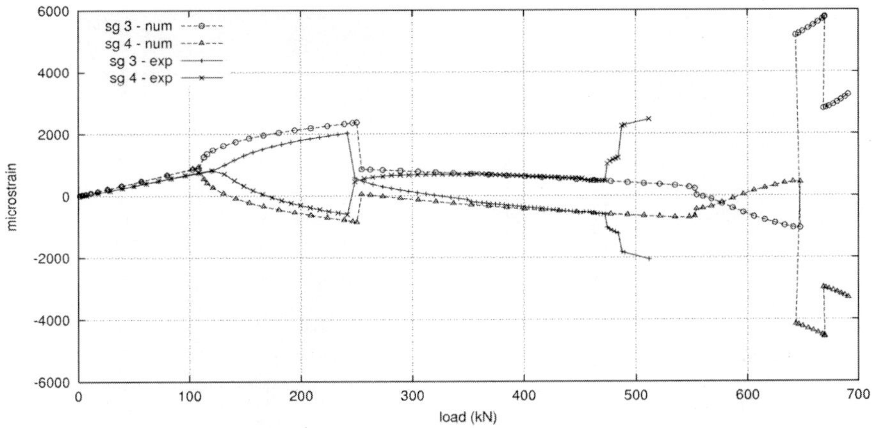

Fig. 3.17. Back-to-back strain gauge pair 3-4 on panel I-P1 (location shown in Fig. 3.13).

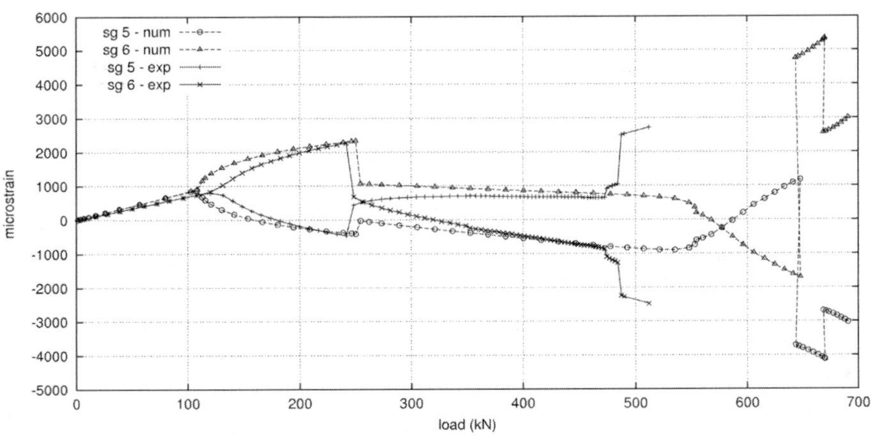

Fig. 3.18. Back-to-back strain gauge pair 5-6 on panel I-P1 (location shown in Fig. 3.13).

(a) (b)

Fig. 3.19. (a) Secondary bonded hat-stiffened panel and (b) mounted in a 250T compression testing machine.

A finite element analysis of this panel was undertaken and the displacements and strains, obtained experimentally at various locations on the panel, were compared with those extracted from the analysis. Figure 3.20 shows the out-of-plane displacement measured at the centre of the panel, marked 'A' on the inset diagram. What is significant is that the finite element analysis predicted a mode-jump at a loading of 385 kN which is remarkably close to the load at which failure of the panel occurred. This close agreement is also evident from the back-to-back strain gauge results in Fig. 3.21, obtained at the location marked 'B' on the inset diagram.

This study confirmed earlier indirect observations which suggested the possibility of structural failure occurring at a mode-jump. It is through the development of an accurate and robust numerical finite element solution scheme that this trigger to structural failure could be confirmed. Further investigations into the secondary bonding of this panel revealed that there were regions at the interface where the bonding was defective. The short-duration high displacements and strains, which were shown to occur during the testing of the I-P2 panel using the high-speed DSP system and presented in Section 3.2.2, may initiate failure, particularly if there is a vulnerability in the structure.

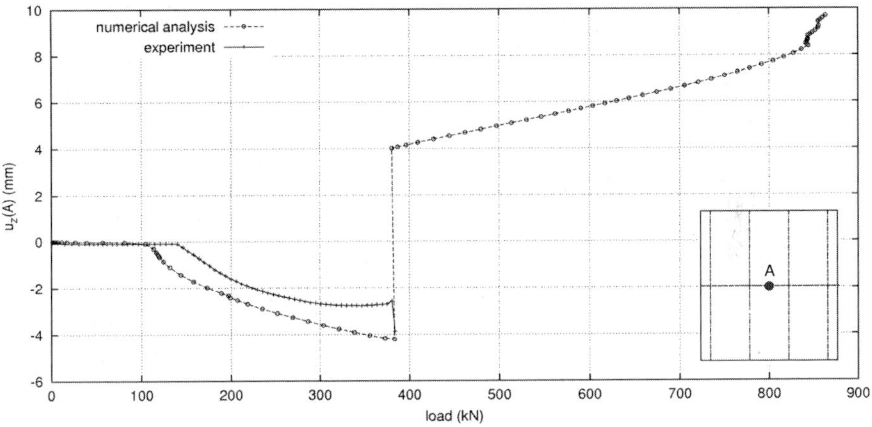

Fig. 3.20. Out-of-plane displacement at the centre of a hat-stiffened panel.

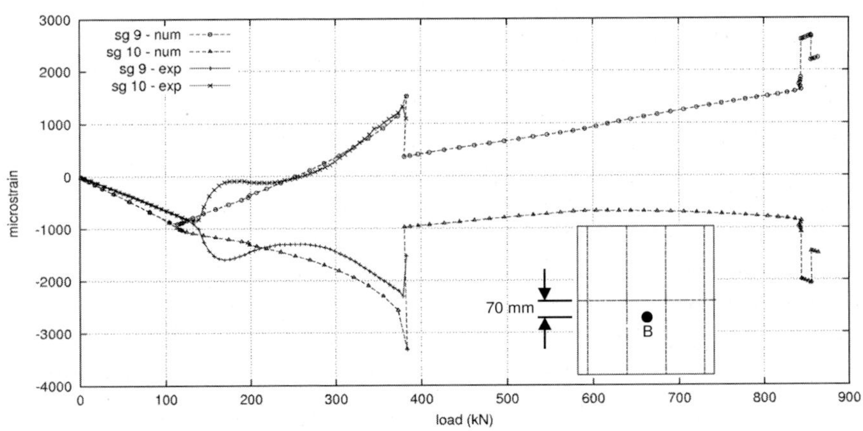

Fig. 3.21. Back-to-back strain gauge results at location marked by point B (gauge 10 is on the stiffener side).

3.5 Concluding Remarks

The mode-jumping observed in postbuckling stiffened panels was shown to be triggered by the high membrane stresses, developed in the postbuckled skin. This phenomenon was further investigated using high speed digital speckle photogrammetry. The mode-jump event, recorded for an I-stiffened panel, was shown to last approximately 106 milliseconds. The out-of-plane displacement and in-plane longitudinal strain responses were characterized by high-frequency oscillations with peaks which were considerably higher than the static values obtained just after the mode-jump when the transient behaviour had subsided.

It was postulated, and subsequently shown, that mode-jumping may release enough energy to cause failure, particularly in vulnerable regions within a structure. The high interlaminar stresses at the skin-stiffener interface in a co-cured panel, for example, are only resisted by a relatively weak resin layer. A secondary-bonded or co-bonded stiffened panel may also be vulnerable if certain regions of the bond are defective.

The design of bonded or co-cured postbuckling structures requires a reliable means of predicting, quantitatively, the likelihood of a secondary instability. Current non-linear finite element solution schemes, readily available in commercial finite element packages, may not converge when modelling this highly non-linear response. A robust scheme was presented which was capable of switching from a standard non-linear quasi-static solution algorithm to a pseudo-transient one when a critical point, along an equilibrium path, was encountered.

The experimental results of an I-stiffened and a hat-stiffened panel were compared with the results predicted by the finite element model and good agreement was achieved in predicting the mode-jump. The hat-stiffened panel was shown to have failed at a load corresponding to a mode-jump and confirmed earlier indirect evidence of this type of structural failure. Deep into the postbuckling regime, the inevitable matrix-cracking in the tested I-stiffened panel, not accounted for in the composite material's behaviour in the finite element model, resulted in poorer correlation between the experimental and numerical results. Further work is currently underway to include in-plane and out-of-plane damage models to enable a more accurate representation of the structural response of composite panels under failure-inducing loads.

3.6 Acknowledgement

The author is indebted to Dr Marco Cerini who coded the algorithms and developed some of the finite element models as part of his PhD research. The assistance and advice of Mr Dennis Hitchings, the original architect of the in-house finite element code in which these new algorithms were implemented, is also greatly appreciated. Most of the work presented in this chapter was funded by the Engineering and Physical Sciences Research Council (EPSRC) under contract no. GR/R32147/01. The use of the high-speed digital speckle photogrammetry at DLR, Braunschweig, was made possible through the award of a European Union grant under the 'Access to Research Infrastructures' scheme. A special thanks to Mr Hans Christian Goetting and Mr Ary Zipfel at DLR's Institute of Structural Mechanics for their considerable expertise in assisting with planning the test programme and in post-processing the data.

3.7 References

1. B. G. Falzon and G. P. Steven, *Compos. Struct.,* 37 : 253 (1997).
2. B. G. Falzon, K. A. Stevens and G. A. O. Davies, *Compos. Part A: Appl. Sci. Manuf.,* 31: 459 (2000).
3. K. A. Stevens, R. Ricci, G. A. O. Davies *Composites,* 26: 189 (2001).
4. B. G. Falzon, *Compos. Part A: Appl. Sci. Manuf.,* 32: 1255 (2001).
5. B. G. Falzon and D. Hitchings, *Compos. Struct.,* 60: 447 (2003).
6. B. G. Falzon and M. Cerini, *Compos. Struct.,* 73: 186 (2006).
7. F. Caputo, R. Esposito, P. Perugini and D. Santoro, *Compos. Struct.,* 55: 347 (2002).
8. Group of Personalities, *European Aeronautics: A Vision for 2020,* Office for Official Publications of the European Communities, Belgium, (2001).
9. G. Romeo and G. Frulla, *Int. J. Solids Struct.,* 31(6): 763 (1994).
10. G. A. Wempner, *Int. J. Solids Struct.,* 7: 1581 (1971).
11. E. Riks, *Int. J. Solids. Struct.,* 15: 529 (1979).
12. M. A. Crisfield, *Comput. Struct.,* 13: 55 (1981).
13. E. Riks, C. C. Rankin and F. A. Brogan, *Comput. Meth. Appl. M.,* 136: 59 (1996).
14. W. T. Koiter, *Colloquium on the Post-buckling of Plates in Metal Structures,* (University of Liege, Begium, 1963).
15. H. L. Cox, *Aeronautical Research Council R & M No. 2042,* (London, 1945).
16. H. Chai, *J. Mech. Phys. Solids,* 49: 209 (2001).
17. J. Singer, J. Arbocz and T. Weller, *Buckling Experiments:Experimental Methods in Buckling of Thin Walled Structures, Vol. 1* (John Wiley & Sons, Chichester, 1998).
18. D. Bushnell, C. C. Rankin and E. Riks, *NASA/CP-1998-206280,* 105 (1998).

19. GoM mbH, *ARAMIS – Deformation measurement using the grating method,* (Braunschweig, 2001).
20. K-J Bathe, *Finite Element Procedures,* (Prentice Hall, New Jersey, 1996).
21. B. G. Falzon and D. Hitchings, *An Introduction to Modelling Buckling and Collapse,* (NAFEMS Ltd, Glasgow, 2006).
22. J. L. Batoz and G. Dhatt, *Int. J. Numer. Meth. Eng.* 14: 1262 (1979).
23. M. A. Crisfield, *Non-linear Finite Elements Analysis of Solids and Structures, Vol. 1,* (John Wiley & Sons, Chichester, 1991).
24. C. Bisagni, *Compos. Part B-Eng.,* 31: 655 (2000).
25. M. A. Crisfield, *Non-linear Finite Elements Analysis of Solids and Structures, Vol. 2,* (John Wiley & Sons, Chichester, 1997).
26. F. Fujii and H. Noguchi, *Commun. Numer. Meth. En.* 18(7): 459 (2002).
27. D. Hitchings, *FE77 User Manual – Version 2.58,* (Imperial College, London, 2002).
28. B. G. Falzon, D. Hitchings and T. Besant, *Compos. Struct.,* 53: 29 (1999).
29. M. Cerini, *Investigation of Secondary Instabilities in Postbuckling Stiffened Composite Structures,* (PhD Thesis, Imperial College London, 2005).

CHAPTER 4

THE DEVELOPMENT OF SHELL BUCKLING DESIGN CRITERIA BASED ON INITIAL IMPERFECTION SIGNATURES

Mark W. Hilburger

NASA Langley Research Center
Hampton, Virginia 23681-2199, USA
E-mail: m.w.hilburger@larc.nasa.gov

An analysis-based approach for developing shell-buckling design factors for cylindrical shells that accurately accounts for the effects of initial geometric imperfections is presented. To develop this approach, measured initial geometric imperfection data from six laboratory-scale graphite-epoxy shells are used to determine a manufacturing-process-specific imperfection signature for these shells. This imperfection signature is then used as input into nonlinear finite-element analyses. The imperfection signature represents a "first-approximation" mean imperfection shape that is suitable for developing preliminary-design data. Comparisons of test data and analytical results obtained by using several different imperfection shapes are presented for selected shells. These imperfection shapes include the actual measured imperfection shape of the test specimens, a "first-approximation" mean imperfection shape with plus or minus one standard deviation in the shape, and a linear-bifurcation-mode imperfection shape. In addition, buckling interaction curves for composite shells subjected to combined axial compression and torsion loading are presented that were obtained by using the various imperfection shapes in the analyses. A discussion on the development of experimentally validated nonlinear finite-element analyses is also presented. Overall, the results indicate that the proposed analysis-based approach presented herein, for developing reliable preliminary-design criteria, has the potential to provide improved, less conservative buckling-load estimates, and to reduce the weight and cost of developing buckling-resistant shell structures.

4.1 Introduction

High-performance aerospace shell structures are inherently thin-walled because of weight and performance considerations and are often subjected to destabilizing loads. In addition, fiber reinforced composite materials are becoming more common in aerospace vehicle structures because of their high strength, high stiffness, and amenability to structural tailoring that permits design solutions that are not possible with metals. Thus, buckling is an important and often critical consideration in the design of these structures and reliable, validated design criteria for thin-walled shells are needed, especially for shells made of advanced composite materials.

As one might expect, shell-buckling design criteria have a history steeped in empiricism.[1,2] From approximately 1930 to 1967, many shell-buckling experiments were conducted on metallic shells. Typically, the experiments yielded buckling loads that were substantially lower than the corresponding analytical predictions, which were based on simplified linear bifurcation analyses of geometrically perfect shells with nominal dimensions and idealized support conditions. The seminal works by von Kármán and Tsien[3], by Donnell and Wan[4], and by Koiter[5] identified small deviations from the idealized geometry of a shell, known as initial geometric imperfections, as a primary source of discrepancy between corresponding analytical predictions and experimental results. However, it appears that because of the intense computational requirements of the nonlinear analyses used by von Kármán and Tsien and by Donnell and Wan, and the asymptotic analyses used by Koiter, at that time, buckling design allowables were determined by establishing lower bounds to test data. Specifically, empirical factors, that have become known as "knockdown" factors, were determined and were to be used in conjunction with linear bifurcation analyses for simply supported shells; that is, these empirical factors were used to "knockdown" the unconservative analytical prediction. This approach to shell design remains prominent in industry practice, as evidenced by the extensive use of the NASA space vehicle design recommendations.[6-8]

From approximately 1960 to 1989, digital computers emerged and matured as a practical tool for scientists and engineers. With these computational tools, considerable effort was put forth to determine analytically the effects of initial geometric imperfections on the buckling response of shells. These efforts were largely based on Koiter's asymptotic theory[1,5,9] for idealized structural shapes with nominal dimensions and also included investigations of shells made of fiber-reinforced composite materials.[10-13] In addition, in-depth, substantial experimental investigations that focused on fiber-reinforced composite shells were conducted at institutions such as the NASA Langley Research Center and the Wright Aeronautical Laboratories at Wright-Patterson Air Force Base, in addition to various academic institutions worldwide. It was during this time period that the measurement of initial geometric imperfections in test specimens and high-fidelity test measurement techniques and apparatus were recognized as important elements in the development of design criteria for buckling of shells.[14-17]

Another significant development that occurred in parallel to the maturation of digital computers in the 1960s to 1980s was the development of computational structural analysis codes based on finite-difference or finite-element methods. Two prominent examples that stimulated the development of this particular technology are the NASA Structural Analysis code (NASTRAN[18]) and the Structural Analysis of General Shells code (STAGS[19]), which were funded by the United States government. The successors and inheritors of these codes now possess robust static and dynamic nonlinear analysis capabilities, along with extensive graphics capabilities, that enable scientists and engineers to model and visualize experiments or actual flight hardware to a tremendous level of detail. Archetypal examples of the application of these tools are the in-depth, highly detailed studies of cylinder buckling response documented in Refs. 20-22 and the comprehensive and highly detailed nonlinear and buckling analyses of the Space Shuttle super-light-weight external fuel tank that is documented in Refs. 23-27. In addition, the extension of these analysis techniques and codes to other disciplines has helped to enable the development of advanced experimental-response measurement systems. Thus, the tools and understanding now exist for developing high-fidelity shell-buckling design criteria for shells of general shape and material composition and for combined mechanical and thermal loads. Recent advances and recommendations for the development of modern shell-buckling design criteria have been given in Refs. 28 and 29. These references examine the potential benefits of adopting probabilistic and hybrid probabilistic-deterministic approaches for developing design criteria. An important concept that is discussed in Ref. 29 and used in the present study is the establishment of characteristic imperfection signatures that are associated with the different shell manufacturing processes. This concept is based on the idea that each manufacturing process will yield shells that possess essentially similar initial geometric imperfection shapes, with a well-defined range of pointwise amplitudes. With this information established, high-fidelity analyses and selective structural testing can be used to determine refined, reliable design criteria for shell buckling that are not overly conservative like the present lower-bound approach found in NASA SP 8007.[6] To this end, some effort has been put toward defining initial geometric imperfection models to be used in shell analysis for isotropic and composite shells.[30,31] Similarly, various probability-based analysis methods have been proposed to develop reliable design buckling loads for imperfect shells in which approximations to the measured imperfections are used in the analyses.[32,33]

This chapter focuses on the development of a refined shell-buckling preliminary-design criteria that is based on high-fidelity nonlinear finite-element analyses that include the effects of a manufacturing-process-specific geometric imperfection signature. Refined preliminary-design data are important because they substantially reduce the chance of getting "locked-in" to a poor design early in a vehicle-development program and, in the context of stability-critical thin-walled shell structures, can give a more accurate preliminary prediction of the load-carrying capacity of the shell. One main objective of the chapter is to present a method for establishing a "first-approximation" characteristic imperfection signature for selected circular cylindrical shells made of a graphite-epoxy material that can be used to develop data suitable for preliminary design. Another main objective is to describe the development and validation of an analysis-based shell-buckling design criteria that accurately accounts for the effects of initial geometric imperfections by using the imperfection signature defined herein. The final objective is to illustrate the potential improvements in buckling predictions that can be realized, particularly for combined loading conditions. Towards these objectives, the testing and analysis procedures that are used in the development of the new design criteria are discussed in Sections 2 and 3, respectively. Next, the development of experimentally validated analysis models is presented in Section 4. Section 5 describes how measured geometric imperfections for six laminated graphite-epoxy shells are used to determine a manufacturing imperfection signature for use in the new analysis-based design criteria. Then, analysis results and experimental results for selected compression-loaded shells are compared to validate the new procedure for developing analysis-based design criteria. Finally, numerically predicted buckling interaction curves are presented for three different shells subjected to combined axial compression and torsion to illustrate the potential benefits of the new procedure.

4.2 Test Specimens, Imperfection Measurements, and Tests

Six test specimens used in the development of the new shell-buckling design criteria are described in this section. These specimens were previously fabricated and tested at the NASA Langley Research Center in support of a research program to study the effects of initial imperfections on the nonlinear response and buckling of compression-loaded composite shells. Much of this work has been documented in Refs. 20-22 and selected details about the test specimens, imperfection measurements, and tests are summarized in this chapter. Additional details on the cylinder tests can be found in these references.

4.2.1 *Test Specimens*

The specimens were fabricated from 12-in-wide (304.8-mm), 0.005-in-thick (0.127-mm) AS4/3502 unidirectional graphite-epoxy tape material. The nominal unidirectional lamina properties of a typical 0.005-in-thick ply with a fiber volume fraction of 0.62 are as follows: longitudinal modulus E_1 = 18.5 Msi (127.5 GPa), transverse modulus E_2 = 1.64 Msi (11.3 GPa), in-plane shear modulus G_{12} = 0.87 Msi (5.99 GPa), and major Poisson's ratio ν_{12} = 0.30. The material was laid up by hand on a 15.75-in-diameter (400.05-mm) mandrel and vacuum bagged and cured in an autoclave to form six shells with different shell-wall laminates. These shells include shells with an 8-ply axially stiff laminate, an 8-ply circumferentially stiff laminate, an 8-ply quasi-isotropic laminate, a 16-ply axially stiff laminate, a 16-ply circumferentially stiff laminate, and a 16-ply quasi-isotropic laminate. These six shells are referred to herein as shells or specimens C1 through C6, respectively. These specimens had a nominal length of 16.0 in. (406.4 mm) and a nominal radius of 8.0 in (203.2 mm). The 8-ply and 16-ply specimens had a nominal shell-wall thickness of 0.040 in. (1.016 mm) and 0.080 in. (2.032 mm),

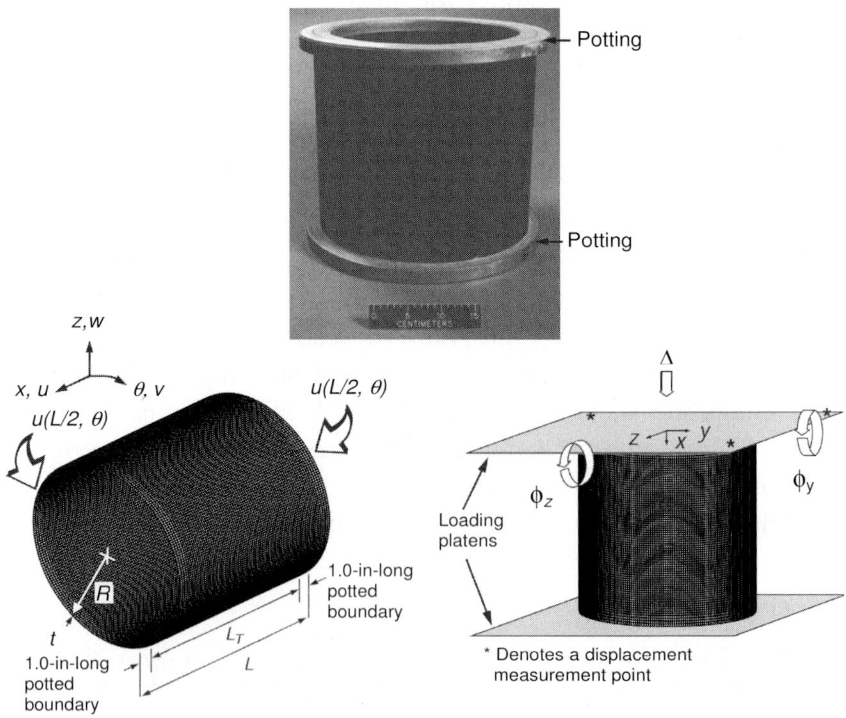

Fig. 4.1. Typical Specimen, finite element model geometry and loading conditions.

respectively, and had shell-radius-to-thickness ratios of 200 and 100, respectively. Both ends of the specimens were potted in an aluminum-filled epoxy resin to ensure that the ends of the specimen did not fail prematurely during the test. The potting material extended approximately 1.0 inch (25.4 mm) along the length of the specimens at each end, resulting in a test section that is approximately 14.0 in. (355.6 mm) long. The ends of the specimens were machined flat and parallel to facilitate proper load introduction during the tests. A photograph of a typical test specimen and the specimen coordinate system used to represent the corresponding geometry is shown in Fig. 4.1. The nominal cylinder length, test-section length, radius, and thickness are designated as L, L_t, R, and t, respectively.

4.2.2 *Imperfection Measurements*

Three-dimensional surveys of the inner and outer shell-wall surfaces of the specimen were made prior to testing to determine the initial geometric shell-wall imperfection shape and the shell-wall thickness distribution. Measurements were taken over a uniform grid with increments of 0.125 in. (3.175 mm) in the axial direction and 0.139 in. (approximately 1° of arc) in the circumferential direction over the exposed surfaces of the specimen. The inner surface measurement was used to determine the initial geometric shell-wall imperfection shape and

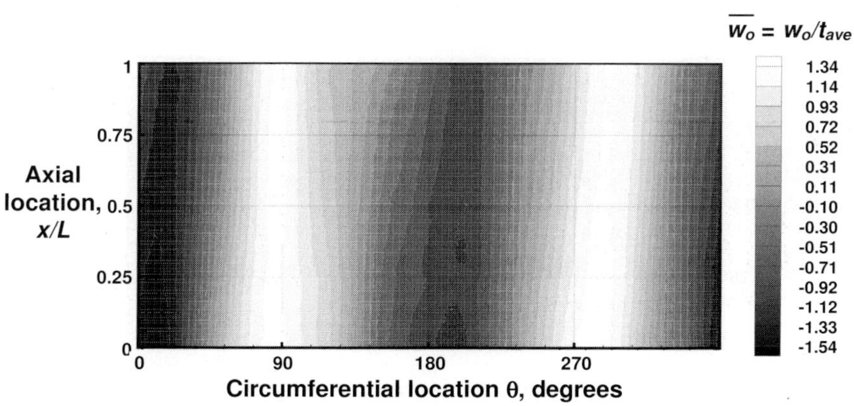

Fig. 4.2. Typical measured inner-surface geometric imperfection shape for shell C3.

the difference between the outer and inner surface measurements was used to determine the shell-wall thickness distribution. A contour plot of the normalized initial geometric shell-wall mid-surface imperfections $w_o(x,\theta)$ for specimen C3 is shown in Fig. 4.2. The measured shell-wall imperfection w_o is normalized by the average measured shell-wall thickness t_{ave} = 0.0381 inches (0.968 mm). These results indicate that the initial geometric shell-wall imperfection is periodic in the circumferential direction and has slight deviations in the axial direction. The maximum amplitude of the imperfection is 1.535 t_{ave}.

A contour plot of the normalized shell-wall thickness variation $\bar{t}_o(x,\theta)$ for the specimen is shown in Fig. 4.3. The measured thickness value t_o is normalized by the average measured shell-wall thickness t_{ave}. These results indicate that the shell-wall thickness, and hence the laminate stiffnesses, varies significantly over a short distance. The thickness varies from 0.928 to 1.321 times t_{ave}. Most of the thickness variation is attributed to local variations in the resin content of the laminate. However, the darker angular pattern in the thickness distribution is caused by gaps between adjacent pieces of graphite-epoxy tape in some of the laminate plies that were generated during the lay-up and curing process. Such a region is referred to herein as a lamina ply-gap or ply-gap. In such a locally thin region, the shell wall is, for example, a 7-ply-thick laminate rather than the nominal 8-ply-thick laminate. In addition, these locally thin regions of the shell wall can

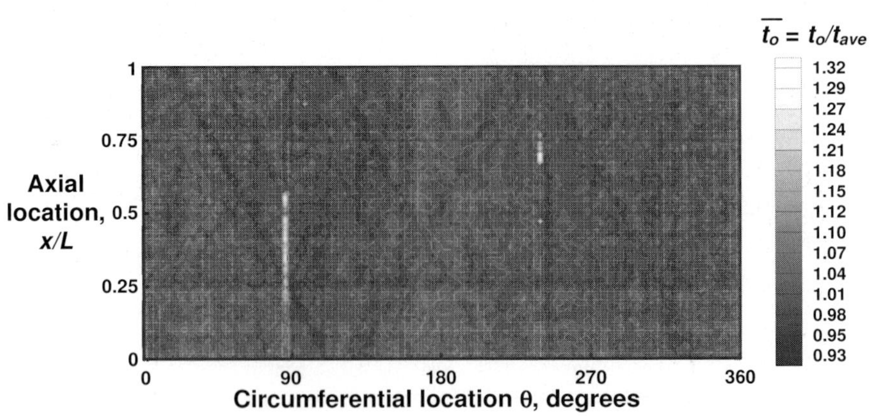

Fig. 4.3. Typical measured shell-wall thickness variation for shell specimen C3.

have, associated with them, a significant mid-surface eccentricity. Typically, a circumferential ply-gap constitutes a gap between two adjacent 90° pieces of tape in the ply, a 45° or helical ply-gap constitutes a gap between two adjacent 45° pieces of tape, and an axial ply-gap constitutes a gap between two adjacent 0° pieces of tape. A magnified cross-sectional view of a ply-gap in a 90° mid-surface ply is shown in Fig. 4.4(a). The width of the ply-gap shown in Fig. 4.4(a) is approximately 0.02 in. (0.51 mm) or approximately half the wall thickness, and the ply-gap depth is between 0.001-in. and 0.005-in. (0.025 to 0.127 mm). However, other observed ply-gaps in the six test specimens have widths as large as 0.1 in. The lighter angular patterns in the thickness contour plot are caused by pinching of the outermost lamina plies of the laminate which forms an outer shell-wall surface ridge. A magnified cross-sectional view of such a region is shown in Fig. 4.4(b).

Measurements of the top and bottom loading surfaces of the specimen were made at every degree around the circumference of the specimen to determine the variation in the shell-end or loading-surface geometry. Typical top and bottom shell-end geometry variations for the specimen are denoted by $\delta_{top}(\theta)$ and $\delta_{bot}(\theta)$, respectively, and are shown in

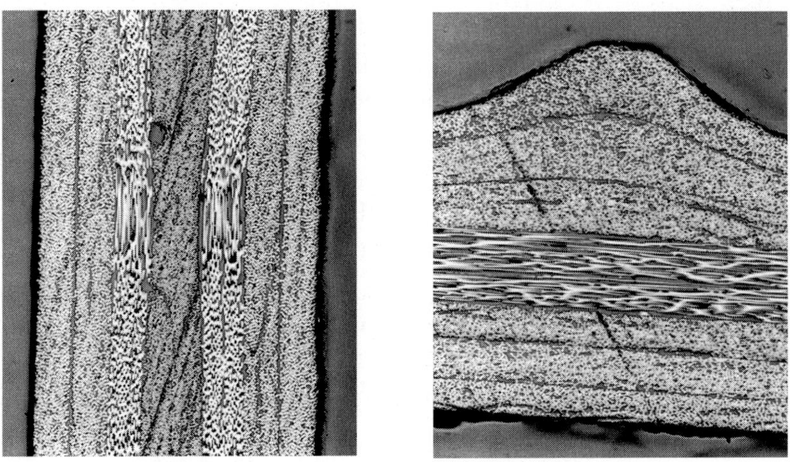

(a) lamina ply-gap (b) outer-surface ridge

Fig. 4.4. Magnified cross-sectional views of a lamina ply-gap and an outer-surface ridge.

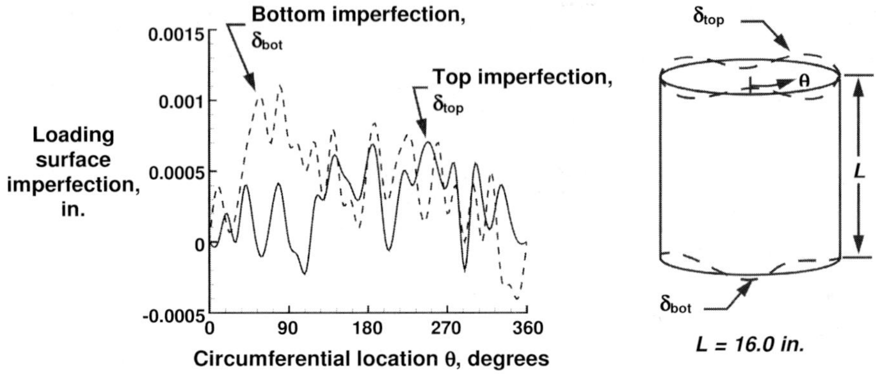

Fig. 4.5. Typical measured shell-end or loading surface imperfections for shell specimen C3.

Fig. 4.5. The maximum amplitude of this shell-end variation is approximately 0.0015 inches, which is approximately 0.01% of the specimen length.

4.2.3 *Test Apparatus and Tests*

The specimens were instrumented with electrical resistance strain gages and direct-current differential transducers (DCDT's) were used to measure displacements. Three non-collinear DCDT's were positioned in three corners of the upper loading platen and used to measure the end-shortening displacement Δ and the rotations ϕ_y and ϕ_z of the loading platen as illustrated in Fig. 4.1. A moiré interferometry procedure was used to qualitatively observe the shell-wall prebuckling, buckling and postbuckling out-of-plane (perpendicular to the shells outer surface) deformation patterns. All data were recorded with a data acquisition system, and the moiré patterns were recorded photographically and on videotape.

The specimens were loaded in compression with a 300 Kip hydraulic universal-testing machine by applying an end-shortening displacement to the loading surfaces of the specimen. To control the load introduction into the specimens, the upper loading platen was aligned with the loading surface of the specimen before each test by adjusting leveling-bolts in the corners of the upper loading platen until strains measured by selected strain gages indicated a uniform axial strain distribution in the shell wall. The specimen was loaded until general instability of the shell occurred.

4.3 Finite-Element Models and Analyses

4.3.1 *Finite-Element Models*

The shells presented in this chapter were analyzed with the STAGS (STructural Analysis of General Shells) nonlinear shell-analysis code.[19] STAGS is a finite-element code developed for the static and dynamic analysis of shells, and includes the effects of geometric and material nonlinearities in the analysis. A typical finite-element model of a cylinder specimen is illustrated in Fig. 4.1. The standard 410 quadrilateral element from the STAGS element library was used in the models and is a flat facet-type element that is based on the Kirchoff-Love shell hypothesis and the nonlinear Lagrangian strain tensor. The element nodes include three translational degrees of freedom and three rotational degrees of freedom. Large element rotations are accounted for by using a corotational algorithm. Typically, the elements of the finite-element mesh are approximately 0.2-in. by 0.2-in. (0.5-mm by 0.5 mm) square. However, some of the elements near the boundary region of the shell are somewhat smaller in order to accurately represent the bending boundary-layer response near the end of the shell. Each element possesses four integration points, which are distributed in such a way as to provide a modeling resolution of approximately 0.1-in. by 0.1-in (0.25-mm by 0.25 mm) square. This integration-point spacing is on the order of the measurement-point spacing used for the initial geometric imperfection measurements of the specimens. Compression-loaded shells and shells subjected to combined axial compression and torsion are analyzed in the present investigation. To simulate the constraints provided by the potting material in a compression-loaded shell, the circumferential and radial displacements v and w were set equal to zero in edge regions of the shell as indicated in Fig. 4.1. For shells subjected to combined axial compression and torsion, the radial displacement w was set equal to zero and a uniform circumferential displacement v was prescribed.

4.3.2 *Nonlinear Analysis Procedure*

The prebuckling, buckling and postbuckling responses of the shells were determined by using the following analysis procedure. The prebuckling responses were determined by using the Riks pseudo arc-length path-following method[34] in STAGS until just before buckling occurred. The unstable buckling response of a shell was predicted by using the nonlinear, transient-analysis option of the code. The transient-analysis option in STAGS uses proportional structural damping and an implicit numerical time-integration method developed by Park.[35] The transient analysis was initiated from an unstable equilibrium state close to the limit point by incrementing the end displacement by a small amount. An initial time step of 1.0E-8 seconds was used in the transient analysis, and the time step was automatically adjusted by the program as a function of the solution behavior. The transient analysis was continued until the kinetic energy in the shell had dissipated to a negligible level, which indicated that the transient response had attenuated. Once the transient analysis had attenuated to a near-steady-state solution, the load-relaxation option of the code was used to establish the stable static equilibrium state.

4.4 Developing Experimentally Validated High-Fidelity Models

The idealized models described in the previous section were modified to more closely represent the actual test configuration and conditions. More specifically, actual measured geometric imperfections, shell-wall thickness variations, nonuniform loading affects, and elastic boundary conditions are included in the models. The modeling approaches and selected results are presented in this section. In addition the predicted results are compared to experimental results.

4.4.1 *High-Fidelity Analysis Models*

Geometrically perfect finite-element models were modified to include the effects of the measured shell imperfections in order to simulate more closely the response of the 'as-tested' specimens. These modeling modifications include the effects of the measured initial geometric shell-wall mid-surface imperfections, shell-wall thickness variations, local shell-wall lamina ply-gaps, thickness-adjusted lamina properties, elastic boundary conditions, shell-end geometric imperfections, and nonuniform end loads. The initial geometric shell-wall mid-surface imperfection $w_o(x, \theta)$ is included in the finite-element models by introducing an initial, normal perturbation to each node of the mesh by using a user-written subroutine with STAGS for that purpose. The user-written subroutine uses a linear interpolation algorithm that calculates the value of the imperfection for the coordinates of each finite-element node based on the measured shell-wall data.

The shell-wall thickness t, mid-surface eccentricity ecz, and lamina material properties E_1, E_2, G_{12} and ν_{12} are adjusted at each integration point of each element in the finite-element models. The shell-wall mid-surface eccentricity is calculated relative to the average shell-wall mid-surface as illustrated in Fig. 4.6; that is, $ecz(x, \theta) = -0.5(t_{ave}-t_o(x,\theta))$. The lamina properties are adjusted by using the rule of mixtures. In the rule of-mixtures calculations, it is assumed that any variation in the lamina ply thickness from the nominal thickness is due to a variation in resin volume only, and that the fiber volume remains constant for each ply. However, there are several assumptions and approximations related to modeling the geometry and stiffnesses of a lamina ply-gap detail that were used in the analysis. First, the finite-element models are limited to

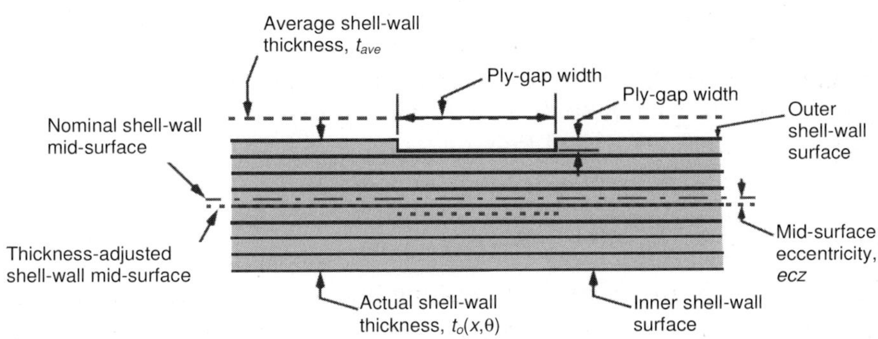

Fig. 4.6. Idealized ply-gap model detail.

modeling the shell-wall thickness variation as discrete step changes, as illustrated in Fig. 4.6, and the resolution of the thickness variation is limited by the finite-element integration point spacing (e.g., 0.1 in. or 2.54 mm for the models presented herein). All ply-gaps detected in the imperfection measurement will be modeled with a fixed width determined by the finite element mesh spacing. Second, as mentioned in Section 2, there is a local reduction in shell-wall stiffness associated with the reduced number of lamina plies. For thick shells it is anticipated that this detail is of little importance, however for thin shells like the ones considered here, the effects of the local stiffness reduction are not known. Fortunately, results from a previous study, illustrating the potential effects of these modeling approximations and mesh refinement on the response of a shell with a ply-gap, indicate that this modeling approximation would cause only slight differences in the magnitude of the local bending response and no more than a 2% variation in the predicted buckling load for a compression-loaded shell.[22] Furthermore, the results indicated that the local shell-wall mid-surface eccentricity is the most important feature of the ply-gap detail for these stability critical problems, and this eccentricity was included in the models for the results presented herein. To provide a better simulation of the constraints provided by the potting material at the ends of the specimens, effective axial and radial potting-support stiffnesses were determined for each shell specimen using a two-dimensional generalized plane-strain finite-element analysis of the potting-material–shell-wall cross-sectional detail shown in Fig. 4.7. Material properties of the potting compound were characterized by Iosipescu tests of the potting material reported by Weiland et al. [17] The nominal properties of the potting material are as follows: Young's modulus E = 1.15 Msi (7.93 GPa), shear modulus G_{12} = 0.36 Msi (2.48 GPa), and Poisson's ratio ν = 0.59. To determine the effective axial and radial potting- support stiffnesses, the following numerical experiments were conducted. In the first numerical experiment, a uniform axial displacement Δ, was applied to an unpotted end of the shell model, and the resulting predicted axial strain response, shown in Fig. 4.8(b), was used to calculate an effective axial stiffness K_A, for the portions of the shell wall supported by the potting material. The results indicate that the axial strain response of the shell wall supported by the potting material is a function of shell-wall laminate modulus. In particular, the decay length of the strain response varies inversely with the axial laminate stiffness. The predicted effective axial potted-shell stiffnesses range from 6.73 to 32.37 Msi (46.4 - 223.2 GPa) for shells presented herein. In the second numerical experiment, a unit concentrated force F_z, was applied to the shell model, and the resulting w (radial or normal) displacement and nodal force at the junction of the shell wall and the potting material was used to calculate an effective radial support stiffness, K_R. The effective radial potting-support stiffness was predicted to be approximately equal to 1.0E5 lbf/in (17.5 N/m).

Nonuniform end loading of a specimen is attributed to initial specimen-end or loading-surface imperfections and to upper loading-platen rotations that are measured during the experiment. First, the measured upper and lower specimen-end or loading-surface imperfections $\delta_{top}(\theta)$ and $\delta_{bot}(\theta)$, respectively, were included in the finite-element model by introducing an initial in-plane axial perturbation to the nodes at the loaded ends of the shell. Then, the compression load was applied to the shell in two parts. The nonuniform specimen-end imperfections, $-\delta_{top}(\theta)$ and $-\delta_{bot}(\theta)$, were applied as displacements to the

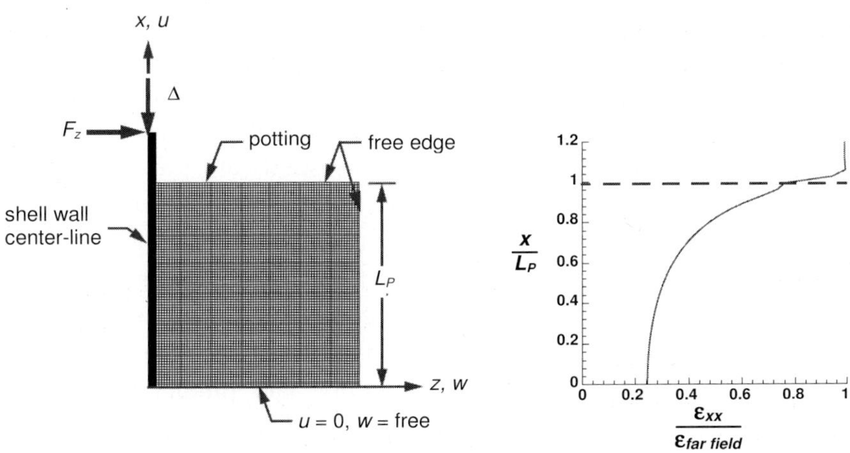

Fig. 4.7. Typical finite element model, geometry and boundary conditions of the potting-shell cross-section detail and typical predicted axial strain distribution at the mid-surface of the shell wall.

upper and lower ends of the shell, respectively, at the beginning of the analysis to simulate a full contact condition between the shell ends and the loading platens. Then, the experimentally measured end-shortening displacement, Δ, and upper loading-platen rotations, ϕ_y and ϕ_z, were applied to the upper shell end or loading surface while holding the lower loading surface fixed as illustrated in Fig. 4.1; that is, $u(-L/2,\theta) = \Delta + R \cos\phi_y \cos\theta + R \cos \phi_z \sin \theta - \delta_{top}(\theta)$ and $u(L/2, \theta) = - \delta_{bot}(\theta)$.

4.4.2 *Typical High-Fidelity Analysis Results*

Analytically predicted and experimentally measured results for the three compression-loaded 8-ply-thick graphite epoxy cylindrical shells considered in this present work are presented in this section. The shell-wall laminates of the three shells include two different orthotropic laminates and one quasi-isotropic laminate. The two 8-ply orthotropic shells, C1 and C2, and the 8-ply quasi-isotropic shell, C3, have shell-radius-to-thickness ratios equal to 200. The predicted results were obtained from finite-element models of geometrically perfect shells and shells that include initial geometric shell-wall mid-surface imperfections, shell-wall thickness variations and thickness-adjusted lamina properties, local shell-wall lamina plygaps, elastic boundary support conditions, and nonuniform loading effects as described in the previous section. These results are presented to illustrate the overall behavior of compression-loaded graphite-epoxy shells and to validate the high-fidelity analysis approach. First, predicted results illustrating the typical nonlinear response of the geometrically perfect 8-ply shells are presented. Next, predicted results illustrating the effects of initial shell geometric and loading imperfections on the buckling response of the 8-ply quasi-isotropic shell C3 is presented. The results include predicted and measured load–end-shortening response curves, predicted and measured load-strain response curves, predicted prebuckling, buckling and postbuckling deformation response patterns, and predicted axial and circumferential stress resultant patterns. The three composite shells considered in this study are interchangeably referred to herein as shells or specimens C1, C2 and C3.

Typical Nonlinear Response of Geometrically Perfect Shells

Results from the nonlinear analyses of selected geometrically perfect shells are presented in this section. The shells analyzed are based on shells C1, C2 and C3 which have orthotropic shell-wall laminates $[-45/+45/0_2]_s$ and $[-45/+45/90_2]_s$, and a quasi-isotropic shell-wall laminate $[-45/+45/0/90]_s$, respectively. First, the load–end-shortening response curves for the shells are compared. Then, results illustrating a typical nonlinear transient collapse response for the quasi-isotropic shell C3 are presented in some detail. Values of the axial load P and the end-shortening Δ presented in this section are normalized by the predicted linear bifurcation buckling load for the quasi-isotropic shell C3, = 42,590 lbs. (189.4 kN), and the nominal shell-wall thickness, t_{nom} = 0.04 in. (1.02 mm), respectively.

The predicted load–end-shortening responses are compared for the three shells in Fig. 4.8(a). The prebuckling load–end-shortening responses are linear up to the general instability point indicated in the figure for each shell. The prebuckling slopes of each response curve vary according to the individual laminate in-plane axial stiffnesses. General instability occurs at normalized axial load values of 0.707, 0.868 and 0.979, for shells C1, C2 and C3, respectively. The general instability response is followed by a sudden reduction in the axial load supported by

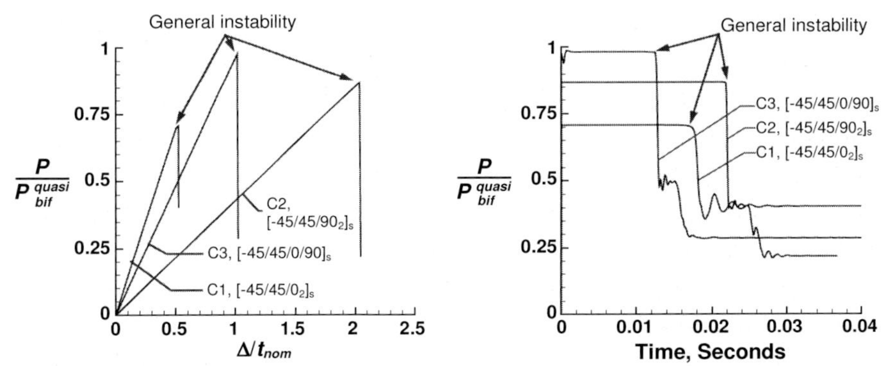

(a) Normalized load—end-shortening response (b) Normalized load-time history during collapse

Fig. 4.8. Numerically predicted nonlinear response of geometrically perfect compression-loaded cylindrical shells.

the shells and is associated with the transient collapse response of the shells. The corresponding load-time histories of the predicted transient collapse response of the shells are shown in Fig. 4.8(b). The initial portion of a typical load-time history curve, e.g., the load-time history curve for shell C1, is associated with the transition from a state just before buckling occurs (time = 0.0) to the state associated with the general instability point indicated in the figure. This initial portion of the load-time history curve only occurs in the analysis of the geometrically perfect shells considered in the present study, and is attributed to the uniformly symmetric character of the initial response. The nonlinear solution algorithm that predicts the structural response of the shell needs a small perturbation in the response to be able to change from the axisymmetric prebuckling response state to an asymmetric general instability and transient collapse response state in the analysis. Since there are no such geometric or loading perturbations in the model for the geometrically perfect shells, it is suspected that, over time small numerical variations in the transient solution accumulate and provide a small perturbation to the equilibrium state of the shell that causes the general instability and transient collapse response to occur. Following the initiation of the general instability response, a typical load-time history curve exhibits a sudden reduction in the axial load supported by the shell which is associated with the transient collapse of the shell. The value of the axial load continues to decrease until the axial load attenuates to a steadystate value. The kinetic energy in the shell dissipates over time and the shell reaches a stable postbuckling equilibrium state after approximately 0.035 to 0.04 seconds have elapsed. Normalized post-collapse postbuckling load values equal to 0.403, 0.218 and 0.286 are obtained for shells C1, C2 and C3, respectively. The results indicate that laminate orthotropy can have a significant effect on the load–end-shortening response of the shells. In particular, while shell C1 has the lowest general instability load value of all three shells, it also has the highest postbuckling load value of all three shells.

The predicted transient deformation responses for selected time steps during the transient collapse of quasi-isotropic shell C3 are presented in Figs. 4.9(a) through 4.9(f). The deformation responses shown in the figures have been scaled for clarity. The shell-wall radial deformations at the general instability point shown in Fig. 4.8 (0.0125 seconds into the predicted collapse response) change in a short period of time from the uniform axisymmetric prebucking deformation pattern shown in Fig. 4.9(a) to the asymmetric transient unstable deformation pattern shown in Fig 4.9(b), at which time the shell begins to collapse. The magnitudes of the shell-wall normal displacements vary between ±0.5 times the shell-wall thickness. The general instability response of the shell is caused by the nonlinear coupling of the localized destabilizing compressive axial and circumferential stresses in the shell and the out-of-plane (radial) deformations that occur in the bending boundary layer near the ends of the shell. The initial buckling deformation pattern indicated in Fig. 4.9(b) is characterized by localized ellipse-like buckles in the bending boundary layer near the ends of the shell. The skewing of the deformation response is attributed to the presence of the small anisotropic bending-twisting coupling stiffness terms of the laminate. After 0.013 seconds have elapsed, the normalized axial load has decreased from 0.983 to 0.479, and the deformation response in the shell has evolved to include additional rows of small ellipse-like buckles as indicated in Fig. 4.9(c). The magnitudes of the shell-wall normal displacements vary from -4 to +2 times the shell-wall thickness. After 0.0134 seconds have elapsed, the normalized axial load has decreased to 0.465, and the local buckles in the deformation pattern have rapidly increased in number to include 24 circumferential half-waves and four axial half-waves as indicated in Fig. 4.9(d). After 0.0152 seconds have elapsed, the normalized axial load has decreased further to 0.426, at which time the small buckles in the shell wall begin to coalesce into larger buckles with displacement magnitudes that vary from -7 to +3 times the shell-wall thickness as indicated in Fig. 4.9(e). After approximately 0.04 seconds have elapsed, the kinetic energy in the shell has dissipated to a negligible level, and the shell has deformed into a stable postbuckling mode-shape that consists of 16 circumferential half-waves and two axial half-waves, as indicated in Fig. 4.9(f). Shells C1 and C2 exhibit similar response characteristics, and are presented in Ref. 22.

(a) Prebuckling deformations, time = 0.0 s.

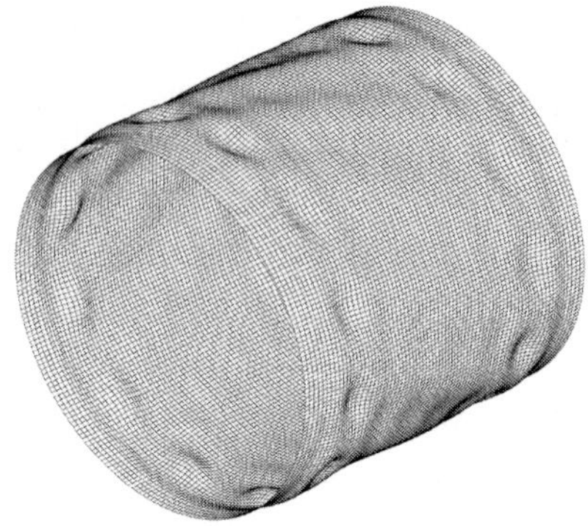

(b) Initial collapse deformations, time = 0.012 s.

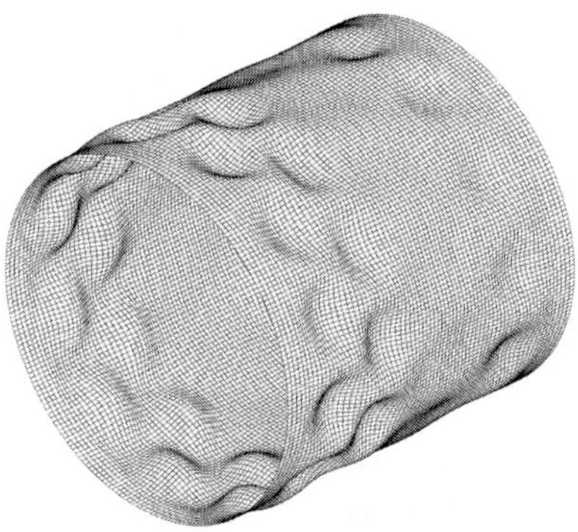

(c) Transient collapse deformations, time = 0.013 s.

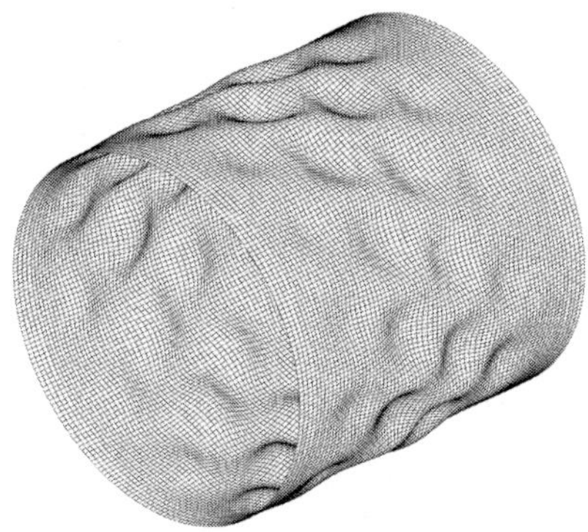

(d) Transient collapse deformations, time = 0.014 s.

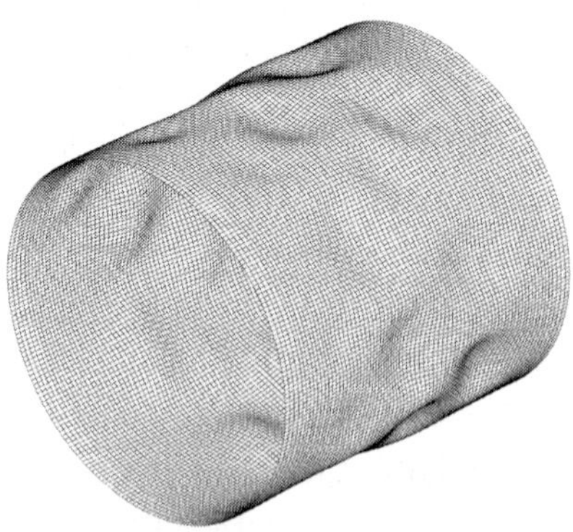

(e) Transient collapse deformations, time = 0.015 s.

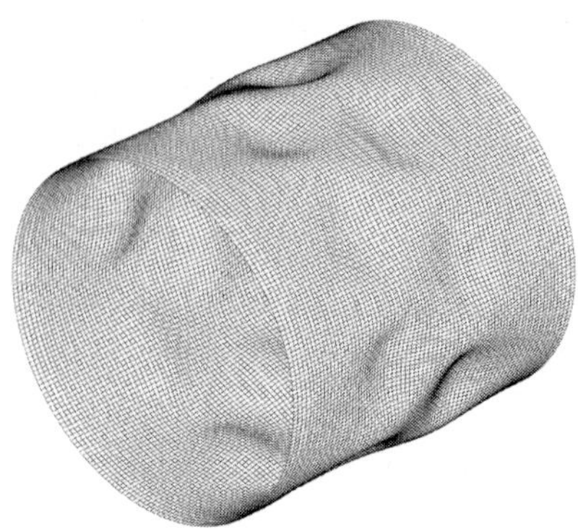

(f) Post-collapse deformations, time = 0.04 s.

Fig. 4.9. Numerically predicted collapse response for a geometrically perfect compression-loaded quasi-isotropic shell C3.

Effects of Initial Imperfections on the Buckling Response

Typical analytically predicted transient response deformations that occur during the collapse of shell C3 with the measured imperfections included in the analysis are shown in Figs. 4.10(a) through 4.10(d). Just before buckling occurs, the shell wall deformations are characterized by several localized ellipse-like deformation patterns as indicated in Fig. 4.10(a). The localization in the deformation pattern is caused by the combination of a local geometric shell-wall imperfection that is in the form of a significant variation in the shell-wall mid-surface geometry, and the

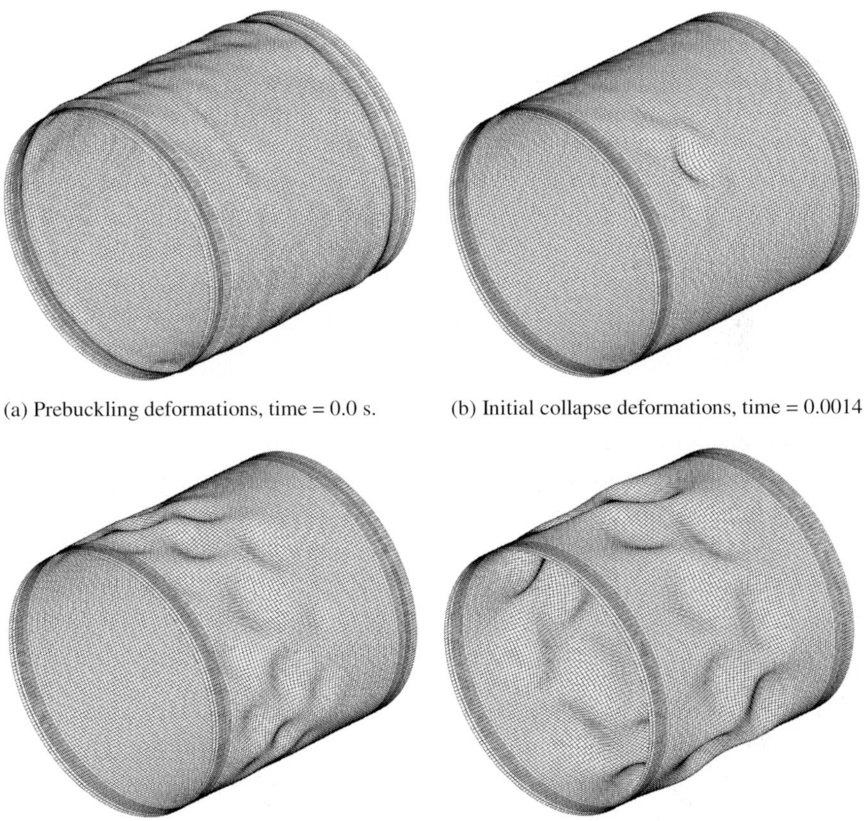

(a) Prebuckling deformations, time = 0.0 s. (b) Initial collapse deformations, time = 0.0014 s.

(c) Transient collapse deformations, time = 0.0024 s. (d) Post-collapse deformations, time = 0.02 s.

Fig. 4.10. Numerically predicted collapse response for an imperfect compression-loaded quasi-isotropic shell C3.

intersection of a helical ply-gap and a circumferentially aligned ply gap in the shell at $x/L_T = 0.25$ and q = 210°. The localized deformations occur in regions with destabilizing compressive axial and circumferential stresses. After approximately 0.00143 seconds have elapsed in the transient response, a single ellipse-like buckle has grown in amplitude and couples with the destabilizing stresses in the shell wall to cause the general instability and collapse of the shell. After approximately 0.00238 seconds have elapsed in the transient response, additional local buckles have formed around the circumference and along the length of the shell as indicated in Fig. 4.10(c). As the buckling process continues, the deformation pattern in the shell wall continues to evolve, and the ellipse-like buckles in the shell begin to coalesce into larger diamond-shaped buckles. After approximately 0.02 seconds have elapsed in the transient response, the kinetic energy in the shell has dissipated to a negligible level, and the shell has deformed into a stable postbuckling mode-shape as indicated in Fig. 4.10(d).

These results indicate that the collapse response of the imperfect shell is initiated by a localized response which leads to the over-all collapse of the shell. These results are different from the collapse response results exhibited by the corresponding geometrically perfect shell which are shown in Figs. 4.9(a)-4.9(f). The transient collapse response occurs after 0.00143 seconds have elapsed in the transient analysis in the shell with imperfections, and this collapse response occurs much earlier in the load-response history of the shell than for the corresponding geometrically perfect shell collapse response. This earlier collapse time is attributed to the localized prebuckling shell-wall deformations and destabilizing in-plane compressive stresses in the shell, which result in a rapid transition from the stable prebuckling state to the unstable transient collapse response. Results for orthotropic shells C1 and C2 indicate similar response characteristics associated with the transient collapse response of these shells.

Predicted and Measured Response Comparisons

Selected results from nonlinear analyses of the orthotropic and quasi-isotropic shells are compared to the experimentally measured results in this section. The nonlinear analysis results are for shell models that included the effects of the measured initial shell-wall geometric and thickness imperfections, thickness-adjusted material property variations, shell-wall lamina ply-gaps, measured loading variations, elastic radial support conditions, and selected specimen parametric uncertainties. A value of the elastic radial support stiffness equal to 1.0E5 lbf/in (17.5 N/m). is used in the models to simulate the boundary conditions provided by the potted end supports of the shell specimens. Upper and lower response bounds were determined based upon the results of a traditional combinatorial analysis of the effects of selected specimen parametric uncertainties. Uncertainties in the measured thickness imperfection, lamina fiber volume fraction, and applied load distribution were considered in the analysis. More specifically, the shell-wall imperfection measurement values and the shell-end-shape imperfection measurement values used in the present study are assumed to be accurate to within ±0.0006 in (0.015 mm), the shell-wall thickness values are accurate to ±0.0012 in (0.030 mm), and the fiber volume fraction specified by the manufacturer of the material to be equal to 0.65 ±0.03 for the 0.005-in-thick graphite-epoxy material. Predicted and measured load–end-shortening response curves, selected load-strain responses, and observed and predicted post-collapse normal displacements are presented in this section.

Analytically predicted and experimentally measured load–end-shortening response curves for shells C1, C2 and C3 are presented in Fig. 4.11. Values of the axial load P and the end-shortening Δ are normalized by the linear bifurcation buckling load of quasi-isotropic shell C3, = 42,590 lbs. (189.4 kN), and the nominal shell-wall thickness t_{nom} = 0.04 in. (1.016 mm), respectively. The solid and dashed lines in the figure represent experimentally measured and analytically predicted results, respectively. Each shell has two predicted response curves representing analytically predicted upper and lower bounds to the response based on specimen parametric uncertainties. The measured results indicate that shells C1, C2 and C3 exhibit the general instability points shown in the figure at normalized load values of 0.652, 0.749 and 0.803, respectively, and are 7.8, 13.7 and 17.6% lower than the predicted loads for the geometrically perfect nominal shells shown in Fig. 4.8(a), respectively.

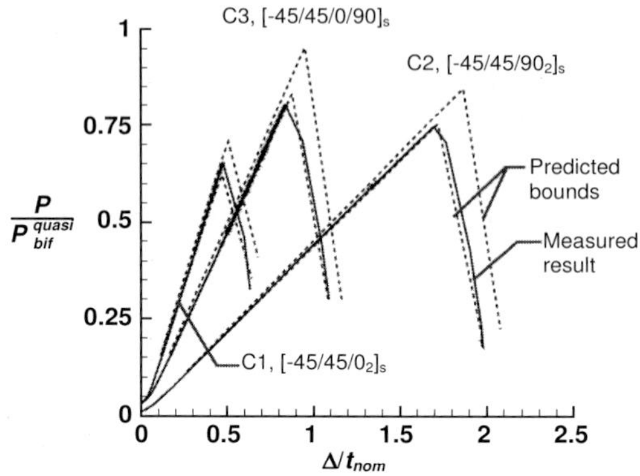

Fig. 4.11. Numerically predicted and experimentally measured load—end-shortening response curves for quasi-isotropic shell C3; predicted results represent response bounds due to uncertainty.

These results indicate that, for the most part, the measured load–end-shortening response curves for the shells fall on or within the analytically predicted response bounds. These results indicate that the measured response curves tend to correlate well with the analytically predicted lower bounds for each shell.

Analytically predicted and experimentally measured strain response curves for a selected point on shell C3 are presented in Fig. 4.12 to illustrate the typical strain response correlation and sensitivity to modeling variations. Values of the axial load P and the axial strain ε_x are normalized by the linear bifurcation buckling load of the quasi-isotropic shell C3, = 42,590 lbs. (189.4 kN), and the corresponding axial strain at buckling ε_{cr} = 0.00257, respectively. The strain results in the figure represents back-to-back surface strain measurements within the bending boundary layer near the ends of the shell. Strain gages 83 and 84 are outer and inner surface strain gages located at $x/L_T = 0.071$ and $\theta = 180^{\circ}$. The bending boundary layer behavior is evident in the divergent character of the back-to-back predicted and measured strain results. These results indicate that the measured prebuckling strain responses, for the most part, fall on or within the predicted response bounds. The results also indicate that the predicted post-collapse strain response can be very sensitive to specimen parametric uncertainties. Similar behavior and response sensitivities were observed for shells C1 and C2.

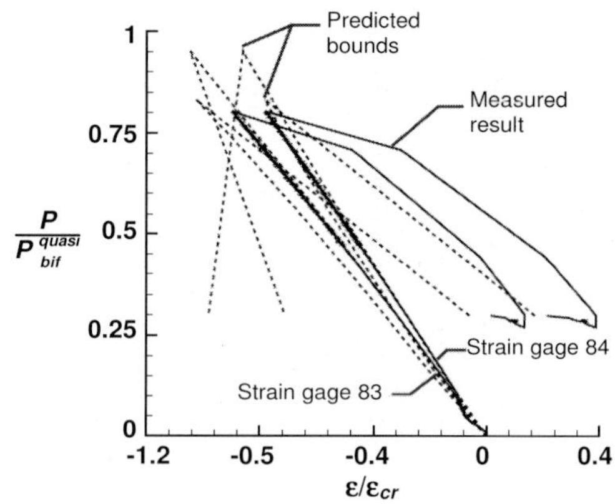

Fig. 4.12. Numerically predicted and experimentally measured load-strain response curves for quasi-isotropic shell C3; predicted results represent response bounds due to uncertainty.

Predicted initial post-collapse normal displacement contours and the corresponding observed moiré fringe patterns for specimen C1 are shown in Fig. 4.13. The dashed contour lines in the predicted displacement contour plots represent inward displacements and the solid lines represent outward displacements. The density of the contour lines indicates the severity of the displacement gradients in the specimen.

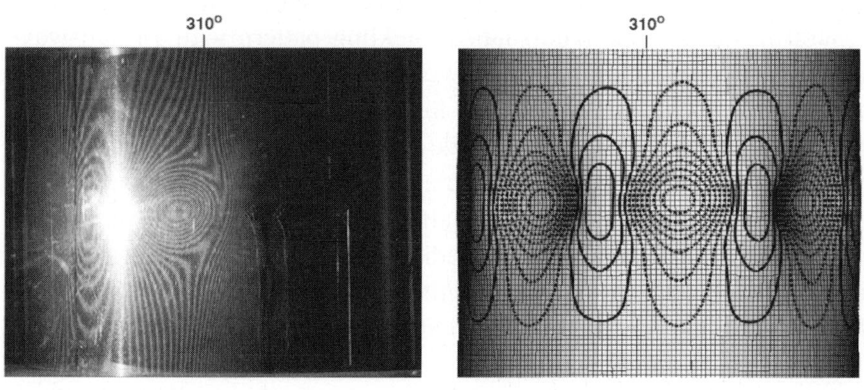

a) Observed post-collapse pattern.

b) Numerically predicted post-collapse deformation response.

Fig. 4.13. Observed and predicted initial post-collapse radial displacement patterns for specimen C1.

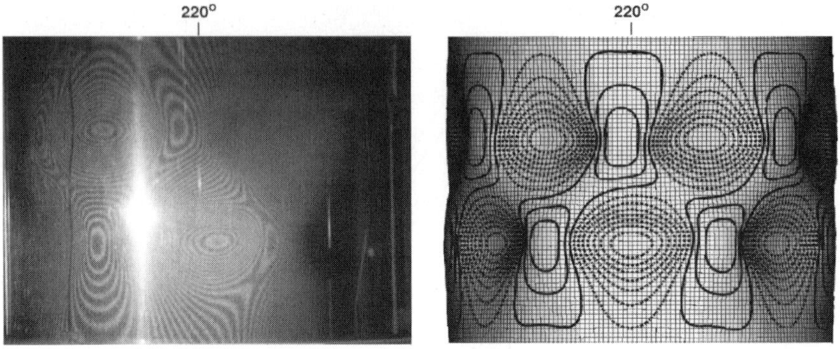

a) Observed post-collapse pattern.

b) Numerically predicted post-collapse deformation response.

Fig. 4.14. Observed and predicted post-collapse radial displacement patterns for specimen C2.

These results indicate that the specimen collapses into a general-instability diamond-shaped buckling pattern with 16 half-waves around the circumference and one half-wave along the length, as predicted by the transient analysis. The analytical results predict that the mode-shape pattern is 15-20° out-of-phase with the observed mode-shape pattern. Predicted initial post-collapse normal displacement contours and the corresponding observed moiré fringe patterns for specimen C2 are shown in Fig. 4.14. These results indicate that the specimen collapses into a general instability diamond-shaped buckling pattern with 14 half-waves around the circumference and two half-waves along the length, as predicted by the transient analysis. In addition, the predicted mode-shape is in-phase with the observed mode-shape. Similar results for shell C3 indicate that the shell collapses into a general-instability diamond-shaped pattern with 16 circumferential half-waves and two axial half-waves. However, the analytical results predicted that the mode-shape pattern is approximately 15° out-of phase with the observed moiré fringe pattern. This discrepancy explains, in part, the large discrepancies in the predicted post-collapse strain response indicated in Fig. 4.12.

4.5 Analysis-Based High-Fidelity Design Criteria

The finite-element models and analysis methods used to develop the new design criteria are based on experimentally validated models presented in the previous section. The results presented in the previous section indicate that highly accurate buckling responses can be predicted for compression-loaded shells if initial geometric imperfections, material property variability, nonuniform thickness distribution, and other features are modelled to a high degree of accuracy. Such validated models that include the effects of these details are generally deemed necessary to develop the most accurate, analysis-based, high-fidelity design criteria. In line with the objectives of the current study, only the effects of the initial geometric imperfection are included in the numerical models. However, the model details, such as the mesh refinement and idealized boundary conditions, and the analysis methods are the same as those used in the previous studies and, as a result, the solutions are expected to be converged. In addition, the validated finite-element models used for compression-loaded shells are also used for the analysis of the torsion-loaded shells and the shells subjected to combined torsion and compression presented herein, and are expected to be converged and adequate because of the finite-element models' extensive refinement.

4.5.1 *Manufacturing Imperfection Signature*

Shell-wall geometric imperfections that had been measured previously for the six laminated-composite shells considered herein were used to identify a manufacturing-process-specific "first approximation" imperfection signature for this type of structure. First, a reference or "best-fit" cylinder for the initial imperfection measurement data for each shell was determined. Specifically, a data reduction program that uses the method of least squares was used to compute the rigid-body translations and rotations and the mean radius of each cylinder. Then, the measured radial imperfections, deviations from a perfect cylinder, were recalculated with respect to the corresponding new best-fit cylinder. The measured imperfection data was then represented by a two-dimensional half-wave Fourier cosine series given by

$$w_o(x,\theta) = t_{nom} \sum_{k=0}^{N} \sum_{l=0}^{N} \cos\left(\frac{k\pi x}{L}\right)\left(A_{kl}\cos\left(\frac{l\pi x}{L}\right) + B_{kl}\sin\left(l\pi\theta\right)\right) \qquad (1)$$

where R, L, and t_{nom} are the shell radius, length and wall thickness; x and θ are the axial and circumferential coordinates; and k and l are integers that denote the number of axial half-waves and circumferential full-waves, respectively. Using a standard representation of this type for the measured imperfection enables convenient analysis and comparison of different imperfection distributions from different shells and different manufacturing processes.

A plot of the coefficient distribution of a half-wave Fourier cosine representation of the measured geometric imperfection shape for shell specimen C3 is presented in Fig. 4.15. The results indicate that the largest-magnitude component of the imperfection is the $k = 0$ and $l = 2$ component and that the magnitudes of the other components decrease rapidly as the wave numbers k and l increase. The $k = 0$ and $l = 2$ component corresponds to a cylinder with an elliptical cross-section that has a magnitude of $1.2t_{nom}$. The next largest component is associated with $k = 1$ and $l = 2$ and has a magnitude of approximately $0.2t_{nom}$. In general, the dominant coefficients are associated with the long wavelength imperfections in the circumferential and axial directions. Coefficients for values of k and $l > 10$ have a negligible effect on the overall imperfection shape. However, the short wavelength components are retained in the imperfection model since these modes may be associated with high

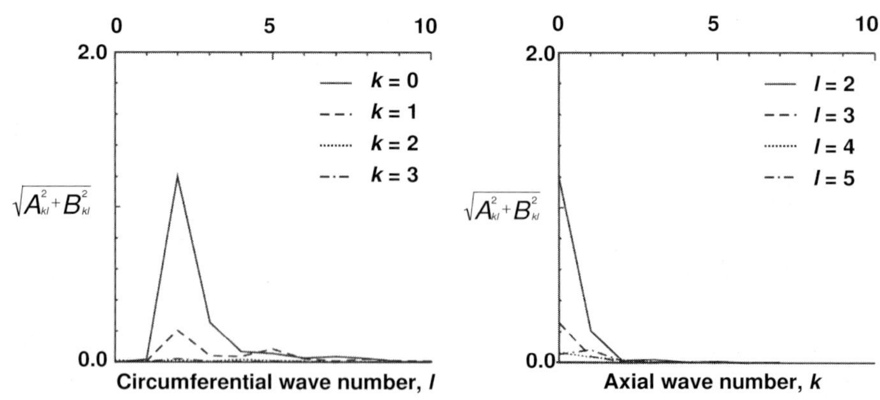

Fig. 4.15. Coefficient distribution of a Fourier series representation of the measured imperfection shape for shell specimen C3.

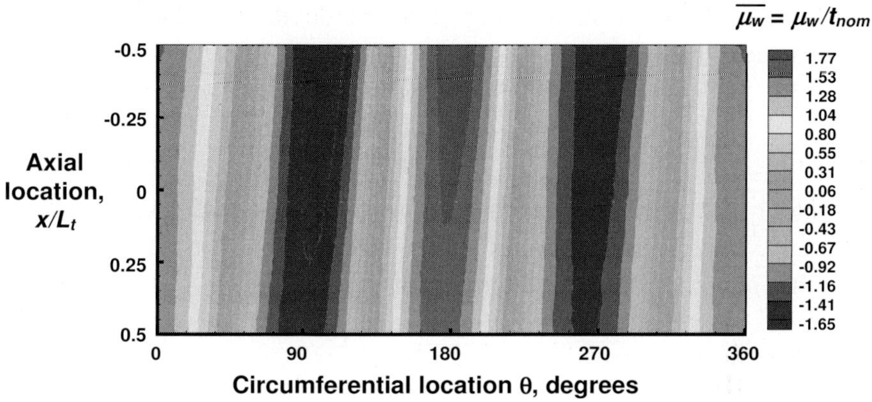

Fig. 4.16. Calculated "first-approximation" mean imperfection shape.

imperfection sensitivity for the shells considered herein even though their magnitudes are generally very small.[37] Results for shells C1, C2, and C4-C6 indicate similar coefficient distributions and suggest that a manufacturing-process-specific imperfection signature for graphite-epoxy laminated-composite shells can be established in terms of a characteristic Fourier coefficient distribution and range. To this end, a mean imperfection shape and a corresponding standard deviation to the imperfection shape were determined based on the measured imperfection data. The mathematical expectation or mean value of $w_o(x, \theta)$ is given by

$$\mu_w(x,\theta)=\langle w_o(x,\theta)\rangle = t_{nom}\sum_{k=0}^{N}\sum_{l=0}^{N}\cos\left(\frac{k\pi x}{L}\right)\left(\langle A_{kl}\rangle\cos\left(\frac{l\pi x}{L}\right)+\langle B_{kl}\rangle\sin(l\pi\theta)\right) \tag{2}$$

where $\langle\ldots\rangle$ denotes the mean value. A contour plot of the normalized mean geometric imperfections, $\mu_w(x, \theta)$, is shown in Fig. 4.16. The mean shell-wall imperfection μ_w is normalized by the nominal shell-wall thickness t_{nom} = 0.04 inches. These results indicate that the initial geometric shell-wall imperfection is periodic in the circumferential direction, exhibits slight variations in the axial direction, and is similar in character to the measured imperfection shape for shell C3 shown in Fig. 4.2. The amplitude of the imperfection μ_w varies from $+1.77t_{nom}$

to -1.65 t_{nom}. The corresponding standard deviation of the geometric imperfection shape $\sigma_w(x, \theta)$ is defined by

$$\sigma_w^2(w_o(x,\theta)) = \left\langle \left(w_o(x,\theta) - \langle w_o(x,\theta)\rangle\right)^2 \right\rangle \tag{3}$$

A contour plot of the corresponding normalized standard deviation of the geometric imperfection shape is shown in Fig. 4.17. The results indicate that the maximum standard deviation is equal to 2.172 t_{nom} and is primarily associated with long wavelength components of the imperfection shape.

The coefficient of variation was calculated for each component of the imperfection and gives the relationship between the mean and the standard deviation. The coefficient of variation, denoted by V, is given by

$$V = \frac{\sigma(X_{kl})}{\mu(X_{kl})} \tag{4}$$

where

$$X_{kl} = \sqrt{A_{kl}^2 + B_{kl}^2} \tag{5}$$

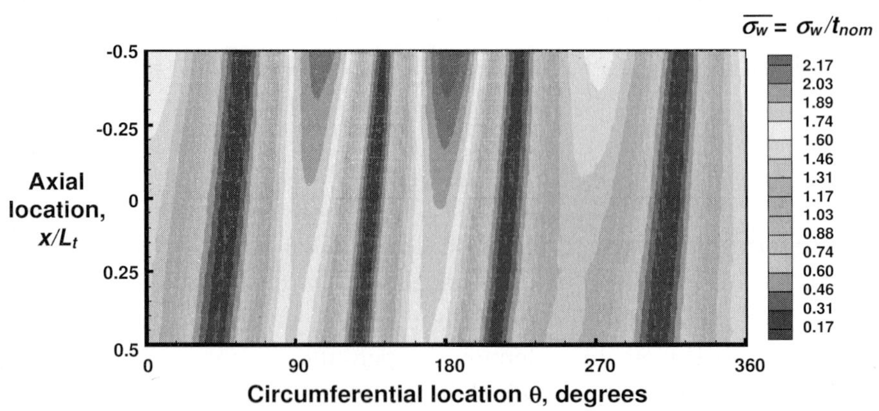

Fig. 4.17. Calculated standard deviation of the imperfection shape.

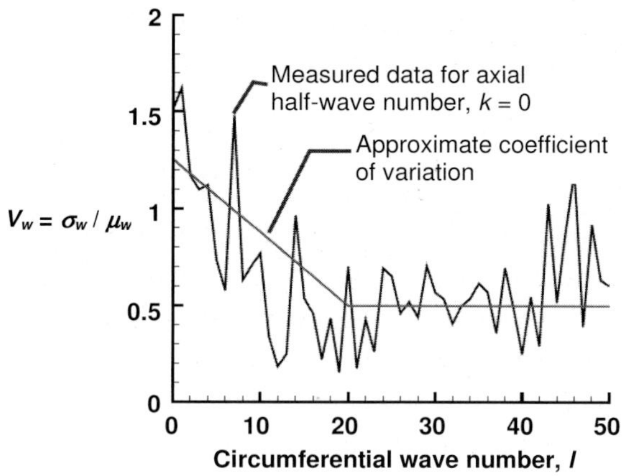

Fig. 4.18. Coefficient of variation V for the axial half-wave number $k = 0$ as a function of the circumferential wave number l.

A typical plot of V for an axial half-wave number $k = 0$ is shown in Fig. 4.18. The results indicate that V is relatively high for long wavelength imperfections; e.g., $V = 1.6$ for $k = 0$ and $l = 2$, and decreases rapidly to a value of approximately 0.5 for $l < 20$. Results for the other values of k were also obtained and indicate similar trends. This information suggests that an approximate expression that relates the mean coefficient values with the corresponding standard deviation can be developed. For example, an approximate V relationship for this particular case that is applicable for all values of k is given by

$$V = \begin{cases} 1.25 - 0.75/20l & for \quad 0 \le l \le 20 \\ 0.5 & for \quad 20 \le l \le 50 \end{cases} \tag{6}$$

This equation is used in subsequent sections when defining "simulated" imperfections in place of the exact standard deviation function given in Eq. 3 to simplify the calculations.

4.5.2 *Response of Compression-Loaded Shells*

The effects of five different imperfection shapes on the response of the three compression-loaded 8-ply shells C1-C3 are described in this section. Specifically, the effects of actual measured shell-wall imperfections, simulated imperfections that are based on the "first approximation" imperfection signature, and linear-bifurcation-mode imperfections were analyzed. The simulated "first approximation" imperfection signature includes the mean imperfection with and without plus or minus one standard deviation. Load-shortening response curves and shell deformations are presented subsequently for the quasi-isotropic shell C3 that illustrate the overall response characteristics for the compression-loaded shell with the five different imperfection shapes. In addition, predicted buckling loads for the two orthotropic 8-ply shells C1 and C2 with the five different imperfection shapes are also presented. Moreover, the predicted buckling loads are compared to the actual experimental buckling loads and to the design load one would get by extrapolating the results in NASA SP-8007[6] for isotropic shells to the composite shells.

Four predicted load-shortening response curves are presented for shell C3 in Fig. 4.19 and include results for a shell with the actual measured imperfection shape; a shell with the mean imperfection shape, denoted by μ_w; a shell with the mean imperfection shape plus one standard deviation, denoted by $\mu_w + \sigma_w$; and a shell with the mean imperfection shape minus one standard deviation, denoted by $\mu_w - \sigma_w$.

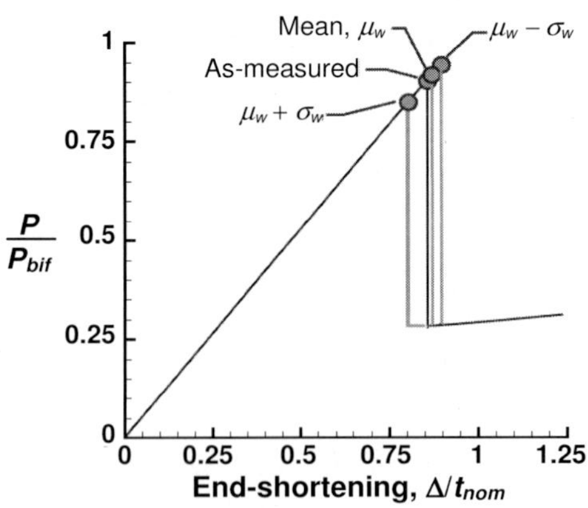

Fig. 4.19. Predicted load—end-shortening response curves for the compression-loaded quasi-isotropic shell C3 with various imperfection shapes.

The axial load P and end-shortening Δ are normalized with respect to the linear bifurcation buckling load of the geometrically perfect nominal shell, P_{bif} = 42,590 lbf (189.4 kN), and the nominal shell-wall thickness, t_{nom} = 0.040 in (1.016 mm), respectively. The load–end-shortening curves clearly indicate a linear prebuckling response. The general instability occurs in the shell at normalized axial load levels that range from P/P_{bif} = 0.855, for a shell with $\mu_w + \sigma_w$ imperfection shape, to 0.943 for the shell with a $\mu_w - \sigma_w$ imperfection shape. The general-instability response of the shell is followed by a sudden reduction in the axial load supported by the shell that is associated with the transient collapse response of the shell. The results indicate that the postbuckling load level is insensitive to the different types of imperfections considered herein.

Deformation results were obtained for shell C3 with the mean imperfection shape, the $\mu_w + \sigma_w$ imperfection shape, and $\mu_w - \sigma_w$ imperfection shape and indicated similar response characteristics to those exhibited by the shell with the actual measured imperfection. In particular, predicted initial buckling deformations for shell C3 are presented in Figs. 4.20(a) through 4.20(c). These results indicate that the initial buckling deformation response of a shell with the mean imperfection μ_w and the $\mu_w + \sigma_w$ imperfection are similar to that exhibited by the shell with the actual measured imperfection and are characterized by a localization of a single buckle near one end of the shell. In contrast, the shell with the $\mu_w - \sigma_w$ imperfection shape exhibits distributed initial buckling deformations shown in Fig 4.20(c) that are similar to those exhibited by the corresponding geometrically perfect shell. These results

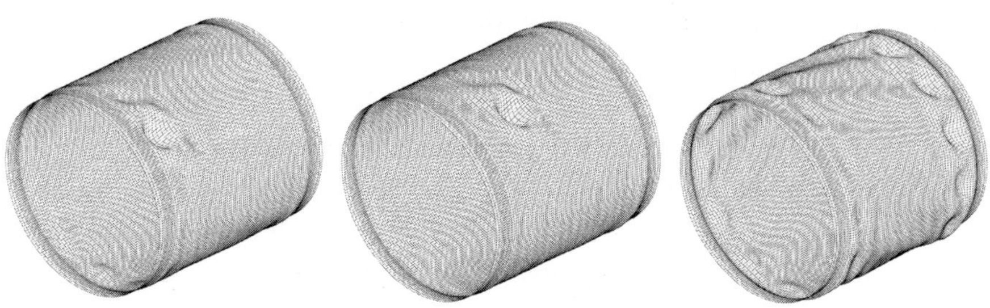

(a) Mean imperfection, μ_w (b) $\mu_w + \sigma_w$ imperfection (c) $\mu_w - \sigma_w$ imperfection

Fig. 4.20. Typical initial buckling deformation responses for shell C3 with selected simulated initial imperfections.

are significant because they suggest that the use of a mean imperfection shape and the corresponding standard deviation can bound the experimental buckling loads as well as predict physically meaningful response characteristics such as the buckling deformations. Results were also obtained for the two 8-ply orthotropic shells C1 and C2. Most notable, is the fact that the predicted buckling loads for the shells with the actual measured imperfection are bounded by the results for the shells with the $\mu_w + \sigma_w$ imperfection shape, and $\mu_w - \sigma_w$ imperfection shape. In addition, the experimental results agree well with the predicted results that include the effects of the actual measured geometric imperfection. The correlation would improve still further if other measured imperfections were included in the finite-element model; e.g., measured thickness variation, material property variation, and loading nonuniformities, as discussed in the previous sections. Specifically, the measured thickness imperfection has been shown to account for an additional 5-10% reduction in the buckling load of the shells. In contrast, the results for shells with linear-bifurcation-mode imperfections with an imperfection amplitude equal to 0.5 times the nominal wall thickness are ultraconservative. Similarly, the lower-bound design recommendation given in NASA SP-8007 for the corresponding isotropic shell is equal to $P/P_{bif} = 0.47$, which is also ultraconservative, as expected.

4.5.3 *Response of Shells Subjected to Combined Axial Compression and Torsion*

A numerical study of the effects of the four different imperfection shapes on the buckling load of the 8-ply shells C1-C3 subjected to combined axial compression and torsion was also conducted. Normalized buckling interaction curves that include curves for a geometrically perfect shell, a shell with the actual measured imperfection, a shell with a $\mu_w + \sigma_w$ imperfection, a shell with a $\mu_w - \sigma_w$ imperfection, a shell with an linear-bifurcation-mode imperfection, and the corresponding NASA SP-8007 design curves are shown for shells C1-C3 in Fig. 4.21, respectively. The measured experimental buckling load for the compression loading is indicated on the figure by the solid circular symbol and is included for comparison. The critical compression and torsion loads, P and T, are predicted from a geometrically nonlinear analysis, and are normalized with respect to the corresponding linear bifurcation buckling loads, and of a geometrically perfect shell. The superscript "+" indicates results for a positive torsion load. The predicted results shown in Fig. 4.21 indicate that the buckling interaction curves for the geometrically perfect and imperfect shells are skewed with respect to the plane given by $T = 0$. The skewing of the predicted interaction curves is caused by the fact that the diagonal compression caused by the shear load is aligned with the outer 45-deg fibers for the case of negative shear and is perpendicular to the fibers for the case of positive shear. Thus, for a negative-shear load the shell is stiffer. In addition, the buckling interaction curves are concave and are highly dependent on the laminate stacking sequence. This behavior is similar to the buckling behavior of anisotropic plates that are subjected to combined compression and shear loads presented in Refs. 38 and 39. In contrast, the NASA SP-8007 design guideline assumes a linear relationship between the critical compression buckling load and the torsion-buckling load. Moreover, the bilinear curve is symmetric about $T = 0$ because the results in NASA SP-8007 are for isotropic shells.

The results in Fig. 4.21 show that shells C1 through C3 are sensitive to the 'as-measured' imperfections and the $\mu_w + \sigma_w$ imperfection when the shell is subjected to compression dominated loading. However, as the torsion load increases, the predicted buckling loads exhibit significantly less imperfection sensitivity for these particular imperfection shapes. The regions of the buckling load interaction curves for which the imperfection sensitivity exists is strongly dependent on the laminate stacking sequence. For example, the buckling load interaction curves for shell C1 exhibit a relatively large region of imperfection sensitivity for normalized torsion loads that ranges from -0.9 to 0.5 and normalized compression loads P/P_{bif} that range from 0.65 to 0.97, as shown in Fig. 4.21(a). However, the regions of imperfection sensitivity in shells C2 and C3 are significantly smaller than that exhibited by shell C1, as shown in Figs. 4.21(b) and 4.21(c), respectively. In contrast, the results indicate that a shell with a linear-bifurcation-mode imperfection exhibits a significant reduction in the buckling loads for all load cases. The large difference in the response trends is caused by a fundamental difference in

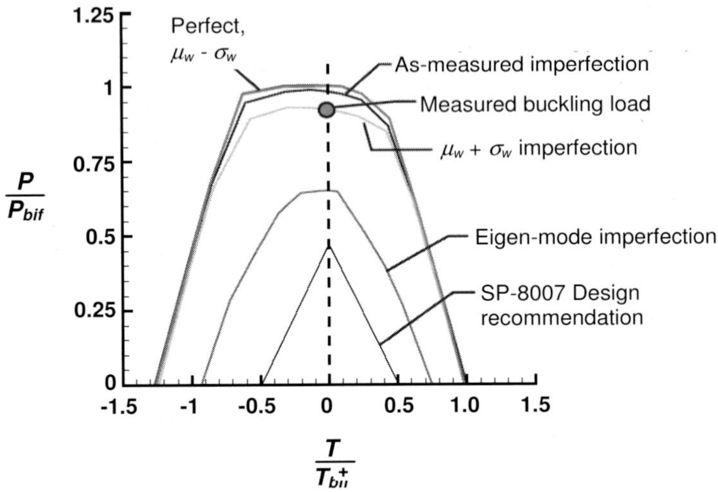

(a) Axially stiff shell C1

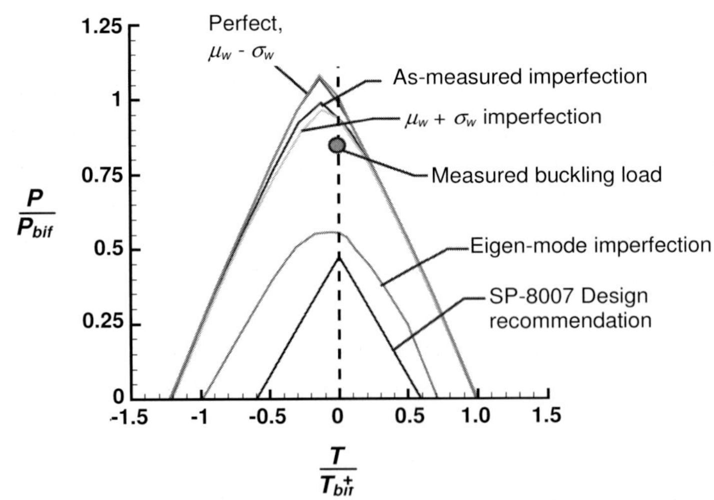

(b) Circumferentially stiff shell C2

Fig. 4.21. (continued) Predicted buckling interactions curves for the 8-ply-thick shells subjected to combined axial compression and torsion.

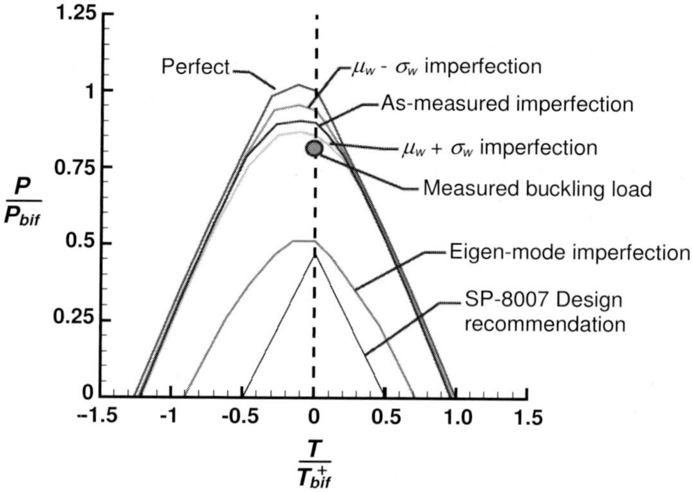

(c) Quasi-isotropic shell C3

Fig. 4.21. (concluded) Predicted buckling interactions curves for the 8-ply-thick shells subjected to combined axial compression and torsion.

the shape of the imperfections in the shell and how they couple with the initial buckling deformation response of the shell. Specifically, the linear-bifurcation-mode imperfection is similar in character to the initial buckling response in the shell and, thus, exhibits a very strong coupling with the initial buckling deformation of the shell, as expected. In contrast, the actual measured imperfection and the imperfection signature, with imperfection amplitudes as large as $w_o/t = 1.92$, are primarily a long-wave imperfection and these imperfection shapes do not exhibit an artificial strong coupling with the buckling deformations in this shell; thus, the reduction in the buckling load imperfection sensitivity is much less pronounced. In general, the results for shells with linear-bifurcation-mode imperfections are, for the relatively small value of the imperfection amplitude considered here, ultraconservative. Similarly, the lower-bound design recommendation obtained by using NASA SP-8007 for the corresponding isotropic shell is also ultraconservative, as expected, and misrepresent the behavioral trends. Finally, it is important to point out that although no test data were obtained for a C3-type shell subjected to torsion or combined torsion and compression, it is expected that the results for the shell with the actual measured imperfection will be very accurate. This expectation is based on the fact that torsion-loaded shells are not compression dominated like the corresponding compression-loaded shells and, as a result, will not exhibit as high of a degree of imperfection sensitivity.

4.6 Concluding Remarks

The approach presented herein for developing shell-buckling design criteria appears to offer a means for obtaining preliminary-design estimates of buckling loads for laminated composite cylindrical shells subjected to combined loads, without having to resort to extensive testing programs. In addition, this approach offers a relatively affordable alternative to relying on historical test data for shells that do not represent the configuration, material system, or fabrication process associated with the design of interest. This design-criteria-development approach could be used as a tool in the early stages of a design process to determine refined estimates of the buckling loads for the actual as-built shell structure under consideration. To insure success for problems of interest where refined design criteria are nonexistent, this approach should always be used with a selected number of carefully conducted tests that should be used to validate the finite-element analyses and provide confidence in the design results. The validation tests should be carefully conducted so that representative analysis results can be generated to correlate with test results. The provenience and pedigree of each test specimen should be determined or measured to assure that the appropriate shell-wall geometry and thickness distribution, material properties, and fabrication process-specific effects are included in the analysis used to develop the design criteria. The design-criteria-development approach described herein has wider implications; that is, it can be used to generate buckling design curves for commonly used structural forms and material systems that account for manufacturing-process imperfections and provide reliable estimates of buckling resistance that are suitable for preliminary design. Design curves of this fidelity will essentially provide a scientifically-based replacement for the overly conservative empirical knockdown factors typically used in design today to account for imperfection sensitivity of shell structures. Once a database of manufacturing-process-specific imperfections is established by the technical community, a designer will be able to determine which various imperfections are representative of the manufacturing procedure or fabrication process that is appropriate for his or her design and then use the relevant design curves. This approach will enable industry to develop a given design without having to repeatedly pay for extensive development tests to develop design knockdown factors. In addition, the refined design criteria should yield substantial weight savings in a given design.

4.7 Acknowledgements

This chapter is dedicated to my good friend and mentor Dr. James H. Starnes, Jr. His career spanned more than 33 years in which time he became nationally and internationally recognized as an expert in structural mechanics and particularly in the field of composite structures technology. His contributions to the understanding of the stability and strength characteristics of aircraft and spacecraft structures has led to structural design criteria that are used throughout the aerospace industry today. In addition, his vision of improved analysis and design technologies for stiffness-critical and strength-critical structures played a major role in guiding much of the structures research conducted at the NASA Langley Research Center over the years, including the research presented in this chapter. Dr. Starnes is fondly remembered as an exceptional supervisor, mentor, and technical leader, and will be greatly missed.

I would also like to thank Dr. Michael P. Nemeth for his valuable input over the years on the development of experimentally validated shell buckling analysis and design methods and for composing the historical overview on shell buckling technology contained in the Introduction of this Chapter.

4.8 References

1. Bushnell, D., *Computerized Buckling Analysis of Shells*, (Kluwer Academic Publishers, Boston, 1989) (see also AFWAL-TR-81-3049, Wright-Patterson Air Force Base, 1981).
2. Singer, J., Árbocz, J., and Weller, T., *Buckling Experiments: Experimental Methods in Buckling of Thin-Walled Structures*, Vol. 2, (John Wiley & Sons, Inc., New York, 2002).
3. von Kármán, T. and Tsien, H. S., *Aeronaut Sci J*, Vol. 8, No. 8, 303, (1941).
4. Donnell, L. H. and Wan, C. C., *J Appl Mech*, Vol. 17, No. 1, 73, (1950).
5. Koiter, W. T., *On the Stability of Elastic Equilibrium.* (in Dutch), (H. J. Paris, Amsterdam, Holland, 1945); translation available as AFFDL-TR-70-25, (Wright-Patterson Air Force Base, 1970).
6. "Buckling of Thin-Walled Circular Cylinders," NASA Space Vehicle Design Criteria, NASA SP-8007, (1965, Revised 1968).
7. "Buckling of Thin-Walled Truncated Cones," NASA Space Vehicle Design Criteria, NASA SP-8019, (1968).
8. "Buckling of Thin-Walled Doubly Curved Shells," NASA Space Vehicle Design Criteria, NASA SP-8032, (1969).
9. Budiansky, B, in *Advances in Applied Mechanics*, Vol. 14, (Academic Press, 1974), pp. 1-65.
10. Árbocz, J. and Hol, J. M. A. M., "ANILISA - Computational Module for Koiter's Imperfection Sensitivity Theory," Report LR-582, (Delft University of Technology, Delft, 1989).
11. Khot, N. S. and Venkayya, V. B., "Effect of Fiber Orientation on Initial PostbucklingBehavior and Imperfection Sensitivity of Composite Cylindrical Shells," AFFDL-TR-70-125, (Wright-Patterson Air Force Base, Ohio, 1970).
12. Cohen, G. A. and Haftka, R. T., *Comput Struct*, Vol. 31, No. 6, 985, (1989).
13. Tennyson, R. C., *Composites*, Vol. 6, No. 1, 17, (1975).
14. Árbocz, J., Babcock, C. D., Jr., *J Appl Mech*, Vol. 36, No. 1, 28, (1969).
15. Árbocz, J. and Williams, J. G., *AIAA J*, Vol. 15, 949, (1977).
16. Árbocz, J., Babcock, C. D., Jr., "Computerized Stability Analysis Using Measured Geometric Imperfections," *Proceedings of the 12th Congress of the International Council of the Aeronautical Sciences*, ICAS-80-20.0, 688, (1980).
17. Árbocz, J., in *Buckling of Shells- A State-of-the-Art Colloquium*, E. Ramm, ed., (Springer Verlag, Berlin, 1982), pp. 535-567.
18. "NASTRAN User's Manual," NASA-SP-222(08),Vol. 2, (1986).
19. Rankin, C. C., Brogan, F. A., Loden, W. A., and Cabiness, H. D., "STAGS Users Manual, Version 4.0," Report LMSC P032594, (Lockheed Martin Missiles & Space Co., Inc., Advanced Technology Center, 1999).

20. Starnes, J. H., Jr., Hilburger, M. W., and Nemeth, M. P., in *Composite Structures - Theory and Practice*, P. Grant and C. Q. Rousseau, eds., (ASTM STP 1383, 2000), pp. 529-550.
21. Hilburger, M. H., and Starnes, J. H., Jr., in *Proceedings of the 42st AIAA/ASME/ASCE/AHS/ASC Structures, Structural Dynamics, and Materials Conference*, AIAA Paper No. 2001-1394, (2001).
22. Hilburger, M.W., and Starnes, J. H., Jr., *Int J Nonlinear Mech*, Vol. 37, 623, (2002).
23. Nemeth, M. P., Britt, V. O., Young, R. D., Collins, T. J., and Starnes, J. H., Jr., *J Spacecraft and Rockets*, Vol. 36, No.6, 788, (1999).
24. Nemeth, M. P., Young, R. D., Collins, T. J., and Starnes, J. H., Jr., *J Spacecraft and Rockets*, Vol. 36, No.6, 828, (1999).
25. Young, R. D., Nemeth, M. P., Collins, T. J., and Starnes, J. H., Jr., *J Spacecraft and Rockets*, Vol. 36, No.6, 820, (1999).
26. Nemeth, M. P., Young, R. D., Collins, T. J., and Starnes, J. H., Jr., *J Spacecraft and Rockets*, Vol. 36, No.6, 812, (1999).
27. Nemeth, M. P., Young, R. D., Collins, T. J., and Starnes, J. H., Jr., *Int J Nonlinear Mech*, Vol. 37, No.4-5, 723, (2002).
28. Nemeth, M. P. and Starnes, J. H., Jr., "The NASA Monographs on Shell Stability Design Recommendations - A Review and Suggested Improvements," NASA TP-1998-206290, (1998).
29. Árbocz, J. and Starnes, J. H., Jr., *Thin Wall Struct*, Vol. 40, 729, (2002).
30. Chryssanthopoulos, M. K., Giavotto, V., and Poggi, C., in *Buckling of Shell Structures, on Land, in the Sea and in the Air*, J. F. Jullien, ed., (Elsevier Applied Science Publishing Co., Inc., New York, 1991), pp. 43-52.
31. Claus, S. J., *Manufacture-Structure-Performance Relationships for Filament-Wound Composite Shells*, Ph. D. Dissertation, The Pennsylvania State University, (1994).
32. Árbocz, J. and Elishakoff, I., *J Appl Mech*, Vol. 107, 122, (1985).
33. Árbocz, J., Starnes, J. H., Hr., and Nemeth, M. P., in *Proceedings of the 41st AIAA/ASME/ASCE/AHS/ASC Structures, Structural Dynamics, and Materials Conference*, AIAA Paper No. 2000-1236, (2000).
34. Riks, E., *J Press Vess*, Vol. 109, 27, (1987).
35. Park, K. C., *J Appl Mech*, Vol. 42, 464, (1975).
36. Riks, E., Rankin, C. C., and Brogan, F. A., *Comput Method Appl M*, Vol. 136 (1-2), 59, (1996).
37. Bushnell, D., in *Collapse Analysis of Structures*, PVP-Vol. 84, ASME, (1984), pp. 30-32
38. Nemeth, M. P., "Buckling Behavior of Long Symmetrically Laminated Plates Subjected to Combined Loadings," NASA TP-3195, (1992).
39. Nemeth, M. P., "Buckling Behavior of Long Anisotropic Plates Subjected to Combined Loads," NASA TP-3568, (1995).

CHAPTER 5

STABILITY DESIGN OF STIFFENED COMPOSITE PANELS – SIMULATION AND EXPERIMENTAL VALIDATION

Alexander Kling

Institute of Composite Structures and Adaptive Systems
German Aerospace Center (DLR)
Lilienthalplatz 7, 38108 Braunschweig, Germany
E-mail: alexander.kling@dlr.de

Allowable compressive loads of thin-walled stiffened panels, made of carbon fibre reinforced plastic (CFRP), are currently limited by the first occurrence of buckling, in most cases a skin-based local instability, which drive the design. Therefore, the remaining structural reserves in the form of load-carrying capacity in the postbuckling regime up to the onset of global stringer-based buckling, are not fully exploited. The efficient design of these structural components requires a novel stability design scenario, to permit postbuckling below ultimate load and at the same time to assure that there is no incidence of structural degradation. This requires accurate, reliable and efficient numerical analysis, commonly accomplished by nonlinear finite element calculations.

After introducing the main aspects of the stability design scenario, the design process of the test structures is detailed, followed by an overview of experimental investigations of two nominally identical panels. Finally the nonlinear finite element (FE) analysis and the validation process at different levels of detail is described.

5.1 Introduction

The industrial strategic research objectives[1] demand a reduction in aircraft operation and development costs by 20% and 50% in the short and long term, respectively. A contribution to this effort is to investigate the possibility of a significant weight reduction in stiffened panel component (refer to Fig. 5.23) of primary aerostructures without prejudice to costs and structural life. A possible approach to tackle the demand for lighter fuselage structures, for example, is to utilize carbon fibre reinforced plastic (CFRP) material, which due to its superior properties allows low-weight designs. However, the ambitious goals require at least one further step towards a novel design scenario, which permits stable postbuckling (only small deformations) stages under ultimate load and at the same time to assure that there is no incidence of structural degradation.[2]

To establish the required level of confidence in this progressive scenario presumes systematic and reliable nonlinear numerical analysis up to the deep postbuckling regime, validated with well designed state-of-the-art experimental investigations.

5.2 Stability design scenario

Several experimental and numerical investigations[3-5] revealed for the considered stiffened CFRP panel-type structures, a significant load carrying capacity after first, in most cases local skin-based, buckling. Figure 5.1 depicts two typical load-shortening curves of axially compressed panels of different designs (Weight 1 as solid line > Weight 2 as dashed line). After a nearly linear prebuckling response, local buckling (skin based) is marked with a dot, at a small change in the slope of the curve. Global (stringer-based) buckling at the beginning of a temporary decline in the curve progression is indicated by a cross and the onset of degradation with a rectangle, after some further progression into the deep postbuckling regime. Limit load (the maximum loads to be expected in service, as defined in the Federal Aviation Regulations FAR Sec. 25.30) and ultimate load (limit loads multiplied by prescribed factors of safety, as defined in FAR Sec. 25.301; commonly the factor of safety is 1.5, FAR Sec. 25.303) are depicted as horizontal lines limiting the maximum allowable load. FAR Part 25 specifies that no permanent deformation of the structure is allowed at limit load and the structure must be able to withstand the ultimate load for at least 3 seconds without failure. As a result of the regulation, elastic buckling (a typical behaviour for CFRP material) is allowed under limit load conditions. Design 1 (Weight 1) is constrained by limit load, (in this case, an optimal structure is designed in a way that first buckling is close to limit load), which represents current common practice. Design 2 (Weight 2), with reduced structural weight is constrained by an ultimate load definition, which is forced by:

- No global, stringer-based buckling below ultimate load (to avoid large deformations, even if they are elastic. (Latest experimental investigations for stiffened CFRP panels at DLR revealed that repeated global buckling (up to 2000 cycles) below onset of structural degradation did not show any effect on load carrying capacity – elastic global buckling.)
- No structural degradation below ultimate load.

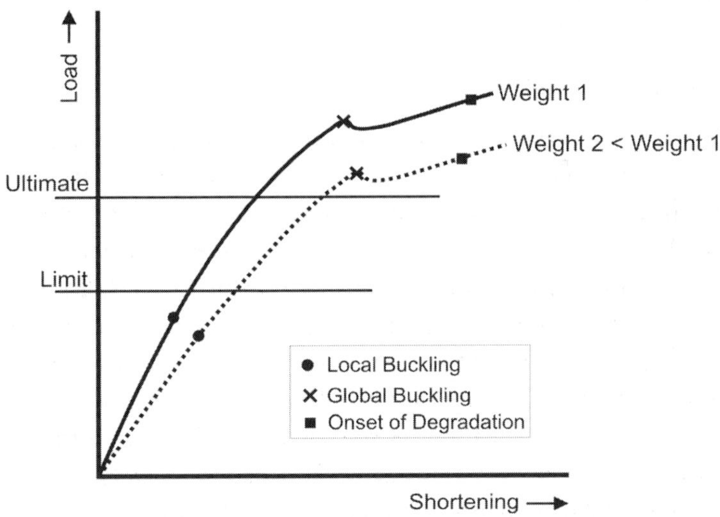

Fig. 5.1. Visualized stability design scenario, based on typical load-shortening runs for the considered structures, comparing two designs: Weight 2 (ultimate load design) < Weight 1 (limit load design).[2]

A stability design scenario, based on the afore mentioned criteria limited by an ultimate load definition, would lead to a significantly higher exploitation of structural reserves and therefore resulting in more weight efficient structures. However, this requires improved and validated analysis procedures, to reliably predict the buckling and postbuckling behaviour of stiffened composite panels based on a sound experimental foundation.

In the subsequent sections an overview on relevant topics with respect to experimental investigation, numerical simulation and validation in the context of the proposed design scenario is given.

5.3 Design of the test structures

The design phase of a test structure plays a decisive role within goal oriented phenomenological or validation experiments (ref. Section 5.4). There are a large number of design parameters for stiffened composite structures, ranging from material selection, panel dimensions, stringer type, pitch and geometry, to the lay-up of stringer and skin. These parameters influence the response and stability characteristics and make the final selection to a challenging task.

Figure 5.2 provides an overview of typical stability-relevant phenomena, including failure modes, using an experimentally measured load-shortening curve of a T-stringer panel including representative pictures from numerical analysis and the test structure after collapse.

Based on experience of former tests as well as intense numerical studies on stiffened panels, an initial design was selected. Some geometric limitations were imposed for testing, using the in-house buckling test facility but similar to real aircraft fuselage structures. The goal was to design a test structure for validation purposes up to the deep postbuckling regime with a noticeable reduction in axial stiffness at the onset of local, stringer-based buckling (N°1 in Fig. 5.2), and a significant load carrying capacity in the postbuckling regime up to global buckling (N°3). Local stringer instabilities as shown in N°2 should be avoided to purify the stability relevant rather punctual phenomena of N°1 and N°2. The onset of degradation should occur far after the onset of global buckling.

Within the design phase ABAQUS/Standard, a commercial multi-purpose finite element program was employed for parametric studies to evaluate which configuration (e.g. stringer spacing, skin and stringer wall thickness or lay-up) show the desired buckling and postbuckling behaviour described in the last paragraph. The design which was finally selected is given in detail in the following section.

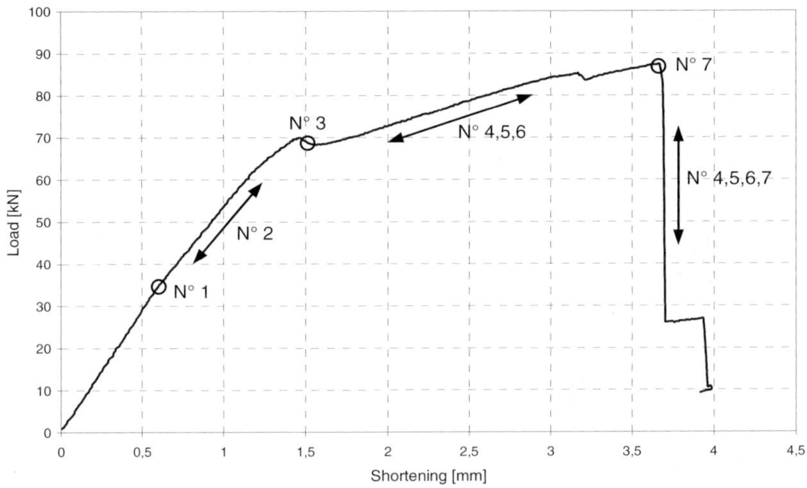

N°1: Local skin buckling

N°2: Local stringer instability and skin buckling

N°3: Global (stringer-based) buckling

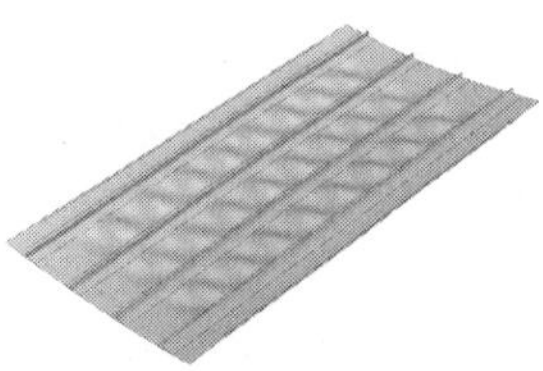

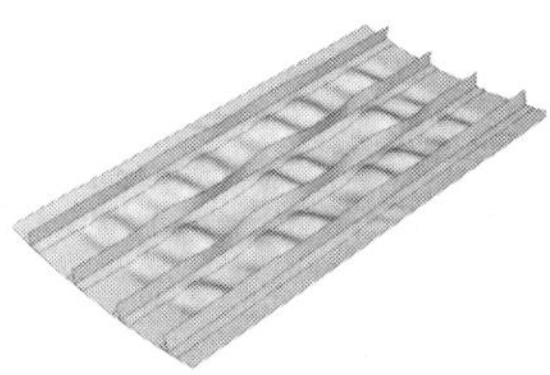

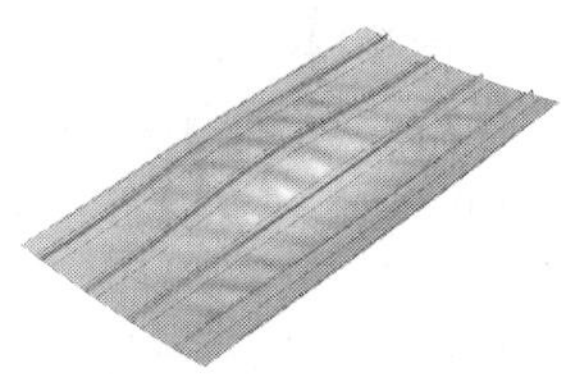

N°4: Fibre fracture
N°5: Inter-fibre fracture

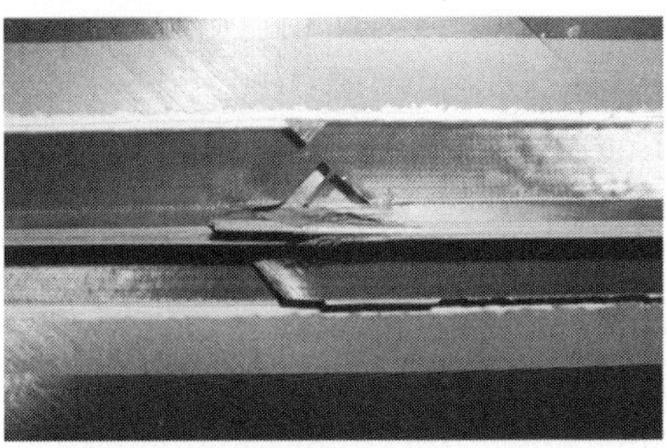

N°6: Delamination
N°7: Stringer-skin separation

Fig. 5.2. Load-shortening curve with typical, stability relevant phenomena including relevant failure modes of a T-stringer stiffened panel.

A by-product of the numerical studies was the examination of the structure's imperfection sensitivity. This was done by introducing normalized buckling modes (from an eigenvalue analysis) as scaled initial geometric imperfections (single and/or superposed). The final design was rather stringer dominant and indicated, as expected, a minor imperfection sensitivity. However, a variation of the numerically applied boundary conditions, especially the ones along the longitudinal edge supports (refer to Fig. 5.7), revealed that the clamping width of the attached metallic supports has a noticeable influence, particularly in the deeper postbuckling regime. As a result of these preliminary numerical findings, modifications were made on the longitudinal edge supports to obtain experimental boundary conditions, which can be appropriately simulated up to the deeper postbuckling regime.

Based on the parametric study to extract a design with the desired buckling and postbuckling behaviour, a fairly good understanding of the nonlinear behaviour was gained. Therefore, first calculations – so called "pre-test" analysis, even in the case of a not fully validated model and numerical approach (however, based on some formerly gained experience with respect to the analysis procedure), are worthwhile, not only for a well designed test structure, but also for a deeper understanding of the complex nonlinear behaviour (e.g. sensitivity with respect to experimentally applied boundary conditions).

5.4 Experiment

Figure 5.3 provides an overview on different stages of experimental investigation for a typical component level test structure (refer to Fig. 5.23 for a classification of structural levels) like the considered stiffened panel. Characterisation experiments on coupon level are usually conducted to extract material properties like elastic modulus, maximum strength values and as exemplarily depicted for composite material the critical energy release rate (GI_c) of Mode I. The data obtained by the material characterisation is mandatory for the certification of the material system as well as an essential input for numerical analysis. Phenomenological experiments on component level focus mainly on the understanding of the complex physical behaviour, as for example stability relevant characteristics, which have been detailed in Section 5.3. The purpose of this type of physical testing can be best described as "understand the physics behind the phenomena" and to subsequently derive a preliminary conceptual model (refer to Fig. 5.19).

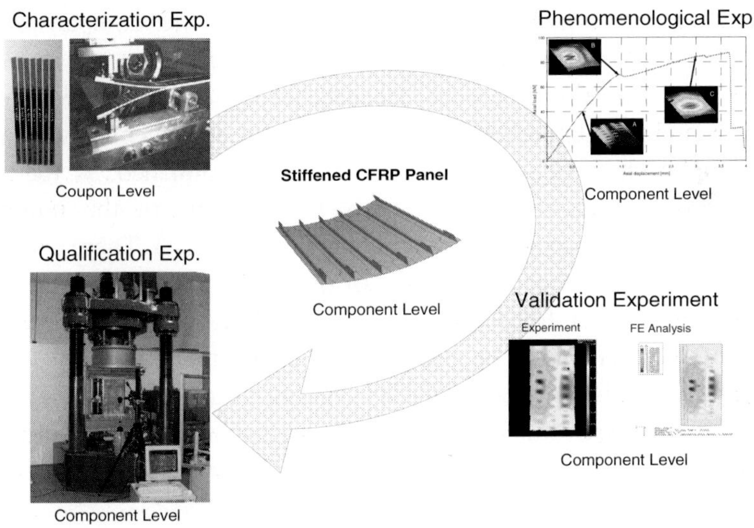

Fig. 5.3. Different stages of experimental investigations on the example of a stiffened CFRP panel.

A fundamentally different purpose lies behind validation experiments.[6] This type of physical experiment is usually conducted after the principle phenomena were understood and numerical methods and models are established, which have to be compared (validated) with reliable test data. Validation experiments are performed to generate high-quality data for the purpose of assessing the accuracy of a virtual model prediction in the domain of interest. To qualify as a validation test, the specimen's material and geometry, initial conditions (e.g. geometric imperfections), boundary conditions, and all other model input parameter must be documented as accurately as possible.[7] During the test, the focus should not only be on gathering data from the applied sensors, but also on possible changes in load introduction or boundary conditions, which might change during the test. All prescribed input, test conditions, measurements and any kind of anomaly have to be fully documented.

Ideally, this approach provides well defined constraints and measurements, requiring only a few, if any, assumptions on the model generation and on the subsequent comparison of numerically predicted and experimentally captured data.

Due to the fact that the experimentally measured data represents the reference against which the numerically results are compared, it is essential to determine possible uncertainties and errors in the measured data as well as test structures manufacturing discrepancies, design tolerances and possible residual stresses imposed during construction.

An important issue, which is not always established, is the close cooperation between the analyst (person in charge of the numerical analysis) and experimenter (person in charge of the test). There should be a common understanding which type of data will be compared and even more importantly is a discussion of possible deficiencies on its prediction or measurement. In addition, the analyst needs to understand the test setup, especially the applied boundary conditions and loading, as well as the testing process. And finally, the analyst should use the predicted sensitivities from the pre-test analysis to point out potential problems with the experimental setup.

Qualification experiments, on component level, focus on the determination of adequacy of the test structure with respect to its intended application. Which means that the experimental setup should "simulate" the structures final application with typical environmental (e.g. temperature, humidity) and loading conditions (load cases or load levels including appropriate boundary conditions) as close as possible. The ultimate goal is not to provide accurate data for validation purposes, rather to ensure that the final design can be certified to fulfil all specifications with the required margin of safety.

The curved arrow in Fig 5.3 indicates a potential loop starting at coupon level, via phenomenological and validation experiments, up to the qualification test. However each type of experiment has its individual self standing justification and motivation within the physical testing process and do not necessarily need to be embedded.

Two extensive experiments on nominally identical stringer stiffened CFRP-panels were conducted to provide a sound database for validation purposes. The following subsection gives an overview on the test structures, the manufacturing, additional information on pre-test analysis, preparation of the specimens, the process of testing as well as a selection of extracted measurements.

5.4.1 *Test structure*

Two stiffened CFRP panels (B0276 and B0281) of nominally identical geometry and material (Hexcel, 6376C-HTA(12K)5-35%) were manufactured.

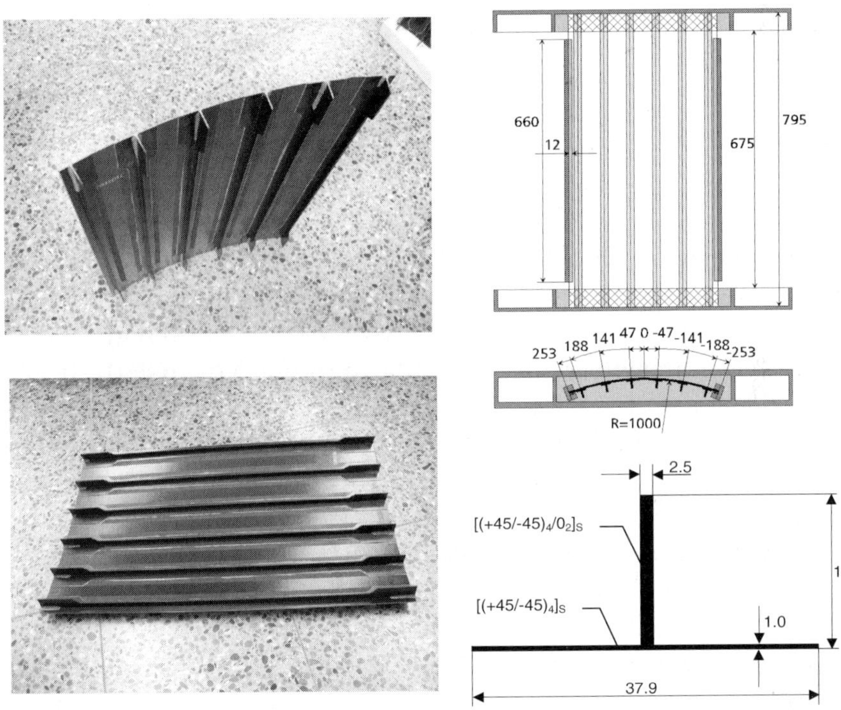

Fig. 5.4. Details of the test structure (geometry and lay-up).

Skin and T-stringer were cured separately and subsequently bonded together using a high strength curing film adhesive (Hexcel, Redux® 312). Figure 5.4 (left) depicts the manufactured test structure and the sketch on the right hand side provides an overview of relevant dimensions, where the panel is shown with installed clamping boxes on the top and bottom for a homogeneous load introduction and the longitudinal edge supports. The stacking sequence of the skin is $[90°/+45°/-45°/0°]_S$ with a single ply thickness of 0.125 mm, where the 0° direction is coincident with the stringer direction. Details with respect to the T-stringer geometry and the lay-up can be found in Fig. 5.4 (right).

Table 5.1. Material properties CFRP prepreg material (Hexcel, 6376C-HTA(12K)5-35%) at room temperature.

E11	[GPa]	131.0
E22	[GPa]	11.2
G12	[GPa]	5.3
G13	[GPa]	5.275
G23	[GPa]	3.950
ν12	[-]	0.301
Tensile strength in fibre direction Xt	[GPa]	2.250
Compressive strength in fibre direction Xc	[GPa]	1.6
Tensile strength normal to the fibre direction Yt	[GPa]	0.064
Compressive strength normal to the fibre direction Yc	[GPa]	0.29
Shear strength	[GPa]	0.098
Inter-laminar shear strength	[GPa]	0.91

Table 5.2. Material properties adhesive layer (Hexcel datasheet, Redux® 312).

E	[MPa]	3000
ν	[-]	0.4
Max. compressive stress	[MPa]	48
Max. shear stress	[MPa]	38
Max. normal stress	[MPa]	8.3

The material properties of the prepreg material are listed in Table 5.1, where index 1 corresponds to the fibre direction, index 2 is perpendicular to the fibre orientation in the plane of the ply. Table 5.2 summarizes the properties of the adhesive layer from the datasheet for Redux® 312.

5.4.2 *Preparation of the test structure*

After manufacturing, a visual inspection was conducted and the main dimensions were checked. Subsequently, both panels were examined using ultrasonic inspection to detect possible, barely visible or invisible damages from manufacturing or transport. The greyscale renderings in Fig. 5.5 depict the outcomes (loss in dB) from ultrasonic inspection. The back wall echo with focus on the stringer, measured from the skin side of the panel, signifies an integral representation of possible failure, whereas the back-wall echo with focus on the skin provide a more detailed insight on the stringer-skin interface. After critical examination, almost no inhomogeneity was detected for the stringer flange, skin and adhesive interface of both panels, which ensured that invisible or barely visible defects (e.g. imperfect adhesive bonding) can be precluded.

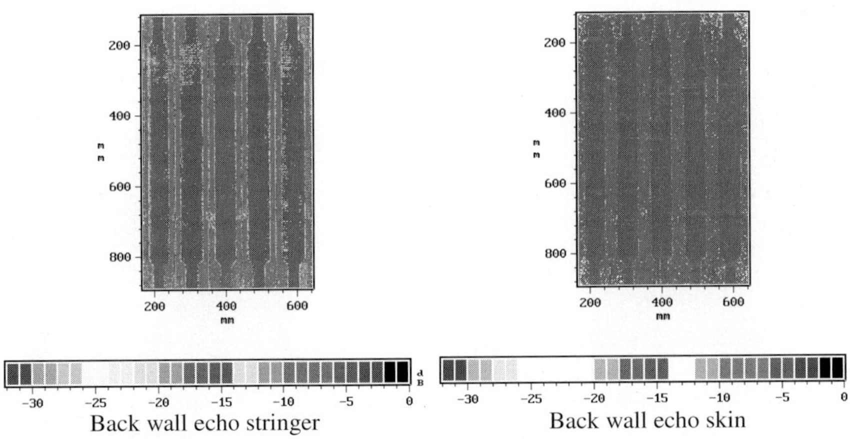

Fig. 5.5. Non-destructive testing with ultrasonic inspection (panel B0281).

This elaborate non-destructive pre-test examination, to ensure high quality test structures, reduces possible uncertainties within the validation process and minimizes post-test speculation, in case of discrepancies between experimental extracted results and numerically predicted data.

In the next step both ends of the panel were embedded into preliminary end-boxes with an epoxy-filler mixture. After curing, the panels were detached from the boxes and subsequently milled to ensure a homogeneous load introduction along the circumferential edge, followed by a final fitting into the test boxes.

Thereafter, an optical 3D digitizing measurement system, based on photogrammetry, was utilized to determine the initial geometric imperfections of the skin, as well as to evaluate a best fit radius, which might differ with respect to the nominal value of 1000 mm due to spring-in effects after demounting from the curing mould within the manufacturing process. This was accomplished by interpolating a best fit cylinder into the measured scatter data of the panel's surface. The pseudo greyscale renderings in Fig. 5.6 represent the aberration with respect to a perfect cylinder of the best fit radius.

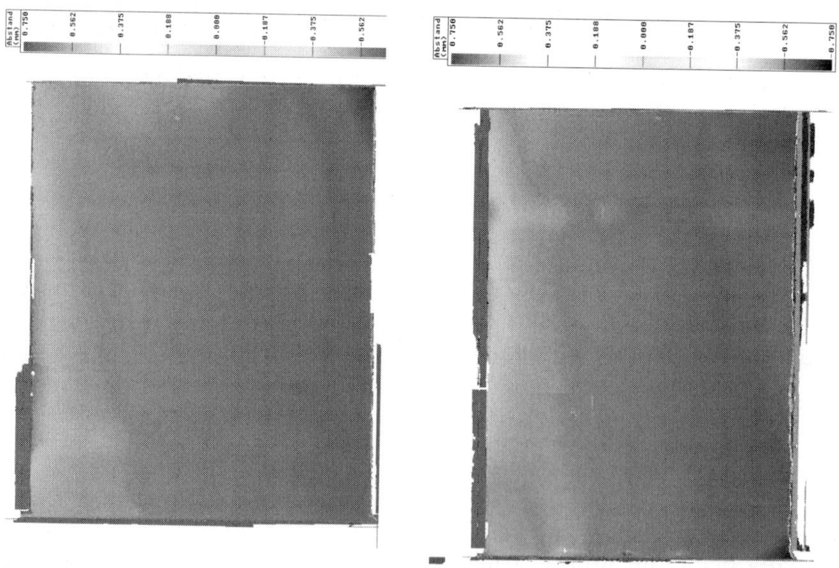

B0276 (measured best fit R=940,8 mm) B0281 (measured best fit R=938.7 mm)

Fig. 5.6. Visualization of the measured initial geometric imperfections. Greyscale renderings represent the aberration with respect to a perfect cylinder of the best fit radius.

Type and location of sensors (e.g. strain gauges, displacement transducers) were determined based on the results from the pre-test analysis and the experience on former panel tests. Overall, 80 strain gauges on skin and stringer side of the panel and 22 displacement transducers (two are in axial/loading direction) were installed for each test.

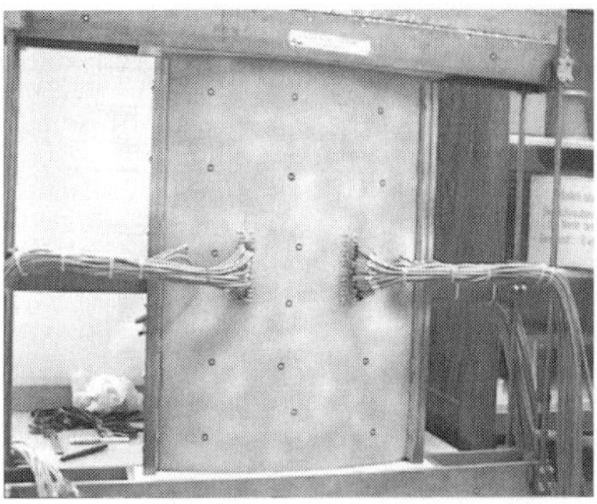

Strain gauges (chains and rosettes)

Displacement transducer and strain gauges

Fig. 5.7. Overview on type and locations of sensors on skin and stringer side of the panel.

Figure 5.7 depicts chains of strain gauges and rosettes as well as 20 displacement transducers attached to the stringer flanges in a regular pattern. In addition, a 3D digitizing measurement system (ARAMIS) based on photogrammetry was utilized to capture deformations on the skin side, therefore an irregular powder spray coating was applied on the skin side as shown in Fig. 5.7 (left). Finally, the panel with the clamping boxes was installed in the buckling test facility.

5.4.3 *Test*

Figure 5.8 (left) depicts the in-house buckling test facility with the build-in panel and in the front one camera pair of the ARAMIS system on a tripod. The schematic sketch on the right provides further insight into the relevant components.

After calibrating and checking of all wiring and sensors, the quasi-static displacement controlled loading process was started. First, three cycles up to about 50% of the expected first/local buckling load were applied in order to compensate for possible settlements and to ensure that data acquisition and sensors work properly. This is followed by the quasi-static loading up to collapse with continuous data acquisition of strain, displacements and axial load, as well as regular measurements of the full scale deformation patterns with the ARAMIS system at 141 load levels. Due to the fact that this rupture test it not repeatable with the same specimen the pre-test planning and final testing process has high significance. Usually the number of nominally identical specimen is small (two, seldom three) and a repetition would be in most cases simply too costly and time consuming.

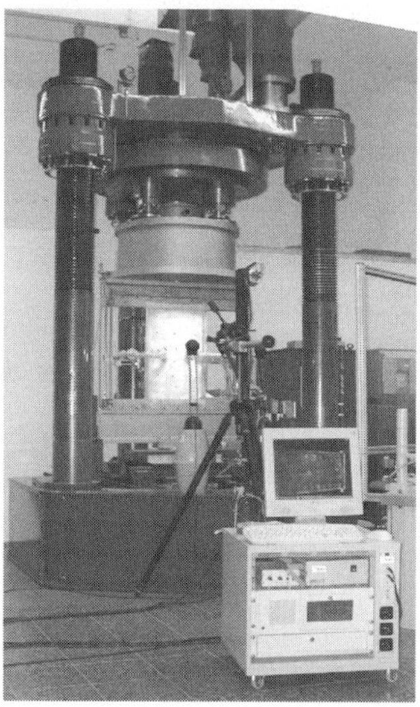

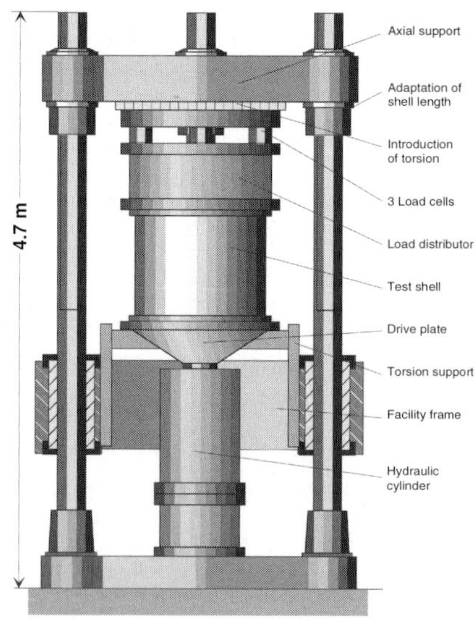

Fig. 5.8. Buckling test facility at DLR with the 3D optical measurement system (ARAMIS) in the foreground (left picture).

5.4.4 *Results*

Within this subsection results from both tests, on the nominally identical panels, will be detailed. The load-shortening curves (scaled to the corresponding load and shortening at local skin-based buckling) of panel B0276 and B0281, depicted in Fig. 5.9, show good agreement up to collapse. Exemplarily, three quantitatively measured deformation patterns at characteristic load levels are depicted using images obtained from the ARAMIS system.

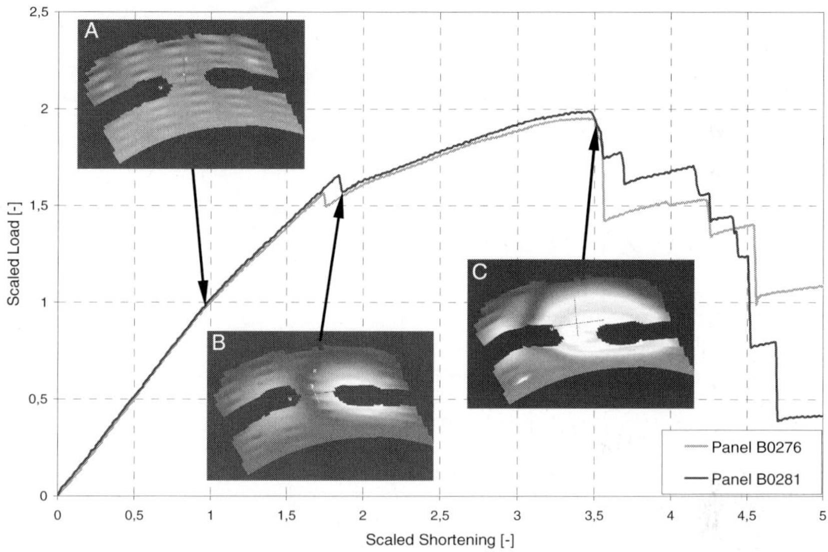

Fig. 5.9. Experimentally extracted load-shortening curve of two nominally identical panels and deformation patterns from the ARAMIS system at characteristic load levels.

In Fig. 5.10, locally measure radial displacements from displacement transducers on the free edge (top) of the stringer blades are depicted. As expected, there was no change in the linear rise of the curves at local skin-based buckling at a scaled shortening of 1.0. Whereas, stringer-bases global buckling coincident with the sharp kink in the load-shortening run caused a sudden rise in the radial deformations, which reproduce the 2/3 to 1/3 buckle (cf. deformation pattern B in Fig. 5.9). The subsequent transition to a central global buckle can be assessed by the run of W88 towards a negative radial displacement and an additional gain in radial direction of W89, W90 and W91. Note, that most of the adhesive connections of the displacement transducer tips and the stringer blades ruptured at collapse (scaled shorting of 3.48), causing significant changes in the radial measurements, however at this stage without any physical relevance.

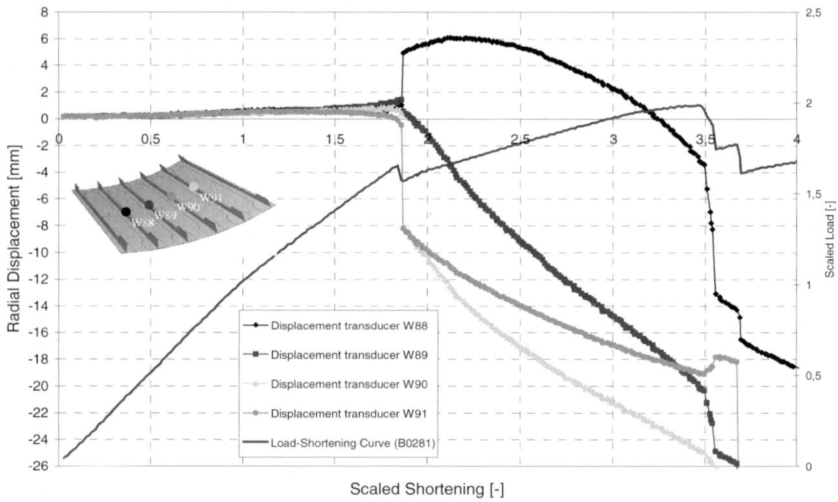

Fig. 5.10. Experimentally measured radial displacements versus scaled shortening and scaled load-shortening curve.

Section of the panel with large stringer-skin separation after collapse

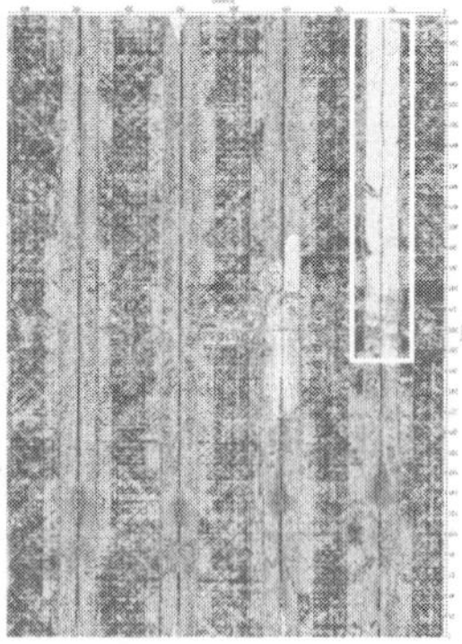

Back wall echo with focus on the stringer-foot

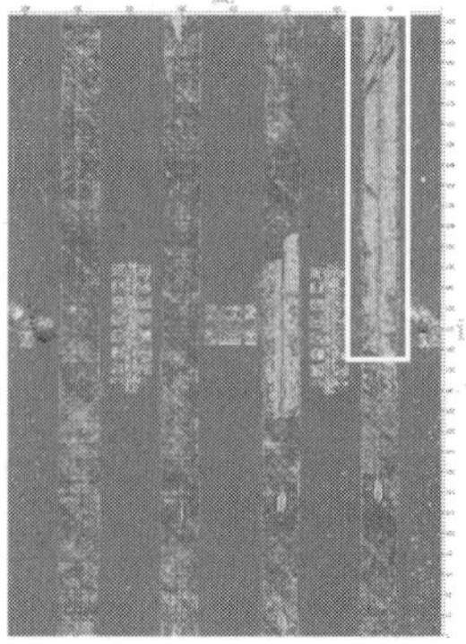

Back wall echo with focus on the skin

Fig. 5.11. Panel after collapse test and results from ultrasonic inspection.

Figure 5.11 (left) depicts a section of the panel after collapse showing a large skin-stringer separation (surrounded by a rectangle). The corresponding findings from the ultrasonic inspection after test, the back wall echo with focus on the stringer-foot in the centre of Fig. 5.11 and the back wall echo with focus on the skin (right) show next to the marked area further visible and invisible damages (cf. to Fig. 5.5 for the pre-test ultrasonic results).

5.5 Analysis

Within this section, modelling aspects as well as the numerical analysis procedure utilizing ABAQUS, a commercial multipurpose finite element software, are detailed.

5.5.1 *Numerical methods*

There are different nonlinear solution methods implemented in ABAQUS, which can be utilized to simulate the buckling and postbuckling behaviour of the considered CFRP panel type structures.[8,9] Next to the established static incremental/iterative solution techniques, like the Newton or arc-length (modified Riks algorithm) method, also nonlinear dynamic methods by implicit or explicit integration are available. In the following the relevant methods implemented in ABAQUS are briefly assessed based on the experience gained at DLR for the considered structures and loading conditions.

Nonlinear static analysis

Newton method: Applicable for load- or displacement controlled loading, with limits (convergence problems) at snap through or onset of global buckling combined with a sudden loss in load carrying capacity. In addition, no decrease in load or displacement is possible for load or displacement controlled loading, respectively.

Newton method (with artificial damping): Applicable for load- and displacement controlled loading, relatively stable (highly dependent on the user selected stabilization factor) also within the transitions to global buckling deformations – most effective and promising results for the considered type of structures and loading conditions (further details with respect to the algorithm is given subsequently). However, it has to be kept in mind that the dissipation energy due to artificial damping, which is taken out of the system, has to be as small as possible to minimize the influence on the results. Analysis of shells with no or minor imperfections are in most cases successful and are, despite the fact that they are nonlinear and incremental/iterative, quite effective with respect to computational costs.

Modified Riks method (arc length): Applicable to follow complex snap-through and snap-back problems (load and simultaneous displacement reduction possible). However, this algorithm has shown a trend to back-tracking near sharp transitions/kinks, like the ones showing up regularly at the onset of global buckling for the considered structures. The use of relatively large imperfections can reduce this effect, however convergence is sometimes hard to establish.

Nonlinear dynamic (quasi-static) method by implicit and explicit integration

Although both methods are truly dynamic procedures, they have a proven value in solving static problems as well. Furthermore, they are predestined for snap through problems with contact. Since a static solution is sought for, the load must be imposed extremely slowly (quasi-static), which causes a considerable increase in computing time.

These short reviews are based on DLR experience in analysing stiffened CFRP panels into the deep postbuckling regime with ABAQUS. The assessments do not claim to be exhaustive and should be always reviewed by the user.

As mentioned before, some additional information will be given with respect to the solution method, which has been selected as the most promising (based on DLR experience with ABAQUS) and effective way to simulate the buckling and postbuckling behaviour of stiffened CFRP panels (Newton method with artificial damping). Basically, there are two ways in which damping can be introduced in a static analysis. Discrete dashpots could be added or alternatively, damping can be introduced through the *STATIC, STABILIZE command.

If damping is used in nonlinear static analysis via the STABILIZE parameter[8], a virtual velocity is calculated to be equal to the displacement increment divided by the corresponding time increment (ABAQUS uses a virtual time for incrementation purposes in static analysis). The damping matrix which is utilized is simply the mass matrix with unit density multiplied by a damping coefficient. The damping coefficient should be as small as possible to minimize the dissipative energy, which is taken out of the system. This has to be checked for each computation to ensure that the viscous dissipation energy (abbreviated in ABAQUS as ALLSD) is a small fraction of the model's strain energy (ALLSE). The default value for the *STATIC, STABILIZE command in ABAQUS is $2x10^{-4}$.

This value is in most cases (based on DLR experience) relatively large, therefore parametric studies should be conducted to use the lowest number possible, which still allows to achieve convergence in the deep postbuckling regime. Figure 5.12 shows the total strain energy and dissipated strain energy as a function of scaled shortening for an artificial damping factor of $2x10^{-6}$. This numerical modification, to introduce artificial damping into the Newton algorithm, has almost no effect prior to global buckling due to the low virtual velocities in this range, however, as soon as instability in the structure develops, the virtual velocity increases rapidly and virtual damping starts to become effective (e.g. large radial displacements at the onset of global buckling as depicted in Fig. 5.18). For a converging solution procedure, the time incrementation usually decreases as velocity increases, which leads to a controlled postbuckling behaviour. All analyses within this report were conducted with the Newton method including artificial damping (*STATIC, STABILIZE) with a factor of $2x10^{-6}$.

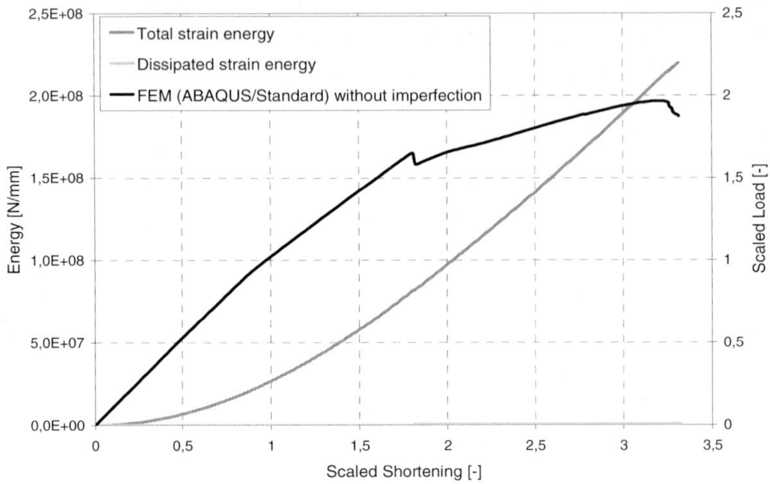

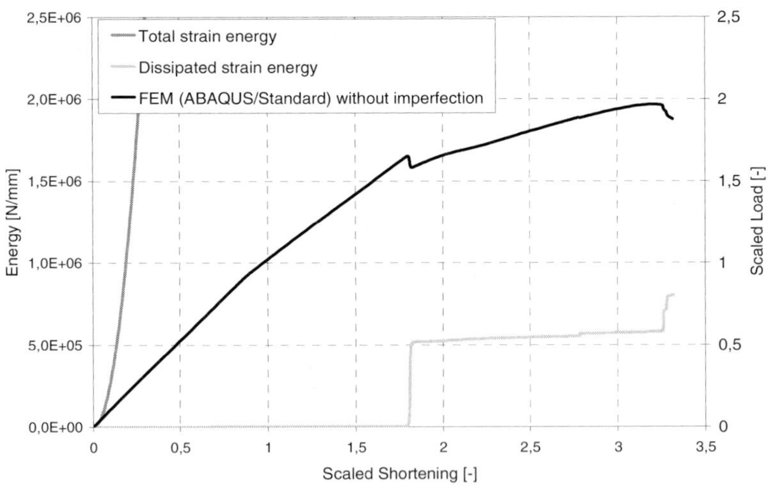

Fig. 5.12. Comparison of dissipated strain energy versus total strain energy with a damping factor of $2x10^{-6}$. Note that the left ordinate (Energy) is scaled forvisualization purposes in the second diagram.

5.5.2 *Analysis procedure*

Figure 5.13 provides an overview on the selected analysis procedure, containing preprocessing, analysis with ABAQUS and the postprocessing. There are several commercially available preprocessing software tools on the market, which can be utilized to generate the desired model for the subsequent analysis. In case parametric studies are planned, an automatic, fully parameterized, mesh generator can significantly reduce the time consumed in tedious model generation.

The dashed path represents an intermediate step (eigenvalue buckling analysis) between preprocessing and the nonlinear analysis in case the sensitivity of initial geometric imperfection should be examined. The intension of this computation is to obtain an estimate of the critical load at which the response of a structure will bifurcate, assuming the behaviour prior to the bifurcation is essentially linear. In addition, a useful estimate of mode shapes is provided, which can be directly introduced as initial geometric imperfections in the nonlinear analysis. Further details with respect to the available eigenvalue extraction methods (Subspace iteration, Lanczos method) can be found in[8,9].

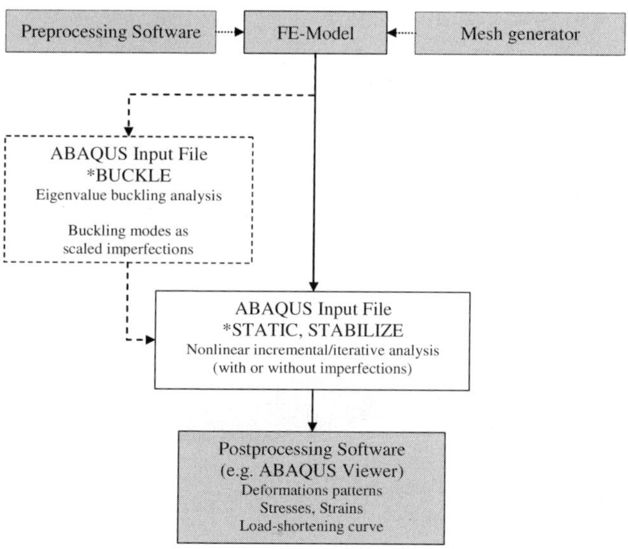

Fig. 5.13. Analysis procedure.

Details with respect to the nonlinear incremental/iterative solution method using the Newton technique with artificial damping have been given in the last section. The required results (e.g. displacements or strains) were extracted within a postprocessing step. In addition, stress based failure criteria, e.g. Maximum Stress, Tsai-Hill, Tsai-Wu and Azzi-Tsai-Hill, currently available in ABAQUS[8] were examined, to identify load levels where first failure has to be expected. These so called "initial failure envelopes" do not distinguish between different failure modes, like fibre or inter-fibre fracture. Extensive information with respect to failure criteria for fibre-reinforced polymer composites can be found in[10].

When failure measures are defined and output is requested, ABAQUS determines the failure index R, which is written to the database. A python script was written to extract the load level and integration-point, at which first-ply failure, for selected failure criteria occurs. Figure 5.14 shows the deformed panel at a load level, for which the Max. Stress failure index first exceeded a value of 1.0, indicating that the stress state in element 7754 (layer 8, integration point 1) traversed the failure envelope.

Fig. 5.14. Visualization of the Max Stress failure index.

5.5.3 *Finite element model*

Based on former extensive studies analyzing thin-walled CFRP structures with ABAQUS, the four-node shell element S4R, (six degrees of freedom at each node, a doubly curved general-purpose element with reduced integration, hourglass control and finite membrane strains) was utilized to discretize skin and stringer. Based on convergence studies a side length of approximately 4 mm was used. This relatively fine mesh ensures that possible local, nonlinear effects are covered within the analysis. Figure 5.15 depicts the discretized panel and an inset, detailing the stringer-skin connection. The adhesive joint was modelled using rigid elements (multi-point-constraints, MPCs) connecting corresponding nodes of skin and stringer-foot shell elements.

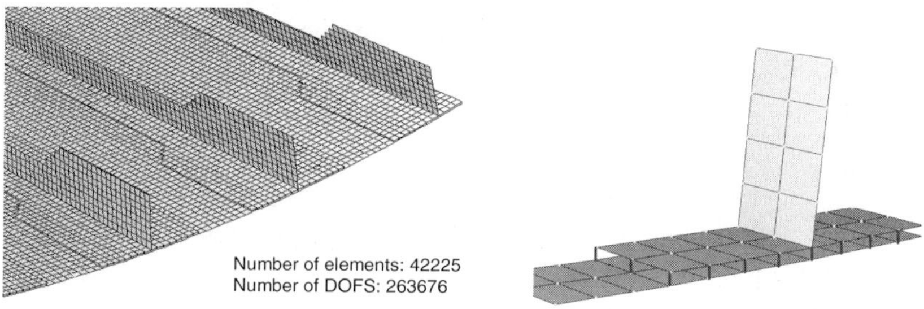

Fig. 5.15. Finite element model and detail of the stringer-skin connection.

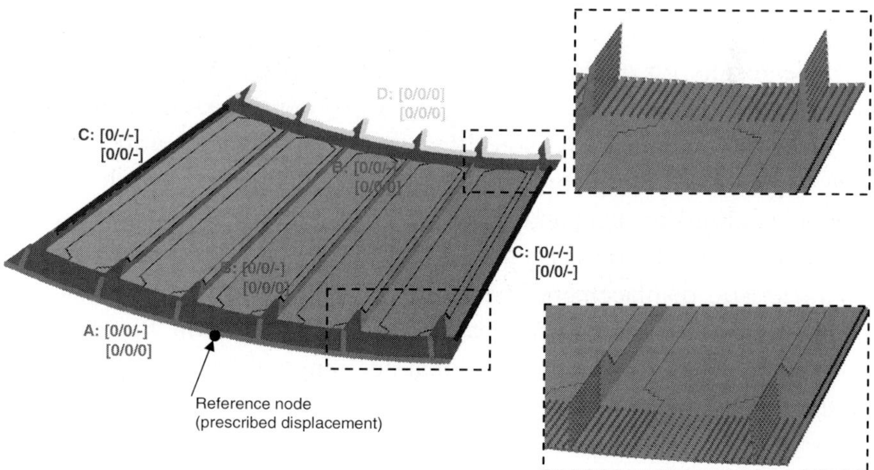

Fig. 5.16. Details of the applied boundary conditions within the FE analysis.

The axial load is applied displacement controlled on a reference node (marked as a dot in Fig. 5.16), which is rigidly connected with all nodes (skin and stringer) on the loaded circumferential edge of the panel (node-set A). The applied boundary conditions for this node-set is [0/0/-], which represent in the first line the displacement degrees of freedom [R/T/Z], and [0/0/0] in the second line characterizing the rotational degrees of freedom [$\varphi_R/\varphi_T/\varphi_Z$], both in a cylindrical coordinate system. The node-set B model the end blocks (refer to the sketch in Fig. 5.4), where is assumed that only axial displacements are possible. The node-sets C stand for the applied boundary conditions on two node lines as a surrogate model of the metallic longitudinal edge supports depicted in Fig. 5.4 or Fig. 5.7. The node line (D) along the circumferential edge represents the fixed end of the panel.

5.5.4 *Results*

Based on the previously detailed analysis procedure and the described finite element model, numerical calculations with ABAQUS/Standard were conducted. Due to the fact that the Newton method with artificial damping was utilized as a solution method additional runs were made to minimize the dissipative energy, which is taken out of the system, whilst still achieving convergence in the deep postbuckling regime.

The results on a global level, the load-shortening curves with and without initial imperfection (scaled low order buckling modes), are summarized in Fig. 5.17. Note, that the scaling factors utilized in the diagram (scaled to the corresponding load and shortening at local skin-based buckling) for load and shortening are identical to the ones used for the experimental findings in Fig. 5.9. The deformation pattern, at characteristic displacement levels, are extracted from the analysis with scaled imperfections, starting with typical "local" skin buckling (A), followed at the sharp kink by a global buckling deformation (B) and one in the transition phase to a central global buckle at a scaled shortening of 3.25 (C). The differences between both curves are essentially small, which confirms a rather stringer dominant design. The marginally smother transition from the prebuckling stiffness to the slightly reduced postbuckling one, at the onset of local skin-based buckling (A), can be explained by scaled initial geometric imperfections from an eigenvalue buckling analysis, triggering this change-over more gradually due to initially superposed skin-buckling-affine deformation patterns. The differences at point B and C are initiated by imperfections and amplified within the incremental/iterative numerical analysis. Based on theses results and additional parametric studies, it can be summarized that this type of structure revealed a minor sensitivity with respect to initial geometric imperfections.

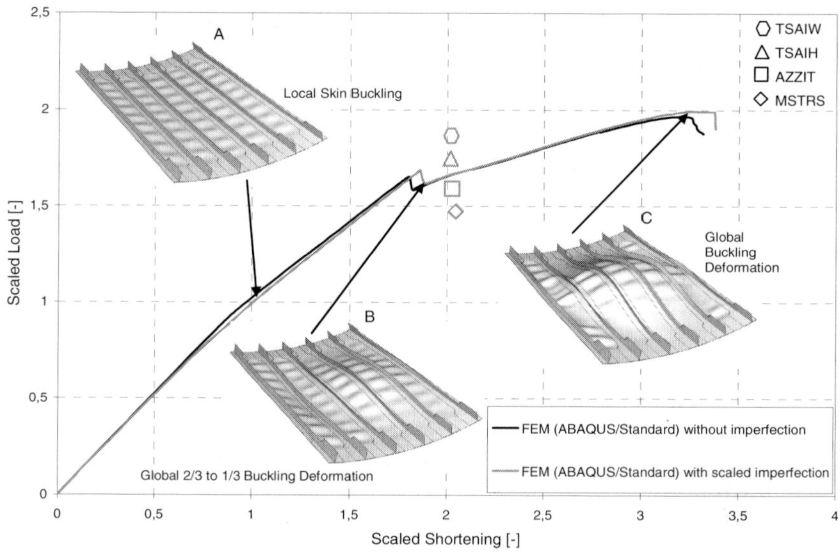

Fig. 5.17. Load-shortening curve from FE analysis with ABAQUS/Standard and deformation patterns at characteristic load levels.

In addition, the results (analysis with scaled imperfections) extracted within a postprocessing step, utilizing four stress bases failure criteria are introduced in Fig. 5.17. Markers indicating first play failure, using different failure criteria (Tsai-Wu, Tsai-Hill, Azzi-Tsai-Hill und Max. Stress) are placed vertically indicating the corresponding scaled shortening. Note, that structural degradation has not been introduced into the nonlinear calculations.

The diagram in Fig. 5.18 focuses on a rather local level, displaying the radial displacements, of four nodes on top of the stringer blade, as a function of the scaled shortening. Up to the onset of global buckling the radial displacements are close to zero, with a sudden change indicating the 1/3 outwards (positive radial displacement) and 2/3 inwards (negative radial displacement) global buckling deformation (refer to deformation pattern B in Fig 5.17). These abrupt local alterations in the radial displacements correspond to the sharp kink in the load-shortening curve. Subsequently, the run of the curves reflect the transition to a single global inwards-oriented buckling deformation (refer to deformation pattern C in Fig. 5.17).

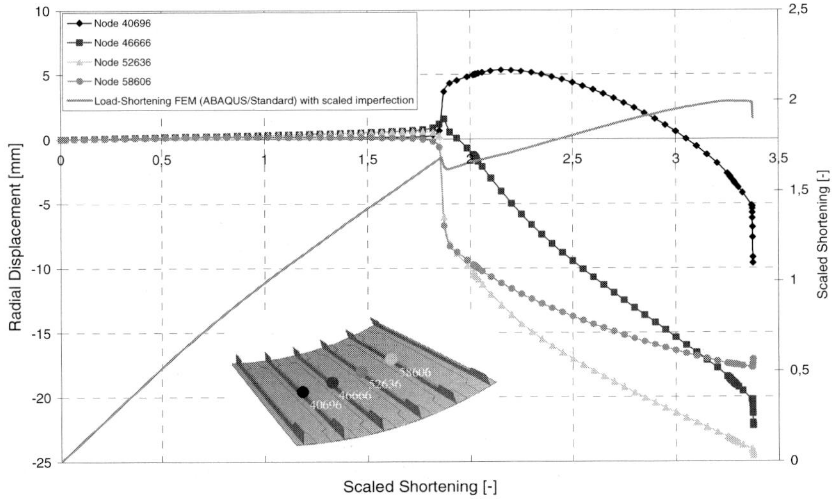

Fig. 5.18. Radial displacement versus scaled shortening and load-shortening curve from FE analysis with scaled imperfections (ABAQUS/Standard).

5.6 *Validation*

This sections provides an overview on different aspects of the validation process, starting with an introduction to give a concise insight in the terminology in modelling and computational simulation, followed by a description of the selected approach to validate the considered structure at different levels of detail, the comparison of numerical results with experimentally extracted data and finally a brief outlook with respect to larger structures.

5.6.1 *Introduction*

Figure 5.19 depicts the basic phases of modelling and simulation as presented by the Society of Computer Simulation[11] or by some other fundamental guidelines for Verification and Validation in the area of computational fluid dynamics.[12] The reality or experiment represents the problem entity to be modelled, which is in the considered case the physical experiments described in Section 5.4 on two thin-walled CFRP panels. The conceptual model is based on the analysis process of the reality or experiment and can be thought of as the sum of all necessary mathematical equations to properly describe the physical behaviour within the domain of interest. Subsequently, these mathematical equations are programmed to obtain the computer model, which is in most cases an operational computer program (e.g. FE software) comprising the conceptual model. Finally, the process of computer simulation connects back the computer model to the reality or experiment. The model qualification provides an idea on the adequacy or fidelity of the conceptual model with respect to the real problem (e.g. physical experiment). Model verification deals with the relationship between conceptual model and the computer model to identify and eliminate possible programming errors. The highlighted area of model

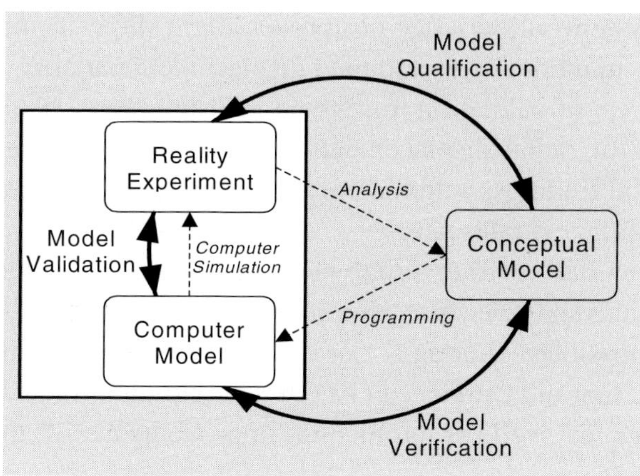

Fig. 5.19. Phases of modelling and simulation.[12]

validation, the main focus within the subsequent sections, determines the confidence in the simulation capability of the computer model in comparison to reality or experimental data.

5.6.2 *Validation approach*

As described in Section 5.4 a significant amount of test data is available to be compared with the numerical results. Based on the examined structures and the focus on their nonlinear buckling and postbuckling behaviour, the selected validation approach focuses on three levels of detail:

On a *condensed global level*, where the overall load-shortening curves are compared between experiment and analysis. The terminology "condensed" means in this context, that global load and shortening are considered to characterize the global behaviour.

This is usually the first step to examine characteristics along the load-shortening run and compare them with experimental findings. For example the axial stiffness in the pre- and postbuckling stage, onset of local buckling (change in axial stiffness) as well as global buckling should map accurately.

In a *full scale level*, where the deformation patterns at different load levels are compared. The applied experimental measurement techniques (ARAMIS system) allow after postprocessing a direct comparison with respect to the numerically determined displacement patterns.

A *local level* of validation, for which measurements, like strains from strain gauges or radial displacements at certain locations are compared with numerical findings, with the purpose to evaluate the correspondence locally, on a rather small scale.

Each level has its individual necessity for a validated numerical model and analysis procedure, due to the fact that with this multi-level approach, possible shortages or discrepancies become obvious. Furthermore, first indications can be extracted to identify possible causes of differences as well as promising improvements of the computer model.

The idea behind the different levels of the problem-oriented (bucking and postbuckling analysis of stiffened thin-walled panels) validation approach, is to raise the awareness that a simple comparison of the load-shortening curve, on a condensed global level, might be not enough to prove the validity of the model and numerical analysis procedure.

5.6.3 *Results*

Within this subsection, each level of the selected validation approach described in the last section will be covered. The condensed global level is met by the load-shortening curves in Fig. 5.20. The experimental results of the nominally identical panels B0276 and B0281 are depicted in comparison with the outcomes from the numerical analysis with and without scaled geometric imperfections. The predicted axial stiffness in the pre- and postbuckling range are very close to the experimentally extracted ones. However, the onset of local, skin-based buckling, represented with a change in the axial stiffness in the load-shortening run occurs, for the numerical results without imperfections, which give a rather defined transition point, slightly earlier in comparison to the experimentally extracted data. The onset of global, stringer-based, buckling shows some variation in the experimental outcomes, however, due to its highly nonlinear behaviour in the postbuckling regime is it still in an acceptable range and corresponds well with the numerical results. In the transition phase from the unsymmetric global 1/3 (outwards), 2/3 buckling (inwards) deformation (cf. Fig. 5.21) to a single central one, the run of all curves are close up to the point of collapse.

The numerical results of the applied stress based failure criteria (cf. Section 5.5) indicate first ply failure shortly after global buckling at a scaled shortening of approximately 2.05. At this load level the global load-shortening run display no detectable loss in load-carrying capacity, which could be explained by this rather local type of structural degradation. The test setup, including the applied sensors, does not allow for detecting this type of failure. In addition, no indication of damage due to audible cracking-noise was observed during the experiment at this load level. As mentioned before, the run of all curves (numerical and experimental ones) are close up to the point of collapse, which would be different in case of accumulated damage in the material of the specimen.

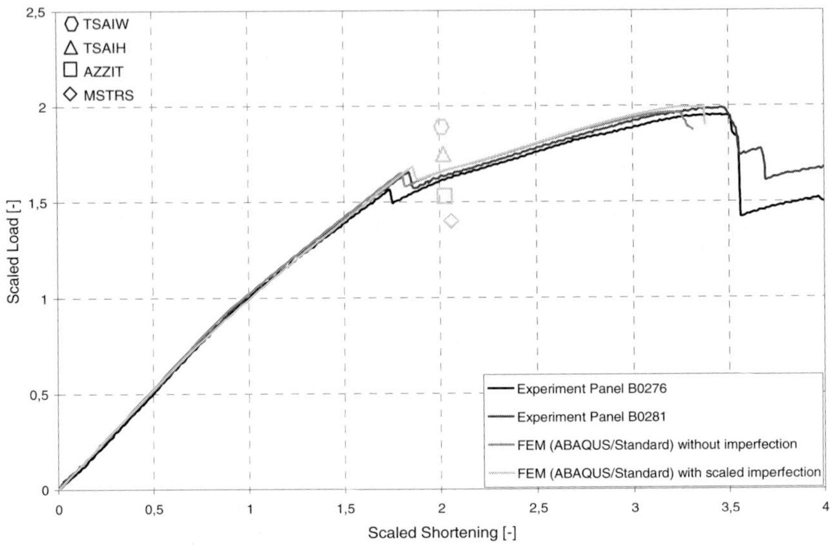

Fig. 5.20. Comparison of the experimentally and numerically extracted scaled load-shortenings curves (condensed global level of validation).

The significant loss in load carrying capacity at 3.5-3.6 scaled shortening from the experimental data are due to structural degradation (e.g. large failure in the stringer-skin interface, major fibre and interfibre failure, extensive delamination), which was located and characterized after the test. The knees in the numerical results at collapse are due to a predicted mode change, followed by the termination of the analysis due to massive convergence problems. It should be emphasized that the numerical and experimental investigations detailed in this chapter focus mainly on the nonlinear buckling and postbuckling behaviour.

Overall, the condensed global level, by comparing the load-shortening run, up to the deep postbuckling regime, revealed a high level of agreement between experimentally extracted data and numerical results.

The side-by-side comparison in Fig. 5.21 of deformation patterns at identical load levels was selected to cover the so called full scale level. The window, which was covered by the ARAMIS system (experimental measurements), has been framed in the greyscale renderings of the numerical analysis.

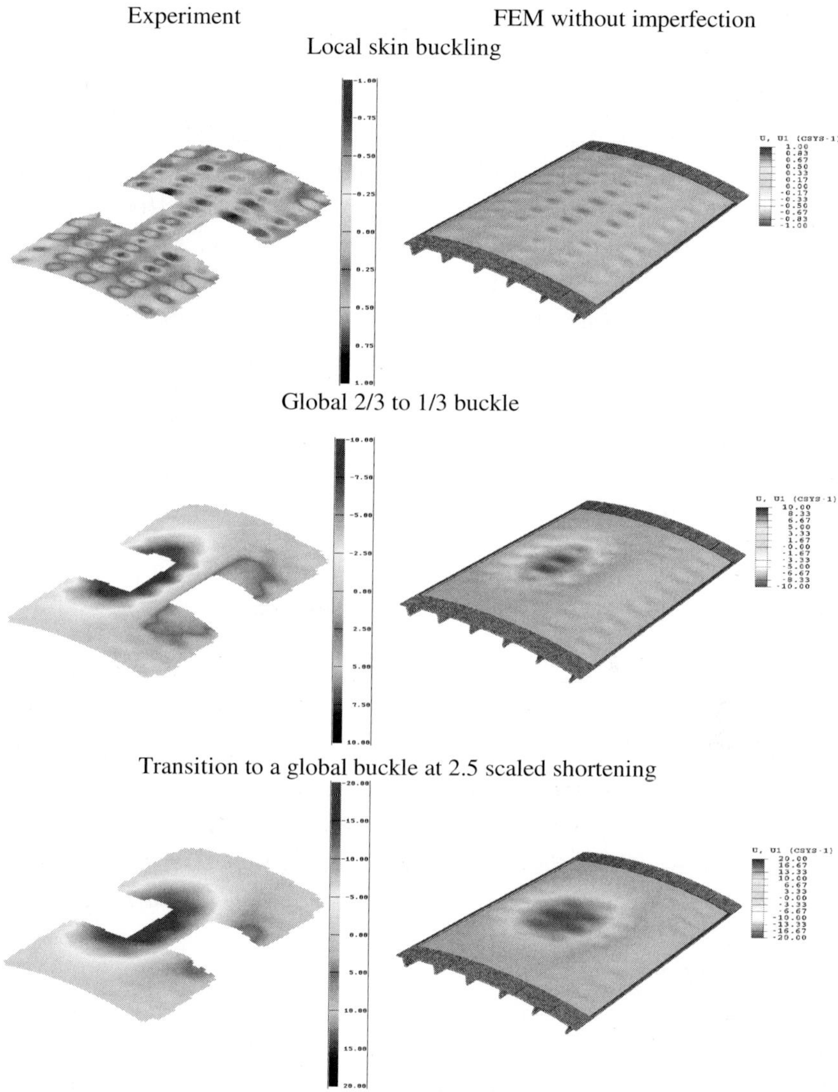

Fig. 5.21. Deformation patterns at characteristic load levels (full scale level of validation).

The measured onset of local skin buckling depicts a slightly more inhomogeneous deformation pattern in comparison to the result form the nonlinear analysis. This could be caused by the influence of the applied boundary conditions and/or some geometric imperfections introduced during the manufacturing process (cf. Fig. 5.6) influencing this highly sensitive stability phenomenon. However, the total number of half-waves in the axial direction is identical for analysis and experiment.

The global, stringer-based, 2/3 to 1/3 buckle depicts a larger expansion in axial as well as circumferential direction for the experimental measurements, however, this is due to different greyscale distribution within the legend of experimental and numerical results. Despite this pure visual discrepancy, further detailed examinations revealed that a high level of agreement was found on the full scale level between the experimental measurements and the numerical results.

To underpin the findings on the local level, experimentally measured and numerically predicted radial displacements at defined locations on top of the stringer blades (refer to Fig. 5.10 or 5.18) were extracted and compared in Fig. 5.22.

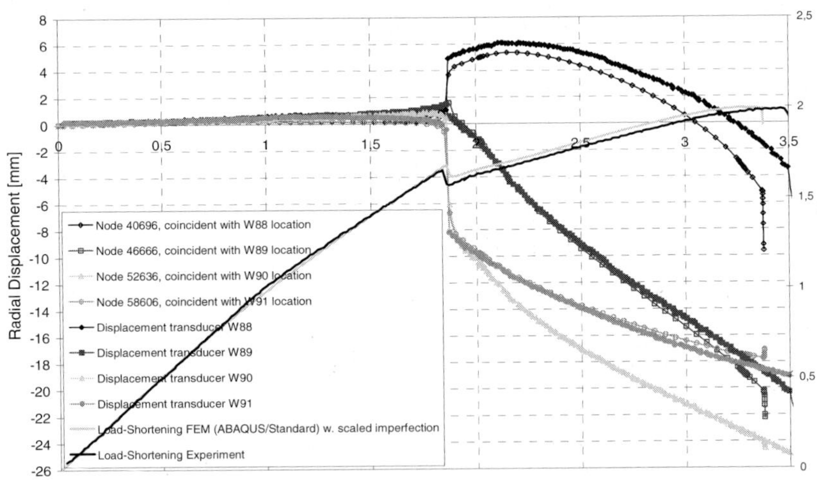

Fig. 5.22. Radial displacement scaled shortening curves from experiment and analysis.

A positive radial displacement corresponds to an outwards deformation from the stringer- to the skin-side. As expected there are almost no deformations prior the onset of global buckling, only small positive radial displacements due to the applied boundary conditions and the axial loading are detected. However, at a scaled shortening of approximately 1.8, which complies with the onset of global buckling, the radial displacements reproduce the unsymmetric global 2/3 to 1/3 buckling deformation (refer to Fig. 5.21). W88 indicates a sudden rise in positive radial direction (1/3). W90 and W91 show an even larger transition in negative radial direction (2/3). The displacement transducer at W89 reflects a linear slope, which can be explained by its initial inflection point location and the subsequent transition to a single central global buckle. This alteration, towards a single global buckle in the centre, explains the run of W88 towards a negative radial displacement and the further negative gain of W89, W90 and W91. The corresponding numerically determined (ABAQUS/Standard with scaled imperfections) displacements revealed a high level of agreement between experimentally extracted data and numerical results

All three levels (condensed global level, full scale level, local level) of the selected validation approach reflected a high level of correlation between the experimentally extracted and numerically determined results, which allows to consider the numerical analysis procedure (model building and solution) as an appropriate methodology to simulate stiffened panel type structures.

5.6.4 *Transferability*

Figure 5.23 depicts characteristic structural levels, in this case for an aircraft fuselage, from coupon level to component or barrel level in a form of a pyramid (inspired by the pyramid of tests introduced by Rouchon[14]). The investigations described before cover both, experiments and analysis, on a lower subcomponent level. For the numerical investigations of stiffened panels, the material properties from coupon level tests (marked by the solid arrow) were utilized.

An important question which arises for the considered stiffened curved CFRP panels is how well the validated numerical analysis procedure is applicable within an extended designs space connecting the subcomponent with the component level, e.g. from the panel level towards larger sections of a whole barrel combined with periodic or symmetric boundary conditions to simulate the periodicity appropriately as well as typical critical loading conditions (cf. dashed arrow).

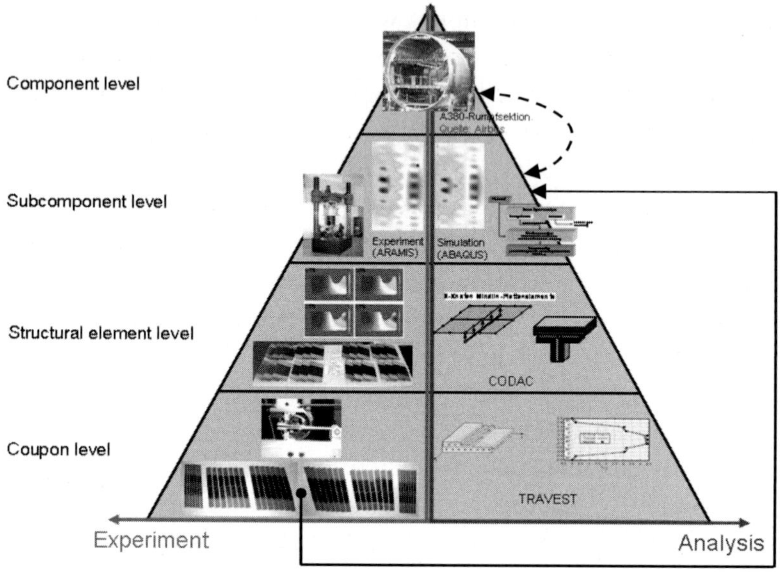

Fig. 5.23. Structural levels (e.g. for an aircraft fuselage).

Preliminary numerical investigations on larger panels with periodic boundary conditions revealed an acceptable level of agreement in comparison to the full barrel. The appropriate load distribution from critical load cases can be obtained via a global-local (submodelling) approach. A rather coarse mesh of the barrel (including all relevant structural components, e.g. floors) is used to determine within a nonlinear analysis nodal displacements, which are subsequently used as prescribed deformations of an appropriately refined mesh in a smaller subsection which was identified to be stability-critical (in most cases lower section of the barrel).

5.7 Conclusions and Outlook

Relevant aspects of simulation and experimental investigation of stiffened composite panels were given. The validation process revealed, at different levels of detail, that the selected numerical analysis procedure (model building and solution) is an appropriate approach to simulate the buckling and postbuckling behaviour of the considered structure. An extension to larger sections (e.g. part of a barrel) was indicated on the basis of preliminary numerical investigations. Based on these findings, with respect to validated virtual testing, the remaining structural reserves in the postbuckling regime may be exploited safely for more weight and cost efficient stiffened panel designs.

This contribution focused on experimental and numerical investigations into the deep postbuckling regime with special focus on validation aspects. Further information with respect to the currently developed methods to consider structural degradation for stiffened CFRP panel type structures can be found in references 15 and 16.

5.8 References

1. Group of Personalities, *European Aeronautics: A Vision for 2020*, Office of Official Publications of the European Communities, Belgium (2001).
2. R. Zimmermann and R. Rolfes, *Compos. Struct.,* 73: 171 (2006).
3. R. Zimmermann, H. Klein and A. Kling, *Compos. Struct.,* 73: 150 (2006).
4. L. Lanzi and C. Bisagni, *AIAA-2003-1795* (2003).
5. H. Abramovich, A. Grunwald, P. Pevsner, T. Weller, A. David, G. Ghilai, A. Green and N. Peker, *AIAA-2003-1793* (2003).
6. W. L. Oberkampf, *Exp. Techniques* 25: 35 (2001).
7. R. G. Sargent, *Winter Simulation Conference*, Wasington D.C. (1998).
8. *ABAQUS Analysis User's Manual*, Vol. 2, Version 6.6 (2006).
9. *ABAQUS Theory Manual*, Version 6.6 (2006).
10. M. J. Hinton, A. S. Kaddour, P. D. Soden: *Failure Criteria in Fibre Reinforced Polymer Composites: The World-Wide Failure Exercise* (Elsevier Ltd, ISBN 008044475X, 2004).
11. S. Schlesinger, *Simulation*, 32(3): 103 (1979).
12. AIAA Guide: *Guide for the Verification and Validation of Computational Fluid dynamics simulation*, ISBN 1-56347-285-6 (1998).
13. P. J. Roache: *Verification and Validation in Computational Science and Engineering*, (Hermosa, ISBN 0-913478-08-3, 1998).
14. J. Rouchon, *Congress of International Council of the Aeronautical Sciences,* 2: 1439 (1990).
15. R. Degenhardt, R. Rolfes, R. Zimmermann and K. Rohwer *Compos. Struct.,* 73: 175 (2006).
16. R. Degenhardt, A. Kling and K. Rohwer, *Congress of International Council of the Aeronautical Sciences*, Hamburg (2006).

CHAPTER 6

ANISOTROPIC ELASTIC TAILORING IN LAMINATED COMPOSITE PLATES AND SHELLS

Paul M. Weaver

Advanced Composites Centre for Innovation and Science (ACCIS)
Department of Aerospace Engineering, University of Bristol, UK
E-mail: paul.weaver@bristol.ac.uk

Significant improvements in structural efficiency are possible using specific, inhomogeneous laminates. These are laminates whose stacking sequence has been optimized against constraints such as buckling, damage tolerance and natural frequencies. Furthermore, there are many design scenarios in which further gains can be made by making use of anisotropy, particularly when buckling or postbuckling is concerned. After a brief introduction to current design practice, buckling and postbuckling formulae are derived for anisotropic plates and shells, subject to different loadings. From these expressions, concepts for anisotropic elastic tailoring are discussed and their implications for design considered.

6.1 Introduction

Aerospace structural design is often driven by the need to prevent or mitigate buckling phenomena. The aim is often to minimise weight. Engineers may use a range of tools to achieve this from simple analytical expressions[1], through bespoke buckling codes[2-3] to detailed finite element analysis (FEA)[4]. Each method is complementary to one other and is most suitable at different stages of the design process. Analytical expressions are most useful at the earliest stages of the design process. Then, design options are fluid and structural idealisation has been used to represent potentially complex structures by relatively simple prismatic beams, plates, curved panels and shells. At this stage, major decisions over structural form, initial sizing and materials selection are often made. Bespoke buckling codes are useful for analysis and design of sub-assemblies and for optimisation runs. At this level of analysis, interactions between structural members, such as joints between stringer and skins may be assessed. Finally, detailed finite element analysis is used to gain more accurate numerical information on sub-assemblies and overall structural performance.

The hierarchical design process outlined above has changed little in recent years despite the increased development of computational tools such as topological optimisation, and high fidelity FEA. However, there remains a need for engineers to have a thorough understanding of basic mechanics and approximate methods. Such knowledge leads to physical insight that may be used to assess the credibility of FEA. As such, approximate methods including simple closed form formulae are useful not only at the initial stages of design but also at every level to provide a credibility check on numerical solutions. This is particularly relevant for laminated composites where there is scope for elastic tailoring. On a basic level, this involves placing fibre orientations along major load paths, or principal directions. Much work has been done to characterise such response, e.g. Refs. 5 and 6. To provide minimum levels of strength in other directions, relatively small percentages of other fibre directions may be used, e.g. 10% rule[7], thus reducing the potential gains in weight saving. In practice, the uncertainty over the nature of loads, multiple load cases, strength and damage tolerance considerations leads to further amelioration of the benefits of elastic tailoring. If these effects are considered sufficiently important the lay-up may be driven to one devoid of any elastic tailoring and will be quasi-isotropic in nature. This lay-up may be initially selected for scoping studies yet it may be considered a responsibility of the engineer to design to a level of performance that is at least as effective as the quasi-isotropic lay-up. A lesson to be learned from such considerations is that the quasi-isotropic lay-up is a good starting lay-up to consider in structural design - or at least it can be considered as the least bad!

Furthermore, the aerospace industry typically manufactures laminates using a limited number of distinct fibre orientations. Typically, these are

0°, 90° and ±45° ply angles. The primary reason for restricting choice appears to be one of cost and manufacturing ease. Actually, Fig. 6.1 shows a lamination parameter diagram[8] for in-plane loading response, noting that a brief introduction to lamination parameters is given in Section 6.2. It is constructed by representing stiffnesses in terms of lamination parameters and material invariants[8-10]. It shows that the boundary of feasible design space is the combination of a locus of single angle ply laminates, that describes a parabola, coupled with an upper limit of $\xi_2 = 1$. All potential lay-ups must lie within this design space. Those based on 0°, 90° and ±45° ply angles must lie within the inscribed triangle that links the vertices of the lamination parameter diagram, and since this triangle covers most of the potential design space it can be said that limiting the fibre angles to 0°, 90° and ±45° plies is not as restrictive as at first might be thought. However, if the design envelope was expanded to include 30° and 60° ply angles then the designs space covered by the inscribed polygon maps very closely to the external parabola of single angle plies. On this basis alone it may be justified to expand the choice of fibre angles to include these additional orientations. Furthermore, by including angles based on 30° and 60° will allow for potentially efficient anisotropic elastic tailoring.

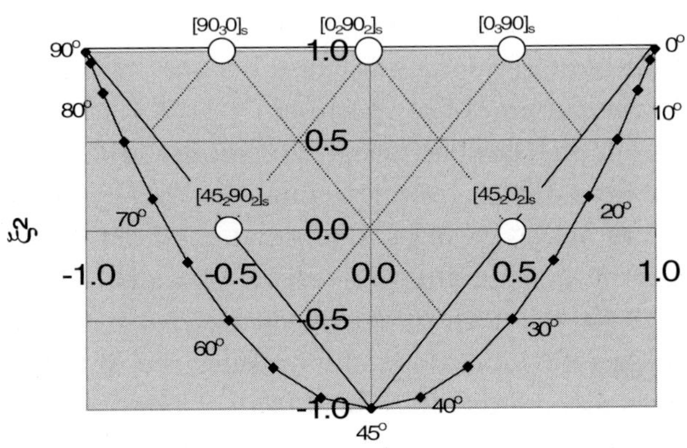

Fig. 6.1. Lamination parameter diagram showing feasible design region.

It is noted that industry practice places further restrictions on lay-up by limiting lay-ups to those that are symmetrically laminated. It is quite possible that such practice arose to minimise warping effects upon cool down from hot temperature cure. Such practice minimises unwanted coupling response (**B**-matrix) between in–plane and out-of-plane behaviour. There is mounting evidence that this practice is unnecessarily limiting, and for cases where buckling is a driver, some optimal lay-ups are antisymmetrically laminated[11].

Further insight into elastic tailoring is gained by realising that the most efficient structures are thin-walled in the sense that one or two of the dimensions are much smaller than the third. Such structures gain their efficiency because loads are transmitted by in-plane response such that the cross-section is uniformly loaded and stress resultants through the thickness of a section are small. Therefore, flexural response of a side wall is minimised. The implications on laminate lay-up are significant because if the structural response is mostly in-plane then stacking sequence effects are minimised because in-plane stiffnesses only depend on proportions of differing ply angles and not their absolute position. Then, for design it is only necessary to select percentages of fibre angles and stacking sequences, which may be chosen to satisfy other considerations such as those on damage tolerance and through-thickness strength. The situation is different where buckling or vibration phenomena or indeed significant bending are design drivers. Then stacking sequences are important because flexural stiffnesses (**D**-matrix) depend on the cubic power of ply position.

In conclusion, the current design situation is one of selection of proportions of fibre angles from a limited set, usually consisting of 0°, 90° and $\pm 45^\circ$ ply angles to meet in-plane strength and stiffness criteria whilst maintaining a minimum proportion, typically 10%, of each ply orientation. Stacking sequences are then chosen to maintain adequate levels of damage tolerance and buckling resistance, if appropriate. This usually involves placing the chief load-bearing plies into the interior of the laminate so that they are offered some protection from damaging through-thickness loads (e.g. impact), by the outer layers. Furthermore, there are constraints on lay-up selection from one structural element to the next; e.g., rib bays, to maintain structural integrity and continuation of load paths. Another pertinent example of such compromise is where a skin cover attaches to a stringer flange. Then, stiffness mismatches must be considered to maintain low levels of interlaminar stress. The design strategy described immediately above may not be the most appropriate when buckling is concerned. Ply angles should not be placed in the major load paths but should be chosen to resist buckling deformations[12]. With such understanding, optimal lay-ups to resist buckling become more intuitive. In the following sections, a classical treatment of plates and shells are evaluated in terms of lamination parameters and possible designs. As such, through-thickness shear effects, damage or interlaminar stresses are not considered here, in any detail. Various scenarios including buckling and postbuckling of flat plates with different boundary conditions subject to either compression or shear loads are considered and draw upon the experiences of the author[11-17]. In addition, the effects of curvature are analysed using the case of a circular cylindrical shell subject to compression loading as an example[18-20].

6.2 Mechanics of Anisotropic Plates

6.2.1 *Introduction*

It is interesting to note that the development of the theories of plates is largely a nineteenth century phenomenon. Kirchhoff[21] presented the first practical plate equations building upon the earlier works of Germain[22] and Lagrange[23]. Indeed, the first anisotropic plate equations were derived by Gehring[24], a student of Kirchhoff. Also, apparently independently of these works, Kelvin[25] published a relatively complete account in 1867, which seems to have largely gone unnoticed. Indeed, Timoshenko[26] (1913) credits Saint-Venant[27] with the derivation of plate equations in his annotated French translation of Clebsch's German pioneering text. The point of mentioning all this is that whilst it could be argued that Kirchhoff was motivated by practical reasons. After all, it was the time of the great industrial age, with wrought iron plate widely available for engineering constructions. However, the same could not be said of

Gehring and Kelvin. At the time, there was no obvious practical application for anisotropic plate equations, their work was apparently done for scientific curiosity. To put these works into context, there remained a philosophical debate on the exact number of elastic constants for isotropic and anisotropic materials. On the one hand, Poisson[28] and Saint-Venant advocated what was known as the rariconstant theory in which it was thought only one elastic constant was necessary to describe an isotropic material (Poisson's ratio was fixed at 0.25) and 15 for a fully anisotropic material. On the other hand, George Green[29] in 1839 using conservation of energy, showed that two and 21 constants, respectively were needed to capture the elasticity of an isotropic and anisotropic material, respectively. This argument between the two camps lasted for many years and was only resolved by careful experimentation by Voigt[30] concerning excitation of single crystals to show that Poisson's ratio could vary. Therefore, the continuum theorists (e.g. George Green) won the day, such that the more accurate description of elasticity encompassed the increased number of independent constants. The next significant advance was made by Huber[31] who summarised many years worth of work in his 1929 treatise on orthotropic plates. He re-derived the orthotropic plates equations (apparently independently of prior work) and was motivated by the new material of its day, namely, reinforced concrete. It appears relatively surprising today that reinforced concrete would be designed as a thin-walled structure such that buckling may outweigh strength as a design driver. Nonetheless, Huber, appears to have published the first treatment of buckling of orthotropic plates. He actually discusses the practical significance of reinforcing concrete either longitudinally or transversely. Based on the mechanics he concludes that either reinforcement results in the same buckling load, a conclusion that endures for lightweight laminated composites of today.

6.2.2 *Initial Buckling of Anisotropic Plates*[13,16-17]

Equilibrium equations

The governing expression for static equilibrium of a flat plate subjected to in-plane loading, as given, for example, by Timoshenko[32] is

$$\frac{\partial^2 M_x}{\partial x^2}+2\frac{\partial^2 M_{xy}}{\partial x\partial y}+\frac{\partial^2 M_y}{\partial y^2}-N_x\frac{\partial^2 w}{\partial x^2}-2N_{xy}\frac{\partial^2 w}{\partial x\partial y}-N_y\frac{\partial^2 w}{\partial y^2}=0 \quad (6.1)$$

where M_x, M_y and M_{xy} are bending and twisting moments, respectively and N_x, N_y and N_{xy} represent in-plane stress resultants. To introduce anisotropy into the problem, the constitutive behaviour of the material is considered.

Constitutive equations

Following Stavsky[33] the constitutive equations are written in a partially inverse form as

$$\begin{Bmatrix}\boldsymbol{\varepsilon}^0\\ \mathbf{M}\end{Bmatrix}=\begin{bmatrix}\mathbf{a} & \mathbf{b}\\ -\mathbf{b}^T & \mathbf{D}^*\end{bmatrix}\begin{Bmatrix}\mathbf{N}\\ \boldsymbol{\kappa}\end{Bmatrix} \quad (6.2)$$

where $\mathbf{a}=\mathbf{A}^{-1}$, $\mathbf{b}=-\mathbf{A}^{-1}\mathbf{B}$, $\mathbf{D}^*=\mathbf{D}-\mathbf{B}\mathbf{A}^{-1}\mathbf{B}$, and where **A**, **B** and **D** are the in-plane, coupling and flexural stiffness matrices, respectively. Note that **b** may thought of as a moment arm or eccentricity matrix because it links in-plane loads, **N**, to bending moments, **M**. It also represents the distance from neutral plane to midplane for a particular stress resultant, N_i. For symmetrically laminated plates $\mathbf{b}=\mathbf{0}$ and $\mathbf{D}^*=\mathbf{D}$. Note that $\mathbf{D}^*$ is the reduced bending stiffness matrix. The A_{ij}, B_{ij} and D_{ij} terms are functions of the material invariants, W_1-W_5 and lamination parameters, ξ_1-ξ_{12} (e.g. Onoda[34], Fukunaga[35]) as

$$\begin{Bmatrix}A_{11}\\ A_{12}\\ A_{22}\\ A_{66}\\ A_{16}\\ A_{26}\end{Bmatrix}=t\begin{bmatrix}1 & \xi_1 & \xi_2 & 0 & 0\\ 0 & 0 & -\xi_2 & 1 & 0\\ 1 & -\xi_1 & \xi_2 & 0 & 0\\ 0 & 0 & -\xi_2 & 0 & 1\\ 0 & \dfrac{\xi_3}{2} & \xi_4 & 0 & 0\\ 0 & \dfrac{\xi_3}{2} & -\xi_4 & 0 & 0\end{bmatrix}\begin{Bmatrix}W_1\\ W_2\\ W_3\\ W_4\\ W_5\end{Bmatrix},$$

$$\begin{Bmatrix} B_{11} \\ B_{12} \\ B_{22} \\ B_{66} \\ B_{16} \\ B_{26} \end{Bmatrix} = \frac{t^2}{4} \begin{bmatrix} 0 & \xi_5 & \xi_6 & 0 & 0 \\ 0 & 0 & -\xi_6 & 0 & 0 \\ 0 & -\xi_5 & \xi_6 & 0 & 0 \\ 0 & 0 & -\xi_6 & 0 & 0 \\ 0 & \frac{\xi_7}{2} & \xi_8 & 0 & 0 \\ 0 & \frac{\xi_7}{2} & -\xi_8 & 0 & 0 \end{bmatrix} \begin{Bmatrix} W_1 \\ W_2 \\ W_3 \\ W_4 \\ W_5 \end{Bmatrix},$$

$$\begin{Bmatrix} D_{11} \\ D_{12} \\ D_{22} \\ D_{66} \\ D_{16} \\ D_{26} \end{Bmatrix} = \frac{t^3}{12} \begin{bmatrix} 1 & \xi_9 & \xi_{10} & 0 & 0 \\ 0 & 0 & -\xi_{10} & 1 & 0 \\ 1 & -\xi_9 & \xi_{10} & 0 & 0 \\ 0 & 0 & -\xi_{10} & 0 & 1 \\ 0 & \frac{\xi_{11}}{2} & \xi_{12} & 0 & 0 \\ 0 & \frac{\xi_{11}}{2} & -\xi_{12} & 0 & 0 \end{bmatrix} \begin{Bmatrix} W_1 \\ W_2 \\ W_3 \\ W_4 \\ W_5 \end{Bmatrix}. \tag{6.3}$$

The lamination parameters may be calculated from the following integrals

$$\begin{aligned} (\xi_1 \quad \xi_2 \quad \xi_3 \quad \xi_4) &= \frac{1}{2}\int_{-1}^{1} (\cos 2\theta \quad \cos 4\theta \quad \sin 2\theta \quad \sin 4\theta)\, du_i \\ (\xi_5 \quad \xi_6 \quad \xi_7 \quad \xi_8) &= \int_{-1}^{1} (\cos 2\theta \quad \cos 4\theta \quad \sin 2\theta \quad \sin 4\theta)\, u_i\, du_i \qquad (6.4) \\ (\xi_9 \quad \xi_{10} \quad \xi_{11} \quad \xi_{12}) &= \frac{3}{2}\int_{-1}^{1} (\cos 2\theta \quad \cos 4\theta \quad \sin 2\theta \quad \sin 4\theta)\, u_i^2\, du_i \end{aligned}$$

$$u_i = \frac{2h_i}{t} \tag{6.5}$$

where h_i is the distance of a particular ply surface from the mid plane, θ is the fibre angle of a particular ply and t is the thickness of the laminate. The material invariants, W_i, are linear functions of the ply stiffness properties, Q_{ij}, as given for example, by Jones[36],

$$
\begin{aligned}
W_1 &= \frac{3Q_{11} + 3Q_{22} + 2Q_{12} + 4Q_{66}}{8} \\
W_2 &= \frac{Q_{11} - Q_{22}}{2} \\
W_3 &= \frac{Q_{11} + Q_{22} - 2Q_{12} - 4Q_{66}}{8} \\
W_4 &= \frac{Q_{11} + Q_{22} + 6Q_{12} - 4Q_{66}}{8} \\
W_5 &= \frac{W_1 - W_4}{2}
\end{aligned} \qquad . \tag{6.6}
$$

It is noted that the material invariants provide useful physical insight into possible laminate response. First, it is noted that all lamination parameters, ξ_1-ξ_{12} are zero for the quasi-isotropic modulus. Secondly, W_1 is an intrinsic measure of the quasi-isotropic Young's modulus for the material. Since Q_{22}, Q_{12} and Q_{66} are typically much smaller than Q_{11} an estimate of its value may be obtained from

$$
W_1 \approx \frac{3Q_{11}}{8}
$$

which allows the designer to obtain the quasi-isotropic stiffness of a laminate from knowledge of its unidirectional modulus. Furthermore, W_4/W_1 is the quasi-isotropic Poisson's ratio and W_5 is the quasi-isotropic shear modulus. It is noted that for larger W_2 (and to a lesser extent W_3) the greater potential there is for orthotropic and anisotropic elastic tailoring.

The curvatures are

$$
\begin{aligned}
\kappa_{xx} &= -\frac{\partial^2 w}{\partial x^2}, \\
\kappa_{yy} &= -\frac{\partial^2 w}{\partial y^2}, \\
\kappa_{xy} &= -2\frac{\partial^2 w}{\partial x \partial y}.
\end{aligned} \tag{6.7}
$$

The immediate goal is to reduce the governing equations into our unknown displacement w, by substituting Eqn. (6.7) into Eqns. (6.2) and (6.1). The governing equation becomes

$$\begin{aligned} & D^*_{11}\frac{\partial^4 w}{\partial x^4}+2(D^*_{12}+2D^*_{66})\frac{\partial^4 w}{\partial x^2\partial y^2}+D^*_{22}\frac{\partial^4 w}{\partial y^4} \\ & +4D^*_{16}\frac{\partial^4 w}{\partial x^3\partial y}+4D^*_{26}\frac{\partial^4 w}{\partial x\partial y^3}-b_{21}\frac{\partial^2 N_x}{\partial x^2}-(2b_{26}-b_{61})\frac{\partial^2 N_x}{\partial x\partial y} \\ & +(b_{11}+b_{22}-2b_{66})\frac{\partial^2 N_{xy}}{\partial x\partial y}-(2b_{16}-b_{62})\frac{\partial^2 N_y}{\partial x\partial y}-b_{12}\frac{\partial^2 N_y}{\partial y^2} \\ & =N_x\frac{\partial^2 w}{\partial x^2}+N_y\frac{\partial^2 w}{\partial y^2}+2N_{xy}\frac{\partial^2 w}{\partial x\partial y}, \end{aligned} \tag{6.8}$$

where b_{ij} are elements of **b**. An expression of this kind, for plates with $b_{ij} = 0$, was first published in the dissertation by Gehring[24], written in Latin and apparently independently, in English, by Lord Kelvin[25]. For initial buckling problems a simplification can be made to Eqn. (6.8) - the variation of uniform resultant loads **N** in the planform, may be neglected. Then, all terms containing derivatives of **N** with respect to x and y disappear. With this assumption, it is noted that buckling expressions for symmetric laminates apply if **D** is replaced by the reduced bending stiffness **D***. The question of whether nonsymmetric laminates buckle or not is briefly discussed in Section 6.3.7. To physically understand the behaviour consider a laminate where **B** = 0 (e.g. symmetric laminates). Each term on the left hand side of Eqn. (6.8) may be thought of as an internal reaction for resisting external loads (N_x, N_y and N_{xy}). These internal reactions reflect individual deformation mechanisms:

- $D_{11}\dfrac{\partial^4 w}{\partial x^4}$: bending along x-axis
- $D_{22}\dfrac{\partial^4 w}{\partial y^4}$:bending along y-axis
- $(D_{12}+2D_{66})\dfrac{\partial^4 w}{\partial x^2\partial y^2}$: twisting and anticlastic contributions

- $D_{16}\dfrac{\partial^4 w}{\partial x^3 \partial y}$: anisotropic twisting due to bending along *x*-axis
- $D_{26}\dfrac{\partial^4 w}{\partial x \partial y^3}$: anisotropic twisting due to bending along *y*-axis

6.2.3 *Significance of Lamination Parameters*

Expressing flexural stiffnesses in terms of lamination parameters and material invariants has important consequences for optimisation and design of composite laminates:

- each set of stiffnesses (in-plane, coupling and flexural) are linear combinations of material invariants and four lamination parameters.

Conventionally, **A**, **B** and **D** stiffnesses, are expressed in terms of transformed stiffnesses, Q^*_{ij}, and ply position, e.g.

$$D_{ij} = \frac{1}{3}\sum_{i=1}^{n} Q^*_{ij}\left(h_j^{\,3} - h_{j-1}^{\,3}\right) \tag{6.9}$$

for a laminate of *n* layers, noting that Q^*_{ij} are fourth order trigonometric function of ply angle[36]. Such expressions are obviously highly nonlinear and have multiple local minima/maxima as a function of ply angle, which makes design and optimisation studies cumbersome. In contrast, by representing the structural stiffnesses as a linear combination, Eqn. (6.3), is simpler. This observation holds because the relative contributions to stiffness between inherent material stiffnesses (material invariants W_i) and ply angle (lamination parameter) have been separated. Importantly, this representation means that materials selection may be done at two levels. The first level concerns choice of material, which is made solely on the relative values of W_i and the second level may be designed solely on the basis of lamination parameters. Furthermore:

- the design space has been reduced from one which is open-ended, consisting of *n* plies, at potentially any fibre orientation to one containing 12 variables - the lamination parameters, ξ_1-ξ_{12}. For symmetric laminates, buckling problems are governed by four variables, ξ_9-ξ_{12}.

- Grenestedt and Gudmundson[37] proved that the feasible region of the 12 lamination parameters is convex. This important feature guarantees that gradient based methods locate a global minimum for linear problems, (statics, buckling, and vibration). Here, linear problems are used to describe those problems in which a variable of interest (e.g. buckling load) is a linear function of **A**, **B** and **D** and hence a linear function of lamination parameters. For example, the plate equation, Eqn. (6.8) for symmetric laminates, shows a linear dependence for buckling loads on D_{ij}. It is noted that the linear dependence does not hold for nonsymmetric laminates due to the presence of nonlinear D_{ij}^* and b_{ij} terms. However, experience has so far shown optimization problems based on symmetric laminates to remain convex.

The plate equation given by Eqn. (6.8) is sufficient for initial buckling studies in which the plate remains flat in prebuckling deformation. If, however, postbuckling studies are undertaken then knowledge of in-plane static equilibrium is needed. The following section details such analysis.

6.2.4 *Postbuckling of Anisotropic Plates*[14-15]

Postbuckling phenomena cannot ignore in-plane deformation contributions as was done for initial buckling. As out of plane deformation arise they must be accompanied by in-plane strains in order to maintain straight edges along the boundaries.

Using von Kàrman plate assumptions, the large-deflection plate equations can be derived in terms of transverse deflection w and a stress function ψ. This approach is adopted because it leads to a system of only two uncoupled differential equations. The stress function ψ is defined by the stress resultants vector **N** :

$$N_x = -\psi,_{yy} \;, \quad N_y = -\psi,_{xx} \;, \quad N_{xy} = \psi,_{xy} \,. \tag{6.10}$$

The vector of reference surface strains $\boldsymbol{\varepsilon}^0$ at $z=0$ and the vector of curvatures $\boldsymbol{\kappa}$ are defined by

$$\varepsilon_x^0 = \frac{\partial u^0}{\partial x} + \frac{1}{2}\left(\frac{\partial w}{\partial x}\right)^2 \quad \varepsilon_y^0 = \frac{\partial v^0}{\partial y} + \frac{1}{2}\left(\frac{\partial w}{\partial y}\right)^2, \qquad (6.11)$$
$$\varepsilon_{xy}^0 = \frac{\partial u^0}{\partial y} + \frac{\partial v^0}{\partial x} + \frac{\partial w}{\partial x}\frac{\partial w}{\partial y}.$$

The differential equation of transverse motion given by Eqn. (6.8) is modified slightly by substituting for stress resultants in terms of force function from Eqn. (6.11) to give

$$\begin{aligned}
&D_{11}^* w,_{xxxx} + 4D_{16}^* w,_{xxxy} + 2\left(D_{12}^* + 2D_{66}^*\right) w,_{xxyy} + 4D_{26}^* w,_{xyyy} + D_{22}^* w,_{yyyy} \\
&+ b_{21}\psi,_{xxxx} + \left(2b_{26} - b_{61}^*\right)\psi,_{xxxy} + \left(b_{11} + b_{22} - 2b_{66}\right)\psi,_{xxyy} \qquad (6.12) \\
&+ \left(2b_{16} - b_{62}\right)\psi,_{xyyy} + b_{12}\psi,_{yyyy} - \left(w,_{xx}\psi,_{yy} + w,_{yy}\psi,_{xx} - 2w,_{xy}\psi,_{xy}\right) = 0
\end{aligned}$$

The other expression that governs the postbuckling of symmetrically laminated plates is the condition of compatibility,

$$\frac{\partial^2 \varepsilon_x}{\partial y^2} - \frac{\partial^2 \varepsilon_{xy}}{\partial x \partial y} + \frac{\partial^2 \varepsilon_x}{\partial y^2} = \left(\frac{\partial^2 w}{\partial x \partial y}\right)^2 - \frac{\partial^2 w}{\partial x^2}\frac{\partial^2 w}{\partial y^2} \qquad (6.13)$$

which is obtained by taking the necessary differentiation and reductions from Eqn. (6.11). By combining Eqns. (6.13) and (6.2), the compatibility equation may be written in terms of w and ψ and it is expressed as:

$$\begin{aligned}
&a_{22}\frac{\partial^4 \psi}{\partial x^4} - 2a_{26}\frac{\partial^4 \psi}{\partial x^3 \partial y} + \left(2a_{12} + a_{66}\right)\frac{\partial^4 \psi}{\partial x^2 \partial y^2} - 2a_{16}\frac{\partial^4 \psi}{\partial x \partial y^3} + a_{11}\frac{\partial^4 \psi}{\partial y^4} \\
&- b_{21}\frac{\partial^4 w}{\partial x^4} - \left(2b_{26} - b_{61}\right)\frac{\partial^4 w}{\partial x^3 \partial y} - \left(b_{11} + b_{22} - 2b_{66}\right)\frac{\partial^4 w}{\partial x^2 \partial y^2} \qquad (6.14) \\
&- \left(2b_{16} - b_{62}\right)\frac{\partial^4 w}{\partial x \partial y^3} - b_{12}\frac{\partial^4 w}{\partial y^4} = \left(\frac{\partial^2 w}{\partial x \partial y}\right)^2 - \frac{\partial^2 w}{\partial x^2}\frac{\partial^2 w}{\partial y^2}
\end{aligned}$$

where a_{ij} are in elements of $\mathbf{a}$. Equations (6.12) and (6.14) constitute an eighth-order system of two equations for transverse deflection w and force function ψ.

6.2.5 *Nondimensional Parameters-Bounds on Values of Parameters*[14-15]

Following a similar approach to Stein[38], it is convenient to introduce the following nondimensional parameters:

$$
\begin{aligned}
&\alpha_A = \sqrt[4]{\frac{a_{11}}{a_{22}}},\ \beta_A = \frac{2a_{12} + a_{66}}{2\sqrt{a_{11}a_{22}}},\ \gamma_A = \frac{a_{16}}{\sqrt[4]{\left(a_{11}\right)^3 a_{22}}},\\
&\delta_A = \frac{a_{26}}{\sqrt[4]{a_{11}\left(a_{22}\right)^3}},\ \eta_A = \frac{a_{12}}{\sqrt{a_{11}a_{22}}},\\
&\alpha_D = \sqrt[4]{\frac{D_{22}^*}{D_{11}^*}},\ \beta_D = \frac{D_{12}^* + 2D_{66}^*}{\sqrt{D_{11}^* D_{22}^*}},\\
&\gamma_D = \frac{D_{16}^*}{\sqrt[4]{\left(D_{11}^*\right)^3 D_{22}^*}},\ \delta_D = \frac{D_{26}^*}{\sqrt[4]{D_{11}^*\left(D_{22}^*\right)^3}},\ \eta_D = \frac{D_{12}^*}{\sqrt{D_{11}^* D_{22}^*}},\\
&\alpha_B = \frac{b_{21}}{\sqrt[4]{a_{11}a_{22}D_{11}^*D_{22}^*}},\ \beta_B = \frac{b_{11} + b_{22} - 2b_{66}}{\sqrt[4]{a_{11}a_{22}D_{11}^*D_{22}^*}},\\
&\gamma_B = \frac{2b_{26} - b_{61}}{\sqrt[4]{a_{11}a_{22}D_{11}^*D_{22}^*}},\ \delta_B = \frac{2b_{16} - b_{62}}{\sqrt[4]{a_{11}a_{22}D_{11}^*D_{22}^*}},\\
&\eta_B = \frac{b_{12}}{\sqrt[4]{a_{11}a_{22}D_{11}^*D_{22}^*}},\ \mu_B = \frac{b_{11}}{\sqrt[4]{a_{11}a_{22}D_{11}^*D_{22}^*}},\\
&\rho_B = \frac{b_{22}}{\sqrt[4]{a_{11}a_{22}D_{11}^*D_{22}^*}},\ \omega_B = \frac{b_{66}}{\sqrt[4]{a_{11}a_{22}D_{11}^*D_{22}^*}},\\
&\nu_B = \frac{b_{16}}{\sqrt[4]{a_{11}a_{22}D_{11}^*D_{22}^*}},\ \tau_B = \frac{b_{62}}{\sqrt[4]{a_{11}a_{22}D_{11}^*D_{22}^*}},\\
&\theta_B = \frac{b_{61}}{\sqrt[4]{a_{11}a_{22}D_{11}^*D_{22}^*}},\ \zeta_B = \frac{b_{26}}{\sqrt[4]{a_{11}a_{22}D_{11}^*D_{22}^*}},
\end{aligned}
\tag{6.15}
$$

where subscripts *A*, *B* and *D* denote, respectively, membrane, coupling and flexural nondimensional parameters. It is noted that the latter have been used extensively by Nemeth[39-40] in parametric studies. The parameters α and β (with either A or D subscripts) denote orthotropic parameters whilst γ and δ denote anisotropic properties associated with angle-ply laminates. In particular, γ_A and δ_A represent membrane anisotropy associated with unbalanced laminates and γ_D and δ_D represent flexural anisotropy effects. Furthermore, the term α_D was originally introduced by Huber[31] and β_D was originally introduced by Seydel[41], as the reciprocal of β_D.

As one might expect, there are many ways of nondimensionalising the plate bending stiffness parameters, but done in this way is particularly useful. For example, for isotropic materials the orthotropy parameters $\alpha_{A,D}$ and $\beta_{A,D}$ take on values of unity and the angle-ply anisotropy parameters $\delta_{A,B,D}$ and $\gamma_{A,B,D}$ are zero valued. Furthermore, for simply-supported and clamped plates the number of flexural-orthotropy parameters has been reduced from four dimensional stiffnesses to two nondimensional stiffness measures - a feature that greatly simplifies laminate design. The nondimensional parameters presented in Eqn.(6.15) and the stiffness ratio $\alpha_D = \sqrt[4]{D_{22}/D_{11}}$ are also useful in the design of special-purpose laminates. This is particularly evident when used in conjunction with the concept of a quasi-isotropic laminate (Tsai and Pagano[42]), as a baseline or starting configuration, which behaves similarly to a corresponding homogeneous, isotropic plate. For this class of laminate constructions, the in-plane stiffness and response characteristics are identical to those for the corresponding isotropic materials. In slight contrast, the bending and twisting stiffness and response characteristics are somewhat different than those for the corresponding isotropic materials. These differences are conveniently represented with the nondimensional parameters given by Eqn. (6.15). Thus, the nondimensional parameters provide a means for quantitatively assessing how different a given quasi-isotropic laminate is from a homogeneous isotropic material. In the context of design, stiffness tailoring may be viewed as perturbing the values of these nondimensional parameters from the values for an isotropic material to obtain a desired response. It is important to note that different laminate constructions that possess identical values for α, β, δ, and γ will exhibit identical response characteristics when these parameters govern the structural behaviour. Thus, it is important to know the practical bounds on α, β, δ, and γ for a wide range of lamina material systems in order to determine the potential for performance enhancements that are possible by using elastic tailoring. Parametric studies made on laminates for which $\mathbf{B} = 0$ indicate that the following nondimensional parameters exhibit the approximate bounds given by

$$\begin{aligned} &0.3<\alpha_A,\alpha_D<3, , \quad -0.8<\beta_A<18, \\ &-3.5<\gamma_A,\delta_A<3.5 \quad -0.9<\eta_A<0.7, \\ &0.05<\beta_D<3 \ , \quad -0.9<\gamma_D,\delta_D<0.9. \end{aligned} \tag{6.16}$$

These bounds are used to determine the importance of the nondimensional terms that appear in the buckling and postbuckling problems defined by the equation of motion and the compatibility equation and to construct approximate closed-form solutions.

6.3 Plate Buckling[13,16-17]

6.3.1 *Introduction*

Approximate expressions for buckling and postbuckling of plates may be obtained using the formulations derived in Section 6.2. Using an approximate mode shape and by ignoring small quantities in the governing equations, simple solutions may be derived for buckling loads and postbuckling stiffnesses. In this section we present solutions for a few load cases of interest. Here, we restrict ourselves to long plates with simply supported edges, which give a conservative estimate of real plated structures. The latter have greater buckling loads due to rotational constraint and finite length. However, it is noted that once a plate is more than three times its width the long plate solution becomes highly accurate. The advantage of providing such solutions is that exact solutions are not necessary from a design perspective. They just need to be sufficiently accurate to cover uncertainty over the exact nature of boundary conditions and material property values. From the author's experience, an error of 10% may typically be allowed, yet the solutions, provided herein, provide accuracies ranging from 0.1% to 10%

depending on load type and lay-up. Furthermore, the solutions are justified from two further perspectives. Firstly, they allow rapid sizing, which may be used in a high-level optimisation process without undergoing expensive computational analyses. Secondly, they provide physical insight which is invaluable to engineers in interpreting their results and aid overall understanding.

6.3.2 *Combined Loading*

Figure 6.2 illustrates a long anisotropic plate subject to combined loadings.

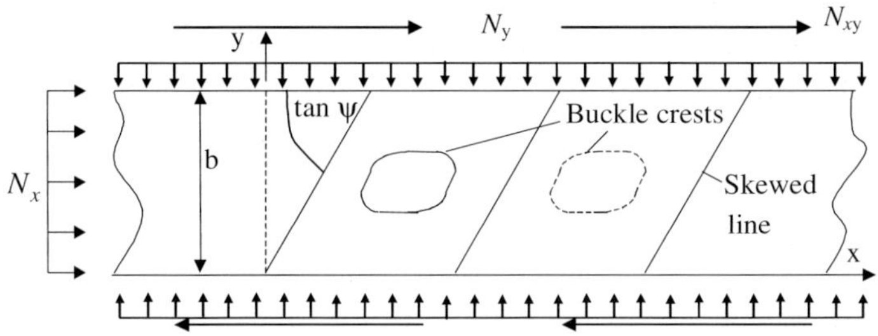

Fig. 6.2. Combined Loads (positive) on a Long Anisotropic Plate.

One method of obtaining a buckling load is to represent the transverse displacement, w, as a complete series of linearly independent basis functions of variable amplitude. A Fourier series, such as the double sine series, is a widely used example, e.g.

$$w = \sum_{m=1}^{\infty}\sum_{n=1}^{\infty} A_{mn} \sin\frac{m\pi x}{a} \sin\frac{n\pi y}{b} \tag{6.17}$$

in which the amplitudes of each mode shape, A_{mn}, are to be found. It is noted that if this expression satisfies both essential (displacements, slopes) and natural (moments, shear forces) on all edges then providing enough terms are retained in the series expansion convergence to the exact solution is guaranteed. However, one reason an accurate closed form solution has not previously been produced before is the reliance on using many (more than three) terms in the Fourier series for representing transverse plate deflections. Once more than a few terms occurs in such a series the resulting expressions become unwieldy and intractable in a closed form. Furthermore, additional trigonometric terms must be included in Eqn. (6.17) to satisfy natural boundary conditions for most anisotropic plates, which is an added complication. Having observed finite element analysis buckling patterns, a one term expression for transverse deflection is proposed. It appears to capture the majority of deformation behaviour, in a useful way, for conceptual design purposes. The effect of exactly satisfying boundary conditions is assessed by comparison with finite element analysis). It transpires that exactly satisfying boundary conditions is secondary in effect for thin and long plates.

The proposed transverse displacement for a plate of length a and width b, at the onset of buckling, is

$$w = w_0 \sin\left(m\pi x/a - ky \right) \sin\left(n\pi y/b \right) \tag{6.18}$$

and is effectively a one-term displacement series. It has the distinct advantage, over an infinite series, that an approximate closed form solution becomes viable. The terms m and n reflect the number of longitudinal and transverse half-waves whilst the k term reflects the influence of flexural/twist anisotropy on buckling patterns i.e. converts a nodal line that is perpendicular to the long edges into one that is slanted (Fig. 6.2). The maximum amplitude of the mode is w_o. It is now convenient to digress slightly and discuss the origins of using Eqn. (6.18) as a waveform in buckling analysis. Timoshenko[26] was the first person to propose such an expression and used it for approximating the buckling mode of a web of an I-section that is subject to shear loading. He realised that such a mode could not exactly satisfy the boundary condition of simple support along the longitudinal edges since the zero bending moment condition is not exactly satisfied. However, he appeared to realise that this was not necessarily important and continued his analysis. He obtained a simple expression for the shear buckling load of an infinitely long isotropic plate and published the result in 1913. Another ten years passed before Skan and Southwell[43] calculated an exact expression from successive approximations until convergence was obtained. Timoshenko's result was 6% too great, yet the simplicity of the result justifies the approximate solution in design.

Later, Sekerzh-Zenkovich[44], following Timoshenko's approach, and using Eqn. (6.18) as the mode form, determined an approximate shear buckling load for an orthotropic plate in terms of undetermined wavenumbers *m* and *n*. Balabuch[45] extended Sekerzh-Zenkovich solution to anisotropic plates and considered combined loading. Although, the framework was set up for developing closed form solutions, buckling loads were once again left in terms of unknown wavenumbers *m* and *n*. Independently of Balabuch's work, Thielemann[46] derived similar equations for buckling loads of anisotropic plates. Surprisingly, these works have not received great attention by other researchers. Perhaps it is because they do not approximate their solutions further to a closed form that they have not received greater use by practising engineers and the fact that the boundary conditions are not obeyed exactly may have hindered acceptance by the academic community. Finally, the work of Green and Hearmon[47] is of merit. They attempted to solve a similar problem using exponential functions, as done earlier by Skan and Southwell on shear buckling of isotropic plates, and once again used the method of successive approximations to obtain numerical results. Convergence was found to be relatively slow.

6.3.3 *Development of Model*

Initially, a similar approach to that taken by Balabuch and Thielemann is followed. Then simplifying procedures, using knowledge of material parameters, are applied. Substituting Eqn. (6.18) into Eqn. (6.8) and making the necessary differentiations and reductions gives

$$\begin{aligned} N_x\lambda_m^{\ 2} + 2N_{xy}\lambda_m\lambda_n + N_y\lambda_n^{\ 2} &= D_{11}\lambda_m^{\ 4} + 2(D_{12}+2D_{66})\lambda_m^{\ 2}\lambda_n^{\ 2} \\ &+ D_{22}\lambda_n^{\ 4} + 2(D_{12}+2D_{66})\lambda_m^{\ 2}k^2 + D_{22}\left(k^4 + 6\lambda_n^{\ 2}k^2\right) \\ &- 4D_{16}\lambda_m^{\ 3}k - 4D_{26}\lambda_m\left[\left(\lambda_n^{\ 2}+k^2\right)k + 2\lambda_n^{\ 2}k\right] \end{aligned} \tag{6.19}$$

with $\lambda_m = \dfrac{m\pi}{a}, \quad \lambda_n = \dfrac{n\pi}{b}.$

Note that to obtain Eqn. (6.19) a Galerkin technique was used[13]. Furthermore, Eqn. (6.19) simplifies to the standard expression, originally developed by Huber[31], and given, for example, by Jones[36] for special orthotropy. In general, the flexural/twist anisotropy terms, D_{i6} that appear in the skew parameter, k and coefficients δ and γ disappear under conditions of special orthotropy. It is possible to determine the minimal buckling load with respect to λ_m, λ_n and k. Obviously, $n = 1$ since Eqn. (6.19) is monotonic in n for all known materials. To determine values for m and k such that a closed form solution to Eqn. (6.19) becomes viable is trickier. However, using a little insight simplifies the solution procedure. Using the nondimensional parameters given in Eqn. (6.15), Eqn. (6.19) may be recast as:

$$K_x = \frac{N_x b^2}{\pi^2 \sqrt{D_{11}^* D_{22}^*}}$$

$$= \frac{1}{1+\kappa\left(\eta^2 + k'^2\right) + 2\chi k'} \left[\frac{1}{\eta^2} + 2\beta_D + \eta^2 + \frac{k'}{\eta^2}\left[2\beta_D k' + k'^3 + 6k'\eta^2 - 4\gamma_D - 4\delta_D\left(3\eta^2 + k'^2\right)\right] \right] \tag{6.20}$$

with $\eta = \alpha_D \dfrac{\lambda_n}{\lambda_m}$, $\kappa = \dfrac{1}{\alpha_D{}^2}\dfrac{N_y}{N_x}$, $\chi = \dfrac{1}{\alpha_D}\dfrac{N_{xy}}{N_x}$ and $k' = \dfrac{\alpha_D k}{\lambda_m}$.

It is possible to obtain a measure of the angle of skew (angle of nodal lines) due to anisotropy (angle ψ in Fig. 6.2). This is found by setting the transverse displacement, w to zero in Eqn. (6.18) and finding an expression for ψ. First, note that ψ is given by

$$\tan\psi = \frac{dy}{dx} \tag{6.21}$$

and that by setting $w = 0$, gives

$$\frac{m\pi x}{a} - ky = n'\pi \tag{6.22}$$

where n' is an integer. Differentiating (6.22) with respect to x and rearranging gives

$$\tan\psi = \frac{dy}{dx} = \frac{m\pi}{ak} = \frac{\lambda_m}{k} = \frac{\alpha_D}{k'} \tag{6.23}$$

as an expression to determine the angle of skew, ψ. As such, k' may be viewed as a nondimensional measure of the angle of skew.

To obtain an approximate closed form solution it is noted that $K_x = K_x(\eta, k'(\eta, k))$, and that the differential dK_x is given by

$$dK_x = \left(\frac{\partial K_x}{\partial \eta} + \frac{\partial K_x}{\partial k'}\frac{\partial k'}{\partial \eta}\right)d\eta + \left(\frac{\partial K_x}{\partial k'}\frac{\partial k'}{\partial k}\right)dk. \tag{6.24}$$

To obtain a buckling load, $dK_x = 0$ or

$$\left(\frac{\partial K_x}{\partial \eta} + \frac{\partial K_x}{\partial k'}\frac{\partial k'}{\partial \eta}\right) = 0 \quad \text{and} \quad \left(\frac{\partial K_x}{\partial k'}\frac{\partial k'}{\partial k}\right) = 0. \tag{6.25}$$

Considering the second term, since $\partial k'/\partial k = \alpha_D/\lambda_m$ then $\partial K_x/\partial k' = 0$ as a necessary condition. However, this leads to $\partial K_x/\partial \eta = 0$ as a necessary condition in the first term. In the following subsections compression loading, shear and biaxial compression are treated separately. This is done as there is no general solution to Eqn. (6.19) and approximations are necessary, and those made were loading specific.

6.3.4 *Compression Loading*

By setting κ and $\chi = 0$ and differentiating Eqn. (6.20) whilst ensuring that $\partial K_x/\partial \eta = 0$ and $\partial K_x/\partial k' = 0$ leaves

$$\begin{aligned} \frac{\partial K_x}{\partial \eta} &= 0 : a\eta^4 + c = 0, \\ \frac{\partial K_x}{\partial k'} &= 0 : fk'^3 + gk'^2 + hk' + i = 0, \end{aligned} \tag{6.26}$$

$$
\begin{aligned}
&a = 1,\\
&c = -1 + 4\gamma_D k' - 2\beta_D k'^2 + 4\delta_D k'^3 - k'^4,\\
\text{with}\quad &f = 1,\\
&g = -3\delta_D,\\
&h = \beta_D + 3\eta^2,\\
&i = -\left(\gamma_D + 3\delta_D \eta^2\right).
\end{aligned}
\tag{6.27}
$$

Eqns. (6.26) are soluble but leave the expressions for K_x in an inelegant form. It is easier to drop all small terms in the coefficients a to i. Noting that the terms γ_D and δ_D have values relatively close to zero, it is proposed to drop terms of order ($\delta_D{}^2, \gamma_D{}^2$) or higher as a first approximation. Furthermore, as it is anticipated that k' is a function of δ_D and γ_D (i.e. the nodal slope is expected to be a function of flexural anisotropy), the value of the c approximates to unity. Then,

$$
\eta = 1 \tag{6.28}
$$

is a solution. This is the expected result for an orthotropic plate as originally found by Huber[31]. Therefore, we have assumed that flexural/anisotropy does not directly affect the value of the longitudinal wavelength. These assumptions are validated later in the Discussion section. Furthermore, the expression for k' is found by neglecting terms of k'^2 or higher in Eqns. (6.26), and is

$$
k' = \frac{\gamma_D + 3\delta_D}{\beta_D + 3} \tag{6.29}
$$

noting that k' is a linear function of δ_D and γ_D. A simple expression for K_x is found by substituting Eqns. (6.29) and (6.28) back into Eqn. (6.20) to give

$$
K_x = 2\left[1 + \beta_D - \frac{(\gamma_D + 3\delta_D)^2}{\beta_D + 3}\right]. \tag{6.30}
$$

The orthotropic solution, originally found by Huber[31] is regained by setting γ_D, $\delta_D = 0$. As the buckling load is an even function of flexural/twist anisotropy (γ_D and δ_D) it is apparent that this form of anisotropy always reduces axial compression buckling loads and confirms the conclusion of other researchers[39,48-49]. Furthermore, it appears that δ_D is three times more effective at reducing buckling loads as is γ_D. This has important consequences for lay-up selection and optimisation studies.

If it is deemed necessary, it is also possible to obtain a more accurate second approximation for the buckling coefficient. Firstly, combining Eqns. (6.26a) and (6.20) gives a simpler form of K_x as

$$K_x = 2\left[\beta_D + \eta^2 + 3k'^2 - 6\delta_D k'\right]. \tag{6.31}$$

Also, note that the last four terms in c are small compared to the first term (unity) when k' is given by Eqn. (6.29), because they are second order in the anisotropy parameters and are also divided by (β_D +3). Thus, a binomial expansion can be used to determine an expression for η^2 that is obtained by taking the square root of minus c. This gives

$$\eta^2 = 1 - 2\gamma_D k' + \left(\beta_D - 2\gamma_D^{\,2}\right)k'^2 + 2\left(\beta_D\gamma_D - \delta_D - 2\gamma_D^{\,3}\right)k'^3, \tag{6.32}$$

noting that if η is required, Eqn. (6.26a) is preferable over taking the square root of the less accurate Eqn. (6.32). Next, by combining Eqns. (6.31) and (6.32), and simplifying, gives a second approximation to the buckling coefficient as:

$$\begin{aligned} K_x = {} & 2(1+\beta_D) - 2\left(\beta_D + 3 + 2\gamma_D^{\,2}\right)\frac{(\gamma_D + 3\delta_D)^2}{(\beta_D+3)^2} \\ & - 4\left(\delta_D + 2\gamma_D^{\,3} - \beta_D\gamma_D\right)\frac{(\gamma_D + 3\delta_D)^3}{(\beta_D+3)^3}. \end{aligned} \tag{6.33}$$

A second approximation for the angle of nodal lines, k', is obtained by substituting the value of η^2 from Eqn. (6.32) into the second expression in Eqn. (6.26) and solving the resulting cubic expression. This is done as follows. There is a closed form solution for the cubic polynomial (Eqn. 6.26b) in k'. It is found from taking the real root from

$$k' = w + \delta_D \tag{6.34}$$

with

$$w = z - \frac{p}{3z},$$
$$z = \left[\frac{-q \pm \sqrt{q^2 + {4p^3}/{27}}}{2}\right]^{1/3}, \tag{6.35}$$
$$p = -3\delta_D^{\ 2} + 3\eta^2 + \beta_D,$$
$$q = -2\delta_D^{\ 3} + \delta_D\beta_D - \gamma_D,$$

noting that there are 6 z roots in 3 sets of complex conjugate pairs.

An alternative approximate solution to k' is found as follows. A more accurate value of k' is found by assuming that k' differs from k_0' by a small amount, such that

$$k' = k_0'(1+e) \tag{6.36}$$

with $e \ll 1$. Then for all integers, p

$$k'^p = k_0'^p(1+pe)$$

is a reasonable approximation. Substituting Eqn. (6.36) into the second expression of Eqn. (6.26), neglecting non-linear (i.e. small) terms in e and rearranging gives

$$e = -\frac{\begin{bmatrix} +2k_0'^3 - 6\delta_D k_0'^2 \\ +2(3\eta^2 + \beta_D)k_0' - 2(3\eta^2\delta_D + \gamma_D) \end{bmatrix}}{\left[6k_0'^{\ 3} - 12\delta_D k_0'^2 + 2(3\eta^2 + \beta_D)k_0'\right]}. \tag{6.37}$$

Equations (6.36) and (6.37) combine to give a second approximation for k', from which a third approximation for η may be from Eqn. (6.26a). Finally, a third (and converged) solution for K_x is found directly from Eqn. (6.20).

The above model is expected to predict the buckling response of flexurally anisotropic plates for finite lengths where $a/b > 3\sqrt[4]{D_{11}/D_{22}}$. At such lengths, the cusps on the buckling curve (K_x vs. $a/b\sqrt[4]{D_{11}/D_{22}}$) converge to constant value[33]. For shorter rectangular plates boundary conditions at the loaded edge is expected to raise buckling loads. Shear deformation is expected to increasingly lower buckling loads for *b*/*t* or *l*/*mt* ratios less than 100. It is noted that Ref. 13 gives details of validation studies.

For tailoring purposes it is noted that laminates with non-zero **B** matrices have reduced buckling loads in comparison with an equivalent laminate for which **B** = 0. This observation is apparent because of the nature of the reduced bending stiffness matrix **D***. Hence, non-symmetric laminates (that buckle conventionally and remain essentially flat at the onset of buckling) will do so at reduced buckling loads. Also, laminates with flexural anisotropy (D_{16} and D_{26}) reduce buckling loads under compression loading. This deduction is made based on the even nature of the functional relationship between K_x and δ_D and γ_D (Eqn. 6.34). The remaining anisotropy (A_{16} or A_{26}) causes coupling between membrane extension and shear. Under compression loading a plate will attempt to shear as well as compress and, if restricted from doing so, will generate induced secondary shear stresses. As such, effectively a combined stress at the onset of buckling is present, which will reduce buckling capacity with respect to a plate under compression loading. Therefore, little anisotropic tailoring benefits can be made under compression loading. However, by changing orthotropic properties significant advantages can be made[a]. To assess the potential for orthotropic elastic tailoring of laminates a material with larger ratios,

---

[a]A similar conclusion may be reached for biaxial loading in which both loads are primary (i.e. not induced)

E_{11}/E_{22} of the basic single ply is chosen as this provides greater scope for tailoring purposes. For this reason P100/AS3502 is chosen with stiffness properties $E_{11} = 369$ GPa, $E_{22} = 5.03$ GPa, $G_{12} = 5.24$ GPa and $\nu_{12} = 0.31$. Before data is shown, appropriate measures of buckling performance must be identified. It is common practice to present graphical data for buckling coefficient, K_x. However, K_x is not an appropriate measure for optimising buckling loads because K_x is itself a function of lay-up, via D_{11} and D_{22}. Therefore, a more objective measure is required. To accomplish this one further non-dimensional parameter is necessary,

$$\varepsilon = \frac{N_x}{N_{x(\text{iso})}} \frac{K_{x(\text{iso})}}{K_x} = \frac{(D_{11}D_{22})^{1/2}}{D_{iso}} \tag{6.38}$$

so that $N_x/N_{x(\text{iso})}$ is used as a more informative measure of buckling performance. It is a relative term as it normalises the buckling load with respect to the buckling load of the quasi-isotropic lay-up of the same material. As such, its value reflects the relative performance of elastic tailoring with respect to the quasi-isotropic lay-up,

$$D_{\text{iso}} = W_1 \frac{t^3}{12}. \tag{6.39}$$

This represents a practical measure for design purposes since the quasi-isotropic lay-up is frequently used and $N_x/N_{x(\text{iso})}$ gives a direct measure of differing lay-ups on buckling behaviour. It is given by

$$\frac{N_x}{N_{x(\text{iso})}} = \frac{\varepsilon K_x}{4} \tag{6.40}$$

noting that

$$K_{x(\text{iso})} = 4 \tag{6.41}$$

Table 6.1. A comparison of buckling loads for uniform uniaxial compression loading.

Laminate	β	γ	δ	ε	K_x	$N_x/N_{x(iso)}$
$(+45)_T$	2.76	0.9	0.9	0.7	1.31	0.28
$(45_2,-45_2)_s$	2.76	0.68	0.68	0.7	4.59	0.81
$(\pm45)_{2s}$	2.76	0.34	0.34	0.7	6.85	1.19
$(\pm45)_{4s}$	2.76	0.17	0.17	0.7	7.35	1.28
$(+45,-45_2,+45)_{a/s}$	2.76	0	0	0.7	7.52	1.31
0	0.28	0	0	0.3	2.47	0.19
90	0.28	0	0	0.3	2.47	0.19
$(\pm45,90,0)_s$	2.21	0.25	0.22	0.8	6.11	1.16
$(90,0,\pm45)_s$	0.27	0.06	0.04	1.2	2.51	0.74

for isotropic materials, which for this case, values of $\beta = 1$, δ, γ, $\kappa = 0$ are substituted into Eqn. (6.30) for K_x.

Table 6.1 shows buckling data for a long plate subject to compression loading. The optimal laminate is an orthotropic 45° angle ply laminate, which performs 31% better than the quasi-isotropic (QI) lay-up. Notably, both unidirectional laminates (0° and 90°) perform poorly, with buckling loads 20% of QI. Interestingly, the orthotropic 45° laminate is antisymmetric $(+45,-45_2,+45)_{a/s}$ in nature. This special stacking sequence eliminates all coupling terms and could be used as a fundamental sublaminate for design purposes. This can be recommended, despite it breaking the over-prescriptive industry rule of only using symmetric laminates.

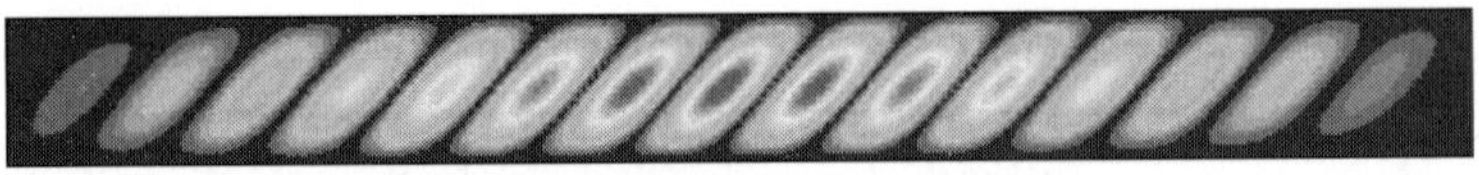

(a) Full plate showing repeitiveness of mode shape and its near independence of length

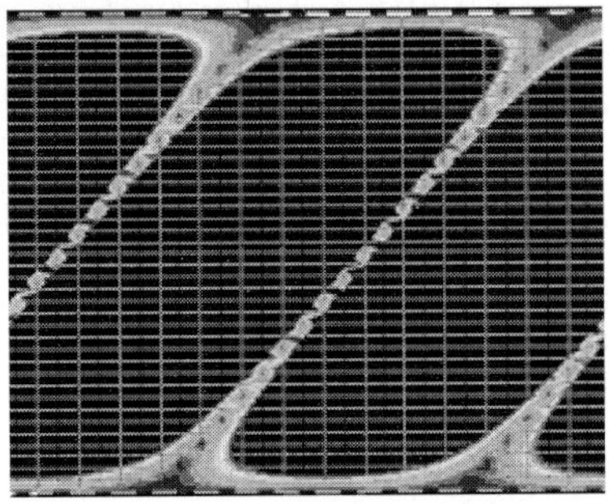

(b) Contours of mode shape showing the smallest 10% of transverse deformation. The straightness and skew angles of the nodal line is confirmed by Eqn. (6.18) and so confirms its validity

Fig. 6.3. FEA results of contours of transverse displacement, at buckling, under compresson loading for [45] laminate.

The validity of the mode shape given by Eqn. (6.18) is proven by comparison with FEA results as shown in Fig. 6.3.

The next section examines biaxial loading. In particular, scope for increasing buckling loads is examined using Poisson's ratio effects to induce secondary (biaxial) stress states, that hinders buckling effects.

6.3.5 *Biaxial Loading*

With $\chi = 0$, Eqn. (6.20) is differentiated to find

$$\begin{aligned} \frac{\partial K_{xb}}{\partial \eta_b} &= 0 : a_b \eta_b{}^4 + b_b \eta_b{}^2 + c_b = 0 \\ \frac{\partial K_{xb}}{\partial k'} &= 0 : d_b k_b'^5 + e_b k_b'^4 + f_b k_b'^3 + g_b k_b'^2 + h_b k_b' + i_b = 0 \end{aligned} \tag{6.42}$$

with

$$
\begin{aligned}
a_b &= 1-\left(2\beta_D - 12\delta_D k' + 5k'^2\right)\kappa \\
b_b &= \left(-2+8\gamma_D k' - 4\beta_D k'^2 + 8\delta_D k'^3 - 2k'^4\right)\kappa \\
c_b &= -1+4\gamma_D k' - (2\beta_D+\kappa)k'^2 + 4(\delta_D+\kappa\gamma_D)k'^3 \\
&\quad -(2\kappa\beta_D+1)k'^4 + 4\kappa\delta_D k'^5 - \kappa k'^6 \\
d_b &= \kappa \\
e_b &= -2\delta_D\kappa \\
f_b &= 2\left(1+\kappa\eta_b^{\,2}\right) \\
g_b &= 2(\kappa\gamma_D - 3\delta_D) \\
h_b &= 2\left(\beta_D + 3\eta_b^{\,2}\right)\left(1+\kappa\eta_b^{\,2}\right) - \kappa\left(1+2\beta_D\eta_b^{\,2}+\eta_b^{\,4}\right) \\
i_b &= -2\left(\gamma_D + 3\delta_D\eta_b^{\,2}\right)\left(1+\kappa\eta_b^{\,2}\right)
\end{aligned}
, \qquad (6.43)
$$

noting that the b subscript here refers to biaxial loading. The term η_b is found directly from Eqn. (6.32) as

$$\eta_b = \left(\frac{-b_b + \sqrt{b_b^{\,2} - 4a_b c_b}}{2a_b}\right)^{1/2} \qquad (6.44)$$

and is a function of k'_b. However, as there is no general solution for a fifth order polynomial, numerical methods (or simplifying assumptions) must be used to determine k'_b. One iteration of the method of successive approximations is used. First, a trial function for k'_b is used and is the value that was determined for uniaxial compression loading[9] i.e. $k'_{b_0} = k'_u$ where the zero subscript is introduced to reflect the trial value of k'_b and the u subscript to clarify its origin, as it was found using uniaxial loading,

$$k'_{b_0} = k'_u = \frac{\gamma_D + 3\delta_D}{\beta_D + 3}. \qquad (6.45)$$

However, its value should reflect the degree of transverse loading since transverse tension tends to decrease the angle of skew whereas transverse compression may actually increase the angle of skew. To reflect this, it is suggested that the trial function, k'_{b_0}, is modified. For significant amounts of flexural anisotropy ($k'_u > 0.2$) and significant transverse compression ($N_y / N_x > 0.1$), then

$$k'_{b_0} = \min\left[\frac{\gamma_D + 3\delta_D}{\beta_D + 3}\left(1 + \kappa/4\right),\ 1\right] \tag{6.46}$$

is suggested. For significant amounts of flexural anisotropy ($k'_u > 0.2$) and significant transverse loading ($N_y / N_x < -0.2$), then

$$k'_{b_0} = \left(\frac{\gamma_D + 3\delta_D}{\beta_D + 3}\right)\left(1 + |\kappa|/2\right) \tag{6.47}$$

is more appropriate. For other values of k'_u and N_y / N_x,, Eqn. (6.45) applies. Then, η_b is found from Eqn. (6.44). Finally, a more accurate value of k'_b is found by assuming that k'_b differs from k'_{b_0} by a small amount, such that

$$k'_b = k'_{b_0}(1 + e) \tag{6.48}$$

with $e << 1$. Then for all integers, p

$$k_b'^{\,p} = k_{b_0}'^{\,p}(1 + pe)$$

is a reasonable approximation. Substituting Eqn. (6.48) into the second expression of Eqn. (6.42), neglecting non-linear (i.e. small) terms in e and rearranging gives

$$e = \frac{-\begin{bmatrix}\kappa k_{b_0}'^{5} - 2\delta_D \kappa k_{b_0}'^{4} + 2\left(1 + \eta_b^2 \kappa\right) k_{b_0}'^{3} + 2\left(-3\delta_D + \gamma_D \kappa\right) k_{b_0}'^{2} \\ + \left(2\left(3\eta_b^2 + \beta_D\right) + \left(5\eta_b^4 - 1\right)\kappa\right) k'_{b_0} - 2\left(3\eta_b^2 \delta_D + \gamma_D\right)\left(1 + \eta_b^2 \kappa\right)\end{bmatrix}}{\begin{bmatrix}5\kappa k_{b_0}'^{5} - 8\delta_D \kappa k_{b_0}'^{4} + 6\left(1 + \eta_b^2 \kappa\right) k_{b_0}'^{3} \\ + 4\left(-3\delta_D + \gamma_D \kappa\right) k_{b_0}'^{2} + \left(2\left(3\eta_b^2 + \beta_D\right) + \left(5\eta_b^4 - 1\right)\kappa\right) k'_{b_0}\end{bmatrix}}. \tag{6.49}$$

Equations (6.46) and (6.49) combine to give k'_b. It is noted that the expression for k'_b from Eqn. (6.44) is appropriate for all values of k'_{b_0}. For many practical lay-ups in composite design Eqn. (6.45) may be substantially simplified by neglecting second-order and higher terms in k_b', a procedure that is appropriate for relatively small values of k_b', such that $\left|k_{b_0}'\right| < 0.3$. Then Eqn. (6.45) simplifies to

$$k_b' = \frac{2\left(\gamma_D + 3\delta_D \eta_b^{\ 2}\right)\left(1 + \kappa \eta_b^{\ 2}\right)}{2\left(3\eta_b^{\ 2} + \beta_D\right) + \left(5\eta_b^{\ 4} - 1\right)\kappa} \text{ for } \left|k_{b_0}'\right| < 0.3. \tag{6.50}$$

The merit in introducing expression (6.50) is the insight that may be gained by the relatively simple functional dependence of k_b' on parameters such as β_D, δ_D, γ_D and κ: and its wide ranging applicability to many practical composites.

As a check k_b' / k_{b_0}' may be evaluated to determine the rate of convergence. If this function (very rarely the case) differs from unity by more than 5% it is recommended to repeat the evaluation of η_b and k_b' from Eqns. (6.44) and (6.46). The approximate closed form solution for the buckling coefficient, K_x is found by substituting Eqns. (6.44) and (6.46) into Eqn. (6.20) for the buckling coefficient. For known composite pre-preg systems δ_D or γ_D rarely exceed values of 0.9 and then only for unbalanced angle ply lay-ups with the principle angle close to 45°. Therefore for most laminates in which δ_D and γ_D are small, it is justifiable to ignore terms higher than k'^2 in the evaluation of η_b. Also, if $k' < 0.1$, then η_b may be given by:

$$\eta_b = \left(\frac{2\kappa + \sqrt{4\kappa^2 + 4 - 8\beta_D \kappa}}{2 - 4\beta_D \kappa}\right)^{1/2}. \tag{6.51}$$

Note, that η_b is written explicitly in terms of known quantities.

Finally, it must be noted that if there is significant transverse loading, i.e. N_y/N_x is large, it is possible that the plate may buckle as a wide column under the influence of N_y only. This happens if the discriminant,

$$b_b^{\ 2} - 4a_b c_b < 0, \tag{6.52}$$

then,

$$N_{bycr} = \frac{\pi^2}{b^2} D_{22}. \tag{6.53}$$

A further point concerns the effect of uniaxial compression loading in which transverse displacement is restricted. This boundary condition induces transverse compressive forces due to Poisson's ratio effects upon application of uniaxial compression loading. It appears quite possible, especially for lay-ups with significant amounts of 45° angle-ply layers, that buckling is triggered by transverse buckling and acts as a wide column (n = 1, m = 1). Although this effect has been recognised by other researchers[40,50], the author is unaware whether this buckling mode has been observed in practice. Presumably, the true longitudinal support condition is one of elastic restraint that lies between the two extremes of free and fully restricted transverse displacement. However, perhaps the elastic restraint could be exploited to improve buckling performance. Such issues are discussed next.

Elastic tailoring under biaxial loading

In this section we once again consider the case of uniform uniaxial compression as the primary load. However, if the longitudinal edges of the plate are restricted from lateral motion then transverse stresses, N_y, will arise. To identify its magnitude, the in-plane constitutive relations for a laminate are considered,

$$\begin{Bmatrix} N_x \\ N_y \\ N_{xy} \end{Bmatrix} = \begin{bmatrix} A_{11} & A_{12} & A_{16} \\ A_{12} & A_{22} & A_{26} \\ A_{16} & A_{26} & A_{66} \end{bmatrix} \begin{Bmatrix} \varepsilon_{xx} \\ \varepsilon_{yy} \\ \varepsilon_{xy} \end{Bmatrix} \tag{6.54}$$

By setting $\varepsilon_{yy} = 0$ and rearranging for N_y in terms of N_x ,

$$\frac{N_y}{N_x} = -\frac{a_{12}}{a_{22}} = -\frac{A_{16}A_{26} - A_{12}A_{66}}{A_{11}A_{66} - A_{16}^{\ 2}} \tag{6.55}$$

where a_{ij} are the in-plane compliance terms. This expression is the effective Poisson's ratio for a material that is loaded in the transverse direction and induces Poisson type strain in the longitudinal direction. It is instructive to examine the conditions for which transverse tension is induced. A sufficient condition is that $A_{16}A_{26} - A_{12}A_{66} > 0$. Note, that current unidirectional materials have elastic properties such that $A_{16}^{\ 2} - A_{11}A_{66} < 0$ and so the possibility of negative Poisson's ratios arises solely from tailoring the elastic properties in the numerator of the above expression. It is therefore a necessary requirement that the plate exhibit membrane anisotropy (i.e. it is unbalanced). A further requirement is that the longitudinal edges of the plate are allowed to shear relative to each other i.e. $\varepsilon_{xy} \neq 0$. The first people to examine the possibility of raising buckling loads by inducing transverse tension using negative Poisson's ratios were Sherborne and Pandey[51]. Further work on these effects, with experimental data, is given in Ref.52.

The data in Table 6.2 represents the buckling response of a number of laminates for which secondary transverse loading has been induced. The last column reflects the buckling response of the laminate with respect to the quasi-isotropic laminate for which there is no induced secondary loading. Consequently, the values in this Table can be directly compared with those in Table 6.1 and used for design studies in tailoring boundary conditions. The first seven laminates all exhibit negative Poisson's ratios. The optimal laminate was found, using the Solver routine in MS Excel[53], to be $[80,24_3]_s$ for the P-100 material. However, it is noted that this lay-up is near-optimal for different unidirectional composite materials. A number of interesting conclusions can be made. First, the optimal buckling load is always less than the buckling load for the quasi-isotropic plate, in which the longitudinal edges is allowed to translate[b] Furthermore, the laminate with the largest negative Poisson's ratio, $[75,30_3]_s$, does not give the largest buckling load. The next four laminates reflect orthotropic angle-ply laminates with differing fibre angles. Note that all perform relatively poorly. Most buckle in a wide column mode under transverse compression. Note that the buckling load of the optimal lay-up for uniaxial compression, $(-45,45_2,-45)_{as}$, is now less than 20% of that of the quasi-isotropic laminate for uniaxial loading. Therefore, by restraining the longitudinal edges from lateral translation, the buckling load decreases from 1.31 (Table 6.1) to 0.19 of the quasi-isotropic case. This is factor that should be considered in designing edge supports of flat plates. Another feature of the results is the apparent similarity of buckling loads amongst the orthotropic angle-ply laminates. Finally, the last two laminates, a quasi-isotropic and a cross-ply laminate are

[b]This is evident because the data in the last column of Table 6.2 are less than unity.

included for comparison. Evidently, both perform relatively well under induced transverse loading. Whilst the cross-ply induces minimal transverse loading the quasi-isotropic generates a state of biaxial compression and buckles in a wide column mode at a load that is 68% of the case for zero induced transverse loading. From these results, it appears that restraining the longitudinal edges from lateral translation provides no improvement in buckling load, over the quasi-isotropic laminate under pure compression.

6.3.6 *Uniform Shear Loading*

For the case where $\kappa_s = 0$, then Eqn. (6.20) simplifies to the buckling coefficient for uniform shear, K_{xy} [8]

$$K_{xy} = 3.42 + 2.05\beta_D - 0.13\beta_D^{\ 2} - 1.79\gamma_D \\ - 6.89\delta_D + 0.36\beta_D(2\gamma_D + \delta_D) - 0.25(2\gamma_D + \delta_D)^2 \tag{6.56}$$

and gives the isotropic value[43], $K_{xy\ (iso)}$ =5.33 for $\beta_D = 1$ and δ_D , $\gamma_D = 0$. The orthotropic expression gives improved accuracy compared with solutions provided by Seydel[41] and listed in Ref. 1. Its accuracy is within 2% for orthotropic plates but reduces to 15% for large negatives values of the anisotropy parameter δ_D, (i.e.< -0.7). Such values only exist for single orientation, unbalanced laminates of extreme orthotropy values such as in materials such as P100/3502. Furthermore, its accuracy is relatively poor for those laminates that are themselves poorly performing in shear, i.e. $K_{xy} < 1$.

Equation (6.56) applies to symmetric plates for which the only anisotropy present originates from flexural/twist coupling, noting that non-symmetric laminates reduce buckling loads in comparison with the equivalent symmetric configuration. The form of Eqn. (6.56) is appealing - it immediately gives the designer a quantitative assessment of the impact of flexural/twist behaviour. Also, it appears that D_{26} coupling behaviour is approximately three to four times more influential on buckling performance than D_{16} as reflected in the coefficients of δ_D and γ_D. Furthermore, it is noted that K_{xy} is primarily an odd function of these parameters and as such this knowledge may be used for elastic tailoring purposes, once it is realised that the sign of δ and γ changes if the sign of angle ply layers in the laminate also changes.

Further elastic tailoring is possible if secondary loading is induced at the support conditions. For instance, if lateral translation of the plate's edges is prevented, then a combined state of primary shear with biaxial tension/compression may be created, which may enhance or diminish buckling loads, depending on the precise lay-up of the laminate.

Table 6.2. A comparison of buckling loads for induced transverse loading.

Laminate	β	γ	δ	ε	N_y/N_x	K_x	$N_x/N_{x(iso)}$
$(75,15_3)_s$	0.48	0.26	0.29	1.1	-0.43	3.08	0.88
$(80,24_3)_s$	0.59	0.37	0.25	1.1	-0.62	3.59	0.96
$(75,25_3)_s$	0.78	0.4	0.35	1.0	-0.67	3.57	0.91
$(75,30_3)_s$	0.97	0.47	0.40	1.0	-0.75	3.63	0.87
$(75,40_3)_s$	1.37	0.58	0.50	0.8	-0.8	3.36	0.67
$(65,30_3)_s$	1.46	0.57	0.58	0.9	-0.58	3.10	0.68
$(76,38_3)_s$	1.25	0.55	0.46	0.8	-0.81	3.50	0.73
$(+45,-45_2,+45)_{a/s}$	2.76	-	-	1.0	0.90	1.12	0.19
$(+30,-30_2,+30)_{a/s}$	2.60	-	-	0.6	0.32	1.20	0.17
$(+72,-72_2,+72)_{a/s}$	1.91	-	-	0.4	3.33	1.67	0.16
(0)	0.28			0.3	0.004	2.47	0.186
(90)	0.28			0.3	0.31	2.47	0.186
$(0_2,90_2)_s$	0.09	-	-	0.9	0.008	2.14	0.48
$[\text{quasi-isotropic}]_s$	1	-	-	1	0.32	2.74	0.68

Lay-up Studies

To assess the potential for elastic tailoring of laminates a series of case studies will be examined using P100/AS3502 with properties given in Section 6.3.5.

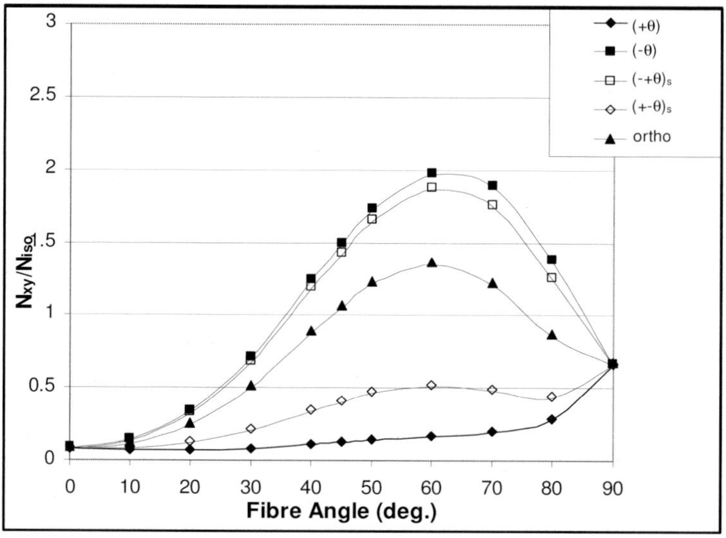

Fig. 6.4. Buckling response of angle ply laminates subject to positive shear.

Primary shear with secondary induced longitudinal and transverse tension and compression, due to elastic restraint is considered and comparisons made with plates under pure shear (no elastic restraint). Finally, the loading case of a shear web of a typical civil passenger aircraft wing will be examined in which positive shear of 2.5g is complemented by negative shear of 1g. It will be shown that the difference in magnitudes of these shear loads allows the possibility of elastic tailoring to take place.

The normalised buckling response, $N_{xy}/\ N_{xy(\text{iso})}$ for a plate under positive shear (positive x and positive y) is shown in Fig. 6.4. Note $N_{xy(\text{iso})}$ is found in a similar way to that shown for compression loading and details are found in Ref. 16. Due to the large negative δ contribution, a single negative ply angle performs best and peaks at angles close to -60°. The following three sections examine the effect of secondary loading induced by membrane anisotropy.

If the plate's long edges are restrained from transverse translation then tensile or compressive stresses may be generated depending on the nature of membrane orthotropy in the laminate.

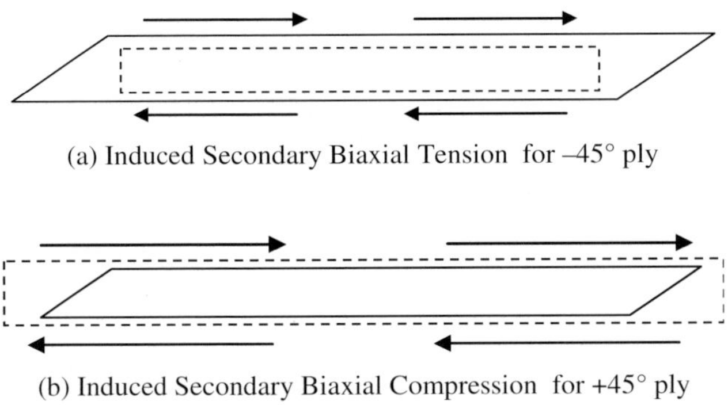

Fig. 6.5. Idealised Representation of Primary Shear Deformation and Induced Secondary Deformation under Positive Shear Loading. Undeformed mesh is rectangular. Deformed mesh is a parallelogram.

Figure 6.5a shows the idealised deformation of a –45° ply subjected to positive shear where all edges are free to translate in the plane of the plate. It is observed that tensile strains in both x- and y- directions have been induced due to membrane anisotropy effects via a_{16} and a_{26} coupling terms. If the plate's edges were restrained then induced secondary compressive stresses arise giving a state of combined loading that significantly diminish buckling performance in comparison with the case of pure positive shear. On the other hand, Fig. 6.5b shows induced compressive strains due to positive shear in a +45° ply. In this case, if the plate's edges were restrained then induced secondary tensile stresses arise giving a state of combined loading that significantly enhances buckling performance. In the analysis that follows N_x and N_y are defined such that they are positive for compressive strains. Since induced loads occur for unbalanced laminates only the single ply laminates will be examined. The origin of these induced loads is considered next.

Primary Load, N_{xy} and induced load, N_y

The induced load, N_y, is determined from the constitutive relations for in-plane loading,

$$\begin{Bmatrix} \varepsilon_x^{o} \\ \varepsilon_y^{o} \\ \gamma_{xy}^{o} \end{Bmatrix} = \begin{bmatrix} a_{11} & a_{12} & a_{16} \\ a_{12} & a_{22} & a_{26} \\ a_{16} & a_{26} & a_{66} \end{bmatrix} \begin{Bmatrix} N_x \\ N_y \\ N_{xy} \end{Bmatrix} \tag{6.57}$$

by setting $\varepsilon_y^{o} = 0$ and by rearranging for N_y in terms of N_{xy} to give

$$N_y = \frac{a_{26}}{a_{22}} N_{xy} \tag{6.58}$$

Primary Load, N_{xy} and induced load, N_x

The induced load, N_x, is determined from Eqn. (6.58) by setting $\varepsilon_x^{o} = 0$ and by rearranging for N_x in terms of N_{xy} to give

$$N_x = \frac{a_{16}}{a_{11}} N_{xy} \tag{6.59}$$

Primary Load, N_{xy} and induced load, N_y and N_x

The induced loads, N_x, and N_y are determined from Eqn. (6.58) by setting $\varepsilon_x^{o} = 0$, $\varepsilon_y^{o} = 0$ and by rearranging for N_x and N_y in terms of N_{xy} to give

$$\begin{aligned} N_y &= \frac{(a_{26}a_{11} - a_{16}a_{12})}{(a_{11}a_{22} - a_{12}^{2})} N_{xy} \\ N_x &= \frac{(a_{16}a_{22} - a_{26}a_{12})}{(a_{11}a_{22} - a_{12}^{2})} N_{xy} \end{aligned} \tag{6.60}$$

As derived in the previous section, induced loads arise from membrane anisotropy and restrictions on in-plane displacements. In reality, a supporting network of ribs and stringers will restrain plate response to an extent that depends on the stiffness of the supporting structure. The idealized case, considered here, represents the response of a plate with infinitely stiff in-plane stiffness of supporting structure. Results are summarized in Fig. 6.6 for three combinations of induced load: N_y; N_x and a combination of N_y and N_x. Buckling coefficients have been calculated using Eqns. (6.20) and (6.56) with induced loads given by the appropriate expressions in Eqns. (6.58), (6.59) and (6.60). Note that response has been calculated for positive shear loading in single ply laminates only. Note that the response for negative loading is found by changing the sign of the ply angle. Single ply laminates have been chosen to maximise membrane anisotropy and also provide upper and lower bounds for buckling response. Solid and hollow markers in the figures refer to positive and negative ply angles, respectively. Diamond and square markers refer to induced N_y and induced N_x only, respectively. The case for combined N_x and N_y is given by a triangular marker, noting that solid triangles are not presented to represent combined induced loading for positive ply angle. This was done because buckling coefficients are in excess of 100, indicating that buckling response is less likely and material failure becomes a strong possibility.

The curves for plates with and without induced loading may be compared between Figs. 6.4 and 6.6. There are notable differences in all cases. Before, it was noticed that unbalanced negative ply angles with positive shear, maximized buckling response. This is no longer the case when considering induced loads. Unfortunately, negative ply angles induce a state of biaxial compression if all the plate's edges are restrained from in-plane translation. As such, the optimal lay-ups are those with positive ply angles that induce relatively large amounts of biaxial tension. Such lay-ups, in P100/AS3502, give very large buckling loads with $K_{xy} > 100$ and $N_{xy} / N_{iso} > 50$. It is noted that ply angles of approximately 10° and 80° maximize induced loads and these ply angle correspond to local peaks in buckling coefficients, in Fig. 6.6.

For design purposes, the current analysis emphasises the important effect, the exact nature of boundary conditions has on buckling of plates, made with composite materials, containing membrane anisotropy. It is noted that it may be possible to prevent shear buckling if the supporting structure is stiff enough to prevent lateral translation of plate edges. If this is the case, it is noted that a web made from a single positive ply angle would be the optimal choice under positive shear loading but would perform poorly under negative shear loading, as indicated by the open triangular markers in Fig. 6.6. One possibility, for making a buckling resistant shear web of minimal weight, is to use a pair of webs each of single ply orientation but of opposite sign. The penultimate section considers the design of a shear web under flight conditions.

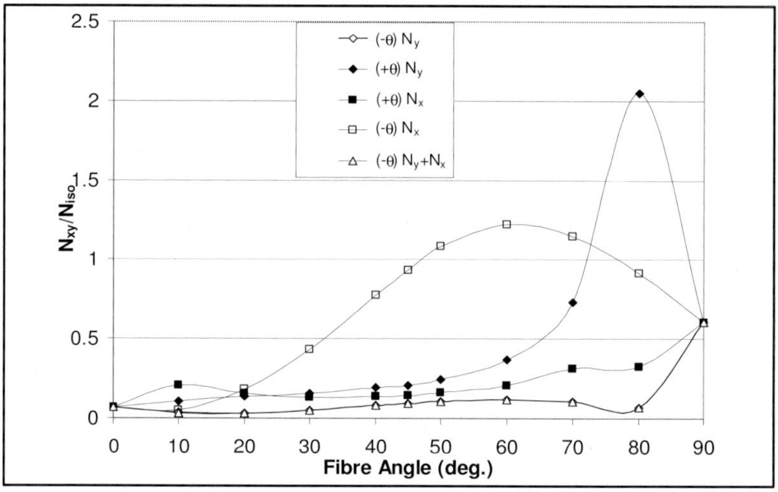

Fig. 6.6. Buckling response of angle ply laminates subject to positive shear and various combinations of induced loads.

Aircraft Loading

Design loads of a shear web in a wing of a passenger aircraft consider both positive and negative shear loads that arise from uplift (flight conditions) and downlift (runway taxiing or flight turbulence). A typical example, is 2.5g under positive shear as one load case and 1g under negative shear, as the other. Our aim is to identify an elastically tailored laminate that is better performing than the optimal orthotropic one with regards to shear buckling. It is assumed that membrane anisotropy, induced loading is not present. As such, the laminate with minimum numerical value of δ_D maximizes buckling under positive shear. Unfortunately, as shown in Fig. 6.4, this laminate performs poorly under negative shear. However, the ratio of positive to negative loading is 2.5, suggesting that elastic tailoring using flexural/anisotropy, via δ is possible. From Fig. 6.4, it is noted that a ply angle of approximately -60° maximises buckling performance for those laminates with flexural anisotropy and orthotropy (noting that the sign of the ply angle is irrelevant for orthotropy). Three laminates are initially considered. The first $(-60, +60)_s$ has large $-\delta_D$ for a balanced symmetric laminate; the second, $(+60, -60)_s$ has large $+\delta_D$ and the last $(-60, +60_2,-60)_{a/s}$ is orthotropic. Values of $N_{xy}/N_{xy(iso)}$ for these lay-ups, taken from Fig. 6.4, are 1.87, 0.51 and 1.34, respectively. Note, that if the direction of shear is reversed then $N_{xy}/N_{xy(iso)}$ for $(-60, +60)_s$ is now 0.51. Since the ratio of 1.87/0.51 is larger than the ratio of positive/negative shear loading, 2.5, no gain in performance is gained through elastic tailoring. Next, $(-60,+60)_{2s}$ is considered. Its values of $N_{xy}/N_{xy(iso)}$ under positive and negative shear loading are calculated from Eqn. (6.20) as 1.64 and 0.97, respectively. Since the ratio between these values is less than 2.5 it is expected that weight may be saved if such a laminate is used. This may be calculated by first considering the extra load carried by this laminate in comparison with the orthotropic 60° laminate. This amounts to 1.64 / 1.34 = 1.22. If the elastically tailored laminate's thickness is reduced such that it has identical buckling loads with the orthotropic laminate, the performance of the elastically tailored laminate will reduce to 0.97 / 1.22 = 0.80 under negative shear. Since this value is greater than 1.34 / 2.5 = 0.54, a sizeable increase in performance may be achieved by using the elastically tailored instead of the orthotropic laminate. This laminate, $(-60,+60)_{2s}$, gives a 22% weight saving and buckles under positive shear but not under negative shear loading. Note, that the theoretically optimal laminate will be designed to buckle under 2.5 g and –1g.

6.3.7 *Postbuckling of Plates under Compression Loading*[14-15]

It is well known that unsymmetrically laminated composite plates under in-plane compression may exhibit prebuckling transverse displacement. In order to obtain an approximate solution for the postbuckling problem, we assume the transverse displacement comprises two terms: the buckling displacement given by Eqn. (6.18) and another that reflects the contribution made from prebuckling deformation due to the nonsymmetric nature of the plate. This second component is found by ensuring the natural boundary condition on the long edges is satisfied. Reference 15 provides details but in nondimensional form is given by

$$\overline{w}_1 = \frac{w}{\sqrt[4]{a_{11}a_{22}D_{11}^*D_{22}^*}} = g\,\frac{y}{b}\left(\frac{y}{b}-1\right) \tag{6.61}$$

where g denotes the amplitude due to zero moment boundary conditions. From the zero moment boundary condition we are able to determine g :

$$g = \frac{K_x\pi^2}{2\alpha_D^2}\eta_B\,. \tag{6.62}$$

In Reference 15 it was shown that plates with $\mathbf{B} = 0$ show two types of prebuckling displacement: one which is a localised edge effect that is maximum close to the short ends of the plate and dissipates over relatively short distances along the plate; the second is that described by Eqn. (6.61) and exists across the width of the plate for plates which have some rotational freedom along the long edges. This important observation highlights that the latter prebuckling displacements are maximum for simply supported long edges and eliminated for clamped plates. As such, only the first type of localized prebuckling displacement occurs in clamped plates. From a tailoring perspective it may be possible to convert the flat plate into a cylindrical strip by ensuring that g is maximized. This would be favourable since cylindrical strips have greater buckling loads than the equivalent flat plate due to circumferential membrane effects. This is done by ensuring that b_{12} is large. This coupling term links direct load with transverse curvature and may be thought of as a Coupling Poisson's ratio. A [90,45] lay-up gives large values of b_{12}. Under axial compression there appear to be little advantage in using nonsymmetric laminates (apart from the possibility of precurving flat plates) or those containing membrane or flexural anisotropy. As such, the remainder of the section describes orthotropic response under compression loading. Instead of a buckling load it is often convenient to characterise the postbuckling of a plate in terms of its stiffness after buckling. Essentially, a Galerkin method is undertaken in which the displacement Eqn. (6.18) and equation of motion, Eqn. (6.12) are combined to produce a third order expression for the amplitude of displacement, w_0 . This solution simplifies to that given by Harris[54], E_{post}, for orthotropic materials,

$$E_{post} = \frac{A_{11}}{t}\left[\frac{1+H^4}{3+H^4}\right] \tag{6.63}$$

where H, is a measure of homogeneity and is given by

$$H = \sqrt[4]{\frac{D_{11}a_{11}}{D_{22}a_{22}}} \tag{6.64}$$

For a homogeneous lay-up the relative postbuckling stiffness is one-half of its initial stiffness since H = 1. It was shown in Ref. 14 that the optimal stiffness is a cross-ply laminate with 87% of 0° plies $(0_{0.87}, 90_{0.13})_s$. For an 8-ply laminate this is almost achieved with $(0_7,90)_s$. To understand this lamination it is informative to consider the deformations that should be considered to raise postbuckling stiffness. Postbuckling is characterised by the increased growth of out-of plane deformations. For a support structure that maintains straight edges this deformation must be accompanied by in-plane stretching. This behaviour is shown, schematically in Fig. 6.7, by the difference in positions between the

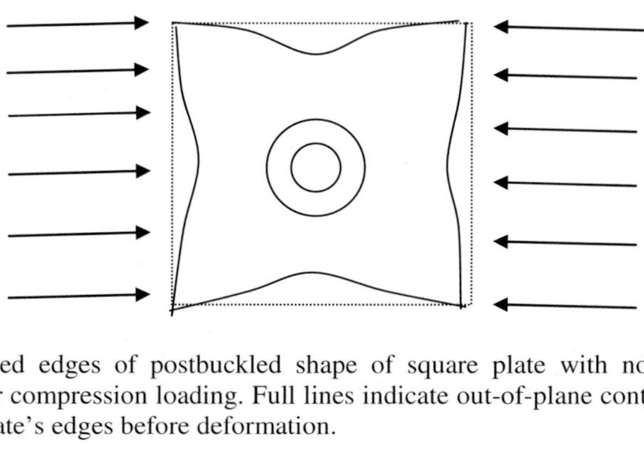

Fig. 6.7. Curved edges of postbuckled shape of square plate with no in-plane edge supports, under compression loading. Full lines indicate out-of-plane contours and dotted lines are the plate's edges before deformation.

curved and straight edges. In practice, so as to maintain straight edges the load is redistributed towards the plate's edges giving support for the concept of an "effective width"[55] design approach. As such the postbuckling stiffness is greatest for the laminate whose in-plane stiffnesses in the plate's principal axes are greatest. Moreover, there is stronger dependence on A_{11} rather than A_{22} reflecting the loading direction.

Design Considerations for Postbuckling of a Flat Plate

It was shown that postbuckling response is optimised by resisting in-plane deformation, particularly that in the loading direction. A cross-ply laminate was shown to be optimal. However, it was also shown previously that to resist buckling deformation it was necessary to design to resist curvatures, particularly anticlastic and twisting curvatures. A $\pm 45^\circ$ laminate was shown to be optimal. However, these optimal laminates each perform badly for the other case, i.e. the optimal laminate for buckling is very poor in postbuckling stiffness and vice-versa. Therefore, the question remains – what is the optimal lay-up when various amounts of postbuckling is considered? This issue was raised in Ref. 14 where differing amounts of postbuckling response are modelled with the design load given as a ratio of the buckling load of the quasi-isotropic lay-up (N_x / $N_{x\ (iso)}$. For loads in excess of 1.5 $N_{x\ (iso)}$, the lay-up that optimises postbuckling stiffness alone was best performing $(0_{0.87}, 90_{0.13})_s$. However, for increasingly lower design loads the optimal lay-up contains an increasing $\pm 45^\circ$ content.

6.4 Cylindrical Shells under Compression Loading[18-20]

Buckling of shells generally occurs at higher loads than flat plates due to the inherent ability of shells to resist external loads by developing membrane stresses in addition to flexural stresses. The accomplish this improvement by means of their curvature, in an analogous mechanism to the way a balloon resists internal pressure by developing surface tensions. A good example of general shell behaviour is found by considering axial compression of circular shells. Not only do they show behaviour characteristic of many shells, they find practical use, e.g. rockets and fuselages under inertial loads (acceleration). The load at buckling, P, may be found from

$$P = \frac{1}{\lambda_m^2}\begin{bmatrix} D_{11}^*\lambda_m^4 - 4D_{16}^*\lambda_m^3\lambda_n + 2\left(D_{12}^* + 2D_{66}^*\right)\lambda_m^2\lambda_n^2 \\ -4D_{26}^*\lambda_m\lambda_n^3 + D_{22}^*\lambda_n^4 \end{bmatrix}$$

$$-\frac{\begin{bmatrix} b_{21}\lambda_m^4 + \left(b_{61} - 2b_{26}\right)\lambda_m^3\lambda_n + \left(b_{11} + b_{22} - 2b_{66}\right)\lambda_m^2\lambda_n^2 \\ + \left(b_{62} - 2b_{16}\right)\lambda_m\lambda_n^3 + b_{12}\lambda_n^4 - \dfrac{\lambda_m^2}{r} \end{bmatrix}^2}{\lambda_m^2\left(a_{22}\lambda_m^4 + 2a_{26}\lambda_m^3\lambda_n + \left(2a_{12} + a_{66}\right)\lambda_m^2\lambda_n^2 + 2a_{16}\lambda_m\lambda_n^3 + a_{11}\lambda_n^4\right)} \tag{6.65}$$

where $\lambda_m = m'n/r$, $\lambda_n = n/r$

and r is the radius, n the number of circumferential waves and m' reflects the angle of skew (analogous to k' for flat plates) due to angle-ply anisotropy or indeed torsion loading. The buckling load is found by iterating through different values of m' and n to obtain the lowest possible value of P. It is noted that the first part of Eqn. (6.65) comprises the buckling expression for a plate. The second part can be further seen as a combination of the extension/flexural couplings terms (those with $[b]$ elements as coefficients) and the membrane terms.

If a compressive load is applied to a laminated cylinder with extension/shear couplings with appropriate boundary condition (circumferential displacements at the ends restrained), the cylinder is actually experiencing a combined loading of compression and torsion, which reduces its buckling load considerably from those predicted by just the compressive load, depending on the magnitude of the extension/shear couplings. However, if the cylinder is subjected to combined compression and torsion as primary loads, extension/shear coupling can be used to induce torsion that opposes the primary torsion and so raise overall buckling loads.

6.5 Conclusions

This chapter has given an overview of recent work on elastic tailoring effects to either mitigate or enhance buckling performance. Emphasis has been given to the physical mechanisms that govern buckling/postbuckling phenomena. This has been partly gleaned by experience but also development of simplified buckling formulae that aid design by capturing the most important deformation phenomena. It is hoped the work will inspire other workers in the field to examine elastic tailoring of thin–walled structures to enhance structural performance.

6.6 Acknowledgements

I would like to thank my co-workers who have helped prove many of the concepts detailed herein. In particular, I would like to thank J. Enrique Herencia for his valuable comments on this chapter.

6.7 References

1. Lekhnitskii, S. G. *Anisotropic Plates*, Gordon and Breach, (1968)
2. Williams, F. W., Kennedy, D., Anderson, M. S. and Butler, R. *AIAA Journal*, **29**, pp1927-1928, (1991)
3. Bushnell, D. *Computers and Structures*,**16** (14), pp167-185, (1983)
4. Hibbitt, Karlsson and Sorenson Inc. *ABAQUS Users Manual,* Version 6.3, Rhode Island, USA, (2003)
5. Datoo, M. H. Mechanics of Composite Materials, Elsevier, (1991)
6. Gurdal, Z. Haftka, R. T. and Hajela, P. *Design and Optimisation of Laminated Composite Materials* Wiley, New York, (1999)
7. Hart-Smith, L. J. *Aerospace Materials*, **5** (2), pp10-16, (1992)
8. Miki, M. Proceedings of the Fourth International Conference on Composite Materials, **1**, pp. 1725-1731, (1982)
9. Tsai, S. W. and Hahn, H. T. *Introduction to Composite Materials*, Technomic, Lancaster, Pennsylvania, (1980)
10. Fukunaga, H. and Hirano, Y., , Proceedings of the Fourth International Conference on Composite Materials, **1**, pp. 565-572, (1982)
11. Weaver, P. M. *Proc. American Society of Composites 18th Technical conference*, paper 142, University of Florida, October (2003)
12. Weaver, P. M *Proc. 47th AIAA/ASME/ASCE/AHS/ASC SDM Conference*, Newport, Rhode Island, 1- 4 May (2006)

13. Weaver P. M. *Pro. Roy Soc* **A 462** (2065), pp. 59-73: (2006)
14. Diaconu, C. G. and Weaver P. M. *AIAA Journal*, **43**(4), pp 906-914 , (2005)
15. Diaconu, C. G. and Weaver P. M., *Int. J. Solids and Structures*, **43**(22-23), pp 6978-6997, (2006)
16. Weaver, P. M. *Proc. of 45th AIAA/ASCE/ASME/AHS SDM conference*, Palm Springs, USA, April (2004)
17. Weaver, P. M. Proc. of 46th AIAA/ASCE/ASME/AHS SDM conference, Austin, USA, April (2005)
18. Wong, K. F. W. and Weaver, P. M. *AIAA Journal*, **43**(12), pp. 2252-2258, (2005)
19. Weaver P. M., *AIAA Journal*, pp101-107, June (2002)
20. Weaver, P. M., Driesen J. R. and Roberts P. *Comp. Sci. Tech.*, **62**, pp91-105, (2002)
21. Kirchhoff, G *J. de Crelle ,(*1850)
22. Germain, S Recherches sur la Théorie des Surfaces Élastiques, Paris, (1821)
23. Todhunter, I. and Pearson, K *A History of the Theory of Elasticity*, Vol. II, Article 1234, Cambridge Universiyt Press, (1893)
24. Gehring, Franciscus De Aequationibus Differentialibus ,quibus aequilibrium et motus laminae crystallinae definiuntur, Ph.D Dissertation, Berlin, (1860)
25. Kelvin, Lord in Thomson and Tait, *Treatise of Natural Philosophy*, article 644, Cambridge University Press, (1867)
26. Timoshenko, S. P., *Annales des Ponts et Chaussees*, 9th Series, 17 (V): Sep-Oct, 372-412, (1913)
27. Clebsch, A *Theorie der Elasticität fester Körper*, Leipzig ,1862 (French translation by Saint-Venant), (1883)
28. Poisson, S. D. *Mem. De l'Acad*, Paris, t. 8, (1829)
29. Green, G. Cambridge Phil. Soc. Trans., **7**, (1839)
30. Voigt, W. *Lehrbuch der Krystallphysik* ,Leipzig u. Berlin, (1910)
31. Huber , M. T. Problem der Statik Technisch Wichtiger Orthotroper Platten, Warsaw, (1929)
32. Timoshenko, S. P. *Theory of Elastic Stability*, McGraw-Hill, (1936)
33. Stavsky, Y. and Hoff, N. J., Chapter 1 in *Composite Engineering Laminates*, Editor G. Dietz, MIT Press, (1969)
34. Onoda, J. *AIAA Journal* **23**(7), p1093, (1985)
35. Fukunaga, H., *AIAA Journal* **29**(4), 641-646, (1991)
36. Jones, R. M. *Mechanics of Composite Materials*, 2nd Edn., Taylor&Francis, (1999)
37. Grenestedt, J. L., and Gudmundson, P., *Proceedings of IUTAM Symposium on Optimal Design with Advanced Materials*, Elsevier Science, Amsterdam, pp. 311–336, (1993)
38. Stein, M., *AIAA Journal*, **12**(12), pp. 1729-1735, (1983)
39. Nemeth, M. P. *AIAA Journal* **24**(11), p1831, (1986)
40. Nemeth, M. P. *NASA TP-3568,* (1995)
41. Seydel, E. *NACA TM 7005*, NACA, Washington D.C., (1933)

42. Tsai, S. W. and N. J. Pagano. *Composite Materials Workshop*,.Technomic Publishing Co., Inc., Stamford, Conn. 06902. pp.223-253. (1968)
43. Skan, S. W. and Southwell, R. V., *Proc. Roy. Soc. London, series A*, **105**, 582-607, (1924)
44. Sekerzh-zen'kovich, Ya. I., *Trudy tsagi* no. 76, 1931
45. Balabuch, L. S. The Stability of Plywood Plates. *Aeronautical Engineering*, U.S.S.R. 11. Translated by British Government under Ministry of Aircraft Production, R.T.P. Translation No. 2199, (1937)
46. Thielemann, W., *NACA Technical Memorandum 1263*,. Translation of "Beitrag fur Frage der Beulung Orthotroper Platten, insbedonderevon sperrholzplatten", (1950)
47. Green, A. E. and Hearmon, R. F. S. *The Philosophical Magazine*, October, 659-688, (1945)
48. Chamis, C. C. *Journal of Structural Division*, ASCE, **95** 2119-2139, (1969)
49. Grenestedt, J. L., *Structural Optimisation,* Vol **3**(7), pp115-120, (1991)
50. Adali, S. in *Buckling and Postbuckling of Composite Plates*, Eds. G.J. Turvey and I.H. Marshall, Chapman and Hall, (1996)
51. Sherborne, AN and Pandey, MD, *Composite Structures,* No. 20, pp 73-81, (1992)
52. Weaver, PM, *Proc. of the 44th AIAA/ASCE/ASME/AHS SDM conference*, paper AIAA-2003- 7942, San Diego, April (2003)
53. Microsoft Excel 97, Solver Tool
54. Harris, G. Z., *International Journal of Mechanical Sciences*, **17**(3), pp. 187–202, (1975)
55. von Kàrman, Th., Sechler, E. E. and Donnell, L .H. *Trans ASME* **54** (2), p52, (1932).

CHAPTER 7

OPTIMISATION OF STIFFENED PANELS USING FINITE STRIP MODELS

Richard Butler and Wenli Liu

Department of Mechanical Engineering, University of Bath
Bath, BA2 7AY, UK
E-mail: R.Butler@bath.ac.uk

The key features of the finite strip buckling analysis program VICONOPT are described. A bi-level strategy, for the efficient optimum design of composite stiffened panels is also presented. Panel level optimization finds a minimum weight cross-sectional geometry based on a substitution of equivalent orthotropic plates for laminated plates. Optimization at the laminate level finds stacking sequences satisfying laminate design rules using either manual selection or a Genetic Algorithm. VICONOPT models are validated with ABAQUS finite element models, and with experimental compressive testing of two blade-stiffened panels. The buckling and postbuckling behaviour of the two panels, with initial buckling in the stiffeners and skin, respectively, is investigated in a high load and high strain range. The bi-level strategy is evaluated by design of a Z-stiffened example, which has been manufactured and tested, and also by design of a wing cover panel which is subject to compression and lateral pressure. The optimum wing panel, considering a variety of stiffener shapes, is 17% lighter than an existing datum design.

7.1 Introduction

The primary load-carrying component in the wing of a typical transport aircraft is a box comprising an upper cover, lower cover and front and rear spars. The covers, which consist of thin skin panels with internal stiffening of various shapes (Figs. 7.1 and 7.2), are supported at regular intervals by ribs, which also transmit loads to the spars. Upper surface cover panels carry axial compressive loads, N_x, in normal flight, and are prismatic, whereas lower surface panels are loaded in tension and may have cut-outs for inspection purposes. Traditionally, the compression panels have been manufactured from high strength aluminium alloys, often with some degree of postbuckling capacity allowed for in the thin skins. However, manufacturers are currently designing aircraft with all-new wings made primarily of carbon fibre reinforced plastic (CFRP) laminates, which despite their obvious attraction in terms of weight saving, leading to fuel saving and increased payload, present some serious challenges to the designer. In particular, there is a need to

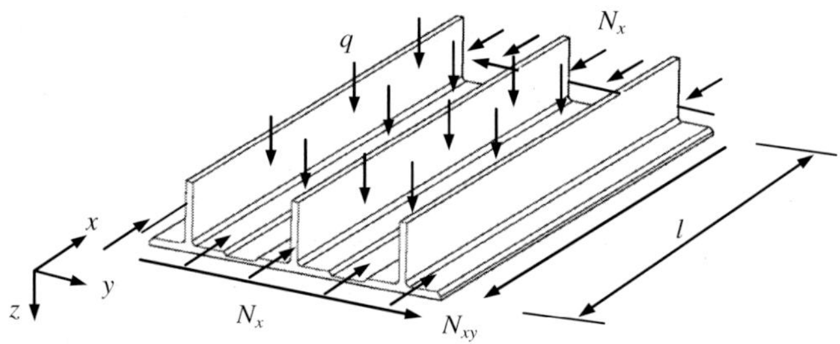

Fig. 7.1. Panel geometry and loading. For the experimental tests, the edges loaded by N_x are fully restrained in rotation and restrained in displacement in y and z. The longitudinal edges are free.

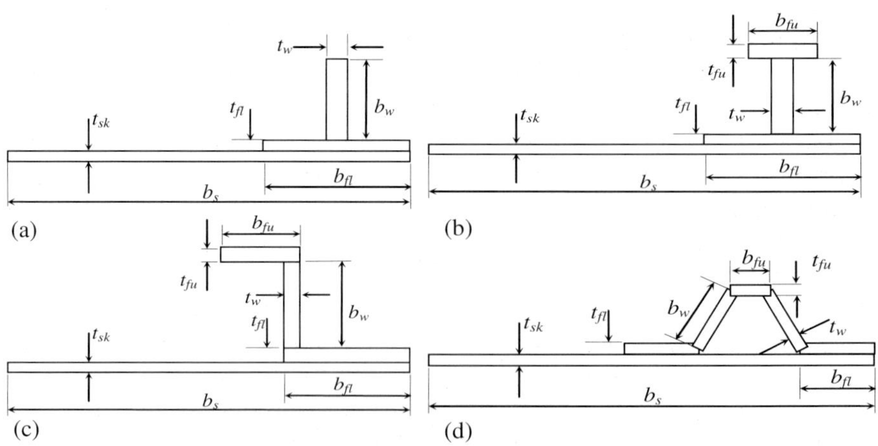

Fig. 7.2. Cross section of stiffened panel repeating portions, showing design variables for (a) blade-stiffened panel; (b) I-stiffened panel; (c) Z-stiffened panel and (d) hat-stiffened panel.

establish modelling and optimisation tools that produce minimum weight designs with sufficient strength, damage tolerance and good manufacturability. The task is complicated by the massive number of potential design variables which include the usual geometric variables as well as a vast range of laminate stacking sequence permutations.

During the design of composite laminates against buckling, the full complexities of detailed modelling, analysis and optimisation are compromised for the sake of efficiency. The optimisation of composite stiffened panels, subjected to buckling constraints, has been considered in many previous studies. Some of these have focused on the simplest modelling method of closed form equations to investigate the structural efficiency of various stiffener shapes for minimum mass and costing. Kollár[1] reported a closed-form equation to determine the local buckling loads of composite structural members when the edges of the webs are rotationally restrained by the flanges. Using closed-form equations as buckling constraints for cost and weight minimization, Kassapoglou[2] found that 'J' stiffeners give the lowest weight configurations while 'T' (blade) stiffeners gave the lowest cost configurations. However the buckling analysis using closed form solutions did not account for coupling between skin and stiffeners or for interaction between overall buckling and local buckling.

Analysis methods which account for interactive effects have been developed using finite strip methods (FSM)[3-5] which divide panels into constant thickness strips and represent the displacement in the strips using polynomial and trigonometric functions. FSMs represent a class of analysis that lies between closed form solutions and finite element (FE) methods in terms of accuracy and computational expense.

With the development of the program PANDA2[6] Bushnell has contributed a substantial volume of work for the design analysis of flat or curved laminated composite panels. Closed-form solutions or discretized models of panel cross sections are applied for calculation of local and general buckling loads.

In recent years there has been growing interest in the application of genetic algorithms (GA)[7] to the optimum design of composite laminates due to the discrete nature of ply thickness. Le Riche and Haftka[8] solved the laminate stacking sequence design problem subject to buckling and strength constraints. Nagendra *et al.*[9] investigated the application of a simple GA to the design of composite blade-stiffened panels combined with a FSM program to evaluate both buckling load and strain constraints. Nagendra *et al.*[10] later presented an improved genetic algorithm that introduced new genetic operators. However, GAs are computationally expensive to use for practical design applications; even with the improved efficiency of the genetic algorithm and the use of efficient analysis methods such as FSM.

This chapter is devoted to the optimisation of compressively loaded prismatic wing cover panels using the program VICONOPT.[11] The key features of the well established strip model, used for buckling analysis are described. A series of recently obtained results are then presented covering validation of the analysis method using experimental tests and nonlinear Finite Element Analysis (FEA), as well as the application of a bi-level strategy for practical composite panel optimisation.

7.2 Buckling Analysis

The particular finite strip model applied here stems from pioneering work done in the late 1960s by Wittrick[12] and the early 1970s by Wittrick and Williams[13] in which the governing differential equations for each constant thickness plate, within the panel cross-section, were solved exactly assuming sinusoidal variations of the buckling deformation in the longitudinal direction (the procedure was also applied to vibration, but this is not covered here). The resulting eigenvalue problem is transcendental in form, and an algorithm[14] was developed to calculate the critical buckling load for any chosen value of longitudinal half-wavelength λ.

The strip theory covers any panel of constant cross section, composed of anisotropic plates each of which can carry any combination of uniformly distributed and longitudinally invariant in-plane stresses. It assumes the panel to be an infinitely long (multi-bay) structure made up of portions, including supporting structures such as ribs, which repeat at longitudinal intervals l. In the simplest form of the buckling analysis[13], the critical buckling load is taken to be the lowest buckling load found for any of a range of values of λ, extending from a value less than the smallest plate width up to the length, l, of the panel. If the panel has no applied shear force and is fabricated from an isotropic material, then the nodal lines of the buckling mode are straight and parallel to the rib supports. Also, since λ values are selected to divide exactly into l, the boundary conditions are consistent with simple supports at the ribs. This analysis gives reasonably accurate results for laminated and other anisotropic panels. However, for a panel carrying a significant shear load, the results are conservative, particularly as λ approaches l; such cases require an alternative analysis[15] that couples responses from a series of λ using Lagrangian multipliers to enforce the end conditions. To allow for continuity over the ribs with adjacent bays, warping of the entire cross-section is allowed at the panel ends. A repetitive geometry feature permits rapid analysis of wide panels. Such analysis is analogous to the infinitely long approach described above and involves modelling a single datum repeating portion by assuming infinite width and using recurrence relations.[16]

The strip theory analysis obtains plate stiffness matrices assuming classical plate theory satisfying the Kirchoff–Love hypothesis, i.e. transverse normals remain perpendicular to the deformed mid-surface of the plate. This assumption becomes inappropriate for thick composite plates and so the analysis can optionally include an approximation[17] for first order shear deformation theory.

Most practical stiffened panel applications give rise to a nonlinear interaction between out-of-plane bending over length l and in-plane compression which will precipitate early material failure and early local or torsional stiffener buckling. Such bending may be caused by manufacturing imperfections or by transverse pressure due, for example, to fuel containment. Manufacturing imperfections are considered in the form of a single half sine wave over l whereas pressure acts on the skin and may assume rib restraints that are either clamped or simply-supported. The strip model allows for imperfections and fuel pressure by assuming that the entire panel acts as a beam–column.[18, 19]

7.3 Optimum Design Strategy

The standard optimum design process[20] proceeds as a series of design cycles in which critical and potentially critical buckling and/or material strength constraints are calculated. The sensitivity of these constraints to design variable changes are found and used along with sequential Taylor series approximations and a mathematical optimiser[21] to produce a design of lower mass than the previous cycle. Feasibility is maintained throughout the design process by stabilization, which factors plate thicknesses to satisfy constraints. The extent of such thickening is used to automatically adjust the move limits for subsequent optimisation within a design cycle, so that move limits are increased (reduced) when the adjustment is small (large), and the best use is made of constraints and sensitivities during each design cycle. Design variables may be plate (or layer) thicknesses, plate widths and, for composites, ply angles. The program can handle various practical features, such as multiple load cases and design variable linking.

The goal of optimisation of laminated CFRP panels is to find the stacking sequence, i.e. the sequence of layer orientations for each laminate, and the cross-sectional geometry of minimum weight panels that will not buckle or fail due to excessive strains for a particular set of design loads and boundary conditions. The four orientations +45°, -45°, 0° and 90° are generally used in design, where a 0° layer has its fibres parallel to the direction of axial load. Practical laminate design rules are normally imposed, based upon well-established experience[22] of laminate design, and have the following typical format.

(i) The 10% rule design is applied to prevent direct loading of the matrix in any direction. As a result, there are at least 10% of 90° plies within skin laminates and stiffener laminates;
(ii) A target of 40% of +45° and -45° plies is used in the skin (and 30% in the stiffeners) to maximize bearing strength for assembly by mechanically fastened joints;
(iii) Grouping of more than 4 plies of the same orientation together is avoided to minimize edge splitting and maintain a homogeneous stacking sequence. (For heavily loaded thick laminates, it is preferred that no more than 3 plies are grouped together.);
(iv) The outer plies of the skin and stiffener laminate are ±45° plies to take damage tolerance aspects under compressive loading into consideration.

For standard VICONOPT design, stacking sequences are always selected before optimisation. However, this approach will invariably produce an optimum design that does not satisfy the laminate design rules. For example, an initial stacking sequence chosen as $[45/\text{-}45/90/0]_S$ might produce the optimum laminate $[45/\text{-}45/90/0_5]_S$, which has 10 plies of 0° orientation grouped in the middle of the laminate, and does not therefore satisfy the rules (i) - (iii).

A bi-level optimisation strategy for practical optimum design of composite stiffened panels has been developed[23] to separate the continuous design problem from the discrete variable problem and solve each using appropriate methods. The strategy is illustrated in Fig. 7.3.

Panel level optimisation finds a minimum weight cross-sectional geometry based on a substitution of equivalent orthotropic plates for laminated plates. Optimisation at the laminate level finds stacking sequences satisfying the above laminate design rules. These two levels are described in detail in the following sections.

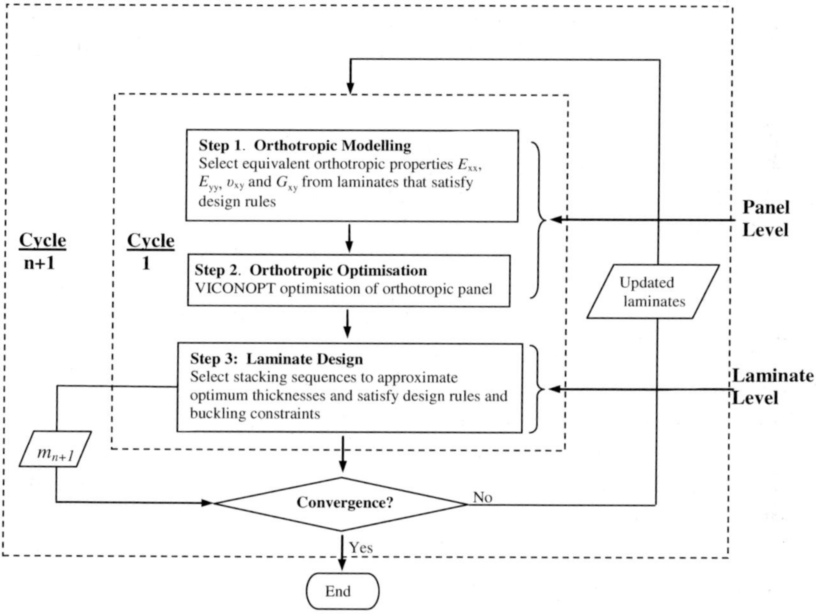

Fig. 7.3. Bi-level optimisation strategy.

7.3.1 *Panel level optimisation*

At this level, the optimisation is based on the VICONOPT sizing strategy to obtain the value of continuous design variables giving minimum mass of stiffened panels subject to buckling and material strength constraints.

The design problem is a constrained minimisation problem of the form

$$\text{Minimize} \quad W(\boldsymbol{X}) \tag{7.1}$$

$$\text{Subject to} \quad G_m(\boldsymbol{X}) < 0 \quad m = 1,2,3,\ldots,n_{pc} \tag{7.2}$$

where $\boldsymbol{X}$ is a vector of design variables, $W(\boldsymbol{X})$ is panel mass, the $G_m(\boldsymbol{X})$ are inequality constraint functions and n_{pc} is the number of inequality constraints at panel level.

In this case, the inequality constraints include the buckling and material strength constraints, as well as side constraints applied to geometric parameters. Design variables (see Fig. 7.2) are the thickness and width of skins, stiffeners and flanges.

The side constraints are applied to the following geometric parameters:

1) stiffener spacing, b_s;
2) thickness of the skin laminate, t_{sk};
3) thickness of the stiffener laminates, t_w, t_{fl}, t_{fu};
4) width of the top flange laminate, b_{fu};
5) width of the bottom flange laminate, b_{fl}, for repair considerations.

The panel level optimisation shown in Fig. 7.3 treats laminates as orthotropic plates of continuous thickness to find the optimum panel cross-section. The orthotropic material properties can either be calculated from the [A] matrix or the [D] matrix of initially selected laminates satisfying the design rules. Both approaches are covered in the following results.

7.3.2 *Laminate level optimisation*

Once the optimum orthotropic panel design with continuous thickness values has been found, the laminate level optimisation is started by rounding-off each laminate thickness to achieve a discrete number of layers. The aim of this level is to substitute the orthotropic plates with laminated plates with minimum mass penalty, whilst also satisfying the buckling constraints, material strength constraints as well as the above laminate design rules.

The optimisation problem at the laminate level is formulated as finding the stacking sequences of the skin, and of the stiffener laminates in order to maximise the buckling factor (i.e. reserve factor) λ_b and material strength reserve factor λ_s obtained (over all strain components) as the smallest ratio of the strain allowable ε_{ij}^{all} to the corresponding strain component ε_{ij}. It is, however, converted to a minimisation problem as follows.

$$\text{Minimise} \quad \frac{1}{\lambda_b}+\frac{1}{\lambda_s} \tag{7.3}$$

$$\text{Subject to} \quad P_i = P_{all,i} \tag{7.4}$$

$$g(\theta) \le 4 \tag{7.5}$$

where P_i are the actual values of orientation percentages in the laminate, and $P_{all,i}$ are the target values. The condition of $P_i = P_{all,i}$ represents the satisfaction of the ply percentage rules upon which the orthotropic material properties are based. $g(\theta)$ represents the grouping constraint, i.e. rule (iii) above.

Laminate selection is carried out using two approaches. These are manual laminate selection and automated laminate selection using a newly developed GA program. The two approaches are described in the following sections.

Manual laminate selection

Each laminate thickness is rounded down to its nearest discrete value. The stacking sequence of each laminate is then selected by hand based on the design rules. (Manufacturing constraints could also be applied at this stage. For example, such constraints could maintain continuity of plies for adjacent panels.) If the design can carry the design load without buckling or material failure, which is checked using VICONOPT analysis, the laminate design is feasible. Otherwise, the thicknesses are increased ply by ply until a check with VICONOPT analysis confirms that the laminate design satisfies all the constraints. In particular, it is seen in the next section that the number of discrete design variables involved in this method corresponds to the number of laminate thicknesses. Hence, the method has fewer variables than other methods, which treat every layer as design variable. This reduces computational time significantly, so giving a practical advantage to the optimisation of composite skin-stiffener structures.

GA laminate selection

A GA code has also been developed to search a population of designs, where each individual in the population represents a stacking sequence, coded in the form of a string, or chromosome. Each unit, or gene, on the string is encoded to indicate plies. The laminate design rules were incorporated into the objective function via a penalty function to create a fitness function. Genetic operators were coded to change the population of strings based on the fitness function. The GA was performed using FORTRAN 95 with a preprocessor and postprocessor to translate VICONOPT input and output files for buckling analysis. The GA fitness function, genetic coding and operators are described in the following.

$$F = \frac{1}{\lambda_b} + \frac{1}{\lambda_s} + \sum_{i=1}^{n_l} \sum_{j=1}^{3} Max[0, P'_{i,j}] \tag{7.6}$$

where $P'_{i,1}$ is defined to add penalty when $P_i > P_{all,i}$ using the following equation

$$P'_{i,1} = \frac{P_i}{P_{all,i}} - 1 \tag{7.7}$$

whereas $P'_{i,2}$ is defined as the following equation

$$P'_{i,2} = 1 - \frac{P_i}{P_{all,i}} \tag{7.8}$$

to add penalty when $P_i < P_{all,i}$. For $P'_{i,3}$, zero indicates that Eq. (7.5) is satisfied whereas a value of one indicates that it is not satisfied. n_l is the number of laminated plates, e.g. 3 for the blade-stiffened panel, i.e. the skin, stiffener and flange.

In Eq. (7.6), the term $\sum_{i=1}^{n_l} \sum_{j=1}^{3} Max[0, P'_{i,j}]$ ensures that the fitness function is penalized for designs that do not satisfy the laminate design rules. It can be seen that a violated constraint will have a value greater than 0 and therefore only if a constraint is violated will the fitness value of a design be increased. Therefore the best design has the lowest fitness function value.

The design variables are the stacking sequences of skin and stiffener laminates. The ply angles of 0°, 90°, 45° and –45° are used directly to create chromosomes for the GA. However, equal probability of occurrence of each of these ply angles (referred to later as the Basic GA method) is biased because the probability of appearance of a ±45 plies is twice that of other plies. For our problem with ply orientation targets, this bias fits well when the target is to obtain quasi-isotropic laminates. Whereas when the target is, for example, a 60/30/10 (i.e. 60% of 0°, 30% of ± 45° and 10% of 90°) laminate, the bias is not beneficial and can cause convergence on a non-optimal design. Experiments were conducted to deal with this problem. A mutation operator that biases replacement of 90° plies with 0° plies was tested. The results showed that this mutation would cause violation of the contiguity constraint-which prevents grouping of over 4 plies of the same orientation together. In contrast, a method for adjusting the probability of appearance of the plies according to the relevant laminate targets fits well and was adopted for further experiments. This adjustment (referred to later as the Improved GA method) is not only performed for creating the initial population, but also during reproduction and mutation. Comparison of the Basic GA and Improved GA methods will be presented in the design examples of Section 7.4.3.

The skin, web and flange laminates are then modelled as chromosomes using the genes within the GA to indicate the stacking sequences. When the laminate is composed of *N* plies, only *N*/2 plies are required to describe the entire symmetric laminate. In particular, at least four plies in the outmost side of each stiffener laminate (see Fig. 7.2), are controlled to be the same as the flange laminates in order to provide integrity to the joint with the skin.

One-point cross-over operator was selected with a probability of 0.8 and the mutation operator was applied to one random gene at a probability of 0.03. An elitist selection was implemented by keeping the best 3 solutions from the previous generation, and new random selections were added at each generation to keep the total population constant when necessary. For all examples in this chapter, a population size of 30 over a maximum 100 generations is used.

7.3.3 *Convergence test*

The mass m_1 of the laminate design generated in Step 3 (see Fig. 7.3), is calculated at the end of the first cycle. The second cycle then starts with the updated equivalent orthotropic material properties obtained from the feasible design of the first cycle to check for convergence. The reason for this is to ensure that the design is optimum for this design region, i.e. within this region of plate thicknesses. The orthotropic material properties for the second cycle are an average of those calculated from the skin and stiffener laminates of the feasible design of the first cycle. Note that the value of maximum allowable strain may require modification to compensate for the updated value of E_{xx}. Note also that the orthotropic material properties obtained from the [D] matrix converge to those from the [A] matrix when the laminates are around 40 layers thick. This means that the method will become more robust when laminates contain a large number of plies.

After the second orthotropic optimum design is obtained in Step 2, the sequential rounding in Step 3 finds the laminate design satisfying the design rules, and buckling and strain constraints. The mass m_2 of the design is calculated and compared with m_1 of the first cycle. If $m_1 \leq m_2$, the feasible design of the first cycle proves to be optimum. If $m_1 > m_2$, the laminate design from the second cycle can be selected as an optimum design or cycling may continue by averaging the two orthotropic material properties and finding the next possible lighter design. Typically, it takes VICONOPT about 2 minutes to complete an optimisation step, and even less time to analyse a laminate design. Hence several cycles can be finished in a short time with the added advantage of accurate prediction of buckling loads for the optimum design.

7.4 Results

7.4.1 *Validation of strip method for local buckling of composite stiffened panels*

Example panels

The stiffened panels, made of CFRP unidirectional UTS/977-2 prepreg, were subject to axial compressive load, and had free longitudinal edges and clamped transverse edges. The material ply thickness was 0.25 mm, and the following properties were assumed: $E_{11} = 117$ GPa, $E_{22} = 17$ GPa, $G_{12} = 4.6$ GPa, $\nu_{12} = 0.3$, and density $\rho = 1584$ kg/m^3. The final configurations were designed using VICONOPT. The design target was a high strain and high load range, i.e. the strain was around 4,000 microstrain, and the compressive design load per unit width was approximately 4 kN/mm.

The two blade-stiffened panels, shown in Fig. 7.4, were designed according to the design target and laminate design rules. For the 3-blade-stiffened panel, the skin has a stacking sequence of [45/-45/0_2/90/0_2/45/-45/0_2/90/0_2/-45/45], and the stiffener blade [45/0/-45/0_3/90/0_2/90/0_3/-45/0/45]$_S$. For the 2-blade-stiffened panel, the skin has a stacking sequence of [45/-45/0_2/90/0/45/-45/0_2/90/0]$_S$, and the stiffener blade has the same stacking sequence as that of the 3-blade-stiffened panel. The stiffener flanges for both panels have half the stacking sequence of the stiffener blades with an additional two plies of ±45 on the bottom to seal the wedge of filler inside the stiffener base. Both panels satisfied the laminate design rules with around 50% of 0° material in the skin and 60% of 0° material in the stiffeners. The lengths of the 3-blade-stiffened panel and the 2-blade-stiffened panel were l = 500 mm and l = 550 mm, respectively. The two blade-stiffened panels were not optimum since different initial buckling modes were to be investigated, i.e. stiffener buckling in the case of the 3-blade panel and skin buckling in the case of the 2-blade panel.

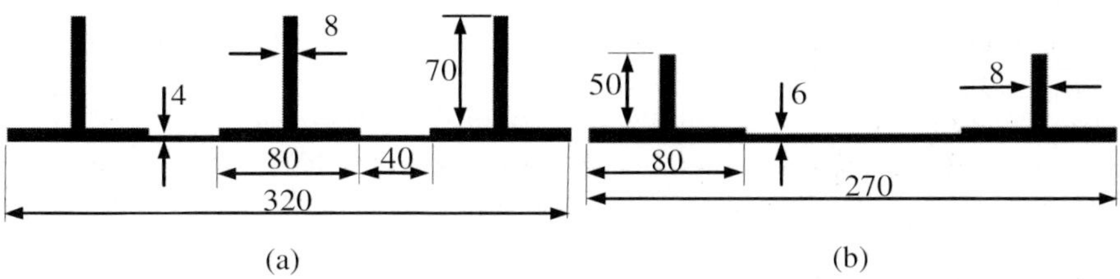

Fig. 7.4. Cross sections of experimental panels with (a) 3-blade stiffeners and (b) 2-blade stiffeners. All dimensions are in mm.

FEA modelling

The general-purpose code ABAQUS[24] was employed to create FE models of the panels using the four-node general element S4R with six degrees of freedom at each node and three integration points along the thickness of each ply. This element can account for transverse shear deformation. With the skin thickness of 4 mm or 6 mm and the stiffener thickness of 8 mm, the transverse shear deformation of the panel is more influential than that of metal panels of the same geometry. The dimension of the element was set to 5×5 mm. Both linear eigenvalue analyses and non-linear static analyses, using a modified Riks method, were performed to study buckling and postbuckling behaviour of the panels.

To model the offsets between the flange and the skin, and between the web and the skin, the skin and stiffeners were meshed as separate components, and then joined together using the multi-point control technique of ABAQUS. The bottom plane of the blade stiffener was located as the reference plane by adding negligibly low stiffness material on the surfaces of the skin and the flange. In this way, the skin nodes and the flange nodes can be moved to the same reference plane, and then constrained with identical translations and rotations. This approach ensures continuity of the skin. In addition, ABAQUS non-linear analysis for the two panels was conducted with a full cosine-wave imperfection over the length of the panel of amplitude 0.5 mm causing increased compression in the stiffener at the panel centre. The reason for this was that the two manufactured panels had a global imperfection due to springback following cure of the composite parts.

Experimental testing and results

The two blade-stiffened panels were manufactured, using an autoclave with controlled pressure and temperature at the University of Bath. An aluminium mould was designed and machined to fabricate the blade stiffeners. The stiffeners and skin were laminated by hand lay-up and then bagged by a standard composite bagging system with a bleeder. They were cured separately in the autoclave and then bonded together using FM-300 film adhesive. Ultrasonic C-scan was employed to check for defects in the panel at various stages during this manufacturing process.

The panels were strain-gauged to monitor the onset and advance of buckling, which was found to be more accurate than load versus end-shortening plots. The upper end and the lower end were potted into a 25 mm thick block of epoxy resin to achieve the clamped end fixture. The skin side opposite the stiffeners was painted white for the application of a Moiré shadow to observe buckling patterns of the panel as they developed during the compression tests. The grating frequency was 2 lines/mm. White lines were drawn on the tips of stiffeners to indicate the buckling shape from the stiffener side. Two high resolution digital cameras were set at each face of the test panel to record the Moiré fringe and stiffener deformation.

Tests were performed using a Dartec 2000 kN testing machine under displacement control. The stroke rate was 0.01 mm/second before buckling and 0.005 mm/second after buckling. To ensure uniform loading, the panels were pre-tested to 500 kN and the cross-head platen adjusted until all strain gauges displayed approximately the same reading. Because the prediction from VICONOPT and ABAQUS indicate that the critical buckling loads occur above 1000 kN, initial buckling loading was performed to 1200 kN for the 3-blade-stiffened panel, and to 1000 kN for the 2-blade-stiffened panel, respectively. After the initial buckling testing, the panels were loaded to failure.

The 3-blade-stiffened panel was designed to buckle in the stiffeners initially. Figure 7.5 presents the experimental results obtained with the strain gauges located on the two side stiffeners at a quarter of the length of the panel from the upper loading edge. The strain gauges were placed back-to-back near the free edge of the stiffener. It can be seen that buckling occurs when the loading reaches 1378 kN, and a strain of 4,400 microstrain. For comparison, the FEA results are also presented in the figure, showing good agreement. The results from VICONOPT, ABAQUS and experimental testing are summarized in Table 7.1. In all cases, the three stiffeners buckled into between two and three half-wavelengths over the panel length during testing.

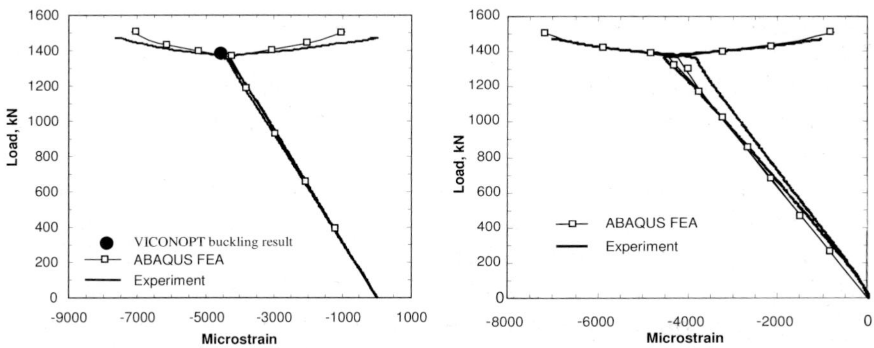

Fig. 7.5. Strain in back-to-back gauges on the two side stiffeners at quarter length distance from end of 3-blade experimental panel.

Table 7.1. Summary of initial buckling results for blade-stiffened experimental panels.

Panels	Results	VICONOPT[a]	ABAQUS		Experiment
			Linear	Non-linear[b]	
3-blade	Initial buckling load (kN)	1397	1396	1362	1378
	Initial buckling strain (310^{-6})	4482	-	4330	4400
2-blade	Initial buckling load (kN)	1259	1205	1150	1160
	Initial buckling strain (310^{-6})	5331	-	4649	4528

[a]No imperfection for VICONOPT results. [b]ABAQUS non-linear results calculated with a full cosine-wave imperfection over the length of the panel of amplitude 0.5 mm causing increased compression in the stiffener at the panel centre.

It was noted that, according to the Moiré pattern, there was no obvious out-of-plane deformation on the skin. This was coincident with FEA results. Both show that the maximum out-of-plane deformation at the anti-nodal line was less than 0.25 mm. It is also shown that the actual imperfection of the panel did not significantly affect the advance of buckling and postbuckling in the stiffener.

The strain data shows that the panel has little postbuckling capacity after such high initial stiffener buckling strain. The panel buckled at 1378 kN and then failed at 1472 kN, which is about 7% above the initial buckling load. There was no evidence of overall buckling until the panel failed. It seems that the buckling deformation tended to pull the stiffeners from the skin since the failed panel indicates separation of stiffeners from skin.

The 2-blade-stiffened panel was designed for initial buckling to occur in the skin. In comparison with the previous panel, this panel has more imperfection sensitivity due to its lower overall bending rigidity, which causes a lower Euler buckling load. This is evident in Table 7.1, which shows that with the same shape and amplitude of global imperfection, the non-linear buckling load of the 3-blade panel is 2% less than the linear result, whereas for the 2-blade panel, the difference is 5%. This was also observed during the pre-testing, where the 2-blade panel presented greater difficulty in achieving uniform strain than the 3-blade panel due to the effect of imperfection. The Moiré pattern, showed in Fig. 7.6 demonstrates obvious out-of-plane deformation of the skin in the form of between two and three half-wavelengths over the panel length, which is

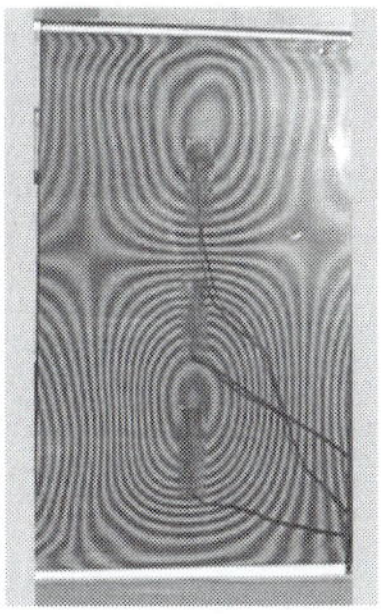

Fig. 7.6. Moiré pattern of post-buckled 2-blade experimental panel.

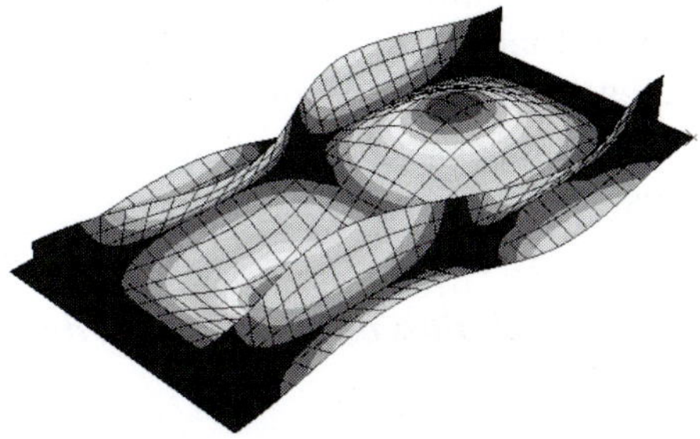

Fig. 7.7. ABAQUS prediction of buckling mode for 2-blade experimental panel.

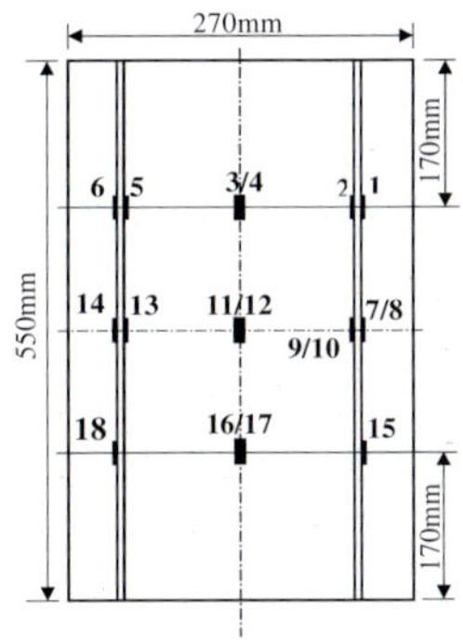

(a)

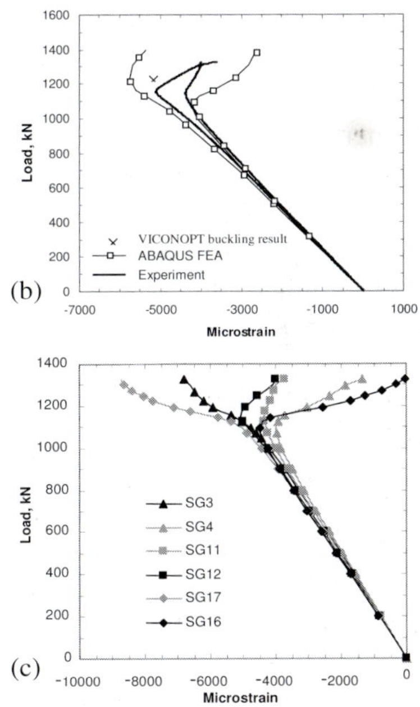

Fig. 7.8. Strain on the 2-blade experimental panel, (a) strain gauge positions, (b) strain from SG11 and SG12, and (c) strain from experiment in the centre of the skin.

coincident with the VICONOPT and ABAQUS analysis results (see Fig. 7.7). The initial buckling loads shown in Table 7.1 are also in good agreement. However, Fig. 7.8 (b) indicates that there are some differences between the postbuckling strains of the experimental test and those predicted by FEA. This is probably due to differences in the boundary conditions of the FEA model, and the experimental test. The boundary conditions assumed in the FEA model are rotational restraints at both loaded edges of the panel, whereas for the test, the upper loading platen is universally jointed, allowing free rotation after buckling occurs. The influence of these differences on postbuckling behaviour has been considered for metallic panels.[25, 26]

7.4.2 *Validation of an optimum design*

Optimum design

The bi-level optimisation strategy was evaluated by the design of a Z-stiffened panel with free longitudinal edges and clamped transverse edges. The material described in Section 7.4.1 was again used. The panel was designed to carry a compressive load of 2.8 kN/mm without buckling and with a maximum allowable mid-surface strain of 5000 microstrain in any plates. The length of the panel was $l = 550$ mm.

The optimisation strategy used the equivalent orthotropic material properties, obtained from the [*D*] matrix of skin and stiffener laminates satisfying the laminate design rules. The variation of properties for the selection of typical laminates given in Table 7.2 is shown in Fig. 7.9. The strategy started with properties based on 10 mm thick laminates. For the skin laminate, this gave $E_{xx} = 68$ GPa, $E_{yy} = 34$ GPa, $G_{xy} = 16$ GPa, and $\upsilon_{xy} = 0.44$. The equivalent orthotropic material properties for the stiffener

Table 7.2. Stacking sequences for typical skin laminates.

Thickness (mm)	Stacking sequence	Percentage of 0/±45/90 material
4.0	$[\pm45/0_2/90/0_2/\pm45/0_2/90/0_2/\mp45]$	50.0% / 37.5% / 12.5%
4.5	$[\pm45/0_3/90/0_2/\pm45/0_2/90/0_3/\mp45]$	55.6% / 33.3% / 11.1%
5.0	$[\pm45/0_2/90/0_2/\pm45/0]_S$	50.0% / 40.0% / 10.0%
5.5	$[\pm45/0_3/90/0_2/\pm45/0]_S$	54.5% / 36.4% / 9.1%
6.0	$[\pm45/0_2/90/0_2/\pm45/0/90/0]_S$	50.0% / 33.3% / 16.7%
6.5	$[\pm45/0_3/\pm45/0_2/90/0/90/0]_S$	53.8% / 30.8% / 15.4%
7.0	$[\pm45/0_3/\pm45/0_2/90/0/90/0/45$ $/-45/0/90/0/90/0_2/\mp45/0_3/\mp45]$	50.0% / 35.7% / 14.3%
7.5	$[\pm45/0_3/\pm45/0_3/90/0/90/0/45$ $/-45/0/90/0/90/0_3/\mp45/0_3/\mp45]$	53.3% / 33.3% / 13.4%
8.0	$[\pm45/0_2/90/0_2/\pm45/0_2/90/0/\pm45/0]_S$	50.0% / 37.5% /12.5%
10.0	$[\pm45/0_2/90/0_2/\pm45/0_2/\pm45/0_2/90/0/\pm45/0]_S$	50.0% / 40.0% / 10.0%

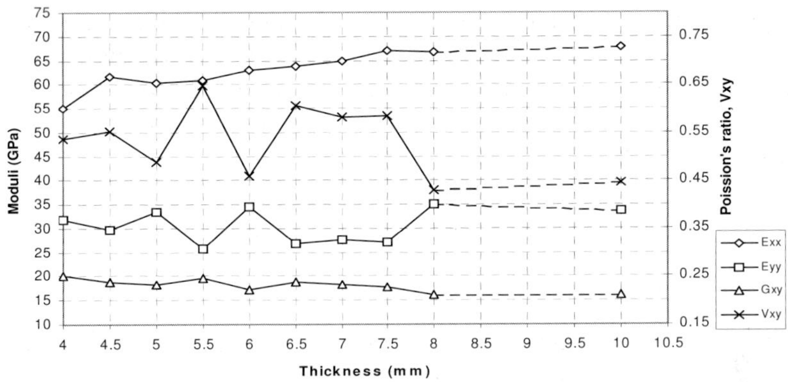

Fig. 7.9. Equivalent orthotropic material properties from [*D*] matrix for laminates in Table 2.

were E_{xx} = 75 GPa, E_{yy} = 33 GPa, G_{xy} = 14 GPa, and υ_{xy} = 0.41. Therefore, the average of these values was used as the input material properties for Step 1 of Cycle 1 and then for the optimisation of Step 2 of Cycle 1. Laminate level optimisation was carried out manually for this problem.

All the design variables used in Step 2 are illustrated in Fig. 7.10. The width of the panel was fixed to 300 mm, and the width of the flange B_1 was constrained to be greater than 40mm. Hence B_1, B_2, B_4, T_1 and T_2 were independent variables, while B_3 was dependent and linked with the width of the panel and B_1. The optimizer could select any set of design points within the feasible domain that satisfy the buckling constraint and the strain constraint. The final optimisation results are summarised in Table 7.3 with the optimum laminate design of the panel (Step 3 of Cycle 1).

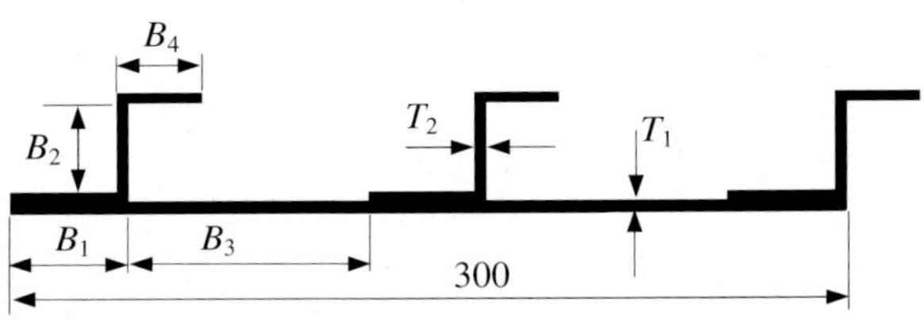

Fig. 7.10. Cross section of Z-stiffened experimental panel with design variables.

When VICONOPT analysis of the optimum design was performed, the initial buckling load N_{cr} was calculated as 2.88kN/mm and the buckling strain was 5053 microstrain, indicating that the design was feasible in terms of strength and stiffness requirements. To ensure the design was optimum, the updated orthotropic material properties were used in a second cycle of optimisation. The updated orthotropic optimum result gave a 2% increase in mass (see Table 7.3) and so the laminate panel design of the first cycle was taken as optimum.

Table 7.3 also presents the N_{cr} for each design after each step. It is noted that when the very different orthotropic material properties were used for the two panels of similar geometry (Step 2 of the first cycle compared with Step 1 of the second cycle), there is only about 6% difference in the N_{cr} of the two panels, although the second cycle E_{xx} of 55 GPa is 24% lower than the 72 GPa used in the first cycle. It is

Table 7.3. Optimisation results for Z-stiffened experimental panel.

	Cycle 1			Cycle 2		
	Step 1[a] Initial design	Step 2[a] Orthotropic optimisation	Step 3[b] Laminate design	Step 1[a] Initial design	Step 2[a] Orthotropic optimisation	Step 3 Laminate design
B_1 (mm)	50	42.1	42	42	42.9	43
B_2 (mm)	50	32.5	33	33	33.7	34
B_3 (mm)	75	86.9	87	87	85.7	86
B_4 (mm)	45	29.3	30	30	30.6	31
T_1 (mm)	5	4.08	$[45/-45/0_2/90/0_2/45/-45/0_2/90/0_2/-45/45]_T$	4	3.82	4
T_2 (mm)	4	3.56	$[45/-45/0_3/90/0]_S$	3.5	3.74	3.5
E_{xx}(GPa)	72.0	72.0	-	55.0	55.0	-
E_{yy}(GPa)	33.5	33.5	-	31.9	31.9	-
G_{xy}(GPa)	15.0	15.0	-	19.9	19.9	-
ν_{xy}	0.42	0.42	-	0.53	0.53	-
Mass(kg)	2.82	2.03	2.00	2.00	2.05	2.04
N_{cr}(kN/mm)	3.50	2.84	2.88	3.01	2.90	2.86

[a]The critical buckling loads N_{cr} for steps 1 and 2 assumed orthotropic properties.
[b]The Step 3 laminate design of Cycle 1 is taken as the final optimum.

assumed that the reason for this is the associated increase in G_{xy} and ν_{xy} values for the design of the second cycle, which results in larger stiffness terms D_{12} and D_{66}, producing similar values of N_{cr} for the two designs. The laminated design N_{cr} of 2.88 kN/mm for Step 3 of the first cycle is 4% lower than the orthotropic design N_{cr} of 3.01 kN/mm for Step 1 of the second cycle. This difference is attributed to the non-vanishing bending-twisting coupled stiffness terms D_{16} and D_{26} in the laminate design. The difference, which is due to the relative position of +45° and -45° material, will reduce for laminates containing more plies.

FE validation

ABAQUS FEA was applied to the Z-stiffened panel to verify the VICONOPT design results. The initial buckling load N_{cr} calculated by linear FEA was 2.85 kN/mm. This was close to the VICONOPT prediction of 2.88 kN/mm. Figure 7.11 shows the initial buckling modes of the panel predicted by VICONOPT and ABAQUS. To illustrate the

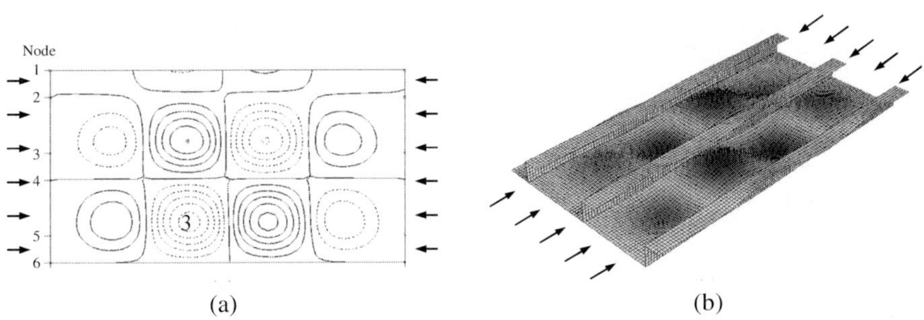

Fig. 7.11. Initial buckling mode of Z-stiffened experimental panel, (a) VICONOPT contour plot of the skin; (b) ABAQUS isometric plot. Cross in (a) shows the location of skin strain gauges.

Fig. 7.12. Z-stiffened panel test specimen.

computational advantage of the strip model, VICONOPT required 5 elements for the skin whereas ABAQUS required over 7,000 elements for the skin. Both indicate that the panel buckles into between four and five half-wavelengths over the panel length.

In order to verify the optimum design of Z-stiffened panel, the panel was manufactured for experimental compression testing (see Fig. 7.12). During manufacture, it was found that the Z-stiffener was more difficult to fabricate than the blade stiffener due to the more complicated tooling with double curvature. It was noted that, due to practical manufacturing constraints, the final specimen has a 5 mm inner radii between the web and flange. However, the optimisation model has a 90° corner at this junction (see Fig. 7.10). After modelling the radii, which increases the exposed width of the skin, it was found that the updated critical buckling loads are, respectively, 13% and 17% lower than the previous straight plate models of VICONOPT and ABAQUS. To correspond to the experimental specimen, the following VICONOPT and non-linear FE results use the updated model with the radii.

During the non-linear analysis, it was found that the Riks arc length analysis of the perfect panel model was not able to pass the first bifurcation point due to severe discontinuity around that point. The end shortening versus load for the perfect model indicated that the critical buckling load was approximately 750 kN (2.5 kN/mm), although the initial buckling mode shape was not coincident with the linear eigenvalue analysis result. It was found that in order to perform non-linear analysis in the postbuckling field with the correct initial buckling mode shape, a

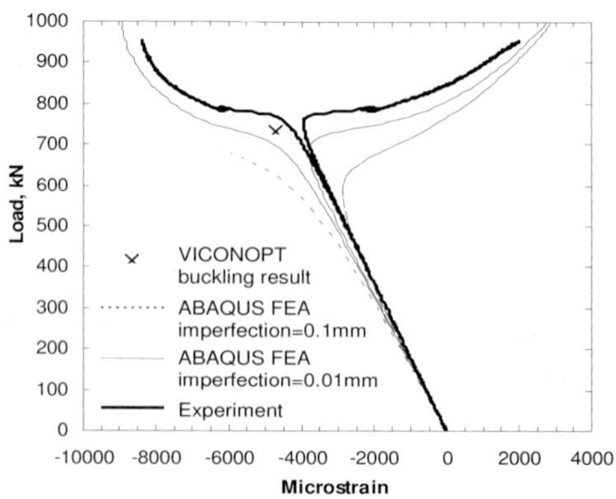

Fig. 7.13. Strain in back-to-back gauges on the skin of Z-stiffened experimental panel.

small initial imperfection in the shape of the (linear) initial mode was required. The lowest amplitude of this mode was found to be 0.01 mm. However, as is well known, imperfections cause reduction of the initial buckling load. For this Z-stiffened panel, the 0.01 mm and 0.1 mm imperfection caused 6% and 10% reduction of the initial buckling load, respectively (see Fig. 7.13). Therefore, it seems that the optimum design of the Z-stiffened panel has more bifurcation paths around the first bifurcation point than the two blade-stiffened panels, for which the analysis easily picked up the first bifurcation path without needing to add the scaled initial imperfection to the non-linear postbuckling analysis. This may be due to the optimisation and due to the increased number of local buckling modes that occur in the Z-stiffened panel.

Although the analysis could follow the equilibrium path beyond the first bifurcation point, the Riks arc length method could not converge around the second bifurcation point. Since the experimental panel failed locally before reaching the second bifurcation point, the Riks arc length method was adequate. If analysis beyond a stable second bifurcation was required, the use of imperfections in modes other than the initial buckling mode might be required. Alternatively, analysis involving stabilization schemes with pseudo damping might be suitable.

Experimental test

The panel was end-potted, and loaded in compression to failure with controlled displacement of 0.01mm/second before buckling and 0.005mm/second after buckling. Fifteen strain gauges were attached to the panel to record initiation of critical buckling and the post-buckled deformation. A Moiré shadow technique was applied to identify the buckling mode, which indicates that the panel buckled into four half-wavelengths over the panel length coincident with the VICONOPT and FEA linear analysis result of Fig. 7.11.

The strain reading from two back-to-back gauges versus the compressive load is presented in Fig. 7.13, where the location of the strain gauges is shown in Fig. 7.11(a). The strain data shows that the panel buckled at 770 kN and then failed at 950 kN, which is about 24% above the initial buckling load. Compared with the blade-stiffened panels, the Z-stiffened panel has an increased postbuckling capacity.

The strain data were also compared with ABAQUS non-linear results in Fig. 7.13. Although the buckling and postbuckling behaviour follows the same bifurcation path, the buckling load from testing is 6% higher than the non-linear FEA result with a 0.01mm imperfection. This difference is attributed to the following two factors. Firstly, the added imperfection will reduce initial buckling loads. Secondly, it is worth pointing out that the material properties used in all VICONOPT and ABAQUS models were obtained from the B-basis statistical procedure, used to derive material properties and allowables. The property value indicated is the value above which at least 90% of the population of values is expected to fall with a confidence of 95%. Compared with mean values, B-basis values are conservative.

Discussion

From the convergence test of the above optimum design example, it is seen that with updated laminate thickness, the design from cycle 2 of the strategy does not provide any more improvement than the design from cycle 1. This means that although the orthotropic material properties calculated from the [D] matrix vary for different laminates, the optimum result obtained from them converges to the one obtained from the orthotropic material properties calculated from the [A] matrix. Hence the [A] matrix is used to calculate orthotropic properties in the results that follow.

7.4.3 *Optimisation of composite wing cover panels*

Optimum design example

The design example presented in this section is an upper wing cover panel between two engines, where fuel is stored within the wing box. Hence a lateral pressure q of 0.109 N/mm^2 is considered, as shown in Fig. 7.1, together with compression N_x of 1.91 kN/mm and shear load N_{xy} of 0.10 kN/mm. The panel is assumed to be infinitely wide and continuous over rib supports spaced at intervals l = 770 mm.

A datum panel design[23] is used as the baseline for optimisation. Based on the blade stiffener concept, the panel mass of the existing datum design was 1.84 kg for a single repeating portion (see Fig. 7.2). The objective of the optimisation is to find a stiffened panel which is lighter than the existing datum design and can carry the combined loads without buckling and with maximum strains below 3,600 microstrain. The panel is made of CFRP IM tape with material properties: E_{11} = 131 GPa; E_{22} = 7.1 GPa, G_{12} = 3.5 GPa; ν_{12} = 0.3, ρ = 1584 kg/m^3 and thickness 0.184 mm. The design laminate target for the percentages of 0°/±45°/90° material is 44/44/12 for the skin and 60/30/10 for the stiffener.

For the example considered here, the panel is assumed to be simply-supported at ribs for buckling analysis, and clamped at ribs for out-of-plane bending due to pressure; causing increased compression in the stiffeners at panel mid-length and increased compression in the skin at panel ends.

The bi-level optimisation strategy described in the previous section was applied to design the composite stiffened panels subject to the combined loads defined above. The following optimum design results are obtained for stiffened panels with blade, I, Z and top hat stiffener shapes. In particular, the GA for automatic selection of laminate stacking sequences at the laminate level was applied and tuned when designing the blade-stiffened panel whereas for the other panels, the laminate selections were carried out manually. Comparison of the two approaches is also presented.

For all the designs, the stiffener spacing b_s was constrained to be equal to 165 mm due to the requirements of manufacturability of the whole wing. A lower bound of 3.312 mm (18 plies) was applied to the thickness of the skin laminates, t_{sk}, to represent a practical design constraint for lightning strike and bolt bearing considerations. Further constraints for each stiffened panel are described in detail in the following.

Blade-stiffened panel

The design process commenced by obtaining the geometry of a repeating portion of blade-stiffened panel from panel level optimisation (see Table 7.4). Two rounded results are also presented in the table, where D_{r1} is the round-up result and D_{r2} is the round-down result. Both designs were taken from panel level to laminate level to select stacking sequences satisfying laminate design rules. The reason for this was because the design D_{r1}, with a laminated design which was easily found manually, was used to tune the GA and its genes.

Table 7.4. Results from panel level optimisation of blade-stiffened wing panel showing rounding.

Design	Mass (kg)	t_{sk} (mm)	t_{sf} (mm)	t_{fl} (mm)	b_{fl} (mm)	b_w (mm)
Panel level	1.61	3.312	8.814	3.918	70	57
Round-up D_{r1}	1.63	3.312 (18 plies)	8.832 (48 plies)	4.048 (22 plies)	70	57
Round-down D_{r2}	1.60	3.312 (18 plies)	8.648 (47 plies)	3.864 (21 plies)	70	57

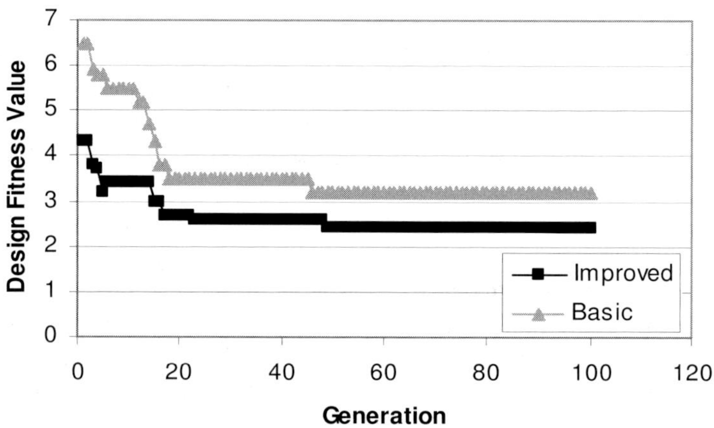

Fig. 7.14. Optimisation history for the Basic GA and Improved BA for blade-stiffened panel.

For the usage of the GA in laminate level optimisation, the comparison of design results of the Basic GA and Improved GA shows that the latter is better at finding a feasible design satisfying the design rules, the buckling constraint and strain constraint. Figure 7.14 shows the history of the objective functions for the two GAs during laminate level optimization.

Table 7.5 gives details of the optimum designs of the blade-stiffened panels. Note that the laminate designs in this table have the same mass as their round-off designs from the panel level. It can be seen that the most optimum design is produced by the GA for the round-down design D_{r2}.

Compared with the datum design,[23] the weight saving of this design is 13%.

I, Z and hat stiffened panels

During panel level optimisation of the I-stiffened problem, a lower bound of 68 mm was applied to the width of the bottom flange b_{fl}. For comparison, the width of the top flange b_{fu} was unconstrained for the design I-a of Table 7.6 and constrained for the design I-b with an upper bound of 15 mm, to account for assembly of the cover panel and ribs.

For the I-stiffened panel designs, it can be seen that the I-a and I-b designs have the same panel mass. Hence only the I-b design has undergone laminate level design. Compared with the existing datum design of the panel mass 1.84 kg, the final laminated I-stiffened panel has a mass of 1.54 kg, i.e. 16% weight saving. This is because the I-stiffened panel has higher bending rigidity than the blade-stiffened panel, reducing the magnitude of strain due to lateral pressure on the top of the stiffener.

For the design of the Z-stiffened panel (see Table 7.6(b)), the optimisation started with design of Z-a, for which the thicknesses of the web, bottom flange and top flange varied. A further constraint was added to the Z-b design to maintain equal laminate thickness, which represents a lower cost manufacturing solution. Because both designs rendered the same panel mass, the Z-b design was considered for laminate level optimisation. However, it was observed that, after rounding-down of the Z-b design, the laminate design (although not shown in Table 7.6(b)), slightly violated the buckling and strength constraints. To overcome this problem, a Z-c design was developed with an upper bound on the thickness of web, top and bottom flange. The laminate design was then achieved without further weight penalty.

For the designs of the hat-stiffened panel, as shown in Table 7.6(c), the buckling load factor for the H-a design is $\lambda_b = 1.69$, i.e. 69% higher than the design load, but the panel has similar mass to I- and Z-stiffened panels. This clearly suggests that the hat-stiffened panel is more efficient than the blade, I- and Z-stiffened panels. The H-a design has a higher buckling load because the stiffener spacing of 165 mm is over-constraining for this example. With unconstrained stiffener spacing, the buckling load of the H-b design equals the design load, and its weight per unit width is 17% less than the datum design.

Table 7.5. Blade-stiffened wing panel optimisation results.

	Design	λ_b	λ_s	Lay-up	Percentage 0/±45/90
Round-up D_{r1}	Manual	1.078	1.009	Skin $[45/-45/0_2/45/-45/0/90/0]_S$ Web $[45/-45/0_3/45/-45/0_3/90/0_2/45/-45/0_2/90/0_2/90/0_2/45/-45/0_2/90/0_2/90/0_2/-45/45/0_2/90/0_3/-45/45/0_3/-45/45]_T$ Flange $[45/-45/0_3/90/0_3/45/0/0/-45/0_3/90/0_3/-45/45]_T$	Skin 44%/44%/12% Web 58%/29%/13% Flange 64%/27%/9%
	GA (Basic)	1.104	0.928	Skin $[45/-45/0/-45/0_2/90/45/0]_S$ Web $[45/-45/90/0/-45/0_3/45/0/-45/0_3/45/45/0/0/45/90/0/90/45/0]_S$ Flange $[45/-45/90/0/0/0/0/-45/0/45/0]_S$	Skin 44%/44%/12% Web 50%/38%/12% Flange 55%/36%/9%
	GA (Improved)	1.113	1.034	Skin $[45/-45/90/0/0/0/-45/0/45]_S$ Web $[45/-45/0/0/45/0/-45/0/0/0/-45/0/0/0/-45/0/45/0/90/0/0/0/0/90]_S$ Flange $[45/-45/0/0/90/0/0/0/-45/0/0]_S$	Skin 44%/44%/12% Web 63%/29%/8% Flange 64%/27%/9%
Round-down D_{r2}	GA (Improved)	1.053	0.991	Skin $[45/-45/90/0/0/45/0/-45/0]_S$ Web $[45/-45/0/0/45/0/90/0/45/0/0/0/0/-45/-45/0/45/0/90/0/0/0/0/\frac{90}{2}]_S$ Flange $[45/-45/0/0/90/0/0/0/-45/0/\frac{0}{2}]_S$	Skin 44%/44%/12% Web 60%/30%/10% Flange 62%/28%10%

Table 7.6. Optimisation results for I-stiffened, Z-stiffened and hat-stiffened wing panels. All thicknesses t and widths b are in mm.

(a) I-stiffened panel

Design	Level	Mass (kg)	λ_b	λ_s	Skin		Stiffener					
					t_{sk}	b_s	Web		Bottom flange		Top flange	
							t_w	b_w	t_{fl}	b_{fl}	t_{fu}	b_{fu}
I-a	Panel	1.55	1.00	1.00	3.312	165	5.888	54	3.884	76	2.418	44
I-b	Panel	1.55	1.00	1.00	3.312	165	5.888	40	3.904	78	12.628	15
	Laminate	1.54	1.01	1.01	3.312	165	5.888	40	3.864	78	12.512	15

Skin $[45/-45/0/0/45/-45/0/90/0]_S$ (44%/44%/12%)

Web $[45/-45/0/0/0/45/-45/0/0/0/90/0/0/90/0/0]_S$ (63%/25%/12%)

Bottom flange $[45/-45/0/0/0/90/0/0/0/45/0/-45/0/0/0/90/0/0/0/-45/45]_T$ (62%/28%/10%)

Top flange $[45/-45/0_3/90/0_2/45/-45/0_3/90/0_2/45/-45/0_3/90/0_2/45/-45/0_3/90/0_2/45/-45]_S$ (59%/29%/12%)

(*Continued*)

Table 7.6. Optimisation results for I-stiffened, Z-stiffened and hat stiffened wing panels. All thicknesses t and widths b are in mm. *(Continued)*

(b) Z-stiffened panel

Design	Level	Mass (kg)	λ_b	λ_s	Skin		Stiffener					
					t_{sk}	b_s	Web		Bottom flange		Top flange	
							t_w	b_w	t_{fl}	b_{fl}	t_{fu}	b_{fu}
Z-a	Panel	1.54	1.00	1.00	3.312	165	4.968	56	5.194	63	3.994	27
Z-b	Panel	1.54	1.00	1.00	3.312	165	5.010	54	5.010	65	5.010	24
Z-c	Panel	1.54	1.00	1.00	3.312	165	4.968	51	4.968	66	4.968	27
	Laminate	1.54	1.01	0.991	3.312	165	4.968	51	4.968	66	4.968	27
		Skin $[45/-45/0/0/45/-45/0/90/0]_S$ (44%/44%/12%)										
		Stiffener $[45/-45/0_3/45/-45/0_3/90/0_2/\frac{90}{2}]_S$ (59%/30%/11%)										

(c) Hat-stiffened panel

Design	Level	Mass (kg)	λ_b	λ_s	Skin		Stiffener					
					t_{sk}	b_s	Web		Bottom flange		Top flange	
							t_w	b_w	t_{fl}	b_{fl}	t_{fu}	b_{fu}
H-a	Panel	1.53	1.69	1.00	3.312	165	3.312	58	3.312	30	4.074	30
H-b	Panel	-	1.00	1.00	3.373	200	8.119	78	3.373	30	4.263	6.9

7.5 Concluding Remarks

Experimental testing, as well as linear and non-linear FEA, was performed on two blade-stiffened panels, which were designed using VICONOPT with initial buckling in the stiffener and skin respectively. The 3-blade panel, with initial buckling in the stiffener, was less sensitive to imperfection than the 2-blade panel, which had initial buckling in the skin. Both panels buckled in the high load and high strain range, and had postbuckling capacity of about 7% and 13% above the initial buckling load for the 3-blade and 2-blade panel, respectively. VICONOPT and FEA predictions were in good agreement with experimental data for both panels. Hence the VICONOPT and FE models were validated by the experimental testing.

A bi-level optimisation strategy for fast design of composite stiffened panels, using VICONOPT and embracing practical composite design rules, has been presented and performed on the design of a Z-stiffened panel. The design was analysed using linear and non-linear FEA within ABAQUS. Detailed modelling of the root radius of the Z-stiffener was required otherwise the VICONOPT prediction is about 13% below the experimentally observed initial buckling load/strain. The bi-level optimization described did not account for this radius, and therefore did not find an accurate optimum design. Hence the efficiency of the Z-stiffened panel, in terms of initial buckling, is lower than that of the blade-stiffened panels (see Table 7.7), whereas its efficiency in terms of postbuckled failure compares favourably with the blade-stiffened panels. It should be noted that all three experimental panels are relatively short and narrow and not therefore representative of aerospace applications such as wing cover panels. It is expected that the efficiency of the Z-stiffened panel, with its increased overall bending rigidity, would be more favourable than blade-stiffened panels for such applications.

The bi-level optimization strategy has also been applied to the design of a wing cover panel. This strategy offers the advantage of considering practical composite design rules via manual laminate selection or an automated GA code. An Improved GA increases the likelihood of 0° ply selection and is shown to improve the chance of convergence to the optimum. The strategy produces weight savings of between 13% and 16% compared with an existing datum design when used to design a typical wing cover panel subject to combined in-plane loads and lateral pressure. This has demonstrated that the strategy is efficient, and reliable for practical optimisation of aircraft wing panels.

Table 7.7. Normalised values of structural efficiency (=load / panel weight) for the three experimental panels.

Panel	Initial buckling[a]	Failure[a]
3-blade	1.00	1.06
2-blade	1.08	1.23
Z	0.98	1.26

[a]The values of loads are those obtained from the experimental test.

Post-buckled strength has not been accounted for because of the relatively large skin thicknesses of these panels. However, some allowance for postbuckling could achieve further weight saving. The VICONOPT strip model has been extended into the post-buckled region,[25, 26] but this simplified analysis does not account for through thickness failure or the effects of damage. Future developments of the methodology will focus on such post-buckled design including damage modelling and multi-panel optimisation.

7.6 Acknowledgements

The authors would like to thank Dr. Martin Gaitonde and Dr. Nihong Yang (Airbus UK Ltd), and Dr. David Kennedy (Cardiff University) for their valuable assistance. This project is sponsored by Airbus UK Ltd and the IdMRC within the Department of Mechanical Engineering, University of Bath. VICONOPT is used with the permission of Cardiff University.

7.7 References

1. L. P. Kollár, *J. Struct. Eng.*, 1202 (2002).
2. C. Kassapoglou, *Compos. Part A-Appl. S.*, 419 (1997).
3. Y. K. Cheung, *Finite Strip Methods in Structural Analysis*. Oxford: Pergamon Press Ltd. 1976
4. T. R. Graves-Smith and S. A. Sridharan, *Int. Mech. Sci.*, 685 (1978).
5. D. J. Dawe, *Int. J. Solids Struct.*,1141-1155 (1977).
6. D. Bushnell, *Comput. Struct.*, 469 (1987).

7. D. E. Goldberg, Genetic Algorithms in Search, Optimisation and Machine Learning, Addison Wesley Longman, Inc. 1989.
8. R. Le Riche and R. T. Haftka, *AIAA J.*, 951 (1993).
9. S. Nagendra, R. T. Haftka and Z. Gürdal, *AIAA-93-1584-CP*, 1993.
10. S. Nagendra, D. Jestin, Z. Gürdal, R. T. Haftka and L. T. Watson, *Comput. Struct.*, 543 (1996).
11. F. W. Williams, M. S. Anderson, D. Kennedy, R. Butler and G. Aston, *NASA-CR-181966*, 1990.
12. W. H. Wittrick, *Aeronaut. Quart.*, 265(1968).
13. W. H. Wittrick and F. W. Williams, *Int. J. Mech. Sci.*, 209 (1974).
14. W. H. Wittrick and F. W. Williams, *J. Struct. Mech.*, 497 (1973).
15. M. S. Anderson, F. W. Williams and C. J. Wright, *Int. J. Mech. Sci.*, 585 (1983).
16. F. W. Williams and M. S. Anderson, *Aspects of the Analysis of Plate Structures - a Volume in Honour of W.H. Wittrick*, D. J. Dawe, R. W. Horstrington, A. G. Kamtekar and G. H. Little (Eds), Oxford University Press, Oxford, 51 (1985).
17. M. S. Anderson and D. Kennedy, 1992. *AIAA-92-22870-CP*, (1992)
18. G. L. Giles and M. S. Anderson, *NASA-TND-6784*, (1972).
19. R. Butler, A. A. Tyler and W. Cao, *Comput. Struct.*, 1107 (1994).
20. R. Butler and F. W. Williams, *Comput. Struct.*, 699 (1992).
21. G. N. Vanderplaats, *NASA-TM X-62*, (1973).
22. M. C. Y. Niu, Composite Airframe Structures: Practical Design Information and Data, Conmilit Press Ltd, Hong Kong, 440 (1992).
23. W. Liu, R. Butler, A. R. Mileham and A. J. Green, *AIAA J.,* 2562 (2006).
24. ABAQUS, Inc., *ABAQUS Theory and User's Manuals*, Version. 6.4, Pawtucket, RI, (2003).
25. M. Lillico, R. Butler, G. W. Hunt, A. Watson, D. Kennedy and F. W. Williams, *AIAA J.*, 996 (2002).
26. M. Lillico, R. Butler, G. W. Hunt, A. Watson and D. Kennedy, *AIAA J.*, 1172 (2003).

CHAPTER 8

STABILITY OF TUBES AND PIPELINES

Hayder A. Rasheed[1] and Spyros A. Karamanos[2]

[1]*Department of Civil Engineering, Kansas State University*
2118 Fiedler Hall, Manhattan, KS 66506, USA
E-mail: hayder@ksu.edu

[2]*Department of Mechanical & Industrial Engineering, University of Thessaly*
Pedio Areos, Volos 38834, Greece
E-mail: skara@mie.uth.gr

An overview of buckling and postbuckling behavior of pipes and tubes is presented, with particular emphasis on nonlinear finite element modeling. Classical bifurcation solutions for isotropic elastic cylinders under external pressure and bending are presented first. Subsequently, the stability of metal tubes and pipes under bending and external pressure is examined, and comparison with available experimental data is conducted. Finally, the behavior of composite cylinders under external pressure is presented, and the important issue of delamination is thoroughly discussed.

8.1 Introduction

The structural analysis of tubes and pipelines has attracted significant attention since the mid nineteenth century, because of its representative applications in all engineering disciplines, including aerospace, automotive, marine, offshore and civil sub and super structures. In those applications the tubes and pipes are subjected to various loading conditions that may cause buckling, and this constitutes a major issue towards safeguarding their structural integrity. Traditionally, the vast majority of tubes and pipes were made of metals. However, the past fifty years have witnessed a steady increase in the use of fiber-reinforced polymers to replace metals in tube and pipe construction. This chapter presents an overview of buckling and postbuckling behavior of pipes and tubular members, for both types of constructs (metal and composite) under various types of load conditions.

8.2 Stability of Elastic Isotropic Cylinders

Classical analytical solutions of stability problems of elastic isotropic cylinders under external pressure and bending are presented, for a better understanding of metal and composite cylinder response. It is worth noting that, recently, the stability of elastic cylinders has received significant attention due to the application of such cylinders in biomechanics, as well as in carbon nanotubes.

8.2.1 *Stability of Elastic Cylinders under Uniform External Pressure*

The problem of buckling of elastic isotropic rings and long tubes under external hydrostatic pressure constitutes a classical Structural Mechanics problem with significant engineering applications. Apparently, Bresse[1] was the first to calculate the buckling pressure of a ring under external pressure. Few years later, Bryan[2] calculated the buckling pressure for a long cylinder, whereas Timoshenko[3] examined the behaviour of pressurized elastic rings in the presence of initial ovality. Those solutions are presented in the classical textbooks of Flügge[4], Timoshenko and Gere[5] and Brush and Almroth[6]. In the following, an outline of the stability solution is presented. Furthermore, the postbuckling behaviour of a uniformly pressurized elastic cylinder is briefly discussed.

Kinematics

A long cylinder is considered, so that the response of every cross-section is identical and free of boundary effects. Under these conditions, each cross section can be treated as a ring with in-plane deformation only (plane strain conditions). Therefore, the solution of the long cylinder under uniform external pressure reduces to the solution of a two-dimensional ring problem, considering plane-strain conditions, and zero out-of-plane deformation. Fig. 8.1 shows the Cartesian and the polar (r, θ) coordinate systems used for the reference and deformed configurations. The displacements in the radial and tangential directions for an arbitrary circumferential fiber at a distance z from the mid surface, are u^* and v^*, respectively. The initial fiber length is dS, and the deformed fiber length is ds. The coordinates of a material point in the reference configuration X_2, X_3 and deformed configuration x_2, x_3 are

$$X_2 = r\cos\theta \;,\; x_2 = (r+u^*)\cos\theta - v^*\sin\theta$$
$$X_3 = r\sin\theta \;,\; x_3 = (r+u^*)\sin\theta + v^*\cos\theta . \tag{8.1}$$

The circumferential strain along the fiber is

$$\varepsilon_\theta^* = \frac{ds - dS}{dS} = \frac{ds}{dS} - 1 \tag{8.2}$$

Rearrangement of Eq. (8.2) and squaring leads to

$$\varepsilon_\theta^* \simeq \varepsilon_\theta^* + \frac{1}{2}\varepsilon_\theta^{*2} = \frac{1}{2}\left[\left(\frac{ds}{dS}\right)^2 - 1\right] = \frac{1}{2}\left[\left(\frac{1}{r}\frac{dx_2}{d\theta}\right)^2 + \left(\frac{1}{r}\frac{dx_3}{d\theta}\right)^2 - 1\right] \tag{8.3}$$

where

$$\frac{dx_2}{d\theta} = \left(\frac{du^*}{d\theta} - v^*\right)\cos\theta - \left(r + u^* + \frac{dv^*}{d\theta}\right)\sin\theta$$

$$\frac{dx_3}{d\theta} = \left(r + u^* + \frac{dv^*}{d\theta}\right)\cos\theta + \left(\frac{du^*}{d\theta} - v^*\right)\sin\theta . \tag{8.4}$$

Substituting Eq. (8.4) into Eq. (8.3) and assuming small hoop strain, one obtains the following expression for the hoop axial strain at any point through the cylinder thickness,

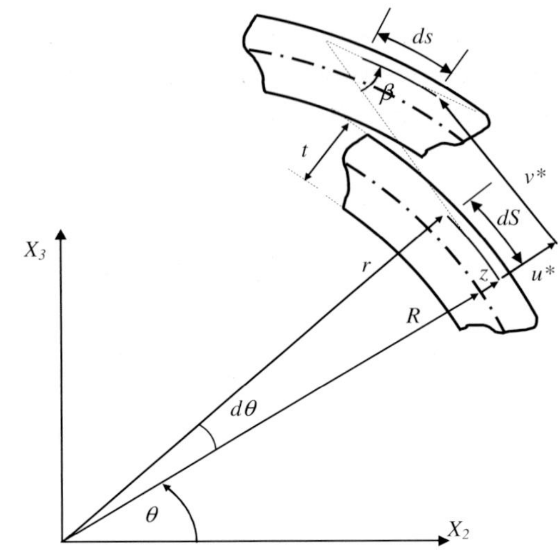

Fig. 8.1. In-plane deformation parameters of a ring.

$$\varepsilon_\theta^* = \left(\frac{v^{*\prime} + u^*}{r} \right) + \frac{1}{2} \left(\frac{v^* - u^{*\prime}}{r} \right)^2 , \tag{8.5}$$

where $()'$ denotes differentiation with respect to θ. Assuming no change in thickness of the ring during deformation, and that fiber rotation, β^*, is the same for the mid-surface as for any other arbitrary fiber aligned in the hoop direction θ, (i.e. plane sections remain plane after deformation and perpendicular to the deformed mid surface), one obtains,

$$u^* = u \ ; \qquad v^* = v + (r - R)\beta$$

$$\beta^* = \frac{v^* - u^{*\prime}}{r} = \beta = \frac{v - u'}{R} , \tag{8.6}$$

where u and v are the displacements of the mid-surface $(r = R)$. Considering thin rings only, the hoop strains are expressed as follows,

$$\varepsilon_\theta^* = \left(\frac{v' + u}{R} + \frac{1}{2} \beta^2 \right) + z \frac{\beta'}{R} = \varepsilon_\theta + z k_\theta , \tag{8.7}$$

where ε_θ and ε_θ^* are the hoop strains along the mid-surface and any other parallel surface, respectively, and k_θ is the circumferential curvature of mid-surface and $z = r - R$.

Constitutive Equations

Assuming isotropic elastic material, characterized by the Young's modulus E and Poisson's ratio, ν, plane strain conditions in the cross-sectional plane $\varepsilon_x^* = 0$, and zero stress in the radial direction $\sigma_r^* = 0$, it is readily obtained from Hooke's law that hoop stress, σ_θ^*, and hoop strain, ε_θ^*, at any material point are related through the following expression

$$\sigma_\theta^* = \frac{E}{1-\nu^2}\,\varepsilon_\theta^*. \tag{8.8}$$

Furthermore, $\sigma_x^* = \nu\sigma_\theta^*$ whereas the out of plane shear stress, $\tau_{x\theta}^*$, and shear strain, $\gamma_{x\theta}^*$, vanish due to the existence of in-plane deformations only. Finally, it is noted that for the case of an isolated ring, the corresponding constitutive equation that relates hoop stress, σ_θ^*, and hoop strain, ε_θ^*, is obtained by replacing $E/(1-\nu^2)$ with E so that $\sigma_\theta^* = E\,\varepsilon_\theta^*$.

Equilibrium Equations

Equilibrium is enforced through the principle of minimum potential energy, Π, which is the sum of the strain energy, U, and the potential of the external pressure, V, equal to the negative of the work W done by the pressure, $V = -W$. The strain energy stored in the system is a function of the hoop stresses and strains only, since either the longitudinal strains vanish in the case of a long cylinder or the longitudinal stresses vanish in the case of a ring. More specifically,

$$U = \frac{1}{2}\int_V \sigma_\theta^*\,\varepsilon_\theta^*\,dV = \frac{1}{2}\int_\theta \int_{Area} \sigma_\theta^*\varepsilon_\theta^*\,dA\; r d\theta. \tag{8.9}$$

Substituting Eq. (8.8) into (8.9), assuming that the ring is thin ($z/R \approx 0$) and conducting the integration the strain energy becomes

$$U = \frac{EAR}{2}\int_0^{2\pi} \varepsilon_\theta^2 \, d\theta + \frac{EIR}{2}\int_0^{2\pi} k_\theta^2 \, d\theta, \tag{8.10}$$

where A and I are the area and the moment of inertia of the cylinder wall. Eq. (8.10) is determined taking into account that $dA = b\,dz$ where b is the length of the pipe segment considered. In the following, a unit length of the long cylinder in the longitudinal direction is considered, so that $dA = dz$ and radial coordinate z is measured from the centroidal cross section axis. Therefore, $A = t/(1-\nu^2)$ and $I = t^3/12(1-\nu^2)$

The work of external fluid pressure loading for a cylinder is,

$$V = p\,\Delta \mathrm{A} = p\left(\mathrm{A}_0 - \mathrm{A}\right), \tag{8.11}$$

where p is the external pressure, $A_0 = \pi R^2$ is the initial cross sectional area of the outer surface, A is the corresponding area in the deformed configuration. Writing A in terms of the displacement components and conducting straightforward calculations[4] one results in the following equation for the potential of pressure.

$$W = \frac{-p}{2}\int_{\theta=0}^{2\pi} \left(2Ru + u^2 + v^2 + uv' - vu'\right) \, d\theta. \tag{8.12}$$

Taking the stationary value of the potential energy ($\delta \Pi = 0$), performing integration by parts, and considering arbitrary virtual displacements $\delta u, \delta v$, the nonlinear equilibrium equations can be obtained in the following form, in terms of the radial and tangential displacements $u(\theta)$ and $v(\theta)$,

$$\left[\left(\frac{v'+u}{R}\right) + \frac{1}{2}\left(\frac{v-u'}{R}\right)^2\right]' + \frac{I}{AR^2}\left(\frac{v-u'}{R}\right)''$$
$$-\left[\left(\frac{v'+u}{R}\right) + \frac{1}{2}\left(\frac{v-u'}{R}\right)^2\right]\left(\frac{v-u'}{R}\right) - \frac{pR}{EA}\left(\frac{v-u'}{R}\right) = 0, \tag{8.13}$$

$$\frac{I}{AR^2}\left(\frac{v-u'}{R}\right)''' - \left[\frac{v'+u}{R}+\frac{1}{2}\left(\frac{v-u'}{R}\right)^2\right]$$

$$-\left[\left(\frac{v'+u}{R}\right)+\frac{1}{2}\left(\frac{v-u'}{R}\right)^2\right]\left(\frac{v-u'}{R}\right)-\frac{pR}{EA}\frac{v'+u}{R}=\frac{pR}{EA}. \tag{8.14}$$

Prebuckling State and Bifurcation

The prebuckling deformation of an externally-pressurized ring is uniform contraction of the ring, where the radial displacement is constant and the tangential displacement is zero, so that $u=u_o,\ v=0,\ \beta=0$ (Fig. 8.2). From simple mechanics of materials, one obtains,

$$u_o=-\frac{pR^2}{EA}. \tag{8.15}$$

Perturbing the prebuckling solution $(u\rightarrow u_0+u_1, v\rightarrow v_1)$ and substitution into equations (8.13) and (8.14), and keeping only linear terms with respect to the perturbation, the linearized equations of equilibrium are obtained,

$$EAR^2\left(v_1'+u_1\right)'+EI\left(v_1+u_1'\right)''=0 \tag{8.16}$$

$$EAR^2\left(v_1'+u_1\right)-EI\left(v_1-v_1'\right)'''+pR^3\left(u_1''+u_1\right)=0. \tag{8.17}$$

Let $\alpha=\dfrac{I}{AR^2}$ and $\gamma=\dfrac{pR}{EA}$. The solution should be periodic and therefore, it is sought in the following form,

$$v_1=c_1\sin n\theta, \qquad n=2,3,\ldots.. \tag{8.18}$$

$$u_1=c_2\cos n\theta, \qquad n=2,3,\ldots.. \tag{8.19}$$

Substituting equations (8.18)-(8.19) into equations (8.16)-(8.17), one readily obtains the following homogeneous algebraic problem

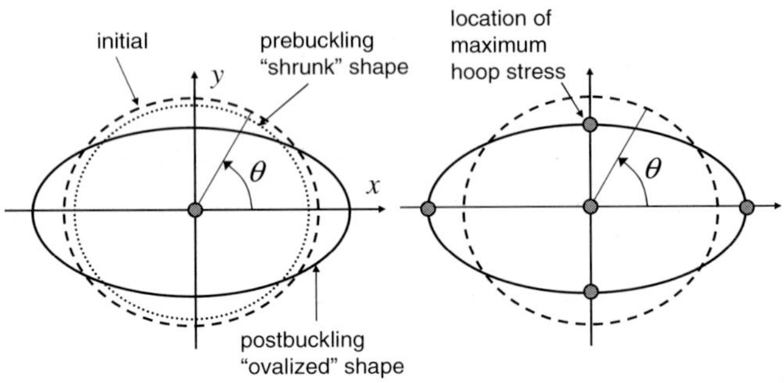

Fig. 8.2. Ovalized cross-sectional shape of a buckled cylinder under external pressure.

$$\begin{bmatrix} n^2(1+\alpha) & n(1+n^2\alpha) \\ n(1+n^2\alpha) & 1+n^4\alpha+\ (1-n^2)\gamma \end{bmatrix} \begin{Bmatrix} c_1 \\ c_2 \end{Bmatrix} = \begin{Bmatrix} 0 \\ 0 \end{Bmatrix}. \tag{8.20}$$

Solution of the above system provides the eigenvalues γ (or p) and the corresponding eigenfunctions,

$$p = p_n = \frac{n^2-1}{1+a}\frac{EI}{R^3}, \tag{8.21}$$

$$c_2 = -\frac{(1+\alpha)n}{1+n^2\alpha}\ c_1, \tag{8.22}$$

where $n \geq 2$. Equation (8.21) for $n = 2$ provides the critical buckling pressure of thin rings, which, for the common case where $\alpha \ll 1$, may simply be expressed as follows,

$$p_{cr} = \frac{3EI}{R^3} = \frac{2E}{1-\nu^2}\left(\frac{t}{D}\right)^3. \tag{8.23}$$

Furthermore, if $\alpha \ll 1$, then $c_2 = -2c_1$, so that the buckling mode becomes

$$u_1 = A\cos 2\theta, \tag{8.24}$$

$$v_1 = -\frac{A}{2}\sin 2\theta, \tag{8.25}$$

where A is an arbitrary constant. The shape of the buckled configuration is shown graphically in Fig. 8.2 and represents an ovalized shape.

Response in the Presence of Initial Imperfections

Suppose that the pipe has initial imperfection of the "inextensional" form,

$$\bar{u}(\theta) = \bar{\alpha}\cos 2\theta, \tag{8.26}$$

$$\bar{v}(\theta) = -\frac{\bar{\alpha}}{2}\sin 2\theta, \tag{8.27}$$

where $\bar{\alpha}$ is an arbitrary constant expressing the imperfection amplitude. It is underlined that this initial deformation corresponds to a stress-free state. Then, it is possible to repeat the above formulation, to obtain the following solution for the additional radial and tangential displacements,

$$u(\theta) = \frac{1}{1 - p/p_{cr}}\bar{\alpha}\cos 2\theta = \frac{1}{1 - p/p_{cr}}\bar{u}(\theta), \tag{8.28}$$

$$v(\theta) = -\frac{1}{2}\left(\frac{1+4\alpha}{1+\alpha}\right)\frac{1}{1 - p/p_{cr}}\bar{\alpha}\sin 2\theta \simeq \frac{1}{1 - p/p_{cr}}\bar{v}(\theta). \tag{8.29}$$

In such a case, the corresponding bending moment around the cross-section is,

$$M(\theta) = \frac{EI}{R^2}(v' - u'')'' = -\frac{3EI}{R^2}\left(\frac{\bar{\alpha}}{1 - p/p_{cr}}\right)\cos 2\theta, \tag{8.30}$$

and its maximum value is located at the four locations $\theta = 0,\ \pi/2,\ \pi,\ 3\pi/2$ (Fig. 8.2),

$$M_{\max} = \frac{pR}{\left(1 - p/p_{cr}\right)}\bar{\alpha}\cos 2\theta. \tag{8.31}$$

Finally, the maximum hoop stress at those four locations is given by the following expression,

$$\sigma_{\theta,\max} = -\frac{pR}{t} + \frac{M_{\max}}{t^2/6} = -\frac{pR}{t} + \frac{pR\bar{\alpha}}{\left(1 - p/p_{cr}\right)} \frac{1}{t^2/6}. \tag{8.32}$$

Postbuckling Behavior

The postbuckling behavior requires a more refined stability theory. Budiansky[7] presented an asymptotic solution, based on the stationary value of the potential energy of a ring under uniform external pressure. It was found that the initial postbuckling behavior is stable, in the sense that the corresponding initial postbuckling path indicates an increase of pressure. Semi-analytical and numerical results have verified Budiansky's asymptotic solution[8].

8.2.2 *Stability of Pressurized Long Elastic Cylinders under Bending*

In numerous practical applications, long cylinders (e.g. pipes and tubes) are subjected to bending. In such a case, together with longitudinal (beam-type) deformation of the cylinder, the response is characterized by a distortion of the cross-section in the form of an oval. This phenomenon, first investigated analytically by Brazier[9], results in a significant reduction of the bending moment capacity and a limit point instability, referred to as "Brazier instability" or "ovalization instability" (Fig. 8.3). Furthermore, the increased compressive stress at the compression side of the cylinder may cause wrinkling of the cylinder wall in the form of a wavy pattern, a bifurcation type of instability. This bifurcation usually occurs before a limit point instability is reached on the ovalization path. Those phenomena, shown schematically in Fig. 8.3, are described briefly in the following.

Ovalization Behavior

A simple and efficient method to model Brazier's ovalizaton phenomenon can be developed. Initially proposed by Wood[10], it is based on a simplified variation formulation that considers both longitudinal and hoop deformation energy. This methodology has been presented in a more general form by Houliara and Karamanos[11], accounting for initial curvature and initial ovality effects. The potential energy is,

$$\Pi = U_L + U_C + V_p - W_p - Mk\,, \tag{8.33}$$

where, U_L is the longitudinal bending strain energy, U_C is the cross-sectional (hoop) deformation strain energy, V_p is the potential due to the applied pressure, W_p is the second-order work of hoop pressure stress $\sigma_p = pR/t$, M is the bending moment and k is the longitudinal applied curvature ($k = 1/R'$ where R' is the radius of curvature in the longitudinal direction). From beam theory, the longitudinal part, U_L, is expressed in terms of strain ε_x in the longitudinal direction as follows,

$$U_L = \frac{EtR}{2}\int_0^{2\pi} \varepsilon_x^2\, d\theta\,. \tag{8.34}$$

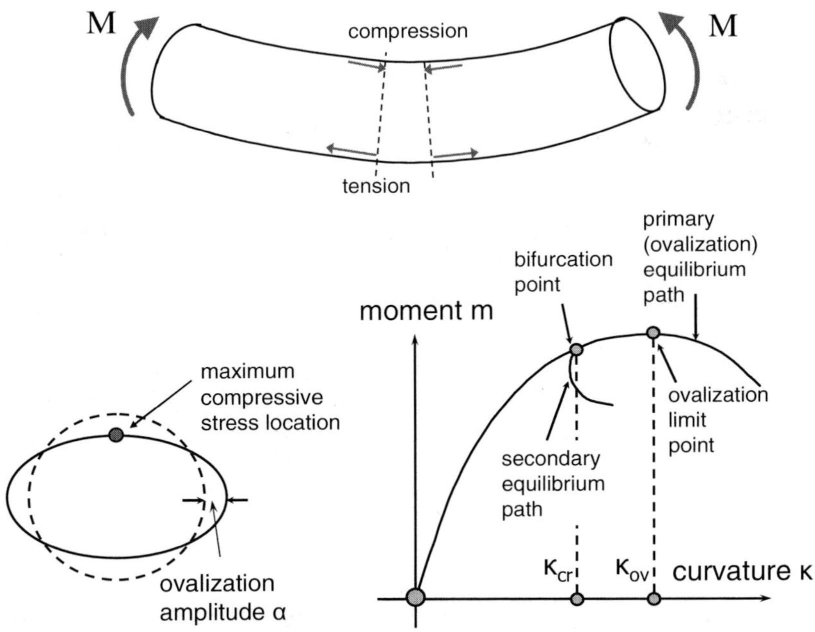

Fig. 8.3. Ovalization of long cylinders under bending.

The strain, ε_x, in the direction of the cylinder's axis is,

$$\varepsilon_x = ky = k[(R+u)\sin\theta + v\cos\theta], \tag{8.35}$$

where R and t are the radius and the thickness of the cylinder's cross-section respectively, y is the distance from the neutral axis, u and v are the radial and tangential displacements respectively.

The hoop strain energy, U_C, is based on ring-theory. From equation (8.9), and for an inextensional displacement field,

$$U_C = \frac{1}{2}\int\int_{A_\theta} \sigma_\theta \varepsilon_\theta R\, d\theta dz = \frac{Et^3}{24(1-v^2)R^3}\int_0^{2\pi} (v'+v''')d\theta. \tag{8.36}$$

A ring subjected to uniform pressure, which remains normal to the deformed surface, is a conservative system. Therefore the potential due to pressure is equal to the product of pressure times the change of the area enclosed by the ring. This can be expressed as follows,

$$V_p = p\Delta A = p\frac{1}{2}\int_0^{2\pi} (2Ru + v^2 - vu' + v'u + u^2)d\theta. \tag{8.37}$$

Finally, the second-order work of hoop pressure stress σ_p is,

$$W_p = \iint_{A_\theta} \sigma_p \bar{\varepsilon}_\theta R\, d\theta dz = \frac{1}{2}p\int_0^{2\pi} (v-u')^2 d\theta, \tag{8.38}$$

where $\sigma_p = pR/t$, and $\bar{\varepsilon}_\theta$ is the second-order (nonlinear part) hoop normal strain defined by:

$$\bar{\varepsilon}_\theta = \frac{1}{2}\left(\frac{v-u'}{R}\right)^2. \tag{8.39}$$

A Ritz discretization, using one trigonometric doubly symmetric term, is assumed for displacements $u(\theta)$ and $v(\theta)$,

$$\begin{aligned} u(\theta) &= \alpha\cos 2\theta \\ v(\theta) &= -\frac{\alpha}{2}\sin 2\theta \end{aligned}, \tag{8.40}$$

where α is the ovalization amplitude. Note that equations (8.40) satisfy the condition of inextentionality. Inserting all the above equations in the expression of potential energy (8.33) and neglecting quadratic terms in the expression of U_L (an assumption also employed by Brazier[9]), one results in the following expression,

$$\Pi(\alpha,k)=\frac{Ek^2R^2t\pi}{2}\left(R-\frac{3}{2}\alpha\right)+\frac{3}{8}\frac{\pi Et^3\alpha^2}{(1-\nu^2)R^3}-\frac{3}{2}\alpha^2 p\pi - Mk\,. \qquad (8.41)$$

The ovalization parameter α can be expressed in terms of the curvature k directly from the minimization of Π in terms of α ($\partial\Pi/\partial\alpha=0$),

$$\zeta=\frac{\alpha}{R}=\frac{\kappa^2}{1-f}, \qquad (8.42)$$

where $\kappa = k/k_N$, $f = p/p_{cr}$, $k_N = t\Big/\left(R^2\sqrt{1-\nu^2}\right)$, and $p_{cr} = Et^3/4R^3(1-\nu^2)$ Subsequently, inserting equation (8.42) into (8.41) and enforcing equilibrium ($\partial\Pi/\partial k=0$), the following expression is obtained for the bending moment,

$$m=\kappa\pi\left(1-\frac{3\kappa^2}{2(1-f)}\right), \qquad (8.43)$$

where $m = M/M_e$ and $M_e = ERt^2\Big/\sqrt{1-\nu^2}$ It is also possible to obtain closed-form expressions for the longitudinal stress, σ_x. Inserting the value of ovalization amplitude, α, in the expressions of displacements $u(\theta)$ and $v(\theta)$, and using (8.35), one obtains a closed form expression for the longitudinal stress $\sigma_x = E\,\varepsilon_x$, as follows,

$$\sigma_x=\sigma_e\kappa\left[\left(1-\frac{3\kappa^2}{4(1-f)}\right)\sin\theta+\frac{\kappa^2}{4(1-f)}\sin 3\theta\right], \qquad (8.44)$$

where $\sigma_e = \mathrm{E}t\Big/(r\sqrt{1-\nu^2})$ is a "reference" stress used for normalization purposes.

Furthermore, analytical expressions for the radius of curvature at the deformed cross-section can be obtained:

$$\frac{1}{r_\theta(\theta)}=\frac{1}{R}+k_\theta=\frac{1}{R}\frac{v'(\theta)-u''(\theta)}{R^2}=\frac{1}{R}+\frac{3\kappa^2\cos 2\theta}{R(1-f)}. \qquad (8.45)$$

Note that the discretized potential energy expression obtained in Eq. (8.41) is compatible with the formulation of Section 8.2.1. More specifically, setting the longitudinal curvature equal to zero, minimization in terms of α yields the buckling pressure of Eq. (8.23).

Bifurcation Instability and Postbuckling Behavior

It is possible to obtain a good estimate for the buckling moment of a bent cylinder, using a simple hypothesis[12]. Buckling would occur at the location where longitudinal stress, σ_x, becomes equal to the buckling stress of a uniformly compressed circular cylinder, with radius equal to the current hoop radius, r_θ, at the critical location,

$$\sigma_x = \frac{E}{\sqrt{3(1-\nu^2)}}\left(\frac{t}{r_\theta}\right). \tag{8.46}$$

In the present case, the critical location is at $\theta = \pi/2$. Using the analytical expressions for the hoop curvature and the longitudinal stress (8.44) and (8.45) at $\theta = \pi/2$, and inserting into the buckling condition (8.46), one results in a third-degree algebraic equation in terms of the buckling curvature κ_{cr},

$$\kappa_{cr}\left(1-\frac{1}{1-f}\kappa_{cr}^2\right)-\frac{1}{\sqrt{3}}\left(1-\frac{3}{1-f}\kappa_{cr}^2\right)=0, \tag{8.47}$$

which can be solved analytically and the critical curvature is:

$$\kappa_{cr} = \frac{1}{\sqrt{3}} - 2\sqrt{\text{abs}\left[\frac{1}{3}(f-2)\right]}\cos\left[\frac{\pi}{3}+\frac{1}{3}\arccos\left(\frac{1}{\text{abs}[f-2]^{3/2}}\right)\right], \tag{8.48}$$

Note that for zero pressure, the solution of equation (8.48) is $\kappa_{cr} = 0.3811$. The corresponding critical moment, m_{cr}, is obtained from equation (8.43) setting $\kappa = \kappa_{cr}$. The location of bifurcation on the pre-buckling path, expressed by equation (8.43) is shown in Fig. 8.4 with an arrow $(\downarrow)$ for a thin-walled elastic cylinder ($R/t = 120$, $\nu = 0.3$). The bifurcation prediction on the ovalization path, using a nonlinear finite element technique[11], is also shown in Fig. 8.4 with an arrow $(\uparrow)$, indicating the above analytical methodology provides good estimates of ovalization and buckling. The buckling shape is also depicted in Fig. 8.4.

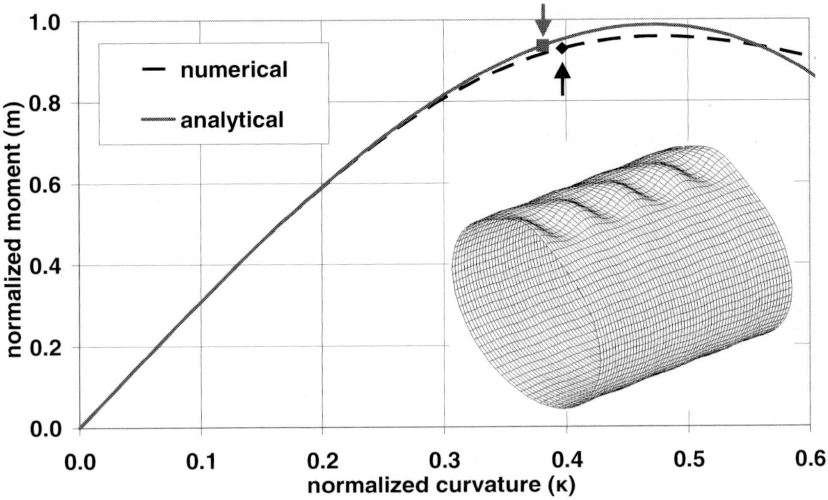

Fig. 8.4. Ovalization and uniform wrinkling of an elastic cylinder under bending loading under zero pressure; numerical results from Houliara & Karamanos[11].

Regarding the nature of postbuckling behavior of elastic cylinders, an asymptotic solution[13] demonstrated that the initial postbuckling path is unstable and symmetric, and it can be approximated by the following asymptotic expression

$$m = m_{cr}(1 - b\xi^2 + ...) \tag{8.49}$$

where b is a positive constant and ξ is a measure of post-buckling deformation. This has been verified recently by more rigorous finite element analysis[11,14].

8.3 Stability of Metal Tubes and Pipelines

The majority of applications are related to pipelines and tubular structures employ metal tubes and pipes. In those applications, the cylindrical members are subjected to a combination of structural loads. In some applications, external pressure is also present, causing additional instability problems and reducing the structural capacity of the member. In the following a numerical finite element formulation is outlined, introduced by Needleman[15], which has been used for the solution of such problems[16,17,18].

8.3.1 *Numerical Finite Element Technique*

A Lagrangian formulation of the tube is adopted, describing the tube as an isotropic continuum using convected coordinates ξ^1, ξ^2, ξ^3. Hypo-elastic constitutive equations are considered[19], which relate the Jaumann rate of Kirchhoff stress to the rate-of-deformation (stretch) tensor,

$$\overset{\nabla}{\boldsymbol{\tau}} = \mathcal{D}^{\text{ep}} \mathbf{d} . \tag{8.50}$$

$\mathcal{D}^{\text{ep}}$ is the fourth-order instantaneous material rigidity tensor, which includes inelastic effects. In terms of components, this equation can be written as follows:

$$\dot{\tau}^{ij} = R^{ijkl} \dot{E}_{kl} , \tag{8.51}$$

where

$$R^{ijkl} = \left(\mathcal{D}^{ep} \right)^{ijkl} - \mathcal{L}^{ijkl} \tag{8.52}$$

$\left(\mathcal{D}^{ep} \right)^{ijkl}$ are the components of $\mathcal{D}^{\text{ep}}$ with respect to the current base vectors, $\dot{\tau}^{ij}$ express the rate of components of Kirchhoff stress tensor with respect to the current base covariant vectors, $\dot{E}_{kl}$ denote the rate of the components of the Green-Lagrange tensor with respect to the reference contravariant vectors and

$$\mathcal{L}^{ijkl} = \frac{1}{2} [g^{ik} \tau^{jl} + g^{jk} \tau^{il} + g^{il} \tau^{jk} + g^{jl} \tau^{ik}] , \tag{8.53}$$

where g^{ij} are the contravariant metric coefficients. In the case of elastic behavior, the material rigidity tensor $\mathcal{D}^{\text{ep}}$ reduces to the elastic rigidity tensor $\mathcal{D}$ with components:

$$\mathcal{D}^{ijkl} = \frac{E}{2(1+\nu)} \left(g^{jl} g^{ik} + g^{il} g^{jk} + \frac{2\nu}{1-2\nu} g^{ij} g^{kl} \right) \tag{8.54}$$

Following classical shell theory[6] the traction component normal to any shell lamina is imposed to be zero and tube thickness is assumed constant. Equilibrium is considered through the principle of virtual work,

$$\int_{V_0} \delta u_{i|j} \tau^{ij} dV_0 = \int_B \delta \mathbf{u} \cdot (-p\mathbf{n}) dB + F\,\delta u + M\,\delta\theta\,, \qquad (8.55)$$

where p is the external pressure (always normal to the deformed shape of the tubular shell wall), $\mathbf{n}$ is the unit normal vector to the outer surface at the current configuration, $\delta u_{i|j}$ denotes the covariant derivative of the components of an arbitrary virtual displacement vector $\delta\mathbf{u}$ with respect to the current base vectors, F and M are the axial force and bending moment respectively, whereas δu and $\delta\theta$ denote the variations of end axial displacement and rotation respectively.

The nonlinear governing equations are solved through an incremental Newton-Raphson numerical procedure. Linearization of equation (8.55) results in,

$$\int_{V_o} \delta U_{i|j} S^{ijkl} \Delta U_{k|l} dV_o = \int_B \delta \mathbf{u} \cdot (-p'\mathbf{n}) dB - \int_{V_o} \delta U_{i|j} (\mathbf{G}^i \cdot \mathbf{g}_k) \tau^{kj} dV_o\,, \quad (8.56)$$

where $\mathbf{g}_k$ are the covariant current base vectors, p' is the pressure at the nearby configuration, $\mathbf{G}^i$ are the contravariant reference base vectors $\Delta U_{k|l}$ and $\delta U_{i|j}$ are the covariant derivatives of incremental displacements $\Delta\mathbf{u}$ and virtual displacements $\delta\mathbf{u}$ respectively with respect to the reference base vectors,

$$S^{ijpq} = (\mathbf{G}^i \cdot \mathbf{g}_k) R^{kjmq} (\mathbf{g}_m \cdot \mathbf{G}^p) + \tau^{jq} G^{ip}\,, \qquad (8.57)$$

are the generalized material moduli and $\tau^{jq} G^{ip}$ is the so-called "geometric rigidity". Galerkin discretization is considered,

$$\Delta\mathbf{u} = \mathbf{N}\,\Delta\hat{\mathbf{U}}\,, \qquad (8.58)$$

$$\delta \mathrm{u} = \mathbf{N}\,\delta\hat{\mathrm{U}}\,, \qquad (8.59)$$

$$\{\Delta U_{k/l}\} = \mathbf{B}\,\Delta\hat{\mathbf{U}}\,, \qquad (8.60)$$

$$\{\delta U_{k/l}\} = \mathbf{B}\,\delta\hat{\mathbf{U}}\,, \qquad (8.61)$$

in which $\mathbf{N}$ is the interpolation matrix that contains shape functions, $\mathbf{B}$ is the matrix of the derivatives of the shape functions, $\Delta\hat{\mathbf{U}}$ contains the increments of nodal degrees of freedom and $\delta\hat{\mathbf{U}}$ are arbitrary. This results in the following discretized incremental problem,

$$[\mathbf{K}]\Delta\hat{\mathbf{U}}=\mathbf{F}_{ext}-\mathbf{F}_{int}, \tag{8.62}$$

where $[\mathbf{K}]$ is the incremental stiffness matrix,

$$[\mathbf{K}]=\int_{V_0}\mathbf{B}^{T}\mathbf{S}\mathbf{B}dV_0 . \tag{8.63}$$

$\mathbf{F}_{ext}$, $\mathbf{F}_{int}$ are the external and internal load vectors respectively.

$$\mathbf{F}_{ext}=\int_{B_q}\mathbf{N}^{T}(-p'\mathbf{n})dB_q \tag{8.64}$$

$$\mathbf{F}_{int}=\int_{V_0}\mathbf{B}^{T}\mathbf{W}^{T}\boldsymbol{\tau}dV_0 . \tag{8.65}$$

$\mathbf{W}$ is a matrix that contains the mixed components of the deformation gradient $\mathbf{G}^{i}\cdot\mathbf{g}_k$ and $\mathbf{S}$ is the matrix corresponding to S^{ijpq} in Eq. (8.57). The incremental procedure is enhanced to enable the tracing of post-buckling "snap-through" or "snap-back" equilibrium paths through an arc-length continuation algorithm. To enable the incremental analysis to follow the post-buckling path, a very small initial imperfection of the tube is imposed. The initial imperfection is considered in the form of the buckling mode, obtained by an eigenvalue analysis just prior to bifurcation, and it is very small yet sufficient to improve convergence near the buckling point and to "trigger" bifurcation.

Shell Element Discretization

A nine-node isoparametric shell element, originally proposed for linear analysis by Ahmad et al.[20], is used. Each node has an associated triplet of orthonormal vectors, $\mathbf{q}_1$, $\mathbf{q}_2$, $\mathbf{q}_3$, (Fig. 8.5). Vector $\mathbf{q}_3$ is oriented through the thickness of the element and

$$\mathbf{q}_3=\mathbf{q}_1\times\mathbf{q}_2 . \tag{8.66}$$

The position of a node (k) in the current configuration is denoted $\mathbf{x}^{(k)}$ and the material coordinates (ξ^1,ξ^2,ξ^3) are chosen so that the element is mapped onto the parallelepiped $-1\le\xi^1\le1$, $-1\le\xi^2\le1$, $0\le\xi^3\le1$. Using quadratic interpolation with respect to ξ^1 and ξ^2, and linear interpolation with respect to ξ^3, the geometry of the element in the current configuration can be expressed as,

$$\mathbf{x}(\xi^1,\xi^2,\xi^3)=\sum_{\kappa=1}^{9}[(\mathbf{x}^{(k)}+t\xi^3\mathbf{q}_3^{(k)})N^{(k)}(\xi^1,\xi^2)], \qquad (8.67)$$

where t denotes the shell thickness and $N^{(k)}(\xi^1,\xi^2)$, $k=1,2,...,9$ are the nodal Lagrangian shape functions. The surface on which the nodes of the element are located is defined by $\xi^3=0$ ($\xi^3=1$ is the opposite surface). There are five nodal degrees of freedom at each node (k): the three components (with respect to a fixed rectangular Cartesian coordinate system) of the displacement increment ($\Delta u_1^{(k)}$, $\Delta u_2^{(k)}$, $\Delta u_3^{(k)}$) and two components of the rotation increment ($\Delta\theta_1^{(k)}$, $\Delta\theta_2^{(k)}$) of $\mathbf{q}_3$, with respect to $\mathbf{q}_1$, $\mathbf{q}_2$, so that the nine-node element has a total of 45 degrees of freedom. The details of the formulation for this element for linear analysis can be found in the work of Ahmad et al.[20]. In Katsounas & Tassoulas[16] and Karamanos & Tassoulas[18] its implementation in the framework of a nonlinear analysis is presented in detail.

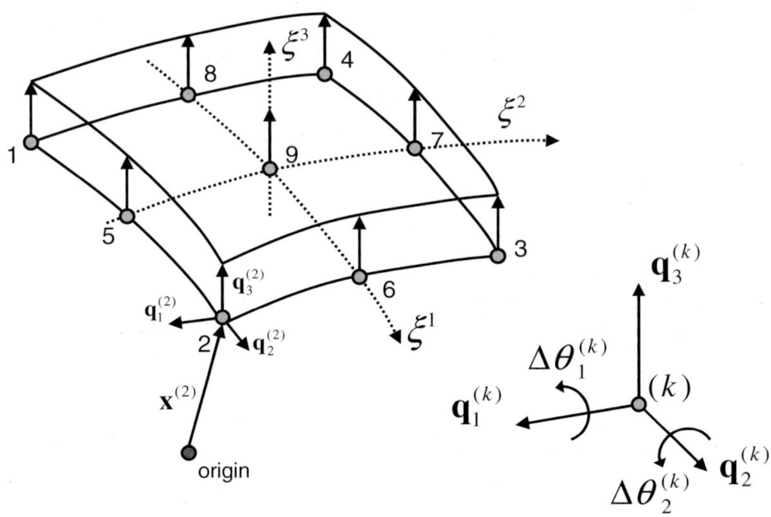

Fig. 8.5. Nine-node shell element; rotational nodal degrees-of-freedom.

Tube Element Discretization

As alternative to the shell element, a three-node special-purpose "tube element"[21,22,23] has been developed, which combines longitudinal (beam-type) with cross-sectional deformation (Fig. 8.6). Convected coordinates are θ, ζ, z initially along the hoop, longitudinal and radial direction respectively. Nodes are located on the tube axis, which lies on the plane of bending, and each node possesses three degrees-of-freedom (two translational and one rotational).

A reference line is chosen within the cross-section at node (k) and a local Cartesian coordinate system is defined, so that the $\overline{x}, \overline{y}$ axes define the cross-sectional plane. The orientation of node (k) is defined by the position of three orthonormal vectors $\mathbf{e}_x^{(k)}$ $\mathbf{e}_y^{(k)}$ and $\mathbf{e}_z^{(k)}$. For in-plane (ovalization) deformation, fibers initially normal to the reference line remain normal to the reference line. Furthermore, those fibers may rotate in the out-of-plane direction by angle $\gamma(\theta)$. Using quadratic interpolation in the longitudinal direction, the position vector $\mathbf{x}(\theta,\zeta,z)$ of an arbitrary point at the deformed configuration is,

$$\mathbf{x}(\theta,\zeta,z) = \sum_{k=1}^{3}\left[\left(\mathbf{x}^{(k)} + \mathbf{r}^{(k)}(\theta) + z\mathbf{n}^{(k)}(\theta) + z\gamma(\theta)\mathbf{e}_z^{(k)}\right)N^{(k)}(\zeta)\right], \quad (8.68)$$

where $\mathbf{x}^{(k)}$ is the position vector of node (k), $\mathbf{r}^{(k)}(\theta)$ is the position of the reference line at a certain cross-section relative to the corresponding node (k), $\mathbf{n}^{(k)}(\theta)$ is the "in-plane" outward normal of the reference line at the deformed configuration and $N^{(k)}(\zeta)$ is the corresponding Lagrangian quadratic polynomial. Using nonlinear ring theory, vector functions $\mathbf{r}^{(k)}(\theta)$ and $\mathbf{n}^{(k)}(\theta)$, can be expressed in terms of the radial, tangential and out-of-plane displacements of the reference line, denoted as $u(\theta)$, $v(\theta)$, $w(\theta)$ respectively.

Functions $u(\theta)$, $v(\theta)$, $w(\theta)$ and the out-plane rotation $\gamma(\theta)$ of the through-thickness fibers are discretized as follows:

$$u(\theta) = a_0 + a_1 \sin\theta + \sum_{n=2,4,6,...} a_n \cos n\theta + \sum_{n=3,5,7,....} a_n \sin n\theta, \quad (8.69)$$

$$v(\theta) = -a_1 \cos\theta + \sum_{n=2,4,6,...} b_n \sin n\theta + \sum_{n=3,5,7,....} b_n \cos n\theta, \quad (8.70)$$

$$w(\theta)=\sum_{n=2,4,6,...} c_n \cos n\theta+\sum_{n=3,5,7,....} c_n \sin n\theta, \tag{8.71}$$

$$\gamma(\theta)=\sum_{n=0,2,4,6,...} \gamma_n \cos n\theta+\sum_{n=1,3,5,7,....} \gamma_n \sin n\theta. \tag{8.72}$$

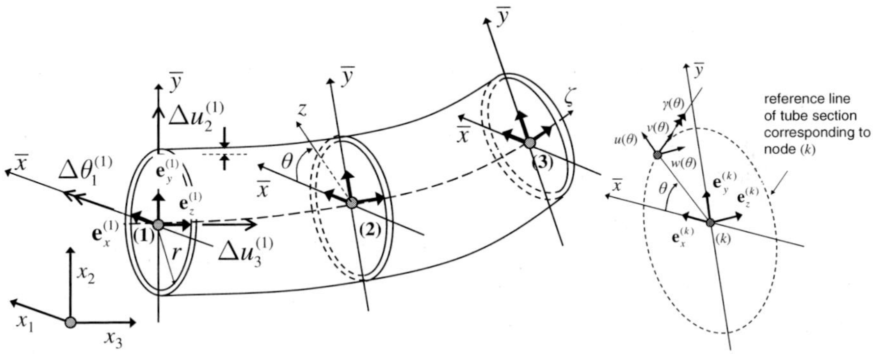

Fig. 8.6. Three-node "tube element"; deformation parameters.

Coefficients a_n, b_n refer to in-plane cross-sectional deformation, and express the ovalization of the cross-section, whereas c_n, γ_n refer to out-of-plane ("warping") cross-sectional deformation. More details on the numerical implementation of this element in the framework of an elastoplastic finite element analysis are described in Karamanos & Tassoulas[21,22,23].

8.3.2 *Buckling of Inelastic Cylinders under External Pressure*

The buckling behavior of elastic cylinders, presented in the previous section, can be used for an initial understanding of metal cylinder buckling. However, the applicability of elastic buckling theory and the corresponding formula (8.23) is limited by the assumption of elastic behavior. Considering that the hoop stress corresponding to the buckling formula, one readily obtains that the elastic buckling theory is valid for,

$$\frac{D}{t}\geq\sqrt{\frac{E}{\sigma'_y(1-\nu^2)}}, \tag{8.73}$$

where

$$\sigma_y' = \sigma_y \Big/ \sqrt{1-\nu+\nu^2}\,, \tag{8.74}$$

is the yield stress under plane-strain conditions, according to von Mises yield criterion This means that for $\sigma_y = 300$ MPa, the D/t ratio should be greater than 27. However, pipes used in numerous applications are quite thicker and buckle in the plastic range. In lieu of a detailed buckling analysis is the plastic range, their buckling pressure may be approximated by the yield pressure formula,

$$p_y = 2\sigma_y \frac{t}{D}, \tag{8.75}$$

so that, in the absence of initial imperfections and residual stresses, the ultimate pressure of a pipe is the minimum pressure calculated from Equations (8.23) and (8.75).

Inelastic behavior has dramatic effect on the post-buckling response of metal cylinders. Consider a relatively thin-walled cylinder that satisfies the criterion (8.73). This means that under external pressure it buckles elastically, at the pressure specified by formula (8.23). The initial post-buckling behavior is stable, with an increasing pressure. However, the ovalized post-buckling shape results in the development of maximum stresses at locations $\theta = 0,\ \pi/2,\ \pi,\ 3\pi/2$, as expressed in equation (8.32). This means that the material yields at those four locations, thus forming a plastic mechanism which is incapable of sustaining further increase of pressure. Under this situation the pipe collapses.

It is possible to approximate the post-buckling behavior of pipes using a four-plastic-hinge kinematic model (Fig. 8.7a.), and assuming rigid-plastic material (no hardening). Equilibrium, shown in Fig. 8.7b, requires that,

$$F_p R\left(\sqrt{2}/2\right) = 2M_p + H\alpha\,, \tag{8.76}$$

$$H = F_p \sin\phi\,, \tag{8.77}$$

where M_p is the plastic moment of the pipe wall, and $F_p = pR\sqrt{2}$. Combining the above two equations and rearranging, the following equation of pressure results,

$$\frac{p}{p_{cr}} = \lambda \frac{1}{2x - x^2}, \tag{8.78}$$

where $x = w(\pi/2)/R$ is the normalized change of length of the vertical pipe diameter and $\lambda = (1-\nu^2)(\sigma_y/E)(D/t)$ is a material/geometric parameter of the pipe. Equation (8.78) shows that the post-buckling behaviour is characterized by a significant drop of the pressure capacity. This means that the post-buckling response is very unstable, and very sensitive to initial imperfections.

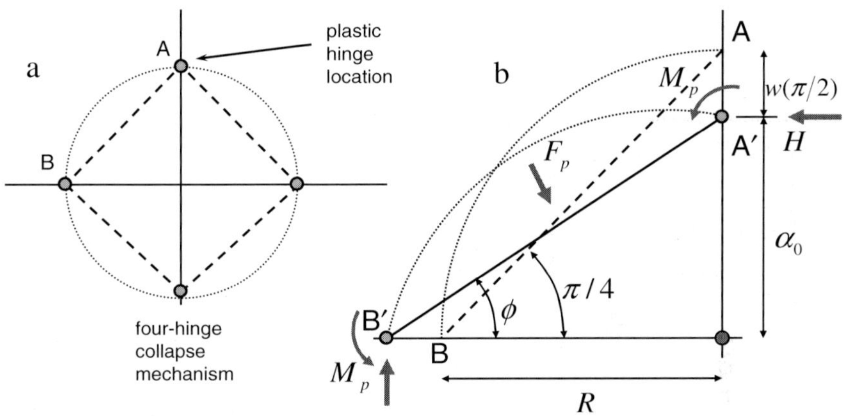

Fig. 8.7. Four-hinge plastic mechanism, simulating the post-buckling inelastic behavior of externally-pressurized cylinders.

Fig. 8.8 shows the corresponding curve for a typical steel tube (σ_y =483 MPa, no hardening and D/t =48.8), analyzed with the "tube element" technique described above. The numerical results in the post-buckling range compare very well with the above simplified formula. Note that an initial ovality equal to 1% (a typical value in pipeline applications), the reduction of the ultimate pressure with respect to the one specified by Equation (8.23) is about 20%.

8.3.3 *Stability of Inelastic Cylinders under Bending and Pressure*

The structural integrity of tubular structures and pipelines is often associated with the stability of metal cylinders under severe structural loads and pressure. The ultimate capacity of offshore pipelines during the installation procedure is a very good example of such behavior. In that case, the combination of bending and external pressure may cause instability and buckling. The basic feature of pipeline response under such loading conditions is cross-sectional ovalization, followed by buckling in the compression side of the cylinder wall. This phenomenon was described in section 8.2.2 for elastic cylinders. However, in the present case, pipes are rather thick and exhibit significant inelastic behavior before reaching an ultimate state.

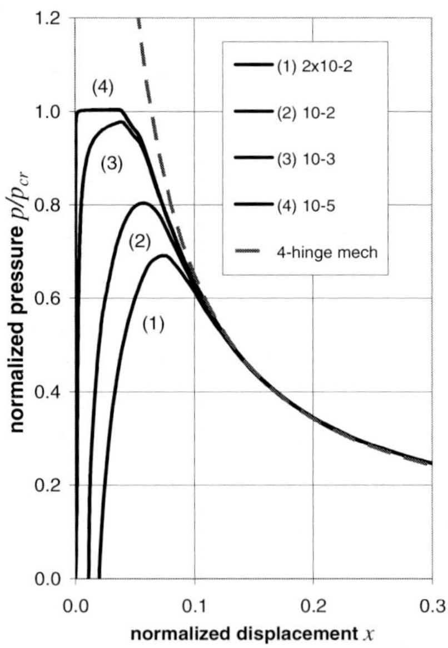

Fig. 8.8. Post-buckling response of an inelastic cylinder under uniform external pressure (D/t = 48.8, σ_y = 483 MPa) for different values of initial ovalization amplitude.

In particular, the response of thick-walled pipes, used in deep-offshore pipeline applications is governed by ovalization instability (a limit-point instability condition), whereas pipes of moderate thickness exhibit buckling in the form of wrinkles and a kink on the compression side of the pipe wall, before reaching a limit point due to ovalization. Fig. 8.9 shows the so-called "collapse envelop" of a relatively thick stainless-steel pipe (denoted as pipe I with $D/t = 34.7$) subjected to a combination of bending deformation and external pressure. The pipes are rather thick and their ultimate state is governed by ovalization instability. The numerical results are obtained through the nonlinear elastoplastic finite element formulation described in section 8.3.1, using the nine-node shell finite element[18]. Kinematic conditions were imposed through special-purpose interface elements to analyze one slice of the pipe with appropriate symmetry conditions. Two loading paths are considered: (a) the $p \rightarrow k$ path, where pressure is applied first, kept constant and, subsequently, bending is applied until collapse; and (b) the $k \rightarrow p$ path, where bending curvature is applied first, kept constant and, subsequently, pressure is raised until instability occurs. The numerical results are in very good agreement with experimental results[24]. Both numerical and experimental results show a strong dependence on the loading path, mainly due to the significant inelastic deformation hat occurs before the critical stage. Apparently, the $p \rightarrow k$ path, which is actually the most likely to occur in a pipeline installation, is the most critical one. Furthermore, it is interesting to note that, in the $p \rightarrow k$ case, the presence of a relatively low pressure level is responsible for a significant reduction of deformation capacity, i.e. the bending curvature at which the maximum bending moment occurs. More specifically, the deformation capacity of the pipe, bent under a pressure equal to 50% of the ultimate pressure is reduced by more than 75% with respect to the corresponding bending deformation capacity under zero pressure.

The ultimate bending behavior of metal cylinders is characterized by a sudden drop of their bending moment capacity, due to wrinkling of pipeline wall at the compression side and subsequent localization of one wrinkle forming a sharp local buckle ("kinking"). In rather thick cylinders, this phenomenon usually occurs after the ovalization limit point, whereas in moderately-thick pipes, local buckling may occur before the maximum ovalization moment is reached. Fig. 8.10 shows the shape of a bent cylinder, obtained by the finite element formulation described in section 8.3.1[23]. For a detailed description and discussion of localized buckling in bent cylinders the reader is referred to the works of Kyriakides & Ju[25], and Ju & Kyriakides[26]. In the presence of external pressure the shape of the local buckle is smoother; however, upon buckle formation, the bending capacity drops quite rapidly[23].

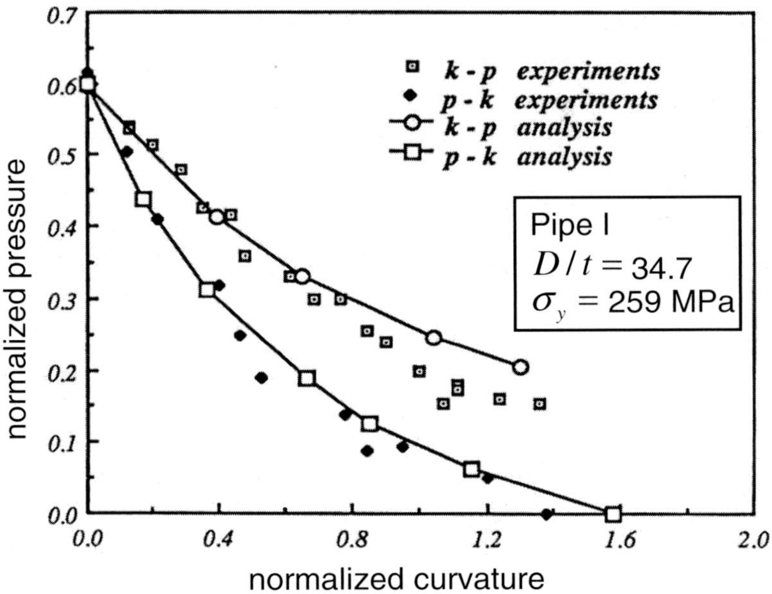

Fig. 8.9. Bending – pressure instability "envelops" for a stainless steel pipe; finite element analysis results[18] compared with experimental data[24].

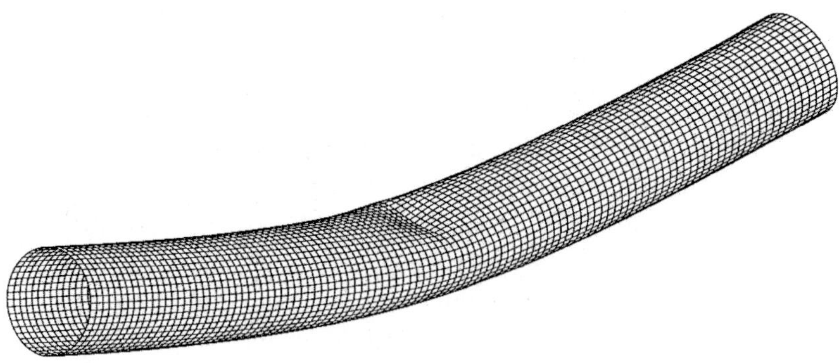

Fig. 8.10. Locally-buckled shape of a bent cylinder[23].

8.3.4 *Propagating Buckles in Metal Pipelines*

A propagating buckle can be described as a state of deformation which, upon initiation, will travel along a pipeline, or other long tubular structure, such as a platform tendon, causing collapse in its passage, as shown in Fig. 8.11. It occurs mainly during pipe-laying operations, where buckling in the form of local damage may occur. Whenever such an accident occurs, if the external ambient pressure is above a certain moderate level, the local damage will be transformed into a "propagating buckle", which will travel along the pipe, flattening a very long pipeline segment[27].

Consider the shape of the buckle, shown in Fig. 8.11. Pressure is always normal to the deformed surface of the pipe and, due to the inclined surface of the deformed pipe, the total force on the surface has a component, which is parallel to the pipe axis. This longitudinal force constitutes the actual mechanism for the propagation of the buckle[17].

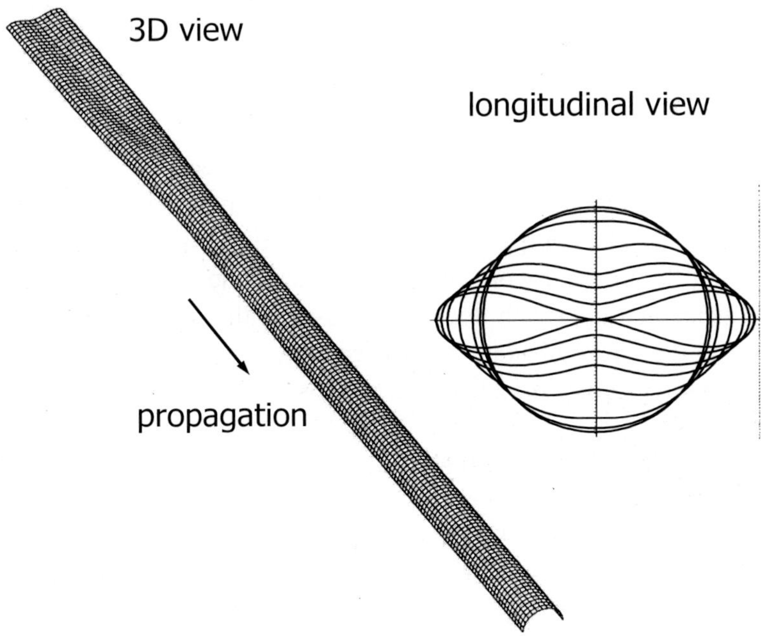

Fig. 8.11. Profile of quasi-static buckle propagation in an inelastic cylinder; finite element analysis[17].

The level of pressure is crucial for the buckle to propagate. There is a minimum level of external pressure that makes the buckle to propagate, called "propagation pressure" p_p and depends on the geometry and the material of the pipe under consideration. If the external pressure is less than this characteristic pressure ($p < p_p$), then propagation does not occur. If the external pressure is equal to the "propagation pressure" ($p = p_p$), then propagation occurs under quasi-static conditions (zero velocity). Finally, if the external pressure is higher than p_p ($p > p_p$), then propagation occurs, with a speed that depends on the level of pressure.

Many early studies have sought to determine the value of the propagation pressure experimentally and develop empirical formulae[28,27]. Palmer & Martin[29] reported a first estimate of the buckling propagation pressure p_p, assuming a four-hinge mechanism for the collapse of the ring with rigid perfectly-plastic material, and considering balance of internal and external work ($W_{int} = W_{ext}$),

$$p_p = \pi \sigma_y \left(\frac{t}{D} \right)^2 . \tag{8.79}$$

Several analytical studies have been based on this concept, considering a more accurate analysis of the collapsing ring in plane strain conditions[30,31]. The energy balance equation $W_{ext} = W_{\text{int}}$ was considered, computing the internal work in an accurate manner. However, the above two-dimensional approach does not recognize the propagating buckle as a three – dimensional state of deformation. Jensen[32] was the first to propose a three-dimensional special-purpose elastoplastic formulation for this problem.

A procedure for nonlinear finite element analysis of buckle propagation was developed at the University of Texas at Austin in early 90's, using in-house nonlinear finite element techniques, based on the elastoplastic large-strain formulation described above and the nine-node shell finite element. Katsounas & Tassoulas[16] analyzed quasi-static buckle propagation, while Song & Tassoulas[17] examined the dynamics of

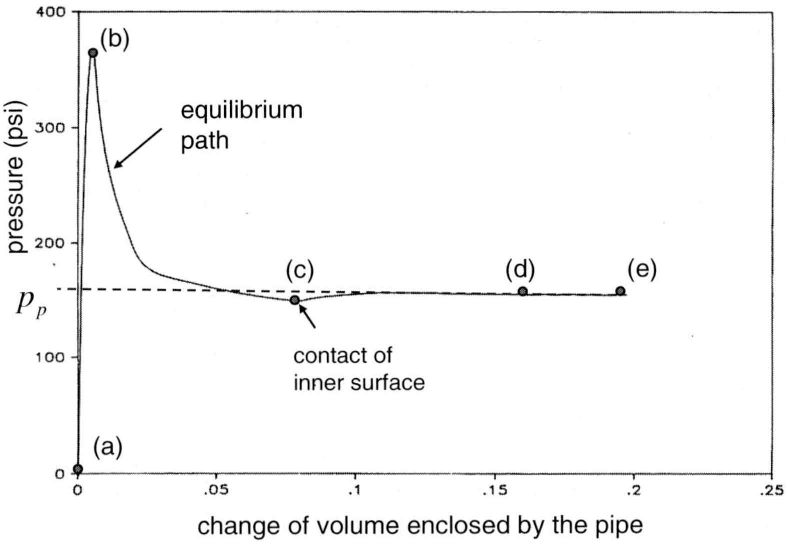

Fig. 8.12. Pressure – change of volume diagram for an externally-pressurized long pipe[17].

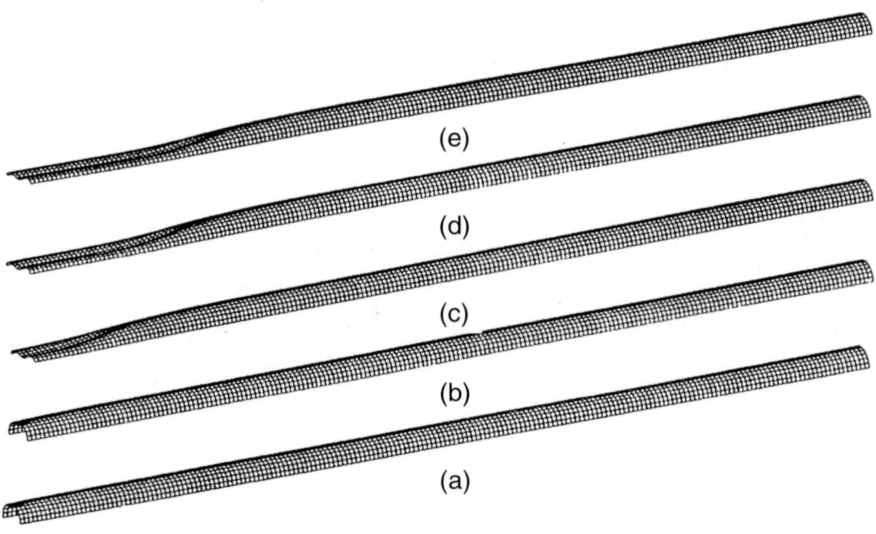

Fig. 8.13. Successive deformation stages of a propagating buckle in a pipeline subjected to external pressure; the labeled stages correspond to the locations indicated in the diagram of Fig. 8.12[17].

the phenomenon. A similar technique was also developed for steady-state finite element analysis of buckle propagation[33]. More recent computational works, on buckle propagation were reported by Mansour & Tassoulas[34], and Kyriakides & Netto[35].

8.4 Stability of Composite Tubes and Pipelines

8.4.1 *Stability of Anisotropic Laminated Rings and Long Cylinders*

The problem of buckling of laminated composite rings and long cylinders has only been recently treated (Rasheed and Yousif[36,37]). This is despite the fact that the formulation follows similar steps to those of section 2. Due to the 3D nature of the constitutive relationships of laminated composites, a consistent static condensation is necessary to accomplish the one dimensional reduction. The formulation is presented below for the general anisotropic laminated lay-up, which adds twisting prebuckling deformation component. The same buckling formula is recovered for the case of orthotropic laminated lay-up constrained by lumping adjacent angle plies $(\pm\alpha^\circ)$ into one layer. This type of stacking sequence is widely used in filament wound rings/long cylinders since this manufacturing process inherently dictates adjacent $(\pm\alpha^\circ)$ layers.

Kinematics

For the same class of intermediate deformations with plane sections before bending remain plane after bending, the kinematic relationships used are the same as those of equations (8.6) and (8.7). In addition, the same linear strain distribution is considered for the axial and shearing strains in thin shells for the sake of the closed form solution,

$$\begin{aligned} \varepsilon_x^* &= \varepsilon_x + z k_x \\ \gamma_{x\theta}^* &= \gamma_{x\theta} + z k_{x\theta} \end{aligned} . \qquad (8.80)$$

Constitutive Equations

Transforming the principal material directions of every composite layer into the ring coordinate directions, the in-plane ply stress-strain relationship becomes,

$$\begin{Bmatrix} \sigma_x^* \\ \sigma_\theta^* \\ \tau_{x\theta}^* \end{Bmatrix} = \begin{bmatrix} \bar{Q}_{11} & \bar{Q}_{12} & \bar{Q}_{16} \\ \bar{Q}_{12} & \bar{Q}_{22} & \bar{Q}_{26} \\ \bar{Q}_{16} & \bar{Q}_{26} & \bar{Q}_{66} \end{bmatrix} \begin{Bmatrix} \varepsilon_x^* \\ \varepsilon_\theta^* \\ \gamma_{x\theta}^* \end{Bmatrix} = \begin{bmatrix} \bar{Q}_{11} & \bar{Q}_{12} & \bar{Q}_{16} \\ \bar{Q}_{12} & \bar{Q}_{22} & \bar{Q}_{26} \\ \bar{Q}_{16} & \bar{Q}_{26} & \bar{Q}_{66} \end{bmatrix} \begin{Bmatrix} \varepsilon_x + z\kappa_x \\ \varepsilon_\theta + z\kappa_\theta \\ \gamma_{x\theta} + z\kappa_{x\theta} \end{Bmatrix} \tag{8.81}$$

where $\bar{Q}_{ij}$ are explicitly defined in standard textbooks on composites in terms of the ply properties and the fiber orientation angle, α, measured with respect to the ring axis[38]. The interaction, among plies in a general anisotropic laminate, requires such laminate stiffness to be assembled first. Then, dimensional reduction by static condensation facilitated by the special structural boundary conditions may be performed. Thus, the constitutive equations are expressed in terms of the force-strain relationships, instead of the stress-strain relationship used for isotropic rings,

$$dN_x = \int_{-\frac{t}{2}}^{\frac{t}{2}} \sigma_x^* \, r \, d\theta dz = \int_{-\frac{t}{2}}^{\frac{t}{2}} \sigma_x^* \left(1 + \frac{z}{R}\right) dz \, R \, d\theta \,, \tag{8.82}$$

Assuming thin shells $(z/R \approx 0)$ and a unit circumferential length $(Rd\theta = 1)$:

$$N_x = \int_{-\frac{t}{2}}^{\frac{t}{2}} \sigma_\theta^* \left(1 + \cancelto{=0}{\frac{z}{R}}\right) dz \,, \tag{8.83}$$

Similarly:

$$N_\theta = \int_{-\frac{t}{2}}^{\frac{t}{2}} \sigma_\theta^* \, dz \;\; ; \;\; N_{x\theta} = \int_{-\frac{t}{2}}^{\frac{t}{2}} \tau_{x\theta}^* \, dz \,, \tag{8.84}$$

$$M_x = \int_{-\frac{t}{2}}^{\frac{t}{2}} \sigma_x^* \, z \, dz \quad ; \quad M_\theta = \int_{-\frac{t}{2}}^{\frac{t}{2}} \sigma_\theta^* \, z \, dz \quad ; \quad M_{x\theta} = \int_{-\frac{t}{2}}^{\frac{t}{2}} \tau_{x\theta}^* \, z \, dz \qquad (8.85)$$

where $N_x, N_\theta, N_{x\theta}$ are the axial, hoop and in-plane ply shearing forces, respectively, $M_x, M_\theta, M_{x\theta}$ are the corresponding bending and twisting moments. Substituting Eq. (8.81) into (8.83)-(8.85) and integrating through the thickness, the standard rigidity expressions of the classical lamination theory are obtained[36].

The Anisotropic Ring Problem

The ring deformation involves a mid-surface axial strain (ε_x) and a shear strain ($\gamma_{x\theta}$) as a result of the lay-up anisotropy (non-zero $\overline{Q}_{16}$ and $\overline{Q}_{26}$). Nevertheless, the axial and shear forces corresponding to these strains ($N_x, N_{x\theta}$) must vanish along the free edges. The bending and twisting moments ($M_x, M_{x\theta}$) are also negligible in the case of thin and narrow rings,

$$\begin{Bmatrix} N_x = 0 \\ N_\theta \\ N_{x\theta} = 0 \\ M_x = 0 \\ M_\theta \\ M_{x\theta} = 0 \end{Bmatrix} = \begin{bmatrix} A_{11} & A_{12} & A_{16} & B_{11} & B_{12} & B_{16} \\ A_{12} & A_{22} & A_{26} & B_{12} & B_{22} & B_{26} \\ A_{16} & A_{26} & A_{66} & B_{16} & B_{26} & B_{66} \\ B_{11} & B_{12} & B_{16} & D_{11} & D_{12} & D_{16} \\ B_{12} & B_{22} & B_{26} & D_{12} & D_{22} & D_{26} \\ B_{16} & B_{26} & B_{66} & D_{16} & D_{26} & D_{66} \end{bmatrix} \begin{Bmatrix} \varepsilon_x \\ \varepsilon_\theta \\ \gamma_{x\theta} \\ k_x \\ k_\theta \\ k_{x\theta} \end{Bmatrix} \qquad (8.86)$$

The 1st, 3rd, 4th and 6th rows of Eq. (8.86) are solved for the strains and curvatures in terms of ($\varepsilon_\theta, k_\theta$) yielding,

$$\begin{bmatrix} A_{11} & A_{16} & B_{11} & B_{16} \\ A_{16} & A_{66} & B_{16} & B_{66} \\ B_{11} & B_{16} & D_{11} & D_{16} \\ B_{16} & B_{66} & D_{16} & D_{66} \end{bmatrix} \begin{Bmatrix} \varepsilon_x \\ \gamma_{x\theta} \\ k_x \\ k_{x\theta} \end{Bmatrix} = - \begin{bmatrix} A_{12} & B_{12} \\ A_{26} & B_{26} \\ B_{12} & D_{12} \\ B_{26} & D_{26} \end{bmatrix} \begin{Bmatrix} \varepsilon_\theta \\ k_\theta \end{Bmatrix} \qquad (8.87)$$

Inverting the matrix on the left hand side of Eq. (8.87),

$$\begin{Bmatrix} \varepsilon_x \\ \gamma_{x\theta} \\ k_x \\ k_{x\theta} \end{Bmatrix} = -[K_{cc}]^{-1}[K_{cr}]\begin{Bmatrix} \varepsilon_\theta \\ k_\theta \end{Bmatrix} \tag{8.88}$$

Substituting Eq. (8.88) back into Eq.(8.86), its 2nd and 5th rows become,

$$\begin{Bmatrix} N_\theta \\ M_\theta \end{Bmatrix} = \begin{bmatrix} A_{ani} & B_{ani} \\ B_{ani} & D_{ani} \end{bmatrix} \begin{Bmatrix} \varepsilon_\theta \\ k_\theta \end{Bmatrix}, \tag{8.89}$$

where

$$\begin{bmatrix} A_{ani} & B_{ani} \\ B_{ani} & D_{ani} \end{bmatrix} = \begin{bmatrix} A_{22} & B_{22} \\ B_{22} & D_{22} \end{bmatrix} - [K_{cr}]^T [K_{cc}]^{-1} [K_{cr}]. \tag{8.90}$$

The Orthotropic Ring Problem

The deformation of orthotropic laminated rings involves uniform out of plane strain with no distortions. This makes ($\gamma_{x\theta}$), ($k_{x\theta}$), ($N_{x\theta}$) and ($M_{x\theta}$) vanish. In addition, the overall axial force (N_x) must also vanish and the bending moment (M_x) is negligible along the free edges reducing equation (8.86) to,

$$\begin{Bmatrix} N_x = 0 \\ N_\theta \\ M_x = 0 \\ M_\theta \end{Bmatrix} = \begin{bmatrix} A_{11} & A_{12} & B_{11} & B_{12} \\ A_{12} & A_{22} & B_{12} & B_{22} \\ B_{11} & B_{12} & D_{11} & D_{12} \\ B_{12} & B_{22} & D_{12} & D_{22} \end{bmatrix} \begin{Bmatrix} \varepsilon_x \\ \varepsilon_\theta \\ k_x \\ k_\theta \end{Bmatrix}, \tag{8.91}$$

The above system of equations may be statically condensed to yield the following hoop force-strain relationship,

$$N_\theta = A_{12}\varepsilon_x \quad + A_{22}\varepsilon_\theta + B_{12}k_x \quad + B_{22}k_\theta = A_{orth}\varepsilon_\theta + B_{orth}k_\theta$$
$$M_\theta = B_{12}\varepsilon_x \quad + B_{22}\varepsilon_\theta + D_{12}k_x \quad + D_{22}k_\theta = B_{orth}\varepsilon_\theta + D_{orth}k_\theta, \quad (8.92)$$

where:

$$\begin{Bmatrix} \varepsilon_x \\ k_x \end{Bmatrix} = - \begin{bmatrix} A_{11} & B_{11} \\ B_{11} & D_{11} \end{bmatrix}^{-1} \begin{bmatrix} A_{12} & B_{12} \\ B_{12} & D_{12} \end{bmatrix} \begin{Bmatrix} \varepsilon_\theta \\ k_\theta \end{Bmatrix},$$

Writing equation (8.92) in a matrix form,

$$\begin{Bmatrix} N_\theta \\ M_\theta \end{Bmatrix} = \begin{bmatrix} A_{orth} & B_{orth} \\ B_{orth} & D_{orth} \end{bmatrix} \begin{Bmatrix} \varepsilon_\theta \\ k_\theta \end{Bmatrix}, \quad (8.93)$$

where:

$$A_{orth} = A_{22} - \frac{D_{11}(A_{12})^2 + A_{11}(B_{12})^2 - 2B_{11}B_{12}A_{12}}{A_{11}D_{11} - (B_{11})^2}$$

$$B_{orth} = B_{22} - \frac{D_{11}B_{12}A_{12} + D_{12}B_{12}A_{11} - D_{12}B_{11}A_{12} - B_{11}(B_{12})^2}{A_{11}D_{11} - (B_{11})^2}$$

$$D_{orth} = D_{22} - \frac{D_{11}(B_{12})^2 + A_{11}(D_{12})^2 - 2B_{11}B_{12}D_{12}}{A_{11}D_{11} - (B_{11})^2}$$

(8.94)

The Anisotropic Long Cylinder Problem

The long cylinder response is known to have in-plane deformation only because the out-of-plane displacements are restrained,

$$\begin{Bmatrix} N_\theta \\ M_\theta \end{Bmatrix} = \begin{bmatrix} A_{22} & B_{22} \\ B_{22} & D_{22} \end{bmatrix} \begin{Bmatrix} \varepsilon_\theta \\ k_\theta \end{Bmatrix} = \begin{bmatrix} A_{ani} & B_{ani} \\ B_{ani} & D_{ani} \end{bmatrix} \begin{Bmatrix} \varepsilon_\theta \\ k_\theta \end{Bmatrix}. \quad (8.95)$$

The Orthotropic Long Cylinder Problem

Similarly, the orthotropic long cylinder response is known to also have in-plane deformation only with zero out-of-plane displacements. Therefore, equation (8.91) reduces to,

$$\begin{Bmatrix} N_\theta \\ M_\theta \end{Bmatrix} = \begin{bmatrix} A_{22} & B_{22} \\ B_{22} & D_{22} \end{bmatrix} \begin{Bmatrix} \varepsilon_\theta \\ k_\theta \end{Bmatrix} = \begin{bmatrix} A_{orth} & B_{orth} \\ B_{orth} & D_{orth} \end{bmatrix} \begin{Bmatrix} \varepsilon_\theta \\ k_\theta \end{Bmatrix}. \tag{8.96}$$

Equilibrium Equations for External Fluid Pressure

It may be shown that the strain energy stored in an anisotropic ring or long cylinder is a function of the hoop stresses and strains only, because either the out-of-plane forces or displacements vanish, see Eq. (8.9). Knowing that $(dA = dz)$ for a unit length of the ring or cylinder in the axial direction and using Eqs. (8.83)-(8.85), (8.89) or (8.95), Eq. (8.9) becomes,

$$U = \frac{R}{2}\int_\theta (N_\theta \varepsilon_\theta + M_\theta k_\theta)\, d\theta = \frac{R}{2}\left\{ \left[A_{ani} \int_{\theta=0}^{2\pi} \varepsilon_\theta^{\,2} d\theta \right] + \left[2B_{ani} \int_{\theta=0}^{2\pi} k_\theta \varepsilon_\theta d\theta \right] + \left[D_{ani} \int_{\theta=0}^{2\pi} k_\theta^{\,2} d\theta \right] \right\}, \tag{8.97}$$

The potential of external fluid pressure loading W for the ring is,

$$\begin{aligned} W &= -p\ \ (V - V_0) = -p\ \ \left((1+\varepsilon_x)A - A_0\right) \\ &= -p\ \ (A - A_0) - \cancel{pA\varepsilon_x}\quad, \varepsilon_x \ll (1 - A_o/A), \end{aligned} \tag{8.98}$$

where p is the external pressure, ($A_0 = \pi R^2$) is the initial cross-sectional area of the mid surface , A is the corresponding area in the deformed configuration, W is the same for long cylinders because $\varepsilon_x = 0$. Writing (A) in terms of the displacement components and manipulating by using the divergence theorem[6] gives,

$$W = \frac{-p}{2}\int_{\theta=0}^{2\pi} \left(2Ru + u^2 + v^2 + uv' - vu'\right) d\theta. \tag{8.99}$$

Minimizing the total potential energy function ($\delta\Pi = \delta U - \delta W = 0$), performing integration by parts, manipulating and considering arbitrary virtual displacements ($\delta u,\ \delta v$), the nonlinear equilibrium equations are found to be the same as those of isotropic rings,

$$\begin{aligned} -N_\theta' R + N_\theta \beta' R - M_\theta' + pR^2\beta &= 0 \\ M_\theta'' - N_\theta R - (N_\theta \beta)' R - p(u+v')R &= pR^2 . \end{aligned} \qquad (8.100)$$

Prebuckling Solution

The pre-buckling solution is obtained by substituting $u = u_o$, $v = v_o$, $\beta = \beta_o = v_o/R$, Eq. (8.7) and (8.89) or (8.95) into Eqs. (8.100),

$$\left(\frac{u_0}{R} + \frac{v_0^2}{2R}\right) = -\frac{pR}{A_{ani}} . \qquad (8.101)$$

The v_o term above is negligible compared to that of u_o. Once u_o is computed from equation (8.101), it can be used to determine v_o from the constitutive equations of the ring by calculating $\gamma_{x\theta}$.

Bifurcation Solution

Substituting equations (8.7) and (8.89) or (8.95) into (8.100), perturbing the displacements u and v ($u \to u_0 + u_1$, $v \to v_0 + v_1$, $\beta \to \beta_0 + \beta_1$) and linearizing,

$$\begin{aligned} &(u_1' + v_1'') + \frac{B_{ani}}{A_{ani}R}(u_1' + 2v_1'' - u_1''') + \frac{D_{ani}}{A_{ani}R^2}(v_1'' - u_1''') \\ &- \frac{v_0}{R^2}\underbrace{(u_1 + u_1'')}_{=0} = 0 \\ &(u_1 + v_1') + \frac{B_{ani}}{A_{ani}R}[v_1' - 2u_1'' - v_1'''] - \frac{D_{ani}}{A_{ani}R^2}(v_1 - u_1')''' + \frac{pR}{A_{ani}}(u_1 + u_1'') \\ &+ \frac{v_0}{R^2}\underbrace{(v_1 + v_1'')}_{=0} = 0 \end{aligned} \qquad (8.102)$$

Note that the above equations differ from those of isotropic rings by having additional coupling terms as well as the extra v_o terms. The latter terms are of higher order and are dropped accordingly. Because of the possible lack of symmetry for this case, the perturbation is allowed to assume the complete Fourier series expansion for *u* and *v*,

$$v_1 = c_1 \sin n\theta + c_2 \cos n\theta$$
$$u_1 = c_3 \sin n\theta + c_4 \cos n\theta \,. \tag{8.103}$$

Letting

$$\alpha = \frac{D_{ani}}{A_{ani} R^2}; \quad \psi = \frac{B_{ani}}{A_{ani} R}; \quad \gamma = \frac{pR}{A_{ani}} \tag{8.104}$$

and substituting equation (8.103) into equations (8.102):

$$\begin{bmatrix} 0 & -\chi & \mu & 0 \\ -\chi & 0 & 0 & -\mu \\ \mu & 0 & 0 & \eta \\ 0 & -\mu & \eta & 0 \end{bmatrix} \begin{Bmatrix} c_1 \\ c_2 \\ c_3 \\ c_4 \end{Bmatrix} = \begin{Bmatrix} 0 \\ 0 \\ 0 \\ 0 \end{Bmatrix}, \tag{8.105}$$

where $\chi = n^2(1+2\psi+\alpha)$, $\mu = n(1+(1+n^2)\psi + n^2\alpha)$, $\eta = 1+\gamma+n^2(\alpha n^2 + 2\psi - \gamma)$. Solving the eigenvalue equations above for γ:

$$\gamma = (n^2 - 1)\left(\frac{\alpha - \psi^2}{1+\alpha+2\psi}\right) \tag{8.106}$$

Eq. (8.106) yields the critical buckling pressure, which may simply be expressed as,

$$p_{cr} = 3\left(\frac{1-\left(\dfrac{\psi^2}{\alpha}\right)}{1+\alpha+2\psi}\right)\frac{D_{ani}}{R^3} = 3\left(\frac{A_{ani}D_{ani} - B_{ani}^{\;2}}{A_{ani}R^3 + 2B_{ani}R^2 + D_{ani}R}\right) \tag{8.107}$$

The above equation reduces to that of Eq. (8.23) for thin isotropic or homogeneous orthotropic rings, because $B_{ani} = \psi = 0$ and $\alpha \ll 1$ in such cases. It is also important to note that if the parameters of the laminated orthotropic ring or long cylinder are used instead (Eq. (8.93) or (8.96)), the eigenvalue problem will noticeably simplify yielding the same equation for critical buckling pressure,

$$\begin{bmatrix} n^2(1+2\psi+\alpha) & n\left(1+(1+n^2)\psi+n^2\alpha\right) \\ n\left(1+(1+n^2)\psi+n^2\alpha\right) & 1+2n^2\psi+n^4\alpha+(1-n^2)\gamma \end{bmatrix} \begin{Bmatrix} c_1 \\ c_2 \end{Bmatrix} = \begin{Bmatrix} 0 \\ 0 \end{Bmatrix} \tag{8.108}$$

$$p_{cr} = 3\left(\frac{1-\left(\dfrac{\psi^2}{\alpha}\right)}{1+\alpha+2\psi}\right)\frac{D_{orth}}{R^3} = 3\left(\frac{A_{orth}D_{orth}-B_{orth}{}^2}{A_{orth}R^3+2B_{orth}R^2+D_{orth}R}\right). \tag{8.109}$$

8.4.2 *Stability of Delaminated Long Cylinders under External Pressure*

Laminated tubular components made of fiber-reinforced composites are widely used in aerospace, automotive, marine, and civil engineering applications today. In addition, the use of such structural components has been considered for deep-water offshore drilling[39]. Despite its several important advantages, composites suffer from layer separation or delamination which leads to loss of integrity and premature failure. Such delaminations may pre-exist in composite tubes due to manufacturing interply bond defects[40], residual radial stresses[41], transportation impacts[42], and environmental effects during the service life of the structural component[43]. It has been observed that thin inner delaminated layers tend to locally buckle inwards in rings under external pressure[44] (Fig. 8.14). In addition, the presence of a small initial out-of-roundness imperfection introduces a deviation from the uniform radial contraction and may affect the delamination buckling and collapse pressure. The present treatment is intended to develop a rigorous numerical procedure for inner delamination buckling of fiber-reinforced laminated composite rings and long cylinders subjected to external fluid-pressure loading[45]. The procedure takes into account the effects of the initial out-of-roundness imperfection, material failure, delamination-face contact and local buckling on the nonlinear geometric response.

Finite Element Model

A multi-angle, laminated, orthotropic, ring finite element model with a partially separated laminate is developed (Fig. 8.14). Symmetry of the cross section about a transverse axis is assumed in the formulation, Fig. 8.14. The geometry and deformation of the outer ring domain is described by Fourier series expansions for radial and tangential displacements u_t and v_t, respectively, along the reference line, as well as for the in-plane rotation ω_t of the through-the-thickness vector, Fig. 8.15.

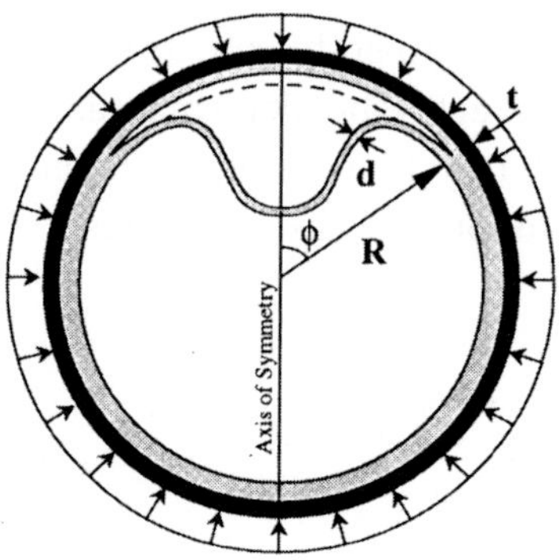

Fig. 8.14. Delamination buckling in rings.

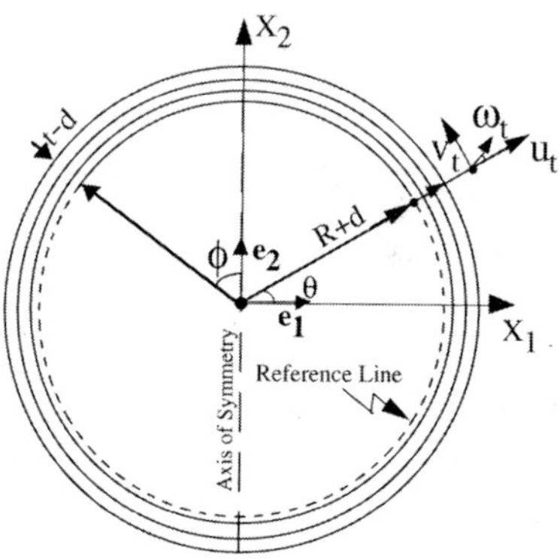

Fig. 8.15. Outer ring domain of element.

The latter accounts for first order shear deformations making the model applicable to moderately thick rings,

$$u_t(\theta) = u_{t0} + a_{t1}\sin\theta + \sum_{n=2,4,\ldots}^{\infty} u_{tn}\cos n\theta + \sum_{n=3,5,\ldots}^{\infty} u_{tn}\sin n\theta\,, \qquad (8.110)$$

$$v_t(\theta) = -a_{t1}\cos\theta + \sum_{n=2,4,\ldots}^{\infty} v_{tn}\sin n\theta + \sum_{n=3,5,\ldots}^{\infty} v_{tn}\cos n\theta\,, \qquad (8.111)$$

$$\omega_t(\theta) = \sum_{n=1,3,\ldots}^{\infty} \omega_{tn} \cos n\theta + \sum_{n=2,4,\ldots}^{\infty} \omega_{tn} \sin n\theta\,, \tag{8.112}$$

where θ is the circumferential coordinate of any material point in the ring wall measured counterclockwise from the $\mathbf{X}_1$ axis, Fig. 8.15. The radial and tangential displacements are common between the inner and the outer rings (Fig. 8.16). The in-plane rotation of the through-the-thickness vector in the inner ring region is expressed by another Fourier series expansion, written in terms of the delamination angle ϕ, Fig. 8.16,

$$\omega_{2t}(\theta) = \omega_{2t\phi} \frac{\pi/2+\theta}{\pi-\phi} + \sum_{n=1}^{\infty} \omega_{2tn} \frac{n\pi(\pi/2+\theta)}{\pi-\phi}, \tag{8.113}$$

where ϕ is the angle between the $\mathbf{X}_2$ axis and the delamination tip in the perfectly circular configuration, Figs. 8.14-8.17. The geometry of the reference line along the delaminated part is defined by equations (8.110) and (8.111) to ensure smooth initial contact. The deformation of the delaminated sub-laminate is also described by Fourier series expansions of the corresponding displacement components, as functions of the delamination angle ϕ, Fig. 8.17,

$$u_p(\hat{\theta}) = u_{p\phi} + \sum_{n=1,3,\ldots}^{\infty} u_{pn} \sin \frac{n\pi(\phi+\hat{\theta})}{2\phi}, \tag{8.114}$$

$$v_p(\hat{\theta}) = v_{p\phi} \frac{\hat{\theta}}{\phi} + \sum_{n=1.}^{\infty} v_{pn} \sin \frac{n\pi\hat{\theta}}{\phi}, \tag{8.115}$$

$$\omega_p(\hat{\theta}) = \omega_{p\phi} \frac{\hat{\theta}}{\phi} + \sum_{n=1}^{\infty} \omega_{pn} \sin \frac{n\pi\hat{\theta}}{\phi}, \tag{8.116}$$

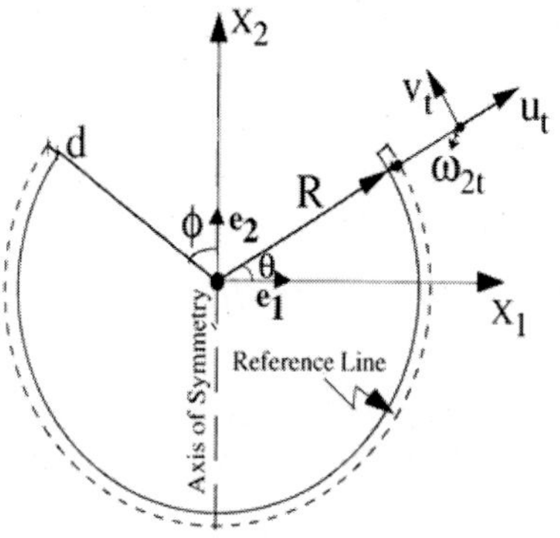

Fig. 8.16. Inner ring domain of element.

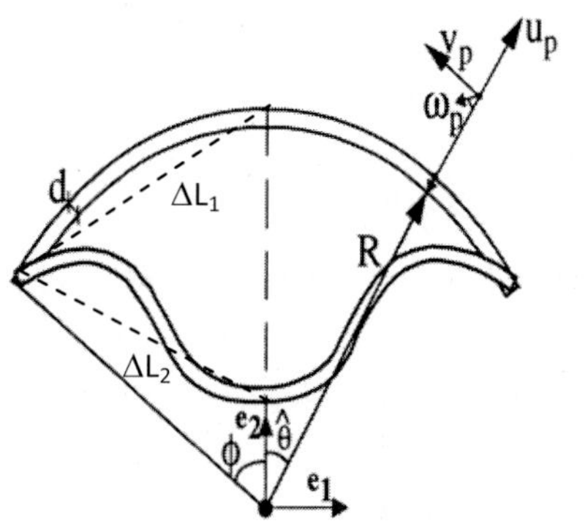

Fig. 8.17. Delaminated region of element.

where $\tilde{\theta}$ is the circumferential coordinate of any material point in the delaminated region measured clockwise from the $\mathbf{X}_2$ axis, Fig. 8.17. The compatibility of deformation between the ring wall and the delaminated layer, at the delamination tip, is maintained by means of three Lagrange multipliers, introduced by imposing the following constraints,

$$u_t(\frac{\pi}{2}-\phi)-u_p(\phi)=0\,, \tag{8.117}$$

$$v_t(\frac{\pi}{2}-\phi)-v_p(\phi)=0\,, \tag{8.118}$$

$$\omega_{2t}(\frac{\pi}{2}-\phi)-\omega_p(\phi)=0\,. \tag{8.119}$$

The degrees-of-freedom of the element are the out-of-plane constant strain ($\mathcal{E}$) and the coefficients of Fourier series in addition to the Lagrange multipliers. Contact between the delamination faces is maintained whenever penetration takes place prior to delamination buckling. Friction is not considered in the analysis, but penetration normal to the delamination faces is prevented through a penalty formulation, as described below.

Equilibrium Equations

The nonlinear set of equilibrium equations is solved incrementally with Newton iterations performed at every load or arc-length step. The tangent stiffness matrix and the internal load vector are derived using the principle of virtual work,

$$\int_V \delta u_{i|j} \sigma^{ij} dV = \int_B \delta \mathbf{u} \cdot \mathbf{T} dB = \int_B \delta \mathbf{u} \cdot (-p\mathbf{n}) dB \,, \qquad (8.120)$$

where σ^{ij} are the contravariant components of the Cauchy stress tensor, $\mathbf{T}$ is the boundary traction vector, p is the external pressure, $\mathbf{n}$ is the unit normal vector to the outer surface at the current configuration, and $\delta\mathbf{u}$ is an arbitrary virtual displacement vector. The above equation is integrated over the reference configuration, V_0, and manipulated to obtain the incremental equilibrium equations[46],

$$\int_{V_o} \delta U_{i|j} S^{ijkl} \Delta U_{k|l} dV_o = \int_B \delta \mathbf{u} \cdot \mathbf{T}' dB - \int_{V_o} \delta U_{i|j} (\mathbf{G}^i \cdot \mathbf{g}_k) \tau^{kj} dV_o \,, \quad (8.121)$$

where $\delta\mathbf{U}$ is the vector of arbitrary virtual degrees-of-freedom, $\Delta\mathbf{U}$ is the vector of incremental degrees of freedom, S^{ijkl} are the components of the tangent stiffness tensor having the material and geometric effects, $\mathbf{T}'$ is the boundary traction vector at a nearby equilibrium configuration, $\mathbf{G}^i$ are the contravariant base vectors in the reference configuration, $\mathbf{g}_k$ are the covariant base vectors in the current configuration, and τ^{kj} are the contravariant components of the Kirchhoff stress tensor[47]. The contact contribution to the left-hand and the right-hand sides of equation (8.121) is,

$$\int_{B_c} (\delta \mathbf{u}_p - \delta \mathbf{u}_t) k_c (\Delta \mathbf{u}_p - \Delta \mathbf{u}_t) \, dB_c \,, \qquad (8.122)$$

and

$$\int_{B_c} (\delta \mathbf{u}_p - \delta \mathbf{u}_t) k_c (\mathbf{x}_p - \mathbf{x}_t) \, dB_c \,, \qquad (8.123)$$

where B_c is the boundary of contact, k_c is the contact spring constant, $(\mathbf{x}_p - \mathbf{x}_t)$ is the penetration vector defined normal to the ring contact surface, $\delta\mathbf{u}_p$ and $\delta\mathbf{u}_t$ are the virtual displacement vectors for the delaminated layer and the ring wall at the contact surface. The equations of constraints (8.117)-(8.119) are also added to equation (8.121) to form the complete incremental equilibrium equations, which may be expressed in a matrix form as follows[46],

$$\delta\mathbf{U}^T \begin{bmatrix} \tilde{\mathbf{K}}_e & \mathbf{G}^T \\ \mathbf{G} & \mathbf{0} \end{bmatrix} \Delta\mathbf{U} = \delta\mathbf{U}^T \left[\int_{B^e} \mathbf{N}^T \mathbf{T}' dB^e - \int_{V_o^e} \mathbf{B}^T \mathbf{W}^T \boldsymbol{\tau} dV_o^e \right]$$

$$- \delta\mathbf{U}^T \left[\int_{B_c^e} k_c \hat{\mathbf{N}}^T (\mathbf{x}_p - \mathbf{x}_t)\, dB_c^e \right] - \delta\mathbf{U}^T \begin{bmatrix} \mathbf{0} & \mathbf{G}^T \\ \mathbf{G} & \mathbf{0} \end{bmatrix} \mathbf{U}, \quad (8.124)$$

$$\tilde{\mathbf{K}}_e = \int_{V_o^e} \mathbf{B}^T \mathbf{S} \mathbf{B} dV_o^e + \int_{B_c^e} k_c \hat{\mathbf{N}}^T \hat{\mathbf{N}} dB_c^e, \quad (8.125)$$

where **N** is the matrix of shape functions and **B** is the matrix of the derivatives of the coefficients of the **N** matrix[46].

Numerical Procedure

Small initial out-of-roundness imperfection is incorporated into the ring geometry for the purpose of global stability analysis under external pressure. Such imperfection is introduced by specifying initial values to the coefficients of the 2θ terms in the radial and tangential displacement expressions above, equations (8.110)-(8.111), to describe a slightly ovalized ring. This drives the solution into the first global buckling mode. The initial out-of-roundness imperfection β is defined as,

$$\beta = \frac{(D+2b)-(D-2b)}{D} \times 100\% = \frac{2b}{R+t} \times 100\%, \quad (8.126)$$

where D is the outside diameter and b is the coefficient defining the initial geometry:

$$u_o = -b\cos 2\theta, \quad v_o = \frac{b}{2}\sin 2\theta, \quad (8.127)$$

Large deformation of the delaminated layer is also permitted for the purpose of local stability analysis. The present algorithm searches for a critical equilibrium by monitoring the eigenvalues of the tangent stiffness matrix. Once a negative eigenvalue is detected, the corresponding eigenvector is used to excite the delaminated part, as detailed below.

The Chord Length Procedure

It has been postulated by Kachanov[48] that the snap buckling process is dynamic in nature. In the present study, several attempts with different arc-length definitions did not produce equilibria near the prebuckling configuration. To overcome this difficulty while still using quasi-static analysis, the chord length procedure is devised. An initial estimate of the delamination buckling configuration is obtained by specifying a reasonable amplitude to the eigenvector of local buckling. This is called the perturbation or the excitation, which serves as a nearby snap buckling configuration to facilitate convergence. The approximate amplitude is evaluated by using the length of the chord, ΔL_1, in the prebuckled position, Fig. 8.17. This chord length is equated to its corresponding value, ΔL_2, at the approximate snap buckled configuration for computing the amplitude λ.

$$\begin{aligned}\Delta\mathbf{L}_1 = \mathbf{x}_{top\ point} - \mathbf{x}_{tip} &= \{-(R+d+u_p(\phi))\sin\phi + v_p(\phi)\cos\phi\}\mathbf{e}_1 \\ &+\{(R+d+u_p(0)) - (R+d+u_p(\phi))\cos\phi - v_p(\phi)\sin\phi\}\mathbf{e}_2\end{aligned} \tag{8.128}$$

$$\begin{aligned}\Delta\mathbf{L}_2 = \mathbf{x}_{bot.pt} - \mathbf{x}_{tip} &= \{-(R+d+u_p(\phi))\sin\phi + v_p(\phi)\cos\phi\}\mathbf{e}_1 \\ &+\{(R+d+\bar{u}_p(0)) - (R+d+u_p(\phi))\cos\phi - v_p(\phi)\sin\phi\}\mathbf{e}_2\end{aligned} \tag{8.129}$$

where $\bar{u}_p(0) = u_p(0) + \lambda\hat{u}_p(0)$ is the radial displacement of the topmost point at the snap buckled configuration, $u_p(0)$ is the same displacement in the prebuckled configuration, $\hat{u}_p(0)$ is the corresponding coefficient of the eigenvector. Thus,

$$\Delta\mathbf{L}_1 \cdot \Delta\mathbf{L}_1 = \Delta\mathbf{L}_2 \cdot \Delta\mathbf{L}_2 \Rightarrow \lambda = -\frac{2\hat{B}}{\hat{u}_p(0)}, \tag{8.130}$$

where

$$\hat{B} = (R+d+u_p(0)) - (R+d+u_p(\phi))\cos\phi - v_p(\phi)\sin\phi. \tag{8.131}$$

Failure Criteria

A macromechanics-based composite material model was implemented. The model considers multi-angle fiber-reinforced polymer plies of orthotropic properties by merging adjacent ($\pm\alpha°$) laminas[46]. Either Tsai-Wu[49] or Hashin[50] ply failure criteria can be invoked[48]. The procedure takes into account nonlinear shear response. However, shear nonlinearity is found insignificant and is thus neglected in the present applications, due to the dominance of hoop layers. Brittle post-failure behavior is assumed for fiber and transverse directions as well as shear.

Numerical Results

IFP and Aerospatiale of France have designed, fabricated, and tested hybrid composite tubular specimens made of high strength carbon/epoxy and S glass/epoxy layers[39]. The specimens were 15 m (49.2 ft) long with a 229 mm (9 in.) inner diameter and a 15.4mm (0.606 in.) structural wall thickness. The S glass/epoxy hoop layers were forming 62% of that thickness. The rest are composed of ±20° helically wound carbon/epoxy composite. The mechanical properties of both layers are given in Table 1. The specimens tested had 13 alternately stacked glass and carbon layers[51], Fig. 8.18. This is referred to as the "Alternating Layup".

Although the specimen design in Fig. 8.18 were not tested for delaminations, their layups are used to parametrically study the effect of delamination buckling on collapse of laminated composite long cylinders under external pressure.

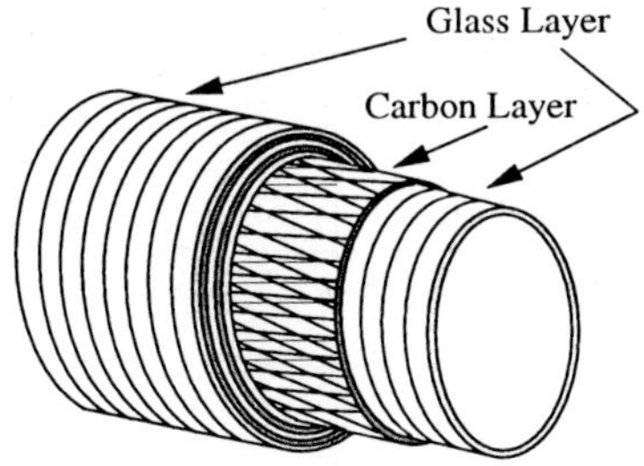

Fig. 8.18. Alternating-layup tube design.

Delamination Length and Through-the-thickness Location

In this section, long cylinders are assumed to have an initial out-of-roundness imperfection of 0.4% with the delamination centered at the region of maximum initial curvature. The effects of delamination length and through-the-thickness depth on local buckling and collapse are studied using the alternating layup.

Fig. 8.19 presents the delamination buckling and collapse pressures for a delamination angle ϕ =40° and a spectrum of inner layer thickness. It can be seen that delamination buckling (for d>0) causes the cylinder to collapse at a lower pressure than that of a defect-free cylinder (d=0). This collapse pressure keeps reducing with the increase in delamination depth until the collapse pressure coincides with the delamination buckling pressure. This defines a critical delamination depth of 2.82mm (0.111 in.). This corresponds to a reduction in collapse pressure of 22.6% compared with a delamination free tube. Longer delaminations, ϕ > 45°, are reported to have no tendency to grow or extend before and after delamination buckling[52]. The initial, delamination buckling and collapse configurations are plotted for the ring with the delamination angle ϕ = 40° and the delamination situated at the carbon/glass interface (d=2.34 mm=0.092 in.), Fig. 8.20.

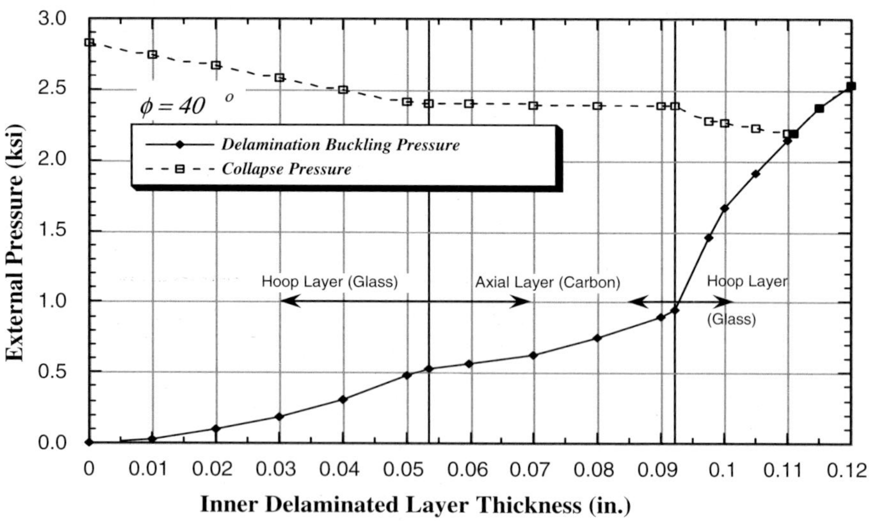

Fig. 8.19. Delamination Buckling and Collapse Pressures for a Delamination Angle of 40° and Varying Delaminated Layer Thickness.

The complete parametric study including the effects of delamination length, depth, orientation with respect to the ovalization imperfection, initial imperfection is presented in detail in Rasheed and Tassoulas[52].

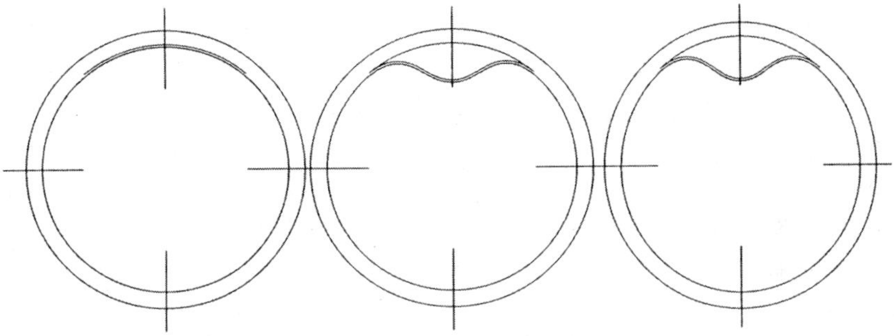

Fig. 8.20. Initial, delamination buckling and collapse configurations of the cylinder with the delamination in the ovalizing zone and ($\phi = 40°$, d/t = 0.152).

Table 8.1. Mechanical properties of composite plies used by IFP and Aerospatiale.

Mechanical Property	S-Glass Layer	Carbon Layer
Fiber Content	66%	60%
E_{11} in GPa (ksi)	57.3 (8300)	140 (20300)
E_{22} in GPa (ksi)	16.4 (2380)	8 (1160)
G_{12} in GPa (ksi)	7.7 (1117)	4 (580)
G_{23} in GPa (ksi)	7.7 (1117)	4 (580)
G_{13} in GPa (ksi)	7.7 (1117)	4 (580)
ν_{12}	0.28	0.28
ν_{23}	0.32	0.42
ν_{13}	0.28	0.28
S^{T}_{fiber} in MPa (ksi)	1600 (232)	1242 (180)
S^{C}_{fiber} in MPa (ksi)	690 (100)	1242 (180)
S^{T}_{matrix} in MPa (ksi)	50 (7.25)	41 (6)
S^{C}_{matrix} in MPa (ksi)	138 (20)	172 (25)
S_{shear} in MPa (ksi)	82 (12)	82 (12)

8.5 References

1. M. Bresse, *Cours de Mechanique Appliquée*, 2nd Edition, Paris, France (1866).
2. G. H. Bryan, *Proc. Cambridge Phil. Soc.*, 6, Cambridge, UK (1888).
3. S. P. Timoshenko, *Applied Mechanics, ASME, APM-55-20*, p. 173 (1933).
4. W. Flügge, *Stresses in Shells*, Springer-Verlag, New York, 525 p. (1960).
5. S. P. Timoshenko and J. Gere, *Theory of Elastic Stability*, McGraw-Hill Book Co., New York, p. 289, (1961).
6. D. O. Brush and B. O. Almroth, *Buckling of Bars, Plates and Shells*, McGraw Hill Inc., USA, 378 p. (1975).
7. B. Budiansky, *Elastic Buckling*, Advances in Applied Mechanics, 14, p. 1, (1974).
8. S. Kyriakides, and C. D. Babcock, *Int. J. of Solids Struct.*, 17 (10), p. 981, (1981).
9. L. G. Brazier, *Proc. Royal Soc.*, series A, 116, p. 104 (1927).
10. J. D. Wood, *J. Appl. Mech., ASME*, 25, p. 453 (1958).
11. S. Houliara and S. A. Karamanos, *Int. J. Nonlinear Mech.*, 41 (4), p. 491 (2006).
12. E. L. Aksel'rad (Axelrad), [in Russian]. Izvestiya Akademii Nauk SSSR, Otdelenie Tekhnicheskikh Nauk, Mekhanika i Mashinostroenie, 4, p. 133 (1965).
13. O. Fabian, *Int. J. Solids Struct.*, 13, p. 1257 (1977).
14. S. A. Karamanos, *Int. J. Solids Struct.*, 39 (8), p. 2059, (2002).
15. Needleman, A., *Finite Elements for Finite Strain Plasticity Problems*, in Plasticity of Metals at Finite Strain: Experiment and Computation, E.H. Lee and R.L. Mallett, Eds, RPI, Troy, New York, p. 387 (1982).
16. A. T. Katsounas, and J. L. Tassoulas, *Finite Element Analysis of Propagating Buckles in Deep-Water Pipelines.* Offshore Technology Research Center, Report No. 1/90B01100, The University of Texas, Austin, Texas. (1989)
17. H. W. Song and J. L. Tassoulas, *Int. J. Num. Meth. Engng.*, 36 (20): p. 3529 (1993).
18. S. A. Karamanos, and J. L. Tassoulas, *J. Engng. Mech., ASCE*, 117 (12), p. 2845 (1991).
19. J. W. Hutchinson, *Finite Strain Analysis of Elastic-Plastic Solids and Structures*, Numerical Solutions of Nonlinear Structural Problems, R. F. Hartung, Ed., *ASME, AMD*, 6, p. 17 (1973).
20. S. Ahmad, B. M. Irons and O. C. Zienkiewicz, *Int. J. Num. Meth. Engng.*, 2, p. 419 (1970).
21. S. A. Karamanos, and J. L. Tassoulas, *Stability of Tubular Members under External Pressure and Structural Loads.*, Offshore Technology Research Center, Report No. 11/93B51100, The University of Texas, Austin, Texas (1993).
22. S. A. Karamanos, and J. L. Tassoulas, *J. Engng. Mech., ASCE*, 122, (1), p. 64 (1996).
23. S. A. Karamanos, and J. L. Tassoulas, *J. Engng. Mech., ASCE*, 122, (1), p.72 (1996).
24. E. Corona and S. Kyriakides, *Int. J. Solids Struct.*, 24 (5), p. 505 (1988).
25. S. Kyriakides and G. T. Ju, *Int. J. Solids Struct.*, 29 (9), p. 1117 (1992).
26. G. T. Ju and S. Kyriakides, *Int. J. Solids Struct.*, 29 (9), p. 1143 (1992).

27. R. E. Mesloh, J. D. Sorenson and T. J. Atterbury, *Gas Magazine*, p. 40 (1973).
28. S. Kyriakides and C. D. Babcock, *J. Pres. Ves. Tech., ASME*, 103 (4),p. 328 (1981).
29. A. C. Palmer and J. H. Martin, *Nature* 254 (5495), p. 46 (1975).
30. S. Kyriakides, M. K. Yeh and D. Roach, *J. Pres. Ves. Tech., ASME*, 106 (2), p. 150 (1984).
31. E. Chater and J. W. Hutchinson, *J. Appl. Mech., ASME*, 51 (2), p. 269 (1984).
32. H. M. Jensen, *Int. J. Solids Struct.*, 24 (1), p. 51 (1988).
33. A. C. Nogueira and J. L. Tassoulas, *J. Engng. Mech., ASCE*, 120 (9), p. 1931 (1994).
34. G. N. Mansour and J. L. Tassoulas, *J. Engng. Mech., ASCE*, 123 (4), p. 359 (1997).
35. S. Kyriakides and T. A. Netto, *Int. J. Solids Struct.*, 37 (46-47), p. 6843 (2000).
36. H. A. Rasheed, and O. H. Yousif, *Int. J. Struct. Stab. Dyn.*, 1 (4), p. 485 (2001).
37. H. A. Rasheed, and O. H. Yousif, *J. of Aero. Engng., ASCE*, 18 (3), p. 129 (2005).
38. R. M. Jones, *Mechanics of Composite Materials*, (Hemisphere Publishing Corporation, New York, 1975).
39. C. P. Sparks, *Proc., 5th Int. OMAE Conf.*, Tokyo, Japan, p. 86 (1986).
40. Y. M. Tarnopolskiy, in G. C. Sih, and V. P. Tamuzs, Eds., Fracture of Composite Materials," Proc. 1st US-USSR Symposium on Fracture of Composite Materials, p.159 (1978).
41. L. M. Kachanov, Translated from *Mekhanika Polimerov*, 2, p. 370 (1974).
42. H. Chai, and C. D. Babcock, *J. Comp. Mat.*, 19, p.67 (1985).
43. F. E. Sloan, and R. J. Seymour, *J. Comp. Mat.*, 24, p.727 (1990).
44. L. M. Kachanov, Translated from *Mekhanika Polimerov*, 6, p.1106 (1975).
45. H. A. Rasheed and J. L. Tassoulas, *J. Engng. Mech., ASCE*, 128 (11), p. 1174 (2002).
46. H. A. Rasheed, *Behavior and Strength of Composite Tubes Considering Delaminations and Other Defects*, Ph.D. Dissertation, The University of Texas at Austin, Austin, Texas, 191p. (1996).
47. L. E. Malvern, *Introduction to the Mechanics of a Continuous Medium*, (Prentice-Hall, Englewood Cliffs, NJ, 713p. 1969).
48. L. M. Kachanov, *Delamination Buckling of Composite Materials,* (Kluwer Academic Publishers, Dordrecht, 95p., 1988).
49. S. W. Tsai, and E. M. Wu, *J. Comp. Mat.*, 5, p.58 (1971).
50. Z. Hashin, *J. Appl. Mech., ASME*, 47, p.329 (1980).
51. C. P. Sparks, P. Odru, H. Bono, and G. Metivaud, Proc. Offshore Tech. Conf., OTC 5797, Houston (Texas), p.467 (1988).
52. H. A. Rasheed, and J. L. Tassoulas, *Int. J. Fracture*, 108 (1), p. 1 (2001).

CHAPTER 9

IMPERFECTION-SENSITIVE BUCKLING AND POSTBUCKLING OF SPHERICAL SHELL CAPS

Seishi Yamada and Manabu Uchiyama

Department of Civil Engineering, Toyohashi University of Technology
Toyohashi 441-8580, Japan
E-mail: yamada@st.tutrp.tut.ac.jp

Partly due to the major discrepancies between observed buckling loads and the prediction from classical theory, the buckling of pressurised spherical caps, as well as axially compressed cylinders, has excited extraordinary interest over the past half century. This chapter reviews the historical change in the understanding of the buckling mechanism of the spherical shell caps. Then fifty-two selected experiments, previously conducted by Yamada are compared with the numerical simulation results recently obtained using an advanced mixed finite element technique. It is shown that since measured geometric initial imperfection distributions are carefully introduced, good agreement between the numerical simulations and the experimental measurements are obtained in not only buckling loads but also postbuckling large deflection behaviour.

9.1 Introduction

In the nonlinear finite element numerical technique, the so-called degenerated finite element approach has been developed in many studies[1-9] over the past two decades. Especially the mixed variational formulation, by which the element does not exhibit locking phenomena even in thin-walled shells, has been pointed out. There are very few reports that have shown good agreement between numerical simulations and experimental results, or which confirm the precision required in the analysis of complex shell buckling behaviour. For this problem, it would be also argued that few comparable precise experiments exist. Recently, Uchiyama and Yamada[10] have applied a nine-node-shell element based on the four-node-shell element by Simo et al.[7] and the interpolation formulation for stress resultant vectors by Sansour and Bufler[8] to the most imperfection-sensitive buckling simulation problems for pressurised shallow spherical caps tested by Yamada[11,12].

The buckling of pressurised shallow spherical caps has excited extraordinary interest over the past 60 years due to the major discrepancies between experimental buckling loads and the predictions from theory. The pioneering research of this area was that of von Karman and Tsien[13]. They attempted to analyse the buckling behaviour of complete spherical shells using a shallow cap model, but recognised the incomplete nature of their results. Since then, this buckling problem has been looked into by a great number of researchers. The solution of axisymmetric (symmetric) snap-through buckling of perfect clamped spherical shells was given by Budiansky[14], however, the experimental results by Kaplan and Fung[15] and Homewood et al.[16] showed that buckling could occur at pressure lower than this numerical solution. In an attempt to resolve this, the non-symmetric (asymmetric) bifurcation buckling analysis was undertaken by Huang[17], but the numerical results were still higher than the above experimental results as shown in Fig. 9.1. In the 1970s, experiments on more accurate specimens were carried out by Tillman[18] and Sunakawa and Ichida[19]; some of their results were shown to be higher than above asymmetric bifurcation buckling pressures. Furthermore, Sunakawa and Ichida[19] cast doubt on the possibility of the asymmetric buckling for the well-known geometric parameter $\lambda > 5.5$, and on the symmetric buckling pressure for $\lambda < 5.5$. In the 1980s, experiments by Yamada[11,12] involving precise measurements for the distributions of initial geometric imperfections and vertical displacements at both prebuckling and postbuckling equilibrium stages, made clear that initial geometric imperfections have serious effects on both the buckling pressure and the deflection mode, and furthermore, type of symmetric initial imperfections cannot only lower but also increase the buckling pressure. It was also shown that for the case of $\lambda > 5.5$, the amplitude of the asymmetric displacement component, having the form of bifurcation buckling wave calculated by Huang[17], becomes large immediately before buckling. It was inferred that Huang's predictions for initially perfect shells were experimentally confirmed by Yamada and Yamada[20] in which five typical experimental results were compared with the solutions from the continuum theory using the nonlinear Marguerre-type differential equations considering imperfections and using the global Galerkin method.

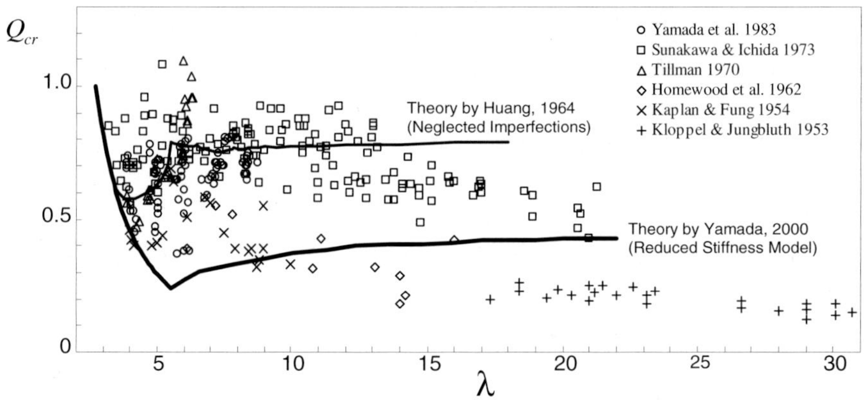

Fig. 9.1. The non-dimensional buckling load normalised by the classical critical pressure versus the non-dimensional geometric parameter for clamped spherical caps under external pressure[21].

9.2 Theoretical background: mixed finite element analytical method

Uchiyama and Yamada[10] adopted the following assumption: (1) strains are small; (2) shell directors are inextensible and straight, but are not necessary to be normal to a shell mid-surface; (3) rotations about the directors are neglected; (4) the material is elastic and homogeneous. Also the singularity-free parameterisation of the rotation field (Simo and Fox[6]) is used.

Figure 9.2 shows the notation and convention for an undeformed shell; φ is a position vector of a point on a shell mid-surface, t is shell thickness, $\mathbf{t}$ is a director, $()_{,\alpha}$ (α=1,2) is a derivative by natural coordinate ξ_α and "$\bullet$" denotes an inner product. We set

$$\mathbf{a}_\alpha = \varphi_{,\alpha},\ a_{\alpha\beta} = \varphi_{,\alpha} \bullet \varphi_{,\beta},\ \gamma_\alpha = \varphi_{,\alpha} \bullet t,\ \kappa_{\alpha\beta} = t_{,\alpha} \bullet t_{,\beta}. \tag{9.1}$$

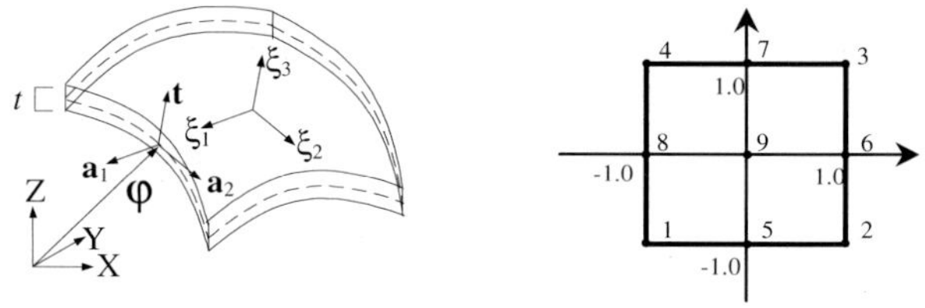

Fig. 9.2. An undeformed shell element.

Then in-plane strains, ε, transverse shear strains, γ, and curvature strains, κ, are defined as follows.

$$\varepsilon = \frac{1}{2}\begin{pmatrix} a_{11} - a_{11}^0 \\ a_{22} - a_{22}^0 \\ 2(a_{12} - a_{12}^0) \end{pmatrix},\ \gamma = \begin{pmatrix} \gamma_1 - \gamma_1^0 \\ \gamma_2 - \gamma_2^0 \end{pmatrix},\ \kappa = \begin{pmatrix} \kappa_{11} - \kappa_{11}^0 \\ \kappa_{22} - \kappa_{22}^0 \\ 2(\kappa_{12} - \kappa_{12}^0) \end{pmatrix}. \tag{9.2}$$

In the expressions, right superscript, $()^0$, denotes unloaded and stress-free configuration. The corresponding stress resultant vectors of in-plane, transverse shear and bending are written as

$$\mathbf{n} = \begin{pmatrix} n_{11} \\ n_{22} \\ n_{12} \end{pmatrix} = \mathbf{C}_n\varepsilon,\quad \mathbf{q} = \begin{pmatrix} q_1 \\ q_2 \end{pmatrix} = \mathbf{C}_q\gamma,\quad \mathbf{m} = \begin{pmatrix} m_{11} \\ m_{22} \\ m_{12} \end{pmatrix} = \mathbf{C}_m\kappa. \tag{9.3}$$

where $\mathbf{C}_n$, $\mathbf{C}_q$ and $\mathbf{C}_m$ are constitutive matrices.

With the help of the Legendre transformations we can give explicit forms of strain energy components (Φ_n, Φ_q, Φ_m) and complementary energy components $(\tilde{\Phi}_n, \tilde{\Phi}_q, \tilde{\Phi}_m)$ when mixed variational principles are constructed. The following relations are obtained;

$$\Phi_n + \tilde{\Phi}_n = \mathbf{n} \bullet \varepsilon, \quad \Phi_q + \tilde{\Phi}_q = \mathbf{q} \bullet \gamma, \quad \Phi_m + \tilde{\Phi}_m = \mathbf{m} \bullet \kappa. \quad (9.4)$$

where

$$\Phi_n = \frac{1}{2}\mathbf{C}_n \varepsilon \bullet \varepsilon, \qquad \tilde{\Phi}_n = \frac{1}{2}\mathbf{C}_n^{-1}\mathbf{n} \bullet \mathbf{n}. \quad (9.5)$$

$$\Phi_q = \frac{1}{2}\mathbf{C}_q \gamma \bullet \gamma, \qquad \tilde{\Phi}_q = \frac{1}{2}\mathbf{C}_q^{-1}\mathbf{q} \bullet \mathbf{q}. \quad (9.6)$$

$$\Phi_m = \frac{1}{2}\mathbf{C}_m \kappa \bullet \kappa, \qquad \tilde{\Phi}_m = \frac{1}{2}\mathbf{C}_m^{-1}\mathbf{m} \bullet \mathbf{m}. \quad (9.7)$$

Denoting a surface element, dA, of the shell mid-surface, M, the functional, π, is defined, due to a hybrid variational principle, as

$$\begin{aligned} \pi &= \int_M \left[\Phi_n + \Phi_q + \Phi_m\right] dA + \Phi_{ext} \\ &= \int_M \left[\mathbf{n} \bullet \varepsilon - \tilde{\Phi}_n + \mathbf{q} \bullet \gamma - \tilde{\Phi}_q + \mathbf{m} \bullet \kappa - \tilde{\Phi}_m\right] dA + \Phi_{ext} \end{aligned} \quad (9.8)$$

where Φ_{ext} is load potential.

The natural coordinates, ξ_α, describing the shell mid-surface are mapped on the bi-unit square of $(\xi, \eta) \in [-1, \ 1] \times [-1, \ 1]$ using standard shape functions, which are used for interpolation functions for the displacement and rotation fields. The stress resultant vectors have the following interpolation functions for nine-node-shell element[8]

$$\mathbf{n} = \mathbf{D}_n \mathbf{S}_n,$$

$$\mathbf{D}_n = \begin{bmatrix} 1 & \eta & \xi & \xi\eta & \eta^2 & \xi\eta^2 & 0 & 0 & 0 & 0 & 0 & 0 & 0 & 0 & 0 & 0 \\ 0 & 0 & 0 & 0 & 0 & 0 & 1 & \eta & \xi & \xi\eta & \xi^2 & \xi^2\eta & 0 & 0 & 0 & 0 \\ 0 & 0 & 0 & 0 & 0 & 0 & 0 & 0 & 0 & 0 & 0 & 1 & \eta & \xi & \xi\eta \end{bmatrix},$$

$$\mathbf{S}_n = \left[S_{n1} \quad S_{n2} \quad \ldots\ldots\ldots \quad S_{n16}\right]^T. \quad (9.9)$$

$$\mathbf{q} = \mathbf{D}_q \mathbf{S}_q ,$$

$$\mathbf{D}_q = \begin{bmatrix} 1 & \eta & \xi & \xi\eta & \eta^2 & \xi\eta^2 & 0 & 0 & 0 & 0 & 0 & 0 \\ 0 & 0 & 0 & 0 & 0 & 0 & 1 & \eta & \xi & \xi\eta & \xi^2 & \xi^2\eta \end{bmatrix},$$

$$\mathbf{S}_q = \begin{bmatrix} S_{q1} & S_{q2} & \cdots\cdots & S_{q16} \end{bmatrix}^T . \tag{9.10}$$

$$\mathbf{m} = \mathbf{D}_m \mathbf{S}_m , \quad \mathbf{D}_m = \mathbf{D}_n , \quad \mathbf{S}_m = \begin{bmatrix} S_{m1} & S_{m2} & \cdots & S_{m16} \end{bmatrix}^T . \tag{9.11}$$

Then we can write

$$\sigma = \begin{bmatrix} \mathbf{n} \\ \mathbf{q} \\ \mathbf{m} \end{bmatrix}, \qquad \mathbf{S} = \begin{bmatrix} \mathbf{S}_n \\ \mathbf{S}_q \\ \mathbf{S}_m \end{bmatrix}, \qquad \mathbf{e} = \begin{bmatrix} \varepsilon \\ \gamma \\ \kappa \end{bmatrix}. \tag{9.12}$$

$$\mathbf{C} = \begin{bmatrix} \mathbf{C}_n & 0 & 0 \\ 0 & \mathbf{C}_q & 0 \\ 0 & 0 & \mathbf{C}_m \end{bmatrix}, \qquad \mathbf{D} = \begin{bmatrix} \mathbf{D}_n & 0 & 0 \\ 0 & \mathbf{D}_q & 0 \\ 0 & 0 & \mathbf{D}_m \end{bmatrix}. \tag{9.13}$$

From Eq. (9.8) the variation of the functional π is obtained as

$$\delta\pi = \sum_1^L \left\{ \int_{Ae} \left(\mathbf{D}^T \mathbf{e} \cdot \delta\mathbf{S} + \mathbf{DS} \cdot \delta\mathbf{e} - \mathbf{C}^{-1}\mathbf{DS} \cdot \mathbf{D}\delta\mathbf{S} \right) \sqrt{a}\, j d\xi d\eta \right\} + \delta\Phi_{ext} \tag{9.14}$$

where L, Ae and j are the total number of element, the mapped shell mid-surface of each element and the Jacobian of mapping, respectively.

Also note that $dA = \sqrt{a}\, j d\xi d\eta$ and $a = \det\left(a_{\alpha\beta}\right)$. The condition $\delta\pi = 0$ yields

$$\sum_1^L \left\{ \int_{Ae} \mathbf{DS} \cdot \delta\mathbf{e} \sqrt{a}\, j d\xi d\eta \right\} + \delta\Phi_{ext} = 0 \tag{9.15}$$

$$\sum_{1}^{L}\left\{\int_{Ae}\left(\mathbf{D}^{T}\mathbf{e}-\mathbf{D}^{T}\mathbf{C}^{-1}\mathbf{D}\mathbf{S}\right)\sqrt{a}\,j\,d\xi\,d\eta\right\}=0 \tag{9.16}$$

Since the stress distributions are allowed to be discontinuous over elements whereas displacements and rotations are continuous, with the following definition

$$\mathbf{A}=\int_{Ae}\mathbf{D}^{T}\mathbf{C}^{-1}\mathbf{D}\sqrt{a}\,j\,d\xi\,d\eta \tag{9.17}$$

we can solve Eq. (9.16) for $\mathbf{S}$ at the element level and obtain

$$\mathbf{S}=\mathbf{A}^{-1}\int_{Ae}\mathbf{D}^{T}\mathbf{e}\sqrt{a}\,j\,d\xi\,d\eta \tag{9.18}$$

Then **S** can be computed from strain using Eq. (9.18), while stresses are computed using Eqs (9.9), (9.10) and (9.11). The linearisation of Eq. (9.16) yields

$$\sum_{1}^{L}\left\{\int_{Ae}(\mathbf{D}^{T}\Delta\mathbf{e}-\mathbf{D}^{T}\mathbf{C}^{-1}\mathbf{D}\Delta\mathbf{S})\sqrt{a}\,j\,d\xi\,d\eta\right\}=0 \tag{9.19}$$

Solving Eq. (9.19) with Eq. (9.18) for $\Delta\mathbf{S}$ on the element level leads to

$$\Delta\mathbf{S}=\mathbf{A}^{-1}\int_{Ae}\mathbf{D}^{T}\Delta\mathbf{e}\sqrt{a}\,j\,d\xi\,d\eta \tag{9.20}$$

On the other hand, the linearisation of Eq. (9.15) yields,

$$\sum_{1}^{L}\left\{\int_{Ae}\mathbf{D}\Delta\mathbf{S}\cdot\delta\mathbf{e}\sqrt{a}\,j\,d\xi\,d\eta+\int_{Ae}\mathbf{D}\mathbf{S}\cdot\Delta\delta\mathbf{e}\sqrt{a}\,d\xi\,d\eta\right\}+\Delta\delta\Phi ext=0 \tag{9.21}$$

Substituting Eq.(9.20) and $\sigma=\mathbf{DS}$ into Eq.(9.21), we obtain

$$\sum_{1}^{L}\{[\int_{Ae}\mathbf{D}^{T}\delta\mathbf{e}\sqrt{a}\,j\,d\xi\,d\eta]^{T}A^{-1}[\int_{Ae}\mathbf{D}^{T}\Delta\mathbf{e}\sqrt{a}\,j\,d\xi\,d\eta]+\int_{Ae}\sigma^{T}\Delta\delta\mathbf{e}\sqrt{a}\,j\,d\xi\,d\eta\}$$
$$+\Delta\delta\Phi_{\text{ext}}=0 \tag{9.22}$$

In this equation the first term is the tangent stiffness, composed of the material and geometric parts. In addition, the second variation of the functional was obtained in the same manner as in Ref.7.

9.3 Experimental background: initial imperfection measurement

Now we deal with a clamped shallow spherical shell subjected to uniform external pressure $\overline{P}$ shown in Fig. 9.3. The shell has the radius curvature R, the thickness t, the base radius a, the distribution of the initial imperfection w^0, Young's modulus E and Poisson's ratio ν and all the quantities were measured for each specimen. The original data of the measurement of E had been scattered with around $\pm 5\%$ of the mean values.

Schematically shown in Fig. 9.4 is the experimental setup at Tohoku University which is similar to that of Sunakawa and Ichida[19]. The test

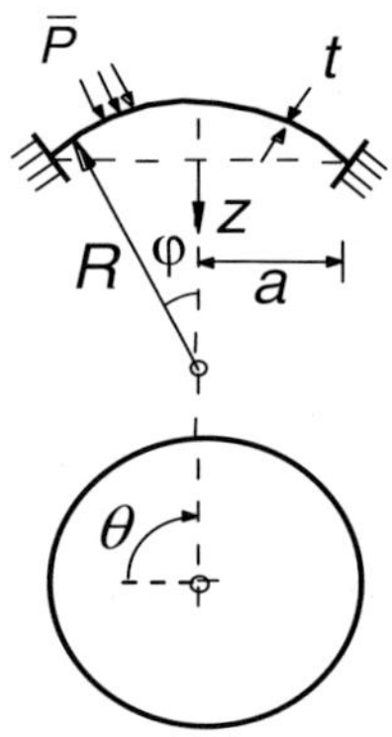

Fig. 9.3. Geometry of a shallow spherical shell.

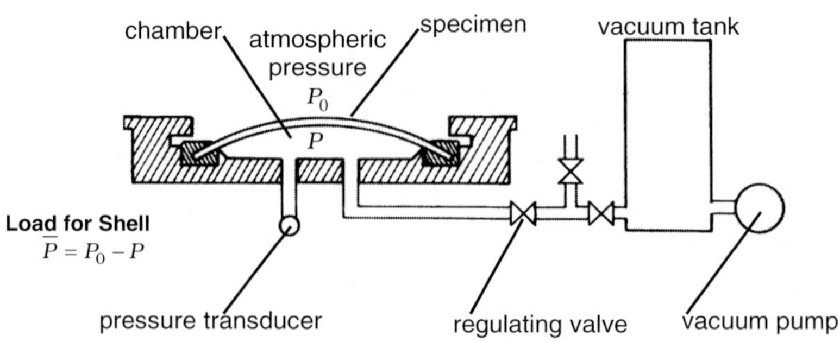

Fig. 9.4. Yamada and Yamada's experimental setup[20].

specimen was bonded, with an epoxy resin adhesive, to the rigid steel supporting ring to provide clamped boundary conditions. The external pressure was applied to the specimens by evacuated the air beneath them. The test system was designed to ensure carefully controlled quasi-static loading of the specimens. A vacuum pump evacuated a vacuum tank and via regulating valves which evacuated the space under the specimen called the chamber. The pressures were measured underneath the specimen by a pressure transducer. Yamada and Yamada[20] employed one linear displacement transducer, movable in the meridional and circumferential directions, to scan the specimens for the initial shapes as well as the deformed prebuckling and postbuckling shapes. The analog measurements of the normal deflection w were recorded on an X-Y recorder and then obtained numerically or graphically from the X-Y plot by means of a graphic reading system and a digital computer. The measurements were to an accuracy of 0.01mm (0.3-1 percent of the shell thickness t, and more precise than t itself, which was within ± 1.5 percent). Selected fifty-two initial imperfections are now available on the web[22].

The specimens in these experiments were formed by thermovacuum molding of rigid polyvinylchloride sheets. Many considerations made it possible to fabricate near-perfect specimens, and to change the type of distribution, the sign and the amplitude of the symmetrical component of initial geometrical imperfections. A strong dependence was observed between the type of distribution of symmetrical initial imperfections and the geometrical dimensions of the specimen. The sign and the amplitude of symmetrical initial imperfections were controlled in the process of forming the specimen using three different processes[11,12].

A total of seventy-nine specimens were tested[11,12], but succeeded in measuring the displacement distributions just before buckling point in fifty-two specimens, to allow consideration of the relationship with Huang's theoretical circumferential buckling wave-number n^H. In this chapter, these fifty-two experiments are selected.

The fifty-two shells were simulated using the mixed finite element method for the nine-node-shell elements. Equilibrium paths were traced by static, incremental analysis until overcoming the buckling point.

Table 9.1. Number of terms.

a mm	$j1$	$j2$	$n2$
150	8	8	12
170	9	9	12
200	10	10	12

Postbuckling paths of a few specimens were also traced. The specimen numbers used were the same as those in Refs.11 and 12; note that "S4", "S5", "S6-1", "S6-2" and "S7" in Ref.20 are "E99", "A87", "B91", "D88" and "C98", respectively.

Initial imperfection distributions towards the vertical direction were measured at $(\mathrm{Nr}\times\mathrm{Na}+1)$ points, where Nr= a mm/10mm was the number of points towards the base radius direction and Na= 24 was the number of points towards θ direction. The base radius of specimens was 150mm, 170mm and 200mm listed in Table 9.1. Initial imperfection distributions towards the vertical direction were approximated by the following Eq. (9.23) using the method of least squares.

$$\frac{w^0}{t} = \sum_{j=1}^{j1} (Ac_j \cos(j\alpha\varphi) + As_j \sin(j\alpha\varphi))$$

$$+\sum_{j=1}^{j2} \sum_{n=1}^{n2} (Bc_{jn} \cos(n\theta) + Bs_{jn} \sin(n\theta)) \cos(2(j-1)\alpha\varphi) \qquad (9.23)$$

$$\alpha = \pi/(2\phi)\,, \qquad \phi = \sin^{-1}(a/R)$$

9.4 Agreement on buckling loads

Three types of meshing $\mathrm{N}\times\mathrm{M}=5\times12$, 10×24 (shown in Fig. 9.5) and 20×48, were preliminary applied[10] to the specimen D89 (a =170 mm, R =1397 mm, t =1.43 mm, E =2.99 GPa, ν =0.36). Here N elements (2N nodes) towards φ direction and M elements (2M nodes) towards θ direction are adopted.

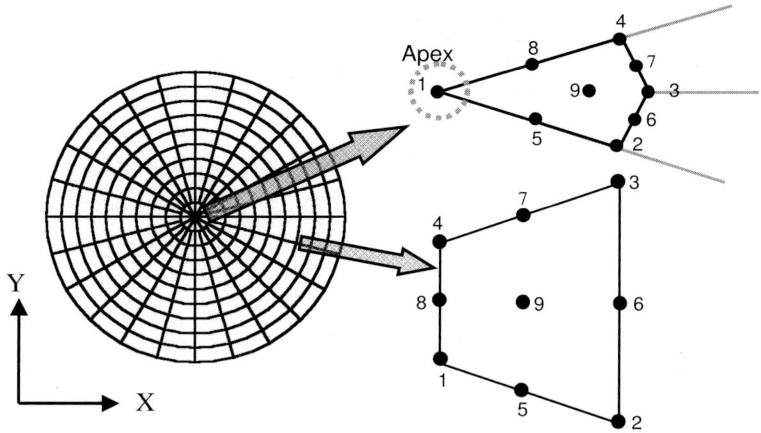

Fig. 9.5. A 10×24 mesh model.

The results are shown in Table 9.2 and Fig. 9.6. The difference in the deflection at the apex was only 3 % between the 10×24 mesh model and the 20×48 mesh model. Note that in nonlinear shell buckling problems the results using a relatively rough approximation (5×12 model in this example) gives lower buckling criteria in comparison to that using a precise model (20×48 model) generally. This is also obtained in the solution using the functional discreteness (Ritz method[23] or global Galerkin method[20]).

The selected fifty-two experiments are compared with the finite element analytical results in Table 9.3. The geometric parameter λ and the classical buckling pressure q_{cl} are as follows,

$$\lambda = \left\{12\left(1-\nu^2\right)\right\}^{1/4}\frac{a}{\sqrt{Rt}} \qquad q_{\text{cl}} = \frac{2E}{\sqrt{3\left(1-\nu^2\right)}}\left(\frac{t}{R}\right)^2 \tag{9.24}$$

In Table 9.3 the average of $Q_{\text{cr}}^{\text{c}}\,/\,Q_{\text{cr}}^{\text{ex}}$ is 1.06 with a standard deviation of 0.06. Taking into account the dispersion of Young's modulus in each specimen made from rigid polyvinylchloride sheets, it can be said that the FEM results are in good agreement with the experimental results. The difference between n^{ex} and n^{c} of A98, 81, A85, B82, C90, D95 is one. It shows difficulties of simulations of shallow spherical shells in which the amplitude of the asymmetric imperfection component was very small, and there were potentially many adjacent buckling modes.

Table 9.2. Buckling pressures and the deflection at the apex δ_0.

mesh	$\overline{P}_{cr}$ kPa	δ_0 mm
5×12	1.132	1.096
10×24	1.175	0.886
20×48	1.178	0.857

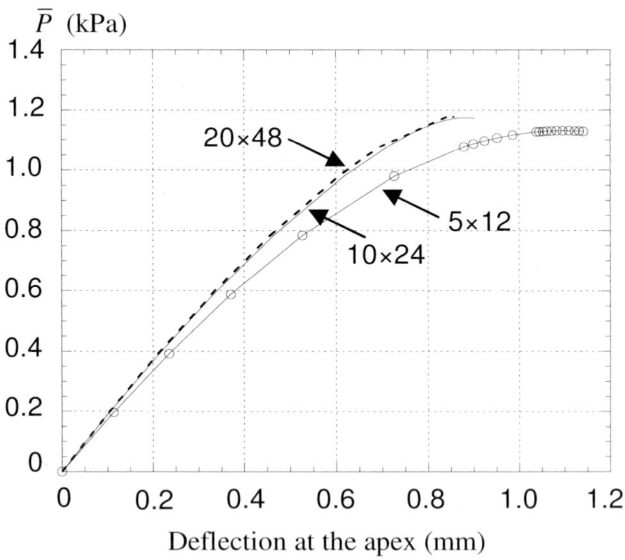

Fig. 9.6. Pressure versus the deflection at the apex.

Table 9.3. Comparisons between experiments and present results.

No.	λ	Q_{cr}^{ex}	n^{ex}	Q_{cr}^{c}	n^{c}	$\frac{Q_{cr}^{c}}{Q_{cr}^{ex}}$	No.	λ	Q_{cr}^{ex}	n^{ex}	Q_{cr}^{c}	n^{c}	$\frac{Q_{cr}^{c}}{Q_{cr}^{ex}}$
E96	3.86	0.71	0	0.78	0	1.12	D89	5.81	0.57	0	0.58	0	1.03
E99	3.87	0.70	0	0.67	0	0.96	B91	5.90	0.66	2	0.71	2	1.07
E95	3.90	0.69	0	0.81	0	1.18	A93	5.92	0.78	2	0.75	1	0.96
E97	3.92	0.74	0	0.82	0	1.11	A99	5.99	0.71	2	0.74	2	1.04
E98	4.00	0.61	0	0.67	0	1.10	A81	6.00	0.38	0	0.40	0	1.04
E92	4.02	0.56	0	0.54	0	0.97	A84	6.00	0.38	0	0.40	0	1.04
E90	4.06	0.55	0	0.56	0	1.03	D83	6.01	0.61	0	0.58	0	0.95
E94	4.06	0.53	0	0.58	0	1.09	D81	6.03	0.52	1	0.55	0	1.06
E91	4.12	0.53	0	0.55	0	1.04	A97	6.07	0.79	3	0.77	3	0.97
B81	4.16	0.53	0	0.54	0	1.02	D88	6.21	0.38	0	0.39	0	1.03
A89	4.18	0.45	0	0.53	0	1.18	F97	6.67	0.65	3	0.65	3	1.00
B80	4.22	0.50	0	0.54	0	1.08	D87	6.78	0.58	2	0.58	2	1.00
F93	4.74	0.75	1	0.81	1	1.08	C93	6.86	0.59	2	0.56	2	0.94
F92	4.75	0.78	1	0.82	1	1.05	F96	6.92	0.65	3	0.69	3	1.07
D85	4.94	0.47	0	0.50	0	1.06	A86	7.04	0.66	2	0.64	2	0.97
D86	4.96	0.45	0	0.50	0	1.10	E81	7.15	0.63	2	0.59	2	0.95
B95	4.97	0.71	0	0.63	0	0.89	A85	7.18	0.66	2	0.66	3	1.00
B94	5.02	0.68	0	0.63	0	0.92	C98	7.29	0.75	3	0.74	3	0.98
A87	5.03	0.56	0	0.55	0	0.99	B82	7.52	0.70	1	0.68	2	0.97
A88	5.04	0.54	0	0.54	0	1.00	C90	7.55	0.79	3	0.71	2	0.90
B93	5.04	0.72	0	0.64	0	0.89	D95	8.22	0.71	3	0.73	4	1.03
D84	5.05	0.52	0	0.49	0	0.95	F90	8.27	0.67	4	0.69	4	1.03
F95	5.66	0.65	1	0.67	1	1.07	F91	8.31	0.70	4	0.71	4	1.01
B90	5.72	0.65	2	0.69	2	1.06	D93	8.34	0.73	4	0.72	4	0.98
F94	5.74	0.65	1	0.69	1	1.07	D99	8.38	0.74	4	0.75	4	1.01
A83	5.76	0.37	0	0.40	0	1.06	D98	8.41	0.75	4	0.75	4	0.99

ex: experiments, c: computed,
Q = buckling pressure normalised by the classical buckling pressure,
n: circumferential wave number of dominate deflection component near the buckling point

9.5 Prebuckling deflection modes near the buckling points

Figures 9.7 and 9.8 show the experimental and FEM results for B91 ("S6-1" in Ref.20) and C98 ("S7" in Ref.20), respectively. The FEM result was in good agreement with the experiment[12] and the numerical analysis of the global Galerkin method[20]. The deflection contour maps for B91 and C98 in Figs 9.7 and 9.8 were also in good agreement with those of experimental results. Numbers of circumferential wave-number of dominate deflection component at the near-buckling point of B91 and C98 are 2 and 3, respectively.

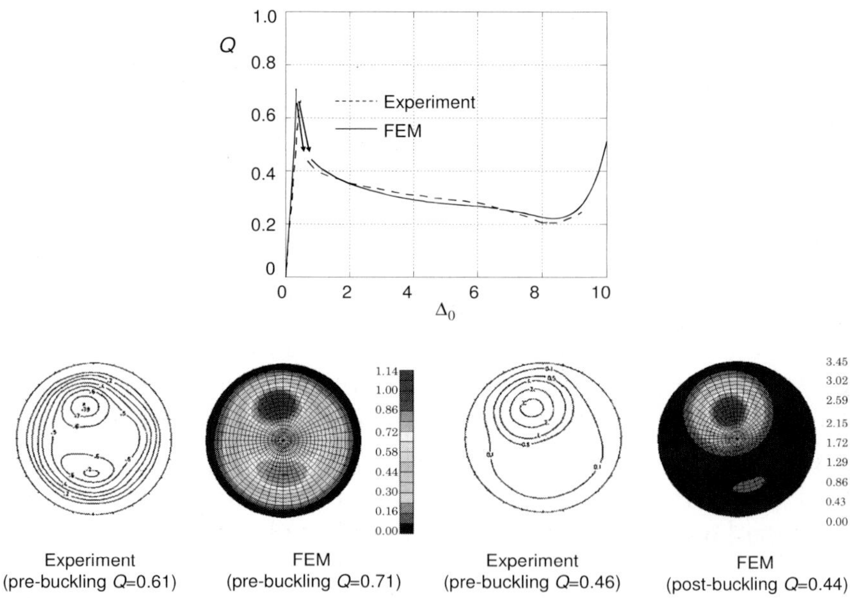

Fig. 9.7. Results for B91 ($Q=q/q_{cl}$, q is pressure, Δ_0: $=\delta_0/t$, and δ_0 is the deflection at the apex).

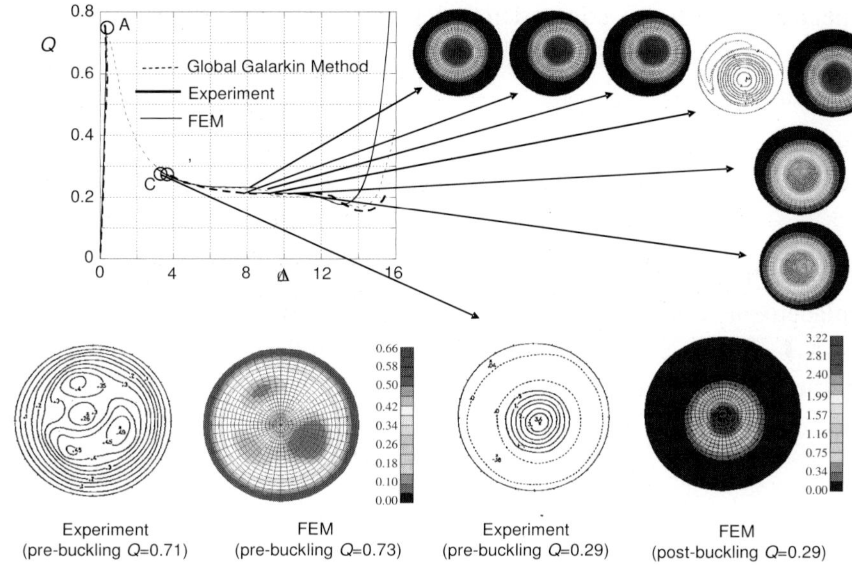

Fig. 9.8. Results for C98 ($Q=q/q_{cl}$, q is pressure, Δ_0: $=\delta_0/t$, and δ_0 is the deflection at the apex).

9.6 Postbuckling deflection behaviour at the static equilibrium state

In Fig. 9.4, the air in the chamber was slowly reduced by a vacuum pump until the shell reached buckling, and then the regulating valve was made tight during the vibration of the shell. In the numerical simulation, the external pressure P_0 was assumed to be constant, and then the internal pressure P or the volume in the chamber V was controlled. During the iteration process in each increment, the following constraint condition was adopted due to the referred experimental condition[12].

$$PV = \text{constant} \tag{9.25}$$

In the incremental calculation process, Yamada et al.[24] controlled the product PV with respect to the chamber, and called this the P-V incremental method. We denote σ, ε, φ and $\mathbf{v}_n$ are stress and strain of shell, the position and normal vector, respectively. Now the region of mid-plane of shell is defined as Se, and number of nodes per shell element and the shape functions are denoted as n and N_{i} $(\mathrm{i}=1,\dots,n)$, respectively. Let the external nodal force and the loading pressure be $\mathbf{f}_{\mathrm{i}}$ $(\mathrm{i}=1,\ldots,n)$ and $\overline{P}=P_0-P$, respectively, then the virtual work may be written as follows,

$$\int_{Se}\sigma^{\mathrm{T}}\,\delta\varepsilon\,\mathrm{d}S=\sum_{\mathrm{i}=1}^{n}\mathbf{f}_{\mathrm{i}}^{\mathrm{T}}\,\delta\varphi_{\mathrm{i}}+\int_{Se}\overline{P}\,\mathbf{v}_{\mathrm{n}}^{\mathrm{T}}\delta\varphi\,\mathrm{d}S \tag{9.26}$$

We set $\Psi^{\mathrm{T}}=[\varphi_1^{\mathrm{T}}\;\;\varphi_2^{\mathrm{T}}\;\;\ldots\;\;\varphi_n^{\mathrm{T}}]$ and $\delta\varepsilon=\mathbf{B}\delta\Psi$ with a strain-displacement operator $\mathbf{B}$. Introducing $\mathbf{F}^{\mathrm{T}}=[\mathbf{f}_1^{\mathrm{T}}\;\;\mathbf{f}_2^{\mathrm{T}}\;\;\ldots\;\;\mathbf{f}_n^{\mathrm{T}}]$ and $\mathbf{b}=[N_1\mathbf{I}\;\;N_2\mathbf{I}\;\;\ldots\;\;N_n\mathbf{I}]$ to Eq. (9.26), the following equation for an arbitrary $\delta\Psi$ is obtained,

$$\int_{S\mathrm{e}}\mathbf{B}^{\mathrm{T}}\sigma\mathrm{d}S-\mathbf{F}-\overline{P}\int_{S\mathrm{e}}\mathbf{b}^{\mathrm{T}}\mathbf{v}_{\mathrm{n}}\mathrm{d}S=\mathbf{0} \tag{9.27}$$

Through linearising Eq. (9.27), the following equation may be obtained using $\Delta\mathbf{b}=\mathbf{0}$ and $\Delta\mathbf{F}=\mathbf{0}$ for this kind of conservative loading problem,

$$\int_{S\mathrm{e}}\Delta(\mathbf{B}^{\mathrm{T}}\sigma)\mathrm{d}S-\Delta\overline{P}\int_{S\mathrm{e}}\mathbf{b}^{\mathrm{T}}\mathbf{v}_{\mathrm{n}}\mathrm{d}S-\overline{P}\int_{S\mathrm{e}}\mathbf{b}^{\mathrm{T}}\Delta\mathbf{v}_{\mathrm{n}}\mathrm{d}S=\text{residual} \tag{9.28}$$

Using a vector $\mathbf{a}_{\mathrm{e}}$ and the element tangent stiffness matrix $\mathbf{K}_{\mathrm{e}}$ which is given from the mixed finite element method, the first and the second terms of Eq. (9.28) are transformed into the following,

$$\int_{S\mathrm{e}}\Delta(\mathbf{B}^{\mathrm{T}}\sigma)\mathrm{d}S=\mathbf{K}_{\mathrm{e}}\Delta\Psi \tag{9.29}$$

$$\Delta\overline{P}\int_{S\mathrm{e}}\mathbf{b}^{\mathrm{T}}\mathbf{v}_{\mathrm{n}}\mathrm{d}S=\Delta\overline{P}\int_{S\mathrm{e}}\mathbf{b}^{\mathrm{T}}\sum_{\mathrm{i}=1}^{n}N_{\mathrm{i}}\mathbf{v}_{\mathrm{n}_{\mathrm{i}}}\mathrm{d}S\equiv\Delta\overline{P}\mathbf{a}_{\mathrm{e}} \tag{9.30}$$

We denote $(\;)_{,\alpha}$ $(\alpha=1,2)$ as differentiation with respect to the natural coordinates ξ_α of an element and $\hat{\varphi}_{,\alpha}$ as a matrix operator of form of $\varphi_{,\alpha}\times$. The linearisation of $\mathbf{v}_{\mathrm{n}}$ becomes following with $\overline{\mathbf{v}}_{\mathrm{n}}=\varphi_{,1}\times\varphi_{,2}$.

$$\Delta\mathbf{v}_{\mathrm{n}}=\frac{1}{|\,\overline{\mathbf{v}}_{\mathrm{n}}\,|}(\mathbf{I}-\mathbf{v}_{\mathrm{n}}\mathbf{v}_{\mathrm{n}}^{\mathrm{T}})\sum_{\mathrm{i}=1}^{n}(-N_{\mathrm{i},1}\hat{\varphi}_{,2}+N_{\mathrm{i},2}\hat{\varphi}_{,1})\Delta\varphi_{\mathrm{i}} \tag{9.31}$$

Then the third term of Eq. (9.28) can be obtained as follows,

$$\bar{P}\int_{S_e} \mathbf{b}^{\mathrm{T}} \Delta \mathbf{v}_{\mathrm{n}} \mathrm{d}S = \bar{P}\mathbf{A}_{\mathrm{e}}\Delta\Psi \tag{9.32}$$

$$\mathbf{A}_{\mathrm{e}} = \left[\mathbf{A}_{\mathrm{e}_{\mathrm{ij}}}\right], \quad \mathbf{A}_{\mathrm{e}_{\mathrm{ij}}} = \frac{N_{\mathrm{i}}}{|\bar{\mathbf{v}}_{\mathrm{n}}|}(\mathbf{I} - \mathbf{v}_{\mathrm{n}}\mathbf{v}_{\mathrm{n}}^{\mathrm{T}})(-N_{\mathrm{j},1}\hat{\varphi}_{,2} + N_{\mathrm{j},2}\hat{\varphi}_{,1}) \tag{9.33}$$

Using **K**, **A** and **a,** obtained by assembling $\mathbf{K}_{\mathrm{e}}$, $\mathbf{A}_{\mathrm{e}}$ and $\mathbf{a}_{\mathrm{e}}$, respectively, Eq. (9.28) becomes, with respect to the incremental displacement $\Delta\mathbf{u}$ and the incremental pressure $\Delta P = -\Delta\bar{P}$,

$$\begin{bmatrix} \mathbf{K} - (P_0 - P)\mathbf{A} & \mathbf{a} \end{bmatrix} \begin{bmatrix} \Delta\mathbf{u} \\ \Delta P \end{bmatrix} = \text{residual} \tag{9.34}$$

On the other hand, the linearisation of the constraint condition, Eq. (9.25), introduces the following equation,

$$\Delta P V + P \Delta V = \text{residual} \tag{9.35}$$

The volume in the chamber V can be considered to consist of variable part V_1 and constant part V_{a}. The volume V_1 is calculated with the vector $\mathbf{v}_{\mathrm{h}}$ which is perpendicular to a base plane of a specimen

$$V_1 = \int_{As} (\varphi - \varphi_{\mathrm{p}})^{\mathrm{T}} \mathbf{v}_{\mathrm{h}} \mathrm{d}A = \int_S (\varphi - \varphi_{\mathrm{p}})^{\mathrm{T}} \mathbf{v}_{\mathrm{h}} \mathbf{v}_{\mathrm{h}}^{\mathrm{T}} \mathbf{v}_{\mathrm{n}} \mathrm{d}S \tag{9.36}$$

$$\Delta V = \Delta V_1 = \int_{As} \Delta\varphi^{\mathrm{T}} \mathbf{v}_{\mathrm{h}} \mathrm{d}A = \int_S \mathbf{v}_{\mathrm{h}}^{\mathrm{T}} \sum_{\mathrm{i}=1}^{n} N_{\mathrm{i}} \Delta\varphi_{\mathrm{i}} \mathbf{v}_{\mathrm{h}}^{\mathrm{T}} \mathbf{v}_{\mathrm{n}} \mathrm{d}S = \mathbf{h}^{\mathrm{T}} \Delta\Psi \tag{9.37}$$

$$\mathbf{h}^{\mathrm{T}} = [\mathbf{h}_1^{\mathrm{T}} \quad \mathbf{h}_2^{\mathrm{T}} \quad \ldots \quad \mathbf{h}_n^{\mathrm{T}}], \qquad \mathbf{h}_{\mathrm{i}} = \int_S N_{\mathrm{i}} \mathbf{v}_{\mathrm{h}}^{\mathrm{T}} \mathbf{v}_{\mathrm{n}} \mathbf{v}_{\mathrm{h}} \mathrm{d}S \tag{9.38}$$

where As means the region of the base plane. From Eqs (9.34) and (9.37), the following equation is obtained.

$$\begin{bmatrix} \mathbf{K} - (P_0 - P)\mathbf{A} & \mathbf{a} \\ P\mathbf{h}^{\mathrm{T}} & V \end{bmatrix} \begin{bmatrix} \Delta\mathbf{u} \\ \Delta P \end{bmatrix} = \text{residual} \tag{9.39}$$

The FEM results obtained by the present P-V incremental method, for the postbuckling deflections of B91 and C98, are depicted in Figs 9.7 and 9.8, respectively. They are in good agreement with the experimental measurement in Ref.12. Care that the FEM results using the so-called arc-length incremental method[10] were not fully able to obtain at some complex postbuckling equilibrium region in which one-dent-type deflection mode at the postbuckling state of the cap is neutral in the circumferential direction.

Shown in Fig. 9.9 are selected three initial imperfection modes[22] in Ref.24. Now the shell having a = 200 mm, R = 540 mm, t = 0.9 mm, E = 3 GPa, $\nu = 0.36$ and the geometric parameter λ $(=[12(1-\nu^2)]^{1/4} a/\sqrt{Rt})$ = 16.3 is adopted here, then its classical buckling pressure $q_{cl} = 2E/\sqrt{3(1-\nu^2)}\cdot(t/R)^2$ is 10.3 kPa.

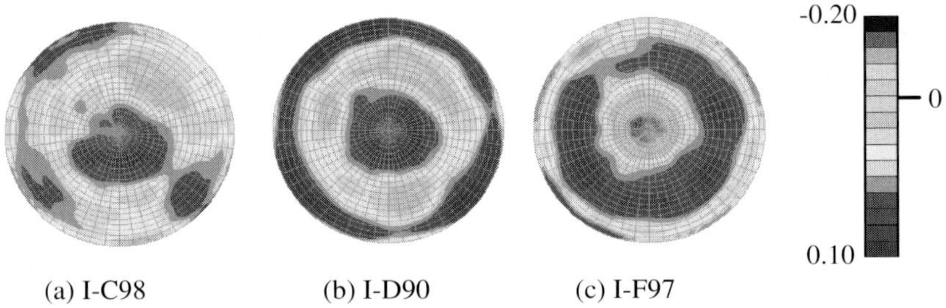

Fig. 9.9. Initial imperfections (in mm).

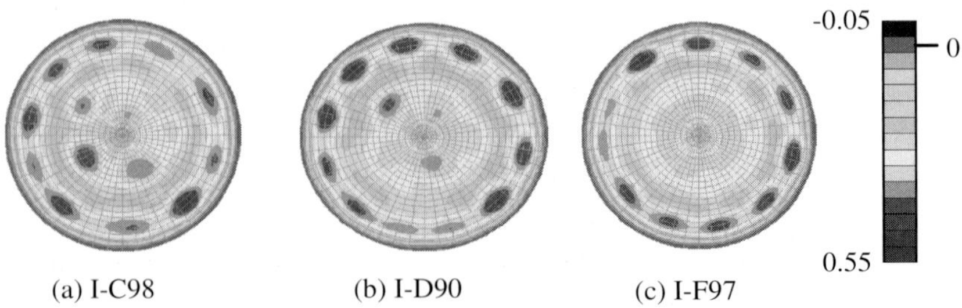

Fig. 9.10. Vertical displacement distribution just before the buckling (in mm).

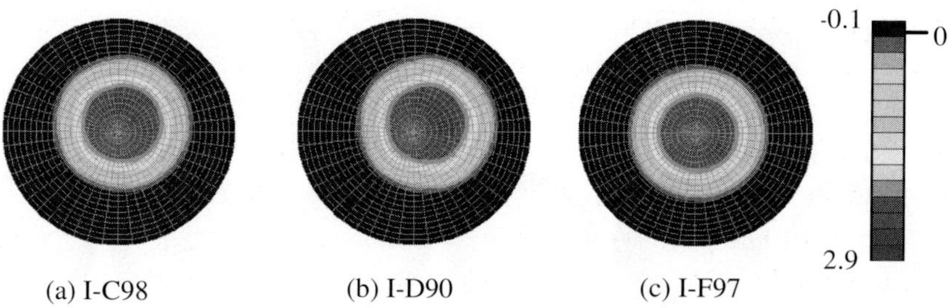

Fig. 9.11. Vertical displacement distribution just after the buckling (in mm).

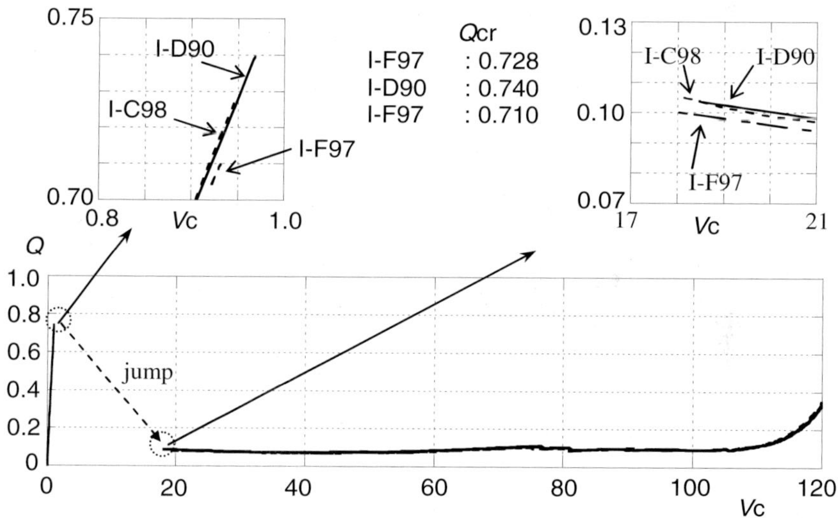

Fig. 9.12. Q versus Vc curves.

Initial shell directors were set as the same as those of a perfect spherical shell because the initial imperfections were small. The models were discretised by 20 elements toward φ direction and 48 elements toward θ direction using nine-node shell elements. The external pressure P_0 was adopted to be 101.3 kPa.

Figures 9.10 and 9.11 illustrate the vertical displacement distribution at pre- and postbuckling states[24]. Figure 9.12 shows the non-dimensional pressure Q $(=\bar{P}/q_{cl})$ versus Vc = (volume change) / (ta^2). From pre-buckling states to post-buckling states $\Delta P V = -101.3$ kPa$\cdot$cm^3 was

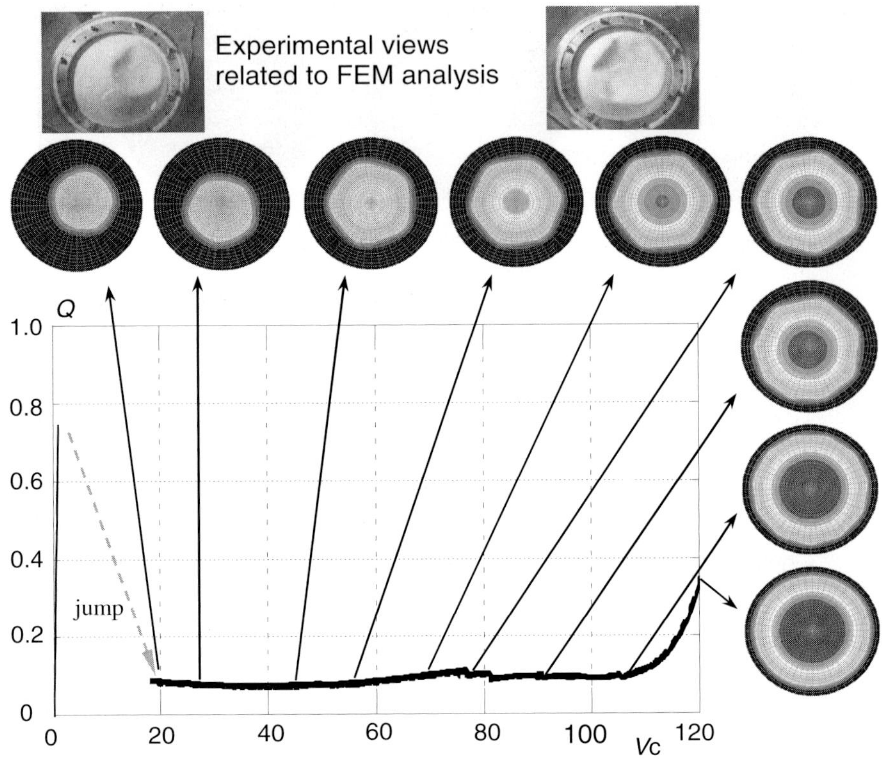

Fig. 9.13. The change of displacement distribution of I-D90.

adopted. At the jumping step associated with snap-through instability phenomena, 60 to 200 iteration steps were needed; however, only three or four steps were required for the prebuckling and postbuckling paths without jumping. The buckling pressure Q in these three cases were 0.728 (I-C98), 0.740 (I-D90) and 0.710 (I-F97). The effect of these initial imperfections on the buckling pressures was slight (4 %). Shown in Fig. 9.10 are the dominantly circumferential wave numbers at the near-buckling point n^B; 10 (I-C98), 11 (I-D90) and 11 (I-F97). Though there are potentially many adjacent buckling eigenmodes, the difference of n^B is only one. In Fig. 9.12, I-C98 and I-D90 show pentagonal displacement distribution at the first equilibrium state after jumping. In Fig. 9.13, the change of displacement distribution of I-D90 is illustrated and it is observed that the displacement distribution mode changes from pentagonal shape to circular one. Many snap-through phenomena may be managed to obtain the solutions by using the P-V incremental method. The use of a traditional arc-length method was also investigated but convergence difficulties.

9.7 Vibration behaviour just after buckling

According to the measurements in Ref.11, the damping factors of a modal damping at no loading state were between 1 % and 2 %. Rayleigh type damping, which was proportional to current stiffness, was used for internal viscous damping, and the damping factor *hi* was set at 1 % for the first eigenfrequency at no loading state.

The air around the specimen was assumed to perform as a dashpot during vibration, that is, its external viscous damping was modelled as a uniform dashpot acting normal to the shell surface. When a_i is used as the area at the node "i", the normal vector $\mathbf{v}_\mathrm{i}$ and the matrix $\mathbf{A}_\mathrm{i}$ can be introduced as follows

$$\mathbf{v}_\mathrm{i} = \begin{bmatrix} v_{\mathrm{i}1} & v_{\mathrm{i}2} & v_{\mathrm{i}3} \end{bmatrix}^\mathrm{T} \tag{9.40}$$

$$\mathbf{A}_\mathrm{i} = \begin{bmatrix} a_\mathrm{i} \,|\, \mathbf{v}_{\mathrm{i}1} \,| & 0 & 0 \\ 0 & a_\mathrm{i} \,|\, \mathbf{v}_{\mathrm{i}2} \,| & 0 \\ 0 & 0 & a_\mathrm{i} \,|\, \mathbf{v}_{\mathrm{i}3} \,| \end{bmatrix} \tag{9.41}$$

Then a damping matrix, $\mathbf{C}_\mathrm{external}$, constructed of dashpots acting normal to the shell surface may be written with a constant γ.

$$\mathbf{C}_\mathrm{external} = \gamma \mathbf{A} \tag{9.42}$$

$$\mathbf{A} = \left[\mathrm{diag}(\mathbf{A}_\mathrm{i}) \right] \tag{9.43}$$

The well-known kinetic equation of free vibration is written as

$$\mathbf{M}\ddot{\mathbf{x}} + \mathbf{C}\dot{\mathbf{x}} + \mathbf{K}\mathbf{x} = \mathbf{0} \tag{9.44}$$

where **M** and **K** are the mass and tangent stiffness matrices, respectively, and

$$\mathbf{C} = \mathbf{C}_{\text{external}} + \mathbf{C}_{\text{int ernal}} \tag{9.45}$$

From Eq.(9.44) the first eigenvector **u** and its conjugate complex vector $\bar{\mathbf{u}}$ are obtained. Since the shell is shallow, vibration normal to the shell surface is much larger than that towards other directions. Then the following constants m and c are set as

$$m = \bar{\mathbf{u}}^{\mathrm{T}} \mathbf{M} \mathbf{u} \cong \rho t \, \bar{\mathbf{u}}^{\mathrm{T}} \mathbf{A} \mathbf{u} \tag{9.46}$$

$$c = \bar{\mathbf{u}}^{\mathrm{T}} \mathbf{C} \mathbf{u} = \gamma \, \bar{\mathbf{u}}^{\mathrm{T}} \mathbf{A} \mathbf{u} \tag{9.47}$$

where ρ and t are the density and the thickness, respectively. Figure 9.14 is a schematic time history of postbuckling vibration. Using the deflections x_n, x_{n+1} and the period T, the following relation is given.

$$\ln\left(\frac{x_n}{x_{n+1}}\right) = \frac{c}{2m} T = \frac{\gamma}{2\rho t} T \tag{9.48}$$

From the measurement on the postbuckling vibration reported in Ref.11, x_n=4.0, x_{n+1}=2.8, T=0.0167sec, $\rho = 1.4\,\text{g/cm}^3$ and t=0.145cm were used. Then $\gamma = 8.671\,\text{g/(cm}^2\,\text{sec)}$ was obtained through Eq. (9.48) on the assumption that the internal viscous damping was sufficiently small during instable condition.

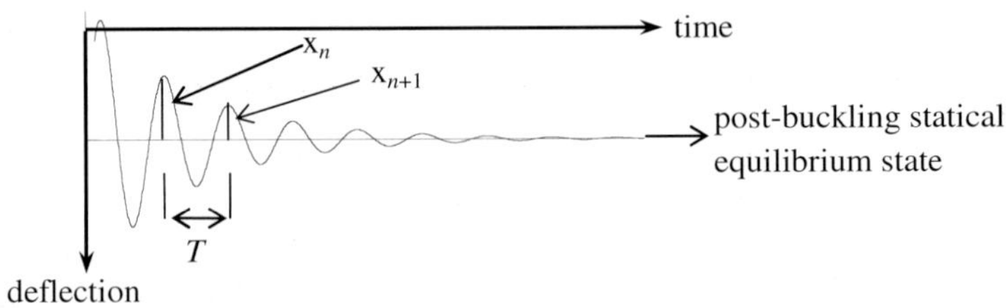

Fig. 9.14. Experimental postbuckling deflection behaviour[11].

Table 9.4 is the input data of C98 reported in Ref.11 and the present results of static analysis are illustrated in Table 9.3 and Fig. 9.8. The FEM results are in good agreement with those of the experiment[11] and the numerical analysis by the Galerkin method[20].

Table 9.4. Input data for C98.

a	170mm
R	1809.62mm
t	0.97mm
E	2.99 GPa
ν	0.36
ρ	1.4 g/cm^3

Here only three vibration cases for C98, Case 1 (*hi*=0%, having no dashpot), Case 2 (*hi*=1.0%, having no dashpot) and Case 3 (*hi*=1.0%, having dashpot), are shown in Fig. 9.15. A time step of 0.0001 seconds was adopted. The first eigenfrequency at no loading state was 909.515 rad/sec. Figure 9.15(a) shows the time history of the deflection at the point, (ϕ=0.449 deg, θ=255.deg). The dynamic behaviours of Case 1 and Case 2 are complex, much similar to those calculated in Ref.11. The result for Case 3, which has the external viscous damping, is shown to have simple vibration characteristics similar to those measured in Ref.11; also the static equilibrium state was obtained in 0.12 seconds.

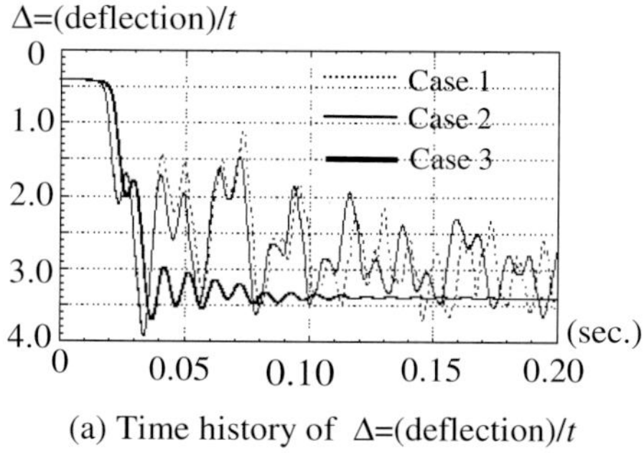

(a) Time history of Δ=(deflection)/t

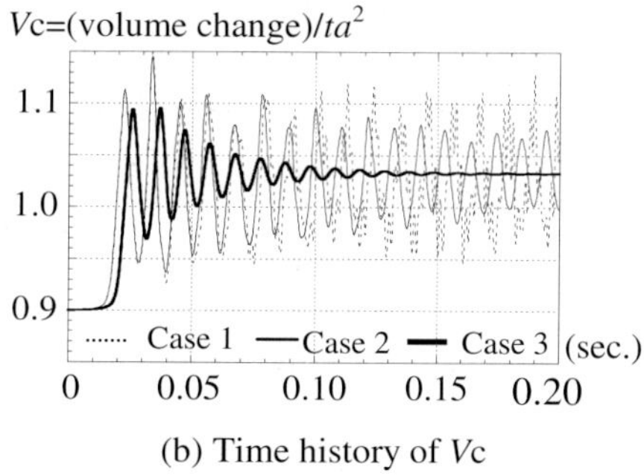

(b) Time history of Vc

Fig. 9.15. Time history.

Figures 9.16 and 9.17 are phase planes with respect to volume change. Even if the damping ratio is very small, the behaviour of Case 2 shows a motion slowly converging at the static postbuckling equilibrium point. The behaviour of Case 3, which has internal and external viscous damping, shows a manner of quick convergence at the postbuckling static equilibrium point. Figure 9.18 illustrates the distributions of postbuckling deflection of the experiment in Ref.11 and the present simulation result, respectively; both are in very good agreement.

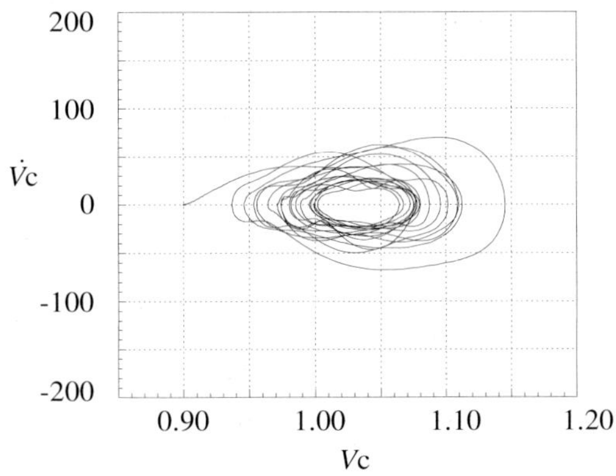

Fig. 9.16. Phase plane (Case 2).

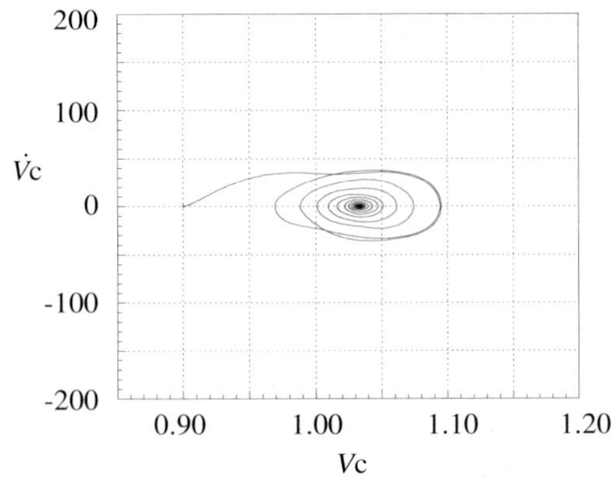

Fig. 9.17. Phase plane (Case 3).

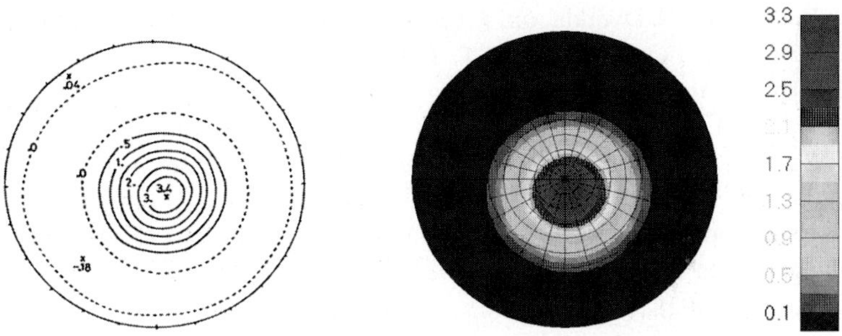

(a) Experiment[11] (b) FEM[10]

Fig. 9.18. Postbuckling deflection in mm.

9.8 Conclusions

For the case of pressurised shallow spherical shells, a nonlinear finite element analysis of elastic pre- and postbuckling deflection has been developed. The mixed variational formulation for a nine-node-shell element has been presented. Fifty-two results selected from seventy-nine experiments by Yamada[11,12] have been compared with the FEM simulations by using each precisely measured initial-shape involving imperfection components. The simulated buckling loads and the deflection behaviour have been seen to be in very good agreements with the complex experimental results. The detailed postbuckling large deflection modes of non-shallow caps, having intrinsically polygonal distribution completely different with the circular one of the shallow caps, have been discovered theoretically. It has also been emphasized that the spherical caps have a high sensitivity to geometric imperfections in their buckling behaviour.

9.9 References

1. S. Ahmad, B.M. Irons and O.C. Zienkiewicz, *Int. J. Numer. Methods Eng.*, 2, 419-451, (1970).
2. E. Ramm, *Formulations and Algorithms in Finite Element Analysis*, MIT Press, (1977).
3. T.J.R. Hughes and W.K. Liu, C*omp. Methods Appl. Mech. Eng.*, 26, 331-362, (1981).

4. K.J. Bathe and E.N. Dvorkin, *Int. J. Computer-Aided Eng. Software*, 1, (1984).
5. W.K.Liu, E.S.Law, D. Lam and T. Belyschko, *Methods Appl. Mech. Eng.*, 55, 259-300, (1986).
6. J.C. Simo and D.D. Fox, *Comp. Methods Appl. Mech. Eng.*, 72, 267-304, (1989).
7. J.C. Simo, D.D. Fox and M.S. Rifai, *Comp. Methods Appl. Mech. Eng.*, 79, 21-70, (1990).
8. C. Sansour and H. Bufler, *Int. J. Numer. Methods Eng.*, 34, 73-115, (1992).
9. H. Noguchi and T. Hisada, *Trans. of the Japan Society of Mechanical Engineers*, 58, A, 943-950, (1992).
10. Manabu Uchiyama and Seishi Yamada, *J. Eng. Mech.*, ASCE, 129, 707-714, (2003).
11. Seishi Yamada, *Experimental investigation of shallow spherical shells*, PhD dissertation, Tohoku University, (1980).
12. Seishi Yamada, Kazuo Uchiyama and Motohiko Yamada, *Int. J. Non-Linear Mech.*, 18, 37-54 (1983).
13. T. von Karman and H.S. Tsien, *J. Aeronatural Sci.*, 7, 43-50, (1939).
14. B. Budiansky, *Proc. Symp. on the Theory of Thin Elastic Shells*, 64-94, (1959).
15. A. Kaplan and Y.C. Fung, *NACA Note* 3212, (1954).
16. R.H. Homewood, A.C. Brine and A.E. Johnson, *Proc. Exp. Stress Analysis*, 18, 88, (1961).
17. N.C. Huang, *J. Applied Mech.*, ASME, 31, 447-457, (1964).
18. S.C. Tillman, *Int. J. Solid Structures*, 6, 37-52, (1970).
19. M. Sunakawa and K. Ichida, *J. Japan Soc. Aeronaut. Space Sci.*, 21, 263-270, (1973).
20. Motohiko Yamada and Seishi Yamada, *Collapse*, Cambridge University Press, 431-441, (1983).
21. Seishi Yamada: *Proceedings IASS Symposium*, Istanbul, 393-402, (2000).
22. http://www.st.tutrp.tut.ac.jp/~yamada/OpenImpData.htm
23. Seishi Yamada and J.G.A. Croll, *J. Applied Mechanics,* ASME, 66, 299-309, (1999).
24. Seishi Yamada, Manabu Uchiyama, Motohiko Yamada: *Proc. 5th Int. Conf. on Computation of Shell and Spatial Structures*, Salzburg, (2005).

CHAPTER 10

NONLINEAR BUCKLING IN SANDWICH STRUTS: MODE INTERACTION AND LOCALIZATION

M. Ahmer Wadee

Department of Civil and Environmental Engineering
Imperial College of Science, Technology & Medicine
London SW7 2AZ, UK
E-mail: a.wadee@imperial.ac.uk

Recent developments in the variational modelling of nonlinear interactive buckling phenomena, found in compression sandwich panels, are presented. Strong interactions between local and global modes are found to lead to localized buckling, in the more compressed face-plate, after the global instability. Important features such as the effect of using orthotropic core materials and face-plates of differing thicknesses, the potential for face–core delamination, panels under combined bending and compression, and the sensitivity to imperfections are discussed. The combination of modal interactions and the introduction of initial geometric imperfections in the struts shows that the critical buckling loads from linear analysis can significantly overestimate their true elastic strength.

10.1 Introduction

Sandwich construction comprising two stiff face-plates separated by a softer core material is popular as a provider of structural strength combined with weight efficiency. It is used extensively in astronautic,[1] aeronautic[2,3] and marine applications,[4] but its efficiency also makes it vulnerable to complicated nonlinear collapse mechanisms.[5–8]

Figure 10.1 shows a sequence of test photographs of a compression sandwich panel in which the different phases of the loading history are represented: (a) pre-buckling followed by (b) overall (Euler) buckling and then (c) interactive buckling where the structure becomes unstable and localization is observed.[9] It is clear from this sequence of photographs that local buckling, which has *localized* to the centre of one face-plate, has contributed significantly to loss of integrity in the overall sense; local failure of the face-

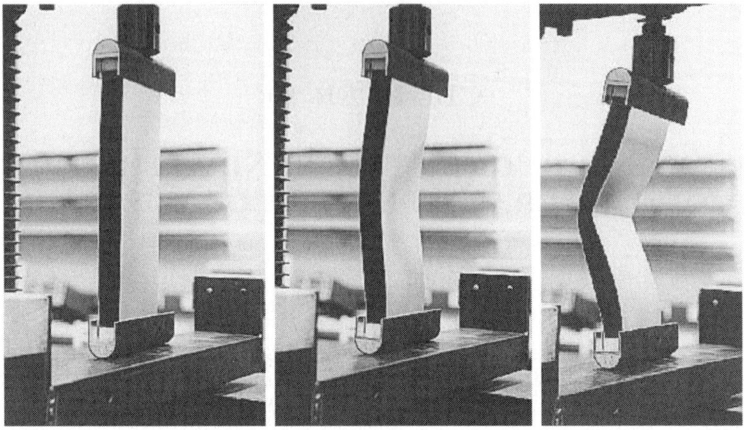

Fig. 10.1. A sandwich panel under axial compression: (a) pre-buckling; (b) critical buckling; (c) interactive buckling.

plate interacts with overall failure as it loses stiffness and shortens. It is well known from classical work[10] that compressed sandwich panels sometimes fail by this combination of overall buckling and local buckling in the face-plates. The nonlinear interaction of overall and local modes of instability, moreover, is common in many component types: examples being stiffened plates,[11] built-up (or reticulated or compound) columns[12] and cylindrical shells.[13] The mechanical response of the compression sandwich panel shows many of the features that these other structures are prone to, such as secondary instabilities and buckle pattern localization. Such phenomena have occupied some attention in the nonlinear buckling literature[14,15] and only until recently, before the advent of modern algebraic manipulation and numerical techniques, analytical treatment leading to practical quantitative information was not viable; the inherent difficulty in such modelling of sandwich structures lying in the basic assumptions connected with bending theory.

Classical sandwich theory,[10] unlike standard Euler–Bernoulli (or engineer's) bending theory, allows shearing displacement in the core; the interactive buckle of Fig. 10.1 clearly could not take place under the constraint of infinite core shear stiffness. The buckling or "wrinkling" of a single face-plate is similar to that of a strut on an elastic foundation, with the core material providing the foundation support. However, the simplest such model, employing the so-called *Winkler* foundation[16] and leading to a fourth order ordinary differential equation, takes the foundation as comprising translational springs alone; shearing resistance is ignored entirely. In the context of a sandwich structure this would deny the system overall integrity, leaving it free to respond as two separate faces buckling, to all intents and purposes, independently. Inclusion of shear in the foundation complicates matters considerably in that at the very least there would be an increase in the order of the underlying governing equation. Such effects are unavoidable and appear naturally in the model formulation presented.

The theoretical development begins with a brief description of the approach to modelling overall buckling of the sandwich panel. To include the possibility of local buckling in conjunction with overall buckling, contributions from both direct and shear stresses are essential in a fully interactive formulation. Overall buckling is represented by two degrees-of-freedom of the long wave mode, and the local buckling profile is determined by applying the calculus of variations to a potential energy functional containing two functions representing respectively the lateral and in-plane displacements of the face under greater compression during overall buckling. The formulation is then enhanced to examine the possibility of face–core delamination, the effects of core orthotropy, the combination of bending and compression, the sensitivity to imperfections and the introduction of panel monosymmetry. It is worth noting that the findings presented currently have been validated through physical experiments.[9]

10.2 Nonlinear buckling model

The most sophisticated sandwich panel considered presently is shown in Fig. 10.2. It shows a panel comprising two thin face-plates of differing thicknesses t_t and t_b representing the "top" face and "bottom" face respectively. The face-plates are assumed to behave linearly elastically with Young's modulus E and Poisson's ratio ν and separated by a 2-dimensionally orthotropic core material with Young's moduli in the x and y directions respectively being E_x and E_y, Poisson's ratios respectively being ν_x and ν_y, and the shear modulus being G_c. The panel is also assumed to be under the combined action of an axial load P and an external couple M, but note that the bending is assumed to be exclusively cylindrical. It is also noteworthy that although ultimately strains would grow to be reasonably large, small strain assumptions are used throughout, on the basis that the qualitative nature of the buckle pattern is unlikely to change significantly by the choice of alternative strain measures.

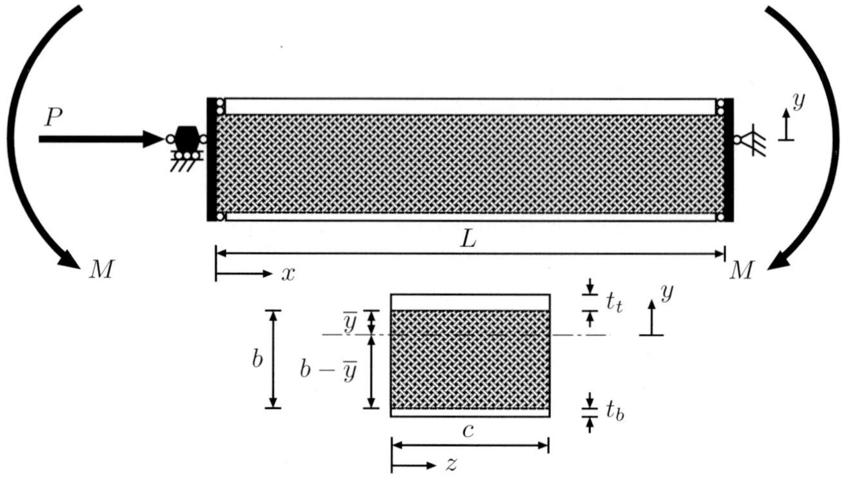

Fig. 10.2. Typical sandwich panel in elevation (top) and cross-section (bottom). Note the separation of the face-plates by the softer core material.

10.2.1 *Overall buckling*

Classical work on compressed sandwich structures showed that buckling first takes place at an initial critical bifurcation point, C, shown in Fig. 10.3 as part of a typical equilibrium diagram for a sandwich strut, in an Euler-type overall mode.[10] The principal difference between Euler strut buckling and overall buckling in sandwich panels is that, unlike in Euler struts, significant shear strains usually develop within the panel's core material. The current formulation allows for this by introducing separate "sway" and "tilt" components of the deformation (Fig. 10.4), whereby a linear combination of these components give the overall mode profile. This two degree-of-freedom formulation for the overall buckling mode ensures that the shear strain is not ignored by basically incorporating a Timoshenko beam assumption.[17] This is an enhancement on the simpler Euler–Bernoulli assumption where plane sections are assumed to remain plane and normal to the neutral surface and would imply that q_s and q_t would be equal. It is worth noting that a similar approach has been used recently to model non-uniform deformation in thin-walled tubes under pure bending.[18] Linear theory shows that the sway component $W(x)$ and the tilt component $\theta(x)$ are successfully modelled by:

$$W(x) = q_s L \sin \frac{\pi x}{L}, \quad \theta(x) = q_t \pi \cos \frac{\pi x}{L}. \qquad (10.1)$$

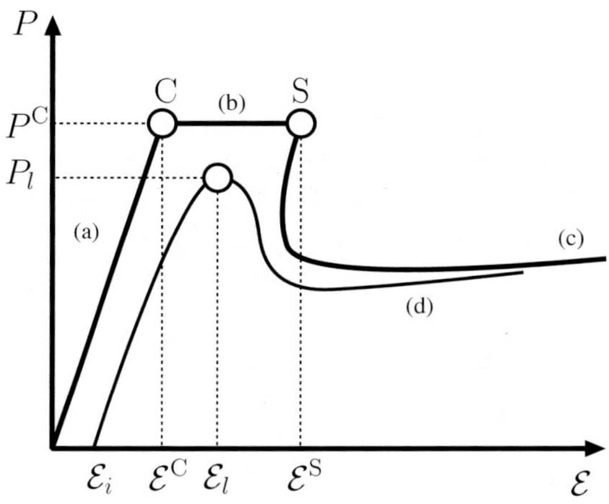

Fig. 10.3. Equilibrium diagram for sandwich struts: applied load P versus total end-shortening $\mathcal{E}$. (a) Fundamental pre-buckling path; (b) Equilibrium path of critical overall buckling; (c) Secondary path of localized buckling. (d) Typical path for an initially geometrically imperfect panel. P^C represents the critical buckling load, P_l represents the limiting load for an imperfect structure and $\mathcal{E}_i$ represents the magnitude of the initial imperfection in terms of end-shortening. Points C and S represent the critical and secondary bifurcations respectively.

Taking primes throughout to represent differentiation with respect to x, the corresponding shear strain γ_{xy} is given by the relationship:

$$\gamma_{xy} = W' - \theta. \tag{10.2}$$

When undergoing overall buckling, for convenience it is assumed that the bottom face-plate always has the greater compression; the extra secondary buckling displacement is thus always confined to that face-plate. With these definitions, the distance of the cross-section neutral axis from the top face–core interface $\overline{y}$ can be found in the conventional way for a section with elements made from different materials and is a key quantity in the model formulation:

$$\overline{y} = \frac{E_c b^2 + E t_b \left[2b + t_b(1 - \mathcal{T}^2)\right]}{2\left[E_c b + E t_b\left(1 + \mathcal{T}\right)\right]}, \tag{10.3}$$

where the ratio of the face-plate thicknesses $\mathcal{T}$, or the monosymmetry parameter, is defined:

$$\mathcal{T} = \frac{t_t}{t_b}. \tag{10.4}$$

When $\mathcal{T} = 1$ the doubly-symmetric case is recovered. Of course if the load P were to be applied elsewhere from the neutral axis then the structure would

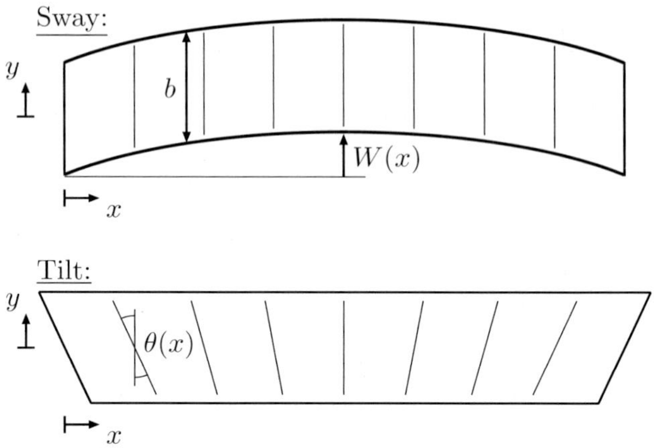

Fig. 10.4. Components of the overall (Euler) mode of buckling. Generalized coordinates q_s and q_t are introduced to represent the magnitude of the sway $W(x)$ and tilt $\theta(x)$ components respectively.

be under a combination of bending and compression, a more complicated modelling problem which is discussed in Sec. 10.3.3.

10.2.2 *Interactive buckling*

Once a sandwich structure buckles in an overall mode, its face-plates carry different compressive stresses. Also, physically, the greatest compressive stress occurs at the midspan of one face. In terms of modelling, this allows the possibility of a secondary bifurcation S (Fig. 10.3) manifesting itself as a local instability occurring at midspan, as a result of the overall mode forcing the face to exceed its local buckling load. The interaction of overall and local modes thus leading to localized buckling. The model therefore describes the corresponding buckling profile and uses this information to obtain an accurate description of the mechanical behaviour; functions $w(x)$ and $u(x)$ are thus introduced to represent lateral and in-plane face-plate displacements respectively as shown in Fig. 10.5.

Bending energy and face-plate imperfections

The components of the strain energy from bending arise from the overall bending of both face-plates, coupled with extra local bending of the bottom face-plate (Fig. 10.5). As stated above, the lower face-plate is always assumed to carry the larger magnitude of compression when the panel un-

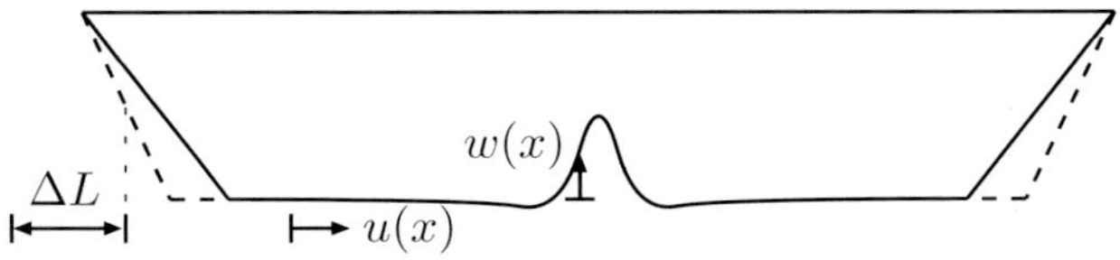

Fig. 10.5. Displacement functions used to model localized buckling in the more compressed face-plate. The displacement ΔL represented the pure squash component of the total end-shortening $\mathcal{E}$.

dergoes overall buckling. Assuming a small rotation curvature relationship, *i.e.* the curvature being equal to the second derivative of the bending displacement with respect to the axial coordinate, the expression for the strain energy stored under bending for the overall mode of buckling is therefore:

$$U_{bo} = \frac{1}{2} E(I_t + I_b) \int_0^L W''^2 \, \mathrm{d}x, \tag{10.5}$$

where I_t and I_b are the local minor axis second moments of area of the top and bottom face-plates respectively:

$$I_t = \frac{ct_t^3}{12(1-\nu^2)}, \qquad I_b = \frac{ct_b^3}{12(1-\nu^2)}. \tag{10.6}$$

The contribution from the bending about the global neutral axis is included in the membrane energy as this relies principally on the tilt mode (see Sec. 10.2.2).

The sandwich panel is also modelled with a generalized imperfection that represents the initial deformation of the more compressed face-plate, w_0. The form of the imperfection closely matches that of an approximate localized buckling mode for a thin strut on a softening Winkler foundation—derived from a first order solution of a double scale perturbation analysis:[19]

$$w_0(x) = A_0 \operatorname{sech}[\alpha(x - L/2)] \cos[\beta\pi(x - L/2)/L], \tag{10.7}$$

where α represents the degree of localization of the imperfection about the strut midspan ($x = L/2$) and β is the number of half-waves in the sinusoidal component of the imperfection. The imperfection is introduced by supposing an initially deformed shape of the more compressed face-plate $w_0(x)$ is then stress-relieved, such that the elemental bending moment, M, and thus stored strain energy of bending U_{bl} drops to zero[20] as shown in Fig. 10.6.

$$\mathrm{d}U_{bl} = \frac{1}{2} M \mathrm{d}\varphi = \frac{1}{2} E I_b (\chi - \chi_0)^2 \, \mathrm{d}x, \tag{10.8}$$

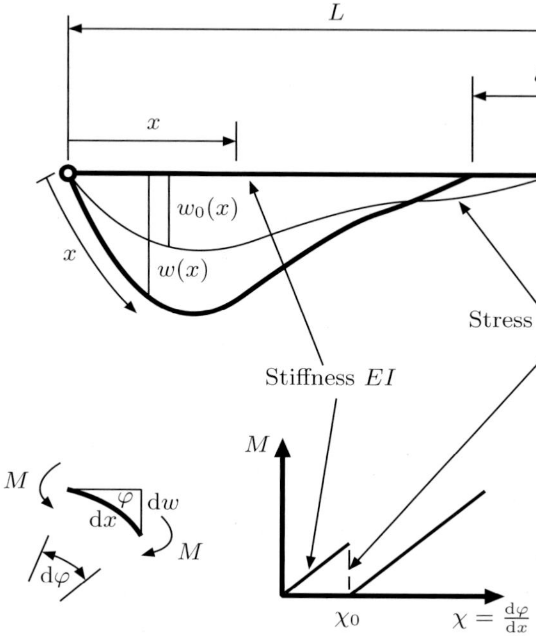

Fig. 10.6. Stress-relieved state of the strut after Thompson and Hunt.[21]

where χ is the curvature of the strut due to w and χ_0 was the curvature of the strut due to w_0. This also assumes that the strain energy in the core is non-zero in the initial state. The total strain energy of bending is therefore:

$$U_b = U_{bo} + \int_0^L \frac{1}{2} EI_b (w'' - w_0'')^2 \, dx. \tag{10.9}$$

Note that a cross-term in $W''w''$ may be expected here but is dropped as its effect is negligible on the governing equations for sandwich panels with the L/b ratios considered currently.

Membrane energy

The membrane strain energy arises from the axial displacement of the face-plates. It is assumed that the release in compression from the overall mode displacement is only a first-order effect and that the nonlinear effects are dominated by w in the bottom face-plate. Therefore, the top face simply contributes the tensile strain resulting from overall buckling, along with a component originating from the pure squash end-shortening Δ,

$$\varepsilon_{xt} = -\overline{y}\theta' - \Delta. \tag{10.10}$$

while the bottom face has, as extra contributions, terms familiar from Von Kármán large-deflection plate theory:[22]

$$\varepsilon_{xb} = (b - \overline{y})\theta' - \Delta + u' + \frac{1}{2}w'^2. \tag{10.11}$$

The total membrane strain energy is therefore the sum of the individual axial strain energy stored in the top and bottom face-plates:

$$U_m = \frac{Et_t c}{2}\int_0^L \varepsilon_{xt}^2 \,\mathrm{d}x + \frac{Et_b c}{2}\int_0^L \varepsilon_{xb}^2 \,\mathrm{d}x. \tag{10.12}$$

Core energy

The core provides the principal resistance to local lateral displacement and shear deformation, and a relatively smaller proportion of resistance to axial displacement when compared to the face-plates. A plane stress assumption is used to model the strain energy of the core. For compatibility with the plane sections remaining plane assumption, it is also assumed that local displacements $w_c(x,y)$ and $u_c(x,y)$ vary linearly through the core breadth b, such that the top face has no local displacement and the bottom face displacements are $w(x)$ and $u(x)$. This is a simplifying approximation to the situation found in the literature[23] where a cubic distribution function is assumed for u_c and a quadratic function for w_c. However, the deviations from the linear profile tend to be confined to the neighbourhood of the face-plates, therefore as long as the local displacement slopes do not become too large, the linear assumption seems to hold well. However, in the far postbuckling range, the model's accuracy reduces—as it would if plasticity effects took over. Thus:

$$u_c(x,y) = \left(\frac{\overline{y} - y}{b}\right) u(x), \qquad w_c(x,y) = \left(\frac{\overline{y} - y}{b}\right) w(x). \tag{10.13}$$

Axial strains in the core are then compatible with axial strains in the face-plates,

$$\varepsilon_x = u_c' + \frac{1}{2}w_c'^2 - y\theta' - \Delta. \tag{10.14}$$

The remaining strains can be written from small deflection theory:

$$\varepsilon_y = \dot{w}_c, \tag{10.15}$$

$$\gamma_{xy} = W' - \theta + w_c' + \dot{u}_c, \tag{10.16}$$

where a dot in this context represents partial differentiation with respect to y. As the core is assumed to be in a state of plane stress ($\sigma_z = \tau_{xz} = \tau_{yz} = 0$), where σ and τ represent direct and shear stresses respectively, the strain energy in the core reduces to:

$$U_c = \frac{1}{2}\int_0^c \int_{\overline{y}-b}^{\overline{y}} \int_0^L (\sigma_x \varepsilon_x + \sigma_y \varepsilon_y + \tau_{xy}\gamma_{xy})\,\mathrm{d}x\,\mathrm{d}y\,\mathrm{d}z. \tag{10.17}$$

Hooke's law for an orthotropic material gives the constitutive relations:[24]

$$\sigma_x = \frac{E_x}{(1-\nu_x\nu_y)}(\varepsilon_x + \nu_y\varepsilon_y), \tag{10.18}$$

$$\sigma_y = \frac{E_y}{(1-\nu_x\nu_y)}(\varepsilon_y + \nu_x\varepsilon_x), \tag{10.19}$$

$$\tau_{xy} = G_c\gamma_{xy}, \tag{10.20}$$

where E_x and E_y are the orthogonal core Young's moduli, and G_c is the core shear modulus; the reciprocal relationship of the constitutive law being:

$$E_x\nu_y = E_y\nu_x. \tag{10.21}$$

where ν_x and ν_y are the orthogonal core Poisson's ratios. The strain energy of the core is thus:

$$U_c = \frac{c}{2}\int_{\overline{y}-b}^{\overline{y}} \int_0^L \left[\frac{1}{(1-\nu_x\nu_y)}(E_x\varepsilon_x^2 + E_y\varepsilon_y^2 + 2\nu_x E_y \varepsilon_x\varepsilon_y) + G_c\gamma_{xy}^2\right]\mathrm{d}x\,\mathrm{d}y. \tag{10.22}$$

Work done

The work done by the axial load is equal to P multiplied by the corresponding total end shortening $\mathcal{E}$. The contribution to $\mathcal{E}$ from local buckling of a single face-plate is clearly dependent on the position of the neutral axis. Together with the contributions from pure squash and sway from overall buckling, the work done by the load P is given as:

$$P\mathcal{E} = P\int_0^L \left[\frac{1}{2}W'^2 - \left(\frac{\overline{y}}{b}\right)u' + \Delta\right]\mathrm{d}x. \tag{10.23}$$

The work done by the external end-moment is equal to M multiplied by the end-rotation Θ, therefore the contributions from the model involve the generalized coordinate of the overall tilt mode q_t and the in-plane face-plate local buckling displacement function u. Thus the work done by the moment M is:

$$M\Theta = M \int_0^L \left(2q_t \frac{\pi}{L} - \frac{u'}{b} \right) \mathrm{d}x. \tag{10.24}$$

Total potential energy and equilibrium equations

The total potential energy of the doubly-symmetric panel ($t_t = t_b = t$ and $\overline{y} = b/2$) is given by the summation of U_c, U_m and U_b, minus the work done terms $P\mathcal{E}$ and $M\Theta$:

$$\begin{aligned} V = \int_0^L \Bigg\{ & C_x \Bigg[\Delta^2 + q_t^2 \frac{b^2\pi^4}{12L^2} \sin^2 \frac{\pi x}{L} + \frac{1}{3} u'^2 + \frac{1}{20} w'^4 - \Delta u' - \frac{1}{3} \Delta w'^2 \\ & + \frac{1}{4} u' w'^2 - q_t \frac{b\pi^2}{6L} \sin \frac{\pi x}{L} \left(u' + \frac{1}{2} w'^2 \right) \Bigg] + \frac{1}{2} k w^2 - \frac{1}{3} k_1 w^3 + \frac{1}{4} k_2 w^4 \\ & + G \Bigg[\frac{u^2}{b^2} - \frac{u w'}{b} + \frac{1}{3} w'^2 + (q_s - q_t)^2 \pi^2 \cos^2 \frac{\pi x}{L} \\ & + (q_s - q_t) \pi \cos \frac{\pi x}{L} \left(w' - \frac{2}{b} u \right) \Bigg] + C_y \Bigg[\nu_x^2 \Delta \left(u' - \Delta + \frac{1}{3} w'^2 \right) \\ & - \nu_x \frac{w}{b} \left(u' + \frac{1}{3} w'^2 \right) \Bigg] + \frac{1}{2} EI \left[(w'' - w_0'')^2 + 2q_s^2 \frac{\pi^4}{L^2} \sin^2 \frac{\pi x}{L} \right] \\ & + D \Bigg[q_t^2 \frac{b^2\pi^4}{2L^2} \sin^2 \frac{\pi x}{L} - \left(2\Delta + q_t \frac{b\pi^2}{L} \sin \frac{\pi x}{L} \right) \left(u' + \frac{1}{2} w'^2 \right) \\ & + u'^2 + \frac{1}{4} w'^4 + u' w'^2 + 2\Delta^2 \Bigg] - P \left(\frac{q_s^2 \pi^2}{2} \cos^2 \frac{\pi x}{L} - \frac{1}{2} u' + \Delta \right) \\ & - M \left(2q_t \frac{\pi}{L} - \frac{u'}{b} \right) \Bigg\} \mathrm{d}x, \end{aligned} \tag{10.25}$$

where:

$$\begin{aligned} I &= \frac{ct^3}{12(1-\nu^2)}, \quad D = \frac{Etc}{2}, \quad G = \frac{G_c cb}{2}, \\ k &= \frac{E_y c}{(1-\nu_x\nu_y)b}, \quad C_x = \frac{E_x bc}{2(1-\nu_x\nu_y)}, \quad C_y = \frac{E_y bc}{2(1-\nu_x\nu_y)}. \end{aligned} \tag{10.26}$$

Parameters k_1 and k_2 are included to adjust the lateral core stiffness properties to simulate real (nonlinear) core situations[25,26] with an initially linear constitutive relationship for small displacement, followed by a destabilization ($k_1 > 0$) and then restabilization ($k_2 > 0$) as displacement grows as represented in Fig. 10.7. At first the cells behave in linearly elastic manner,

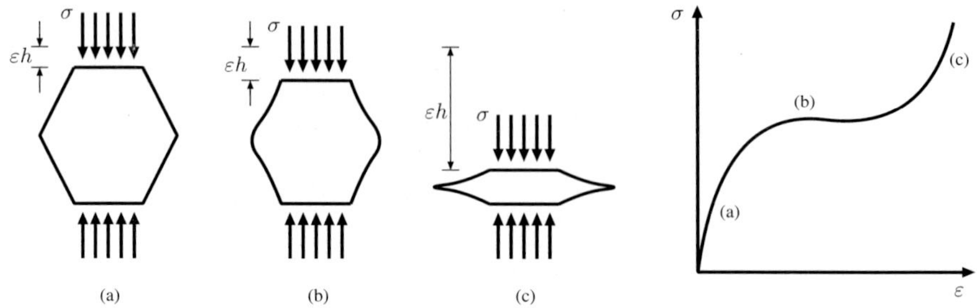

Fig. 10.7. Core material behaviour in compression at the microscopic cell level of nominal depth h. (a) Linear elastic range; (b) Elastic buckling of cell walls causing a plateau in the stress–strain response; (c) Cell wall crushing causing restiffening of the core material.

Fig. 10.7(a). After a certain amount of compressive strain, the cell walls buckle elastically causing a loss of stiffness, sometimes to an extent where it becomes negative, Fig. 10.7(b). However, as the buckles become larger and the opposite walls of the cell are so close that they begin to bear up against each other, Fig. 10.7(c); the stiffness then rises again to values close to the linear elastic value. Although no plasticity effects have been included, for modelling purposes, the lateral resistance of the core, F, comprise a softening quadratic nonlinearity (k_1) to model the initial loss of stiffness and a hardening cubic nonlinearity (k_2) to account for the subsequent increase. The nonlinear lateral resisting force F is therefore:

$$F = kw - k_1 w^2 + k_2 w^3. \tag{10.27}$$

Polystyrene and polyurethane foams exhibit this type of nonlinear elastoplastic constitutive behaviour in compression when strains become moderately large. Longitudinal and shear strains tend to be small in comparison with the lateral core strain; these nonlinearities are therefore included only in the lateral direction. A full elastoplastic formulation of the core is deliberately avoided as the localized buckling characteristics are not unduly affected by this.

The functional form of the potential energy is typified by $\int \mathcal{L}(Q, Q', Q'')\, \mathrm{d}z$ where Q in this case can be $w(x)$ or $u(x)$. Such an integral is extremized if:[27]

$$\delta \int_0^L \mathfrak{L}(Q, Q', Q'')\, \mathrm{d}x = \left[\frac{\partial \mathfrak{L}}{\partial Q''} \delta Q' + \left\{ \frac{\partial \mathfrak{L}}{\partial Q'} - \frac{\mathrm{d}}{\mathrm{d}x} \left(\frac{\partial \mathfrak{L}}{\partial Q''} \right) \right\} \delta Q \right]_0^L + \int_0^L \left\{ \frac{\mathrm{d}^2}{\mathrm{d}x^2} \left(\frac{\partial \mathfrak{L}}{\partial Q''} \right) - \frac{\mathrm{d}}{\mathrm{d}x} \left(\frac{\partial \mathfrak{L}}{\partial Q'} \right) + \frac{\partial \mathfrak{L}}{\partial Q} \right\} \delta Q\, \mathrm{d}x = 0. \tag{10.28}$$

With the assumption of simple supports at the ends of the panel the physical boundary conditions:

$$w(0) = w''(0) = w(L) = w''(L) = 0, \tag{10.29}$$

eliminate most of the boundary terms in Eq. (10.28) when $Q \equiv w(x)$. However, slightly more complicated boundary conditions arise for $Q \equiv u(x)$:

$$\left(1 + \frac{C_x}{3D}\right) u'(x_0) + \left(1 + \frac{C_x}{4D}\right) \frac{1}{2} w'^2(x_0) - \left(1 + \frac{C_x - C_y \nu_x^2}{2D}\right) \Delta + \frac{P}{4D} + \frac{M}{2Db} = 0, \tag{10.30}$$

where $x_0 = 0$ and L. The second part of the minimizing process is to eliminate the integrand in Eq. (10.28). This leads to ordinary differential equations (ODEs) in each of the variables $w(x)$, and $u(x)$ plus three integral constraints for the generalized coordinates q_s, q_t and Δ respectively, as given below:

$$\begin{aligned} EI\,(w'''' - w_0'''') &+ D\left[q_t \frac{b\pi^2}{L} \left(\sin \frac{\pi x}{L} w'' + \frac{\pi}{L} \cos \frac{\pi x}{L} w' \right) + 2\Delta w'' \right] - (2u''w' \\ &+ 2u'w'' + 3w'^2 w'') + G\left[\frac{u'}{b} - \frac{2}{3} w'' + (q_s - q_t) \frac{\pi^2}{L} \sin \frac{\pi x}{L} \right] + kw - k_1 w^2 \\ &+ k_2 w^3 + C_x \left[\frac{2}{3} \Delta w'' - \frac{1}{2} \left(u''w' + u'w'' + \frac{6}{5} w'^2 w'' \right) + q_t \frac{b\pi^2}{6L} \left(\sin \frac{\pi x}{L} w'' \right.\right. \\ &\left.\left. + \frac{\pi}{L} \cos \frac{\pi x}{L} w' \right) \right] + C_y \left[\frac{\nu_x}{b} \left(\frac{2}{3} w w'' + \frac{1}{3} w'^2 - u' \right) - \frac{2}{3} \nu_x^2 \Delta w'' \right] = 0, \end{aligned} \tag{10.31}$$

$$\begin{aligned} \left(D + \frac{C_x}{3}\right) u'' &+ \left(D + \frac{C_x}{4}\right) w'w'' - \frac{G}{b} \left(\frac{u}{b} - \frac{1}{2} w' \right) \\ &- \pi \cos \frac{\pi x}{L} \left[\frac{G}{b} (q_s - q_t) - \left(D + \frac{C_x}{6}\right) q_t \frac{b\pi^2}{2L^2} \right] - \frac{C_y \nu_x}{2b} w' = 0, \end{aligned} \tag{10.32}$$

$$P = \frac{2\pi^2 EI}{L^2} + \frac{2G}{q_s}\left[(q_s - q_t) + \frac{1}{\pi L}\int_0^L \cos\frac{\pi x}{L}\left(w' - \frac{2}{b}u\right)\mathrm{d}x\right], \quad (10.33)$$

$$\begin{aligned} q_s = (1+s)q_t - \frac{2M}{\pi G L} + \int_0^L \left[\frac{1}{\pi L}\cos\frac{\pi x}{L}\left(\frac{2}{b}u - w'\right)\right. \\ \left. - \frac{s}{b\pi^2}\sin\frac{\pi x}{L}\left(u' + \frac{1}{2}w'^2\right)\right]\mathrm{d}x, \end{aligned} \quad (10.34)$$

$$\Delta = \frac{\frac{P}{2} + \frac{1}{L}\int_0^L \left[D\left(u' + \frac{1}{2}w'^2\right) + \left(\frac{C_x - C_y\nu_x^2}{2}\right)\left(u' + \frac{1}{3}w'^2\right)\right]\mathrm{d}x}{(2D + C_x - C_y\nu_x^2)}, \quad (10.35)$$

where:

$$s = \frac{b^2\pi^2}{2GL^2}\left(D + \frac{C_x}{6}\right). \quad (10.36)$$

10.2.3 *Linear eigenvalue analysis*

Apart from Sec. 10.3.3, the rest of Secs. 10.2 and 10.3 deal with the perfectly straight panel under pure compression (*i.e.* $w_0 = 0$ and $M = 0$). The first result can be obtained by linear eigenvalue analysis which gives the critical load for overall buckling P^{C}, arising on the pure squash fundamental path F at which $q_s = q_t = w = u = 0$, see Fig. 10.3. The potential energy V is non-diagonal with respect to q_s and q_t, and so the initial buckling mode is triggered when the matrix

$$\mathbf{V}_{ij}^{\mathrm{F}} = \begin{pmatrix} \left.\frac{\partial^2 V}{\partial q_s^2}\right|^{\mathrm{F}} & \left.\frac{\partial^2 V}{\partial q_s \partial q_t}\right|^{\mathrm{F}} \\ \left.\frac{\partial^2 V}{\partial q_t \partial q_s}\right|^{\mathrm{F}} & \left.\frac{\partial^2 V}{\partial q_t^2}\right|^{\mathrm{F}} \end{pmatrix} \quad (10.37)$$

becomes singular. Rearranging the resulting equation to find P^{C} gives:

$$P^{\mathrm{C}} = \frac{2\pi^2 EI}{L^2} + \frac{2Gb^2\pi^2}{L^2}\left[\frac{\left(D + \frac{C_x}{6}\right)}{2G + \frac{b^2\pi^2}{L^2}\left(D + \frac{C_x}{6}\right)}\right]. \quad (10.38)$$

The expression matches the classical result for the overall buckling load of a sandwich strut.[10]

10.2.4 *Perfect isotropic struts with soft cores*

Having formulated the necessary equations and calculated the overall critical buckling load, the full postbuckling equilibrium solutions are presented.

These are found from solving the fully nonlinear system of ODEs and integral equations given in Eqs. (10.31)–(10.35) subject to the boundary conditions given in Eqs. (10.29) and (10.30) at $x = 0$ and the symmetry condition at the panel midspan:

$$w'(L/2) = w'''(L/2) = u(L/2) = 0, \tag{10.39}$$

numerically with the continuation package *AUTO*.[28] In this section the core material is assumed to be isotropic and soft, assumptions that imply the following relations:

$$\begin{aligned} E_c &= E_x = E_y, \quad E_c \ll E, \\ \nu_c &= \nu_x = \nu_y, \quad G_c = \frac{E_c}{2(1+\nu_c)}. \end{aligned} \tag{10.40}$$

The solutions are presented as follows: first, the position of the secondary bifurcation S on the critical path for overall buckling is presented followed by the full postbuckling results for a particular panel geometry where S was sufficiently close to C (see Fig. 10.3) thereby maximizing the modal interaction.

For the numerical study presented below, the basic dimensions of the panel were the length L, which was varied as a parameter, the breadth $c = 254\,\text{mm}$, the face-plate thickness $t = 0.508\,\text{mm}$ and the core depth $b = 5.08\,\text{mm}$. For the materials, the face-plates had the following properties: $E = 68947.57\,\text{N/mm}^2$ and $\nu = 0.3$; the core had the isotropic properties: $E_c = 198.57\,\text{N/mm}^2$ and $\nu_c = 0.2$. To reflect experimental evidence of the core behaviour described in Fig. 10.7, a destabilizing quadratic term and a restabilizing cubic term was included in the core's lateral resistance F, these were respectively taken as: $k_1 = 6728\,\text{N/mm}^3$ and $k_2 = 1346\,\text{N/mm}^4$.

Location of secondary bifurcation

The secondary bifurcation can be found by linearizing the ODEs. Solutions were first sought at constant q_t by shooting for the symmetry condition, Eq. (10.39), using an initial value problem solver. The results of this were then ported into *AUTO* so that numerical continuation could be used to reduce parametrically the value of q_t until the trivial solution was found. This defined the position of the secondary bifurcation point S numerically (see Fig. 10.3) and the magnitude of the tilt mode q_t^{S}. Moreover, assuming that prior to S, both u and w were both trivial, the following relationship could be deduced from Eq. (10.34):

$$q_s^{\text{S}} = q_t^{\text{S}}(1+s). \tag{10.41}$$

Figure 10.8 shows the relation between the magnitude of the sway mode at the position of secondary bifurcation q_s^S with length, where if $q_s^S = 0$ it would imply that the critical and secondary buckling modes are triggered simultaneously. The graph also shows corresponding results from an earlier periodic Rayleigh–Ritz formulation.[5] In each case the secondary bifurcation that leads to localized buckling occurred well before the secondary bifurcation leading to periodic buckling and this finding remained true throughout the current study, as may be expected from more general work on the strut on a nonlinear (softening) foundation.[29]

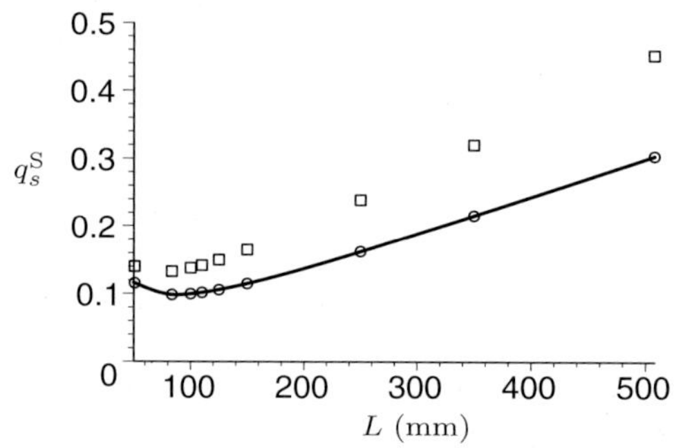

Fig. 10.8. Magnitude of sway mode q_s at secondary bifurcation S versus strut length L. (○): current model, (□): periodic Rayleigh–Ritz model.[5]

Over the range $50 \leqslant L \leqslant 150$ mm, the sway mode at the secondary bifurcation was relatively small, implying a situation where unstable localized buckling could occur early in the loading history. The curve had a minimum at approximately $L = 100$ mm, and therefore this length was chosen for the detailed example below. This panel's ratio of L/b of approximately twenty is not overly slender nor stocky.

Postbuckling response

The postbuckling results for the near-optimum length of $L = 100$ mm are presented by solving the fully nonlinear system. Figure 10.9(a) shows the localized buckling deformation of the face-plate under greater compression. Note how the central amplitude begins to dominate as the load reduces from P^C on an unstable postbuckling equilibrium path after the secondary bifurcation S. The load–end-shortening relationship of Fig. 10.9(b) shows the linear elastic non-trivial fundamental equilibrium path, the neutral critical buckling path—as a small rotation moment–curvature relationship was assumed—and an unstable secondary localized buckling path. The destabilization–restabilization characteristic in the core's lateral force–displacement relationship, as given in Eq. (10.27) and shown in Fig. 10.9(d), reflects the form of the stress–strain curve in Fig. 10.7. The system restabilizes at $P \approx 0.40P^{\mathrm{C}}$, when the influence of the hardening cubic nonlinearity begins to dominate over the softening quadratic term. The main noteworthy point is that a 60% drop in the theoretical structural strength has been observed in this panel and together with realistic imperfections, the situation would worsen, indicating that due care needs to be exercised when designing such panels to avoid interactive buckling early in the loading history.

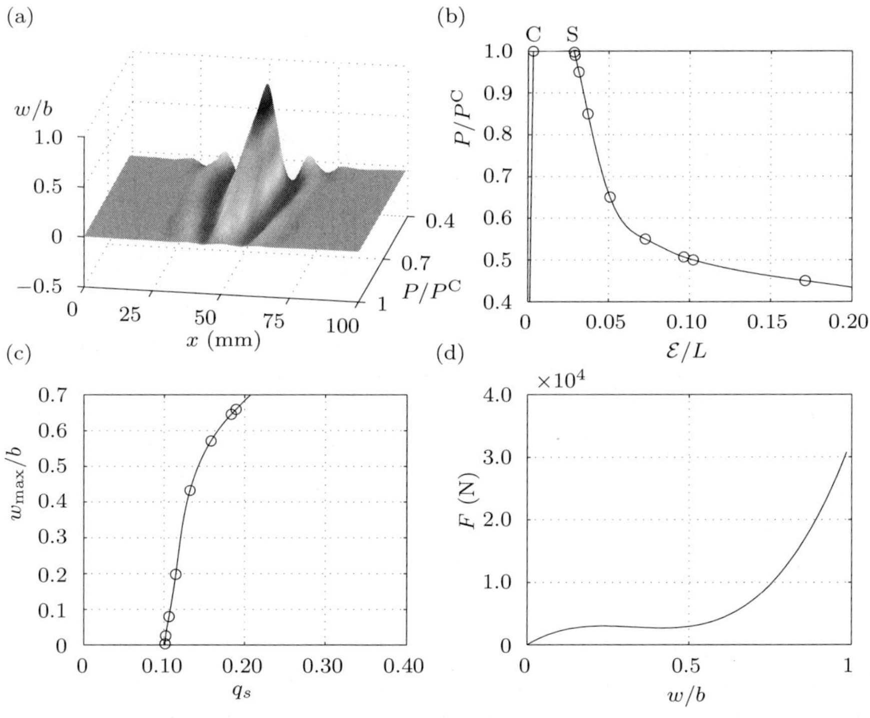

Fig. 10.9. Postbuckling response for the strut of $L = 100$ mm with a cubic core: (a) Localized buckle profiles; (b) Load–end-shortening curve; (c) Relative magnitudes of the overall and localized modal amplitudes; (d) Lateral resistance of the core F plotted against the displacement of the face-plate w into the core.

10.3 Special perfect cases

In this section postbuckling responses are discussed for panels with orthotropic cores, those susceptible to face–core delamination and those subjected to axial compression and bending. Parametric studies, where material properties are varied, are presented with the important cases highlighted.

10.3.1 *Core orthotropy*

For comparison purposes, a sandwich section was chosen which had received attention in the literature.[30] The face-plate properties were the same as in Sec. 10.2.4. For the core: depth $b = 50.8$ mm, width $c = 254$ mm, and the overall strut length $L = 508$ mm. Other properties were initially taken from an isotropic panel and varied later to obtain results for the orthotropic core. The preliminary isotropic core properties were: Young's modulus: $E_c = E_{xi} = E_{yi} = 198.57$ N/mm^2, shear modulus: $G_{ci} = 82.74$ N/mm^2, Poisson's ratio: $\nu_{xi} = \nu_{yi} = 0.2$.

Orthotropy in the core essentially means that the shear modulus G_c of the core is independent of the Young's modulus components E_x and E_y. Varying G_c not only has a substantial influence on the critical load for overall buckling but also an important effect on the proximity of the critical and secondary bifurcations. An illustration of this can be seen in Fig. 10.10(a) which shows the magnitude of sway at the secondary bifurcation q_s^S plotted against G_c. In the isotropic case $q_s^S = 0.073$, whereas when $G_c = 20$ N/mm^2, $q_s^S = 0.023$—reducing G_c in this case having a marked and adverse effect on the postbuckling by bringing the two bifurcations much closer together thereby increasing the sensitivity to initial imperfections as discussed in Sec. 10.4.

By varying the core lateral modulus E_y, while keeping $\nu_x \nu_y = 0.04$ and G_c constant, Fig. 10.10(b) was obtained. The two curves represent the cases of $G_c = G_{ci}$ (upper curve), and $G_c = 20$ N/mm^2 (lower curve). Both curves show q_s^S dropping monotonically with a shallow gradient. Similarly, varying the core axial modulus, E_x while keeping $\nu_x \nu_y = 0.04$ and G_c constant, we obtained Fig. 10.10(c). The two curves again representing two separate values of G_c: $G_c = G_{ci}$ (upper curve) and $G_c = 20$ N/mm^2 (lower curve). It was found that the reduction in the relative displacements required for localized buckling was not significantly affected by either changing E_y or E_x.

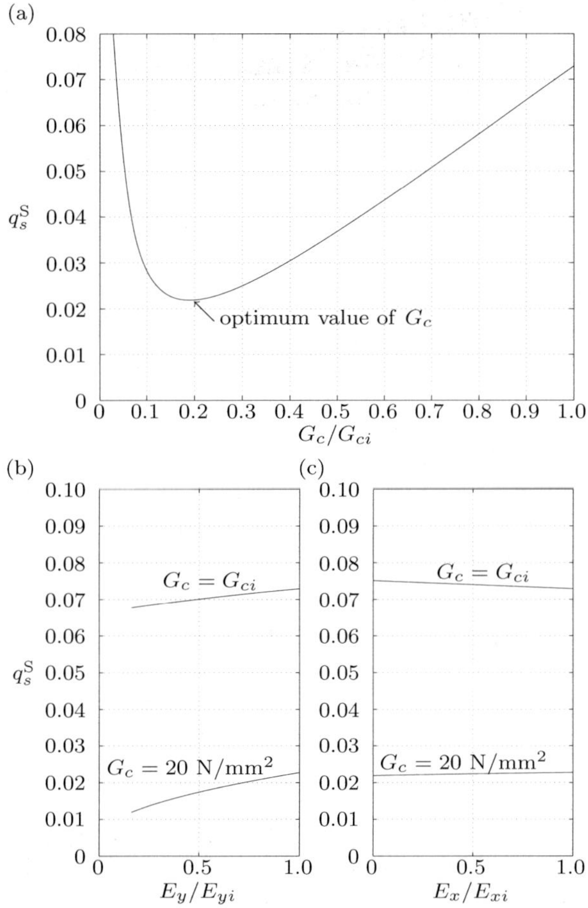

Fig. 10.10. Effect of varying (a) G_c, (b) E_y and (c) E_x on the proximity of critical and secondary bifurcations.

Comparison of isotropy and orthotropy

It is interesting to monitor the change in G_c that maximizes the interaction between local and overall buckling. Figure 10.11(a) compares the maximized interactive effect, measured again by the value of q_s^S, with the equivalent for an isotropic core, as the direct core stiffness is altered. The curves touch at $E_c = 25$ N/mm^2, indicating that, for a system of these particular dimensions and material characteristics, isotropy maximizes the interaction. When $E_c > 25$ N/mm^2, decreasing G_c while holding E_c constant brings the bifurcations together. This is because, as G_c increases in relation to E_c, the strut tends to the engineers' bending theory situation of zero shear strain, denying the shear deformability that is essential for localized buckling to occur.

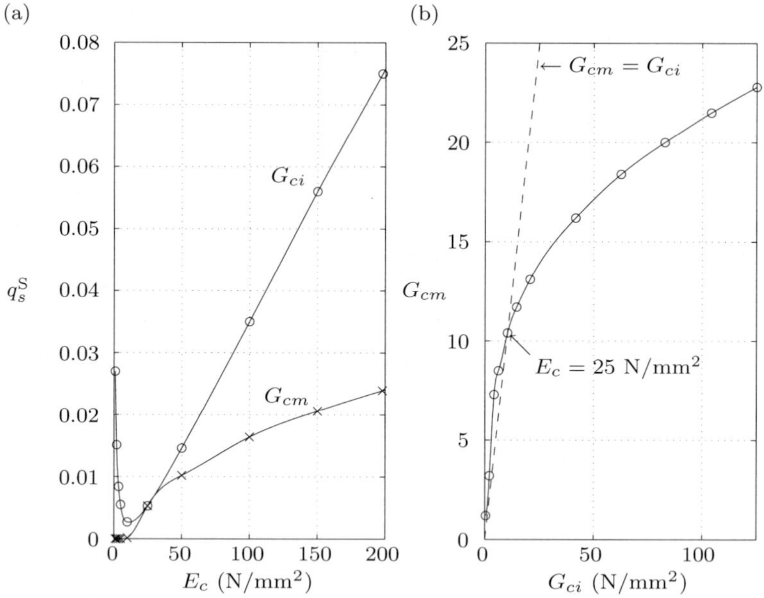

Fig. 10.11. (a) Comparison of the secondary bifurcation position in isotropic cores (○) and orthotropic cores where G_c is varied to maximize the interaction (×). (b) Values of G_c that maximize the interaction.

On the other hand, when $E_c < 25$ N/mm^2 the bifurcations are brought together by increasing G_c; indeed, q_s^{S} tends rapidly to zero as E_c reduces, indicating that primary and secondary bifurcations become virtually coincident, a phenomenon which is often associated with a high degree of imperfection-sensitivity.[21,31] This is the opposite situation: as G_c drops the strut loses shear resistance and the overall mode tends to pure sway with no tilt; this implies that there is little or no overall integrity and the face-plates are acting almost as two independent Euler struts with the result that there is little or no tendency for the response to localize to one face. The maximum interactive effect is thus found for cores that have sufficient shear stiffness for overall integrity.

10.3.2 *Face–core delamination*

As discussed in the previous sections, the usual situation is that the strut buckles first into the overall mode. However, after the secondary bifurcation that leads to interactive localized buckling has occurred, the local deflections relative to the face–core interface take on more importance. If core shear terms are ignored for the time being and it is assumed that local deflections are primarily inwards (positive as shown in Fig. 10.5) the lateral resistance of the core acts as a nonlinear Winkler foundation; this is assumed to take the form of a polynomial function, as described in Eq. (10.27) and Fig. 10.7. If local deflections are negative then the core is in tension there with the consequence that the face–core adhesive potentially becomes the weakest link in the structure; when a certain value of the lateral tensile (peeling) strain $\varepsilon_y = \varepsilon_y^*$ is exceeded the face–core interface breaks and delamination occurs. This is seldom a problem until overall buckling occurs because it is only then that the differential compression in the face-plates, which leads to localized buckling, increases the local peeling stresses very significantly. Shear-driven delaminations as seen in laminated composites[32] are less likely as the adhesive used to bond the interface is usually very stiff as compared to the core and the shear strength is therefore much higher at the interface than in the rest of the core.[33]

The immediate loss of the face–core interface changes the structural response dramatically. Not only does the face-plate start to behave like an Euler strut in the delaminated zone, but the shear stresses in the core from the localized buckling mode also vanish. Therefore, in the delaminated zone the only shear contributions that remain are purely from overall buckling. However, potential energy contributions from the membrane and from bending with the work done by the load P remain the same. To account for these sudden changes in the governing equations, a modified Heaviside function $\mathcal{H}_d(\varepsilon_y)$ is introduced:

$$\mathcal{H}_d = \begin{cases} 1 & \text{for } \varepsilon_y \leqslant \varepsilon_y^* \\ 0 & \text{for } \varepsilon_y > \varepsilon_y^*, \end{cases} \tag{10.42}$$

where ε_y^* is defined as limit of the lateral strain before delamination occurs.

Core energy

Under the assumption of soft cores, Eq. (10.40), and that core stresses from local buckling vanish in the delaminated zone, the remaining core strains can be written:

$$\varepsilon_y = \dot{w}_c \mathcal{H}_d, \tag{10.43}$$

$$\gamma_{xy} = W' - \theta + (w_c' + \dot{u}_c)\,\mathcal{H}_d. \tag{10.44}$$

The strain energy of the core is thus

$$U_c = \frac{E_c c}{2(1-\nu_c^2)} \int_0^L \int_{\overline{y}-b}^{\overline{y}} \varepsilon_y^2 \, \mathrm{d}y \, \mathrm{d}x + \frac{G_c c}{2} \int_0^L \int_{\overline{y}-b}^{\overline{y}} \gamma_{xy}^2 \, \mathrm{d}y \, \mathrm{d}x. \tag{10.45}$$

Substituting the appropriate expressions and integrating with respect to y, we obtain:

$$U_c = \int_0^L \left\{ \mathcal{H}_d^2 \left(\frac{1}{2} k w^2 - \frac{1}{3} k_1 w^3 + \frac{1}{4} k_2 w^4 \right) + G \left[\mathcal{H}_d^2 \left(\frac{1}{3} w'^2 + \frac{u^2}{b^2} - \frac{1}{b} u w' \right) + \mathcal{H}_d (q_s - q_t) \pi \cos \frac{\pi x}{L} \left(w' - \frac{2u}{b} \right) + (q_s - q_t)^2 \pi^2 \cos^2 \frac{\pi x}{L} \right] \right\} \mathrm{d}x. \tag{10.46}$$

and because of the definition of $\mathcal{H}_d$ it can be assumed that

$$\mathcal{H}_d^2 \equiv \mathcal{H}_d. \tag{10.47}$$

Numerical results

Numerical studies from the literature have shown postbuckling results by assuming pre-existing delamination zones at discrete lengths along the panel in compression[34,35] and in bending;[36] they have effectively studied geometrically imperfect structures. Pre-existing delaminations in the current context are dealt with in Sec. 10.4.1. The strut used in the numerical study presented below took the material and geometric properties from the isotropic study in Sec. 10.2.4.

It is assumed that the appearance of the delamination is not affected by the gap between the critical and secondary bifurcations which mark points of the appearance of overall buckling and interactive (localized) buckling (see Fig. 10.3). This is because the delamination is solely based on the displacement of the face-plate relative to the core; the contribution from overall buckling to this relative displacement is small—the significant increase being after the secondary bifurcation. The lateral strain limit, ε_y^*, was chosen to be approximately 8%—a realistic value for a cellular core because of the face–core interface adhesive (Fig. 10.12). Notwithstanding these assumptions, several key points arise that are discussed below.

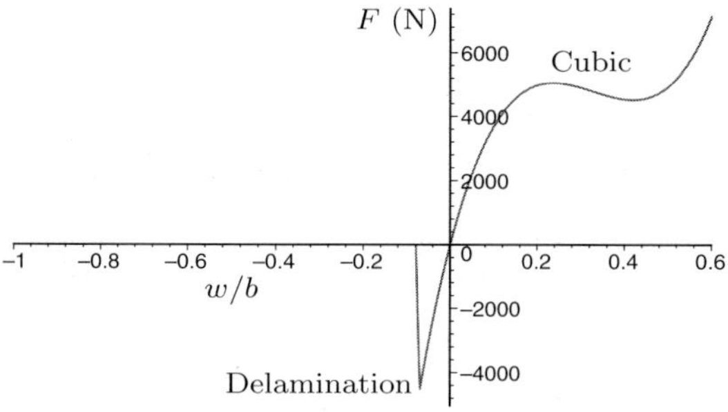

Fig. 10.12. Core material force–lateral strain properties to model the sudden change in constitutive response caused by delamination.

The structural response depends on the shape of the localized buckle pattern; an initial midspan deflection of the more compressed face-plate into the core makes it less likely to delaminate and thus keep structural integrity as shown in experimental studies[9] whereas outward deflections increase the probability of delamination. The results for outward face-plate displacements are shown in Figs. 10.13–10.14. The graphs showing the change in $w(x)$ against the load P also show by implication the highly localized distribution of the peeling strains.

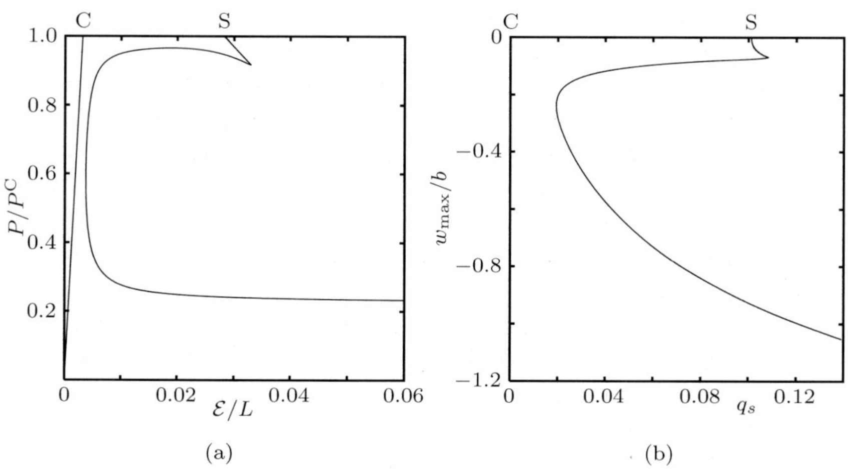

Fig. 10.13. (a) Load P versus total end-shortening $\mathcal{E}$ for the face–core delamination case; (b) shows the maximum peeling strain (w/b) against the magnitude of the overall mode (q_s).

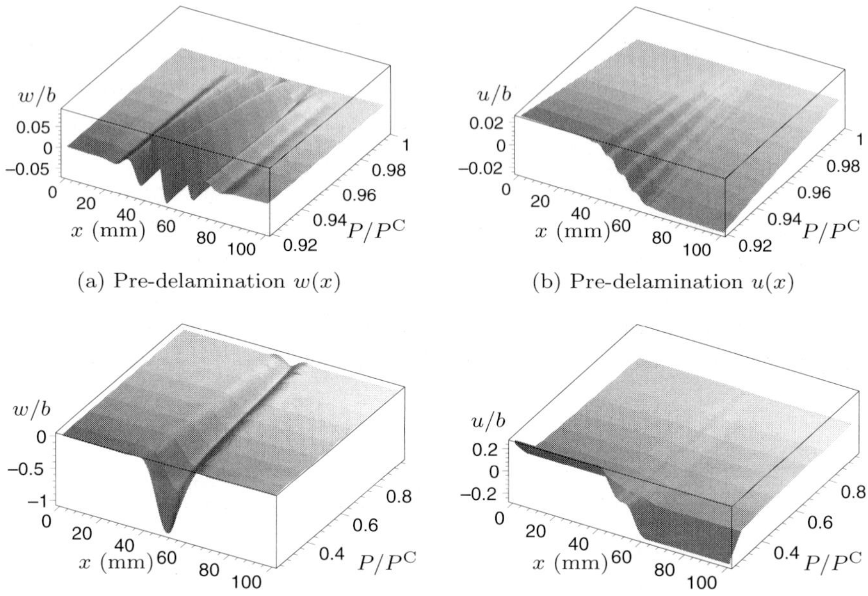

(a) Pre-delamination $w(x)$ (b) Pre-delamination $u(x)$

(c) Post-delamination $w(x)$ (d) Post-delamination $u(x)$

Fig. 10.14. Distribution of the localized buckling pattern for the cubic core case as defined by the relative lateral and in-plane displacements $w(x)$ and $u(x)$ respectively.

The postbuckling load versus end-shortening curves differ significantly for inward and outward ($w(x) < 0$ dominating) face-plate deflections. This is easily explained by the asymmetry of the constitutive law of the core. Although the initial response beyond the secondary bifurcation is slightly stiffer than for inward deflections the effect of $\mathcal{H}_d(\varepsilon_y)$ is highly destabilizing. When this threshold is violated delamination occurs and the stored energy in the core is released very rapidly and the buckle profile has a distinctive single hump; the evolution of the buckle pattern being shown in Fig. 10.14 where the profiles are plotted as functions $w(x, P)$ and $u(x, P)$. In the numerical example presented the system subsequently restabilized under rigid loading at approximately $0.23P^{\mathrm{C}}$. The reason for this post-delamination restabilization is the fact that in the delaminated zone the face-plate is free to deform like an Euler strut. As the postbuckling response of the linearized Euler strut is neutrally stable, when deformations become large enough, this begins to dominate the behaviour culminating in the restabilization. Of course, if the response was measured from a rigidly loaded experiment, the load would have dropped from the limit point at

$0.92P^C$ down to $0.24P^C$—a dramatic loss in load carrying capacity due to the delamination. Moreover, the response was independent of the cubic model for the compression lateral force–displacement relationship as given in Eq. (10.27). The severe snap-back response shown in Fig. 10.13(a) also implies a strong imperfection sensitivity characteristic particularly with the proximity of the fundamental and unstable postbuckling path. Small defects in panel construction would have led to the structure buckling before the theoretical critical load P^C was reached (see Sec. 10.4).

Figure 10.15(a) shows the buckle pattern changing once the system restabilizes under rigid loading at $P \approx 0.24P^C$. Although the amplitudes change very significantly, the localized buckle wavelength λ, as defined in Fig. 10.16, begins to reduce very gradually. If the equations are reduced, taking the case where $\mathcal{H}_d = 0$, it turns out that in the delaminated zone the parameter that controls the buckle wavelength is Δ. In the numerical study it was noted that in the region where the response restabilizes Δ increased

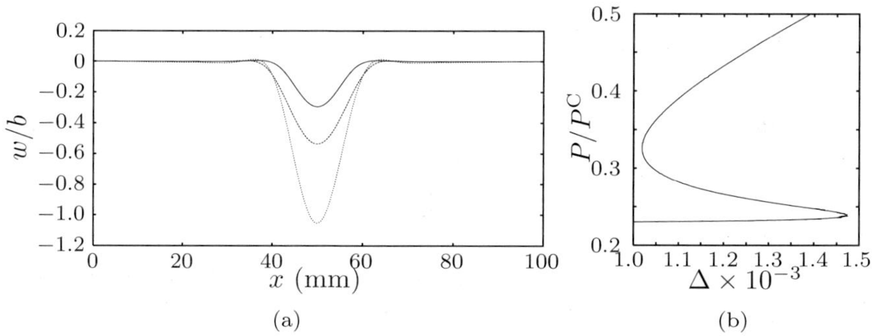

Fig. 10.15. (a) Change in buckle pattern after restabilization in rigid loading, where $P/P^C \approx 0.24$. (b) Note small change in the buckle wavelength and the increase then reduction in Δ as P reduces from $0.3P^C$ to $0.2P^C$.

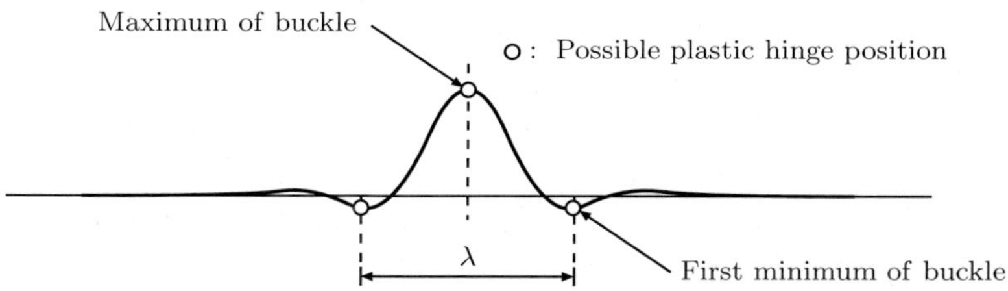

Fig. 10.16. Definition of localized buckle wavelength λ.

briefly and then decreased, as shown in Fig. 10.15(b), for both cases such that the buckle wavelength increased then decreased. However, the rate of reduction was so slow that in reality plasticity (in the form of a plastic hinge) or fracture may well start to govern the response. It is also noteworthy that delamination propagation in laminated composite materials also exhibits arrest in situations of mechanical restabilization.[37]

10.3.3 *Combined loading*

The behaviour of beam-columns can also be discussed quantitatively. A very brief study is presented where two different types of combined loading are considered. The first case is that the axial load P acts at an eccentricity e from the neutral axis, which adds a bending moment on the structure M that is equal to the product of the axial load P and the eccentricity e. The second case considers a panel that has an axial load P coexisting with constant equal and opposite externally applied moments M at both ends—*i.e.* P and M are independent.

Limiting axial loads

Using the same properties as the panel in Sec. 10.2.4 but with an orthotropic soft core, a parametric study was conducted showing the effect of varying the core shear modulus G_c between 30 and 160 N/mm^2. Figure 10.17 shows the distribution of the limiting values of axial loads (P_l as shown

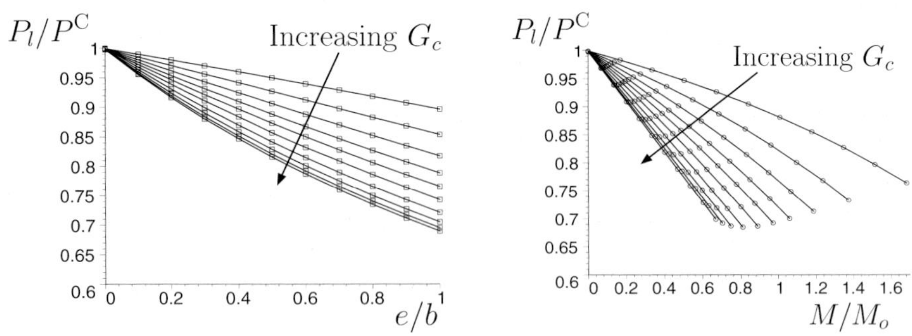

(a) Eccentric load: $M = Pe$ (b) Constant moment: M

Fig. 10.17. Interaction curves of limiting axial loads P versus the combined moment from (a) an eccentricity e, where $M = Pe$, and (b) a constant end moment M for valid G_c values. Note that the benchmark moment M_o is equal to $P^C b$.

in Fig. 10.3) when panels were subjected to combined bending and axial compression. As can be seen, increasing the shear modulus destabilized the panels relatively more as the co-existing moment was increased. Note that if snap-back occurred after the limit point then these would be overestimates of the realistic strength of the panel. In that case it would become necessary to analyse the extent of the snap-back effect, which is beyond the scope of the current discussion.

10.4 Imperfection sensitivity

10.4.1 *Doubly-symmetric panels*

Pre-existing delaminations and geometric imperfections

To account for pre-existing delaminations as well as those caused by excessive peeling strains in the localized buckling process, further Heaviside functions alongside $\mathcal{H}_d(\epsilon_y)$ are introduced by including $\mathcal{H}_p(x)$ and $\mathcal{H}_{pw}(x, w)$. Owing to the fact that a related study[20] indicated that geometric imperfections symmetric about the strut midspan ($x = L/2$) gave the most severely unstable behaviour, all the pre-existing delaminations considered currently are similarly symmetric about the panel midspan:

$$\mathcal{H}_p = \begin{cases} 0 & \text{for } L/2 - x_D/2 \leqslant x \leqslant L/2 + x_D/2 \\ 1 & \text{otherwise,} \end{cases} \tag{10.48}$$

where x_D is a length parameter defining the region of the pre-existing delamination. For the case of the strain energy from core shear, the modified Heaviside function $\mathcal{H}$ is assumed to be the product of $\mathcal{H}_d$ and $\mathcal{H}_p$:

$$\mathcal{H} = \mathcal{H}_d \mathcal{H}_p. \tag{10.49}$$

However, when the lower face-plate is not bonded to the core but still compresses the core purely by contact without adhesion, the behaviour of the face-plate is analogous to a strut on a Winkler foundation model mentioned earlier. In the strain energy expression, this effect is accounted for by the lateral strain ϵ_y. In this case, the Heaviside function only becomes zero when there is no contact between the face-plate and the core ($w(x) < 0$). This adjustment is accounted for in the function $\mathcal{H}_{pw}$ where this is the same as $\mathcal{H}_p$ except for the condition on the sign of $w(x)$:

$$\mathcal{H}_{pw} = \begin{cases} 0 & \text{for } L/2 - x_D/2 \leqslant x \leqslant L/2 + x_D/2 \text{ and } w(x) < 0 \\ 1 & \text{otherwise.} \end{cases} \tag{10.50}$$

Therefore, for the case of the lateral core strain energy, the new modified Heaviside function $\mathcal{H}_w$ is assumed to be the product of $\mathcal{H}_d$ and $\mathcal{H}_{pw}$:

$$\mathcal{H}_w = \mathcal{H}_d \mathcal{H}_{pw}. \tag{10.51}$$

The core strain energy therefore changes to:

$$U_c = \int_0^L \left\{ \mathcal{H}_w^2 \left(\frac{1}{2} k w^2 - \frac{1}{3} k_1 w^3 + \frac{1}{4} k_2 w^4 \right) + G \left[\mathcal{H}^2 \left(\frac{1}{3} w'^2 + \frac{u^2}{b^2} - \frac{1}{b} u w' \right) \right.\right. \\ \left.\left. + \mathcal{H}(q_s - q_t) \pi \cos \frac{\pi x}{L} \left(w' - \frac{2u}{b} \right) + (q_s - q_t)^2 \pi^2 \cos^2 \frac{\pi x}{L} \right] \right\} \mathrm{d}x \tag{10.52}$$

and because of the definitions of $\mathcal{H}$ and $\mathcal{H}_w$ it can be assumed that

$$\mathcal{H}^2 \equiv \mathcal{H} \text{ and } \mathcal{H}_w^2 \equiv \mathcal{H}_w. \tag{10.53}$$

It is worth noting that pre-existing delaminations do not act like imperfections in the sense that they do not convert bifurcations into limit points, but they do tend to increase the sensitivity to geometric imperfections by bringing the two bifurcations much closer together.

Numerical results

Using the technique outlined in Sec. 10.2.2, a system of two ODEs, three integral equations and a set of boundary conditions can be written for both geometric imperfections and pre-existing delaminations.[38,39] In this section, the panel studied in Sec. 10.3.2 now includes a pre-existing delamination of extent $\tilde{x}_D$. Note that $\tilde{x}_D$ is expressed as a proportion of the total length L. The panel is also assumed to have an initially deflected face-plate with imperfection magnitude $\mathcal{E}_0$, which is effectively the first-order approximation of the end-shortening of the localized imperfection (Fig. 10.6):

$$\mathcal{E}_0 = \int_0^L \frac{1}{2} {w_0'}^2 \, \mathrm{d}x. \tag{10.54}$$

The current purpose is to establish the effect of pre-existing delaminations on the equilibrium response. By parameterizing the strut length, the potential sensitivity to geometric imperfections can be deduced. The particular strut configuration was chosen because this had been shown to have only a small sensitivity to geometric imperfections when in a perfectly bonded state—the critical and secondary bifurcations being fairly remote from each other. Therefore, increasing the length of the pre-existing delamination x_D from zero to 10% of the strut length about the midspan, the proximity

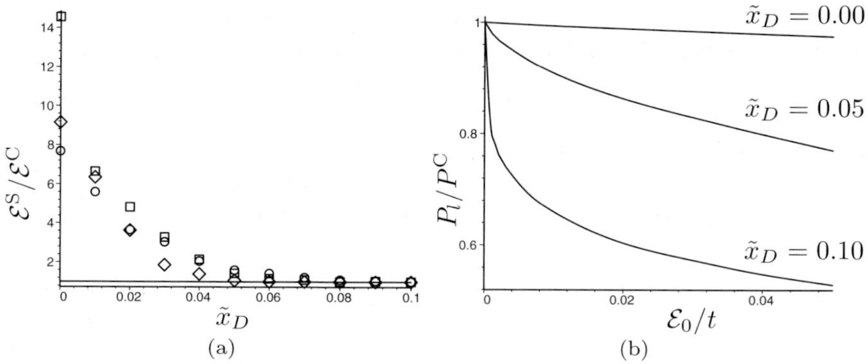

Fig. 10.18. (a) Proximity of bifurcations against the nondimensional pre-existing delamination length $\tilde{x}_D$. Symbols: $L = 50$ mm (□), $L = 75$ mm (○), $L = 100$ mm (◇). As $\tilde{x}_D$ increases, $\mathcal{E}^S/\mathcal{E}^C \to 1$, implying that the critical and secondary bifurcations are basically coincident. (b) Imperfection-sensitivity curves for strut of length $L = 100$ mm with varying degrees of pre-existing delamination $\tilde{x}_D$.

of the critical and secondary bifurcations could be evaluated as shown in Fig. 10.18(a). The graph shows $\mathcal{E}^S/\mathcal{E}^C$ versus $\tilde{x}_D$, where $\tilde{x}_D = x_D/L$, which approaches unity for two out of the three cases shown when $\tilde{x}_D > 0.075$. This means that for $L = 75$ mm and 100 mm, when the pre-existing delamination length is greater than 7.5% of the strut length, the critical and secondary bifurcations are effectively coincident—making a *compound bifurcation point* leading directly to localization. This, in itself, is a strong indication of the existence of severe sensitivity to initial geometric imperfections in conjunction with the delaminated system.

The following study quantifies this further by introducing geometrical imperfections in the more compressed face-plate, as described in Sec. 10.2.2. Combining this geometric imperfection with the pre-existing delamination gives an insight into the equilibrium behaviour postulated to be highly imperfection-sensitive.

As shown in earlier work,[20] the shape of the most severe geometric imperfection—the one that minimizes the limit point load P_l (Fig. 10.3)—changes as the imperfection magnitude $\mathcal{E}_0$ changes. Reducing the complexity of the numerics by fixing β to the value that matched the local buckling profile at the secondary bifurcation, A_0 and α could be varied to find the worst case imperfection shape for a particular value of $\mathcal{E}_0$. The strut example is as before with $L = 100$ mm and for this case $\beta = 15$. Three levels of pre-existing delamination are presented in this example: $\tilde{x}_D = 0, 0.05, 0.10$.

The loci of the relative limit loads P_l/P^{C} plotted against the imperfection magnitudes $\mathcal{E}_0$ give the so-called *imperfection-sensitivity* curves for the component with, in this case, differing levels of pre-existing delaminations and are plotted in Fig. 10.18(b).

The varying nature of the worst case imperfection is shown in Fig. 10.19. The cases shown are for $\tilde{x}_D = 0.05$ and $\tilde{x}_D = 0.10$. The case where there is no pre-existing delamination ($\tilde{x}_D = 0$) is not shown owing to its lack of significant imperfection sensitivity as shown in Fig. 10.18(b), for instance when $\mathcal{E}_0/t = 0.04$, $P_l P^{\mathrm{C}} = 0.98$. A point to note here is that the shape of the worst case localized imperfection becomes less spiked (α decreasing) with increasing $\tilde{x}_D$ as the delaminated region is wider at the beginning.

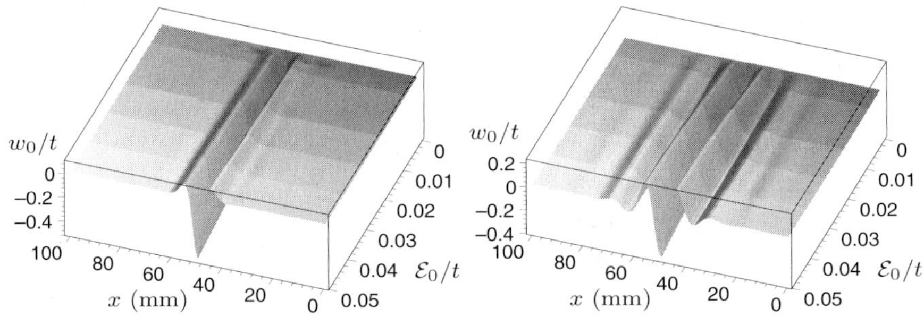

(a) $\tilde{x}_D = 0.05$ (b) $\tilde{x}_D = 0.10$

Fig. 10.19. Variation of the worst case imperfection in struts with a pre-existing delamination.

In the study of the perfect panel in Sec. 10.3.2, the postbuckling behaviour including delamination was considered and this showed a large gap between the critical and secondary bifurcations followed by a severe *snap-back* in the initial subcritical state when $P/P^{\mathrm{C}} \approx 0.92$; the equilibrium configuration of the strut changed very significantly and the load-carrying capacity dropped drastically under controlled end-shortening. The results for the cases of imperfect struts with pre-existing delaminations are presented in Figs. 10.20 and 10.21. Note that for these cases the values of $\mathcal{E}_0/t$ were fixed at 0.02 and that their nondimensionalized amplitudes A_0/t for the worst case imperfection were 0.24 and 0.32 for the respective cases where $\tilde{x}_D = 0.05$ and $\tilde{x}_D = 0.10$. The graphs for both sets of results showed the usual features for imperfect structures; buckle deformation growing while end-shortening was increased and a smoother transition

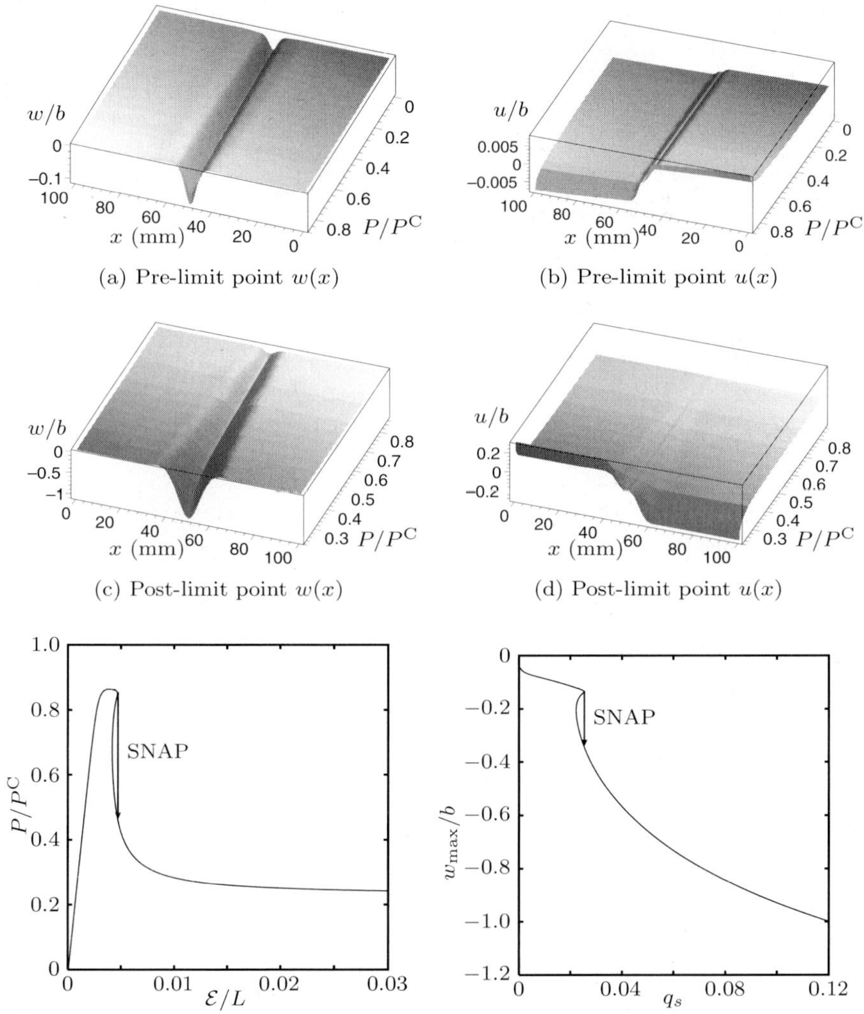

(a) Pre-limit point $w(x)$ (b) Pre-limit point $u(x)$

(c) Post-limit point $w(x)$ (d) Post-limit point $u(x)$

(e) Load vs end-shortening (f) Relative buckle amplitudes

Fig. 10.20. Distribution of the localized buckling pattern for the imperfect case where $\tilde{x}_D = 0.05$ and $\mathcal{E}_0/t = 0.02$.

into larger deformations with greater imperfections (or in this case larger pre-existing delaminations). Moreover, the equilibrium paths showed that only small deformations had occurred when the load passed the limit point ($P = P_l$). For the case of $\tilde{x}_D = 0.05$, the load–end-shortening curve shown in Fig. 10.20(e) also exhibited a slight snap-back after the limit point was

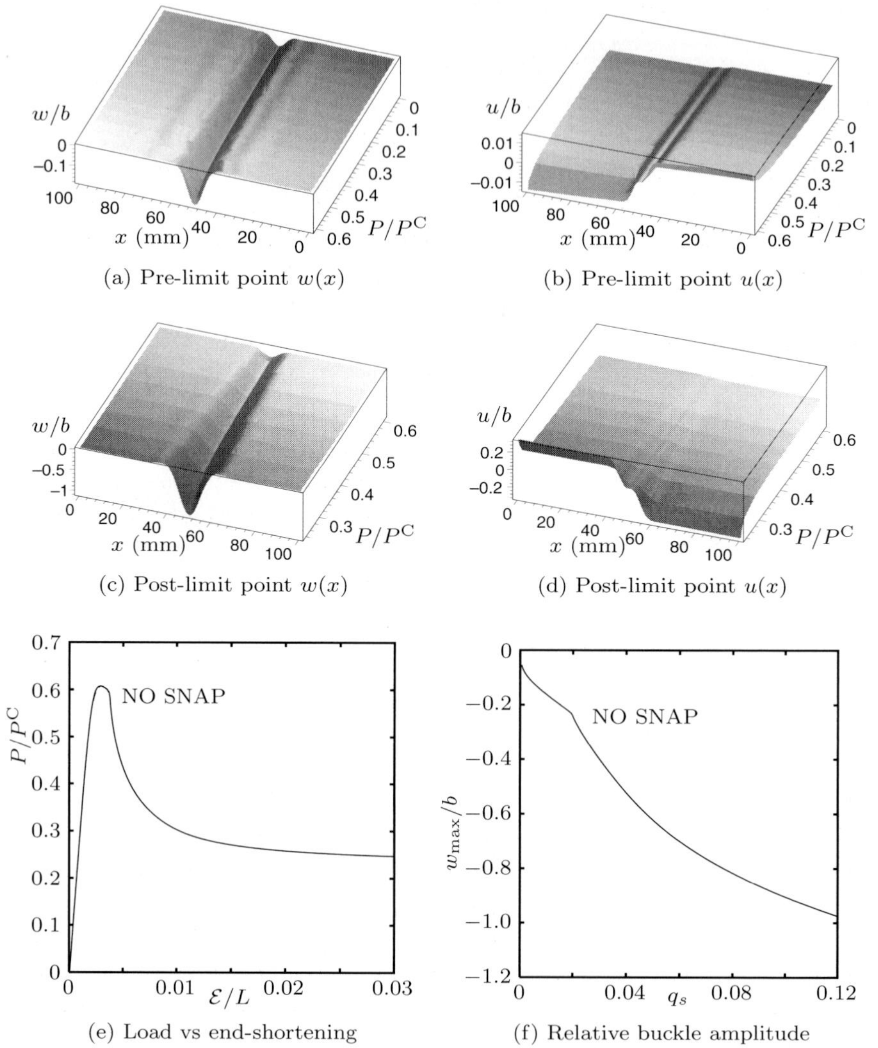

(a) Pre-limit point $w(x)$ (b) Pre-limit point $u(x)$

(c) Post-limit point $w(x)$ (d) Post-limit point $u(x)$

(e) Load vs end-shortening (f) Relative buckle amplitude

Fig. 10.21. Distribution of the localized buckling pattern for the imperfect case where $\tilde{x}_D = 0.10$ and $\mathcal{E}_0/t = 0.02$.

reached. There was no such snap-back for the case of $\tilde{x}_D = 0.10$, as shown in Fig. 10.21(e), highlighting the smoother nature of the equilibrium path with higher values of $\tilde{x}_D$, but with the drawback that the corresponding value of P_l was significantly less.

10.4.2 *Monosymmetric panels*

When monosymmetric panels buckle in the overall mode, the initial sign of the mode amplitude q_s is of paramount importance; this mode is triggered at the same load for both positive and negative values of q_s putting either the thicker or thinner face-plate into extra compression. Therefore, with a face-plate configuration with differing thicknesses, a source of asymmetry in the post-critical buckling response is introduced; the secondary bifurcation that triggers localized buckling occurs at different magnitudes of q_s depending on its initial sign. This leads to the conclusion that the degree of imperfection sensitivity of the strut must be asymmetric.

To demonstrate this, examples are presented such that the relative face-plate thicknesses of initially symmetric struts presented earlier are varied. We then focus on a particular strut found to have a highly pronounced overall–local mode interaction and present the postbuckling characteristics together with a study of its sensitivity to initial geometric imperfections.

Proximity of secondary bifurcation

The basic strut used in the numerical study had the following material properties taken from Sec. 10.3.1. The strut length and core depth were kept constant: $L = 508$ mm and $b = 50.8$ mm respectively; the face-plate material properties were also kept constant: Young's modulus $E = 68947.57$ N/mm^2 and Poisson's ratio $\nu = 0.3$. The face-plate thicknesses and the Young's modulus of the core were varied in the numerical study.

Figure 10.22 shows the effect of changing the monosymmetry parameter $\mathcal{T}$ on the relative gap between the critical bifurcation for overall buckling and the secondary bifurcation for localized buckling ($\mathcal{E}^{\mathrm{S}}/\mathcal{E}^{C}$). Three different cases of core modulus are shown and it can be seen that the degree of monosymmetry can bring together the separate bifurcations, which in turn results in a less stable structure in the postbuckling range as discussed in previous sections. For the cases of $E_c = 300$ N/mm^2 and 198.57 N/mm^2, there is clearly a worst case monosymmetry with a minimum value $\mathcal{E}^{\mathrm{S}}/\mathcal{E}^{C}$ being achieved for a certain value of $\mathcal{T}$. For the third case, $E_c = 50$ N/mm^2, the bifurcations become all but simultaneous for a relatively large range of $\mathcal{T}$, this is most likely due to the local mode becoming critical for those configurations.[20]

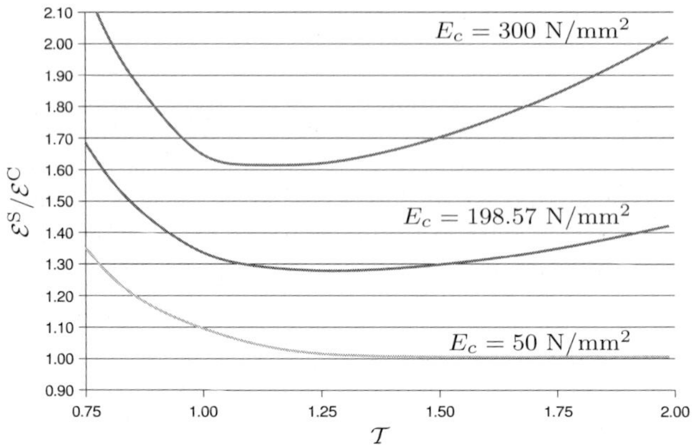

Fig. 10.22. Relative proximity of secondary and critical bifurcations for struts with different cores and monosymmetries.

Postbuckling results

Leading on from the previous section, postbuckling results are presented for a strut that had quite differing relative secondary bifurcation properties depending on the direction of the overall buckling. Figures 10.23, 10.24 and 10.25 show a schematic representation of the selected strut, buckling displacements and equilibrium diagrams for the geometrically perfect strut respectively. The selected strut had the following dimensions and properties: face thicknesses 1.0 mm and 0.8 mm; length $L = 508$ mm; core depth

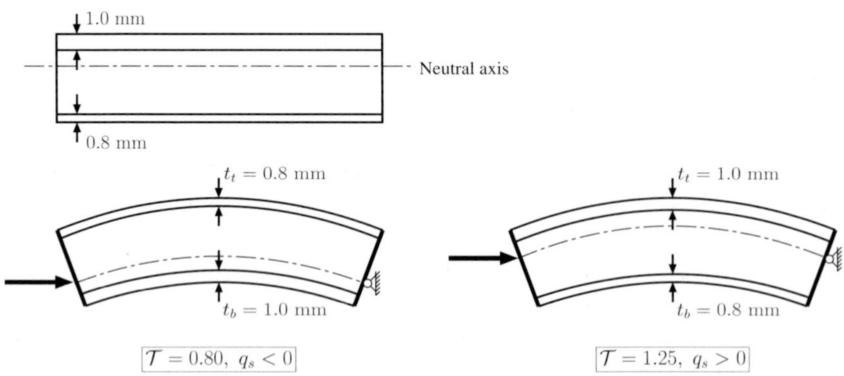

Fig. 10.23. The selected strut: definitions of the thicknesses t_t, t_b, monosymmetry parameter $\mathcal{T}$ and the sign of overall buckling q_s.

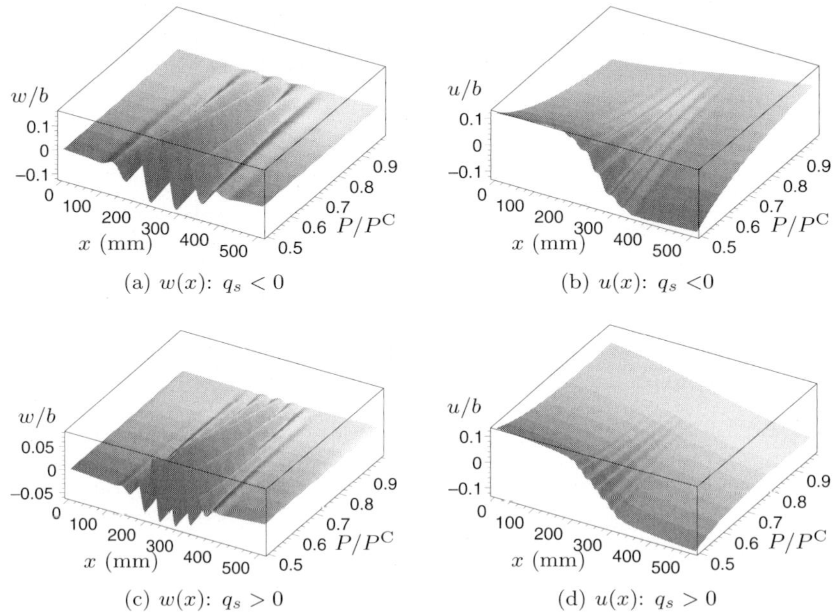

(a) $w(x)$: $q_s < 0$ (b) $u(x)$: $q_s < 0$

(c) $w(x)$: $q_s > 0$ (d) $u(x)$: $q_s > 0$

Fig. 10.24. Postbuckling profiles of the monosymmetric strut with face thicknesses 1.0 mm and 0.8 mm. Other dimensions and properties: $L = 508$ mm, $b = 50.8$ mm, $E_c = 50$ N/mm^2 and $\nu_c = 0.2$.

and properties $b = 50.8$ mm, $E_c = 50$ N/mm^2 and $\nu_c = 0.2$. As shown in Fig. 10.23, if the thicker face-plate was in extra compression after overall buckling had been triggered then q_s would be negative, and if the thinner face-plate was in extra compression after overall buckling then q_s would be positive. For $q_s < 0$ the value of the monosymmetry parameter $\mathcal{T} = 0.8$ with the bifurcation proximity: $\mathcal{E}^{\mathrm{S}}/\mathcal{E}^{\mathrm{C}} = 1.267$, and for $q_s > 0$ the value of the monosymmetry parameter $\mathcal{T} = 1.25$ with the bifurcation proximity: $\mathcal{E}^{\mathrm{S}}/\mathcal{E}^{\mathrm{C}} = 1.016$.

It can be seen in the graphs that the interactive postbuckling modes of w, shown in Figs. 10.24(a) and 10.24(c), have differing localized wavelengths for the face-plates. Moreover, the important result from this is shown in Fig. 10.25 where the proximity between the critical and secondary bifurcations is significant only for $q_s < 0$; for $q_s > 0$ the gap is much smaller. This implies that if the strut had imperfections forcing $q_s > 0$, the strut would be much less likely to reach the linear eigenvalue critical load than if the condition $q_s < 0$ were imposed by the imperfection. This potential asymmetry in the sensitivity to initial imperfections is quantified below.

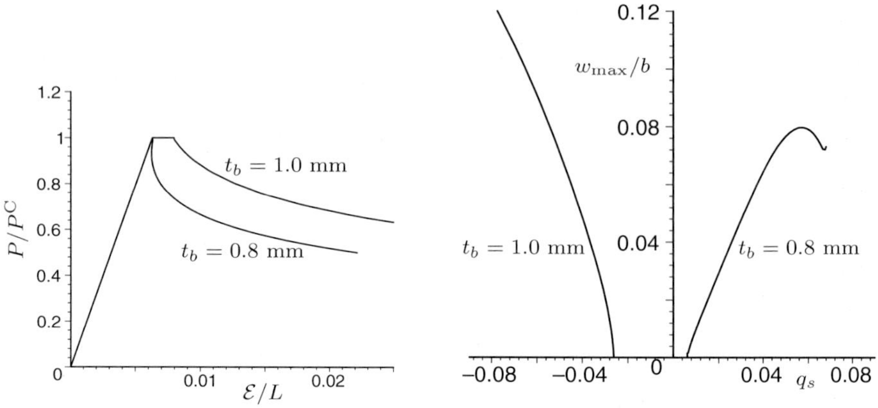

(a) Load vs end-shortening (b) Relative buckle amplitudes

Fig. 10.25. Postbuckling equilibrium diagrams of the monosymmetric strut with face thicknesses 1.0 mm and 0.8 mm. Other dimensions and properties: $L = 508$ mm, $b = 50.8$ mm, $E_c = 50$ N/mm^2 and $\nu_c = 0.2$.

Asymmetry in imperfection sensitivity

The geometric imperfection w_0, expressed in Eq. (10.7), is introduced for the selected strut from the previous section. For each face the imperfection sensitivity for the periodic mode was determined using the following technique: keeping the localization parameter α at zero and the wave number parameter β at the linear eigenvalue solution, the value of the imperfection magnitude $\mathcal{E}_0$ was varied; each variation of $\mathcal{E}_0$ giving an explicit value of A_0. The limit load P_l was then determined and the imperfection-sensitivity curves were plotted. For the localized imperfection the worst case was determined as described in Sec. 10.4.1.

Figures 10.26–10.28 show the imperfection sensitivity curves and the changing profile of the imperfection respectively for the cases where the sole imperfection was confined to either the thicker or the thinner faceplate. As expected from previous work,[20] the localized imperfection gave the more severely unstable equilibrium response for non-trivial values of $\mathcal{E}_0$. Moreover, the monosymmetry of the sandwich panel meant that the relative imperfection sensitivity was higher when the thinner face was imperfect rather than the thicker face; for example, when $\mathcal{E}_0/t_b = 0.025$, the limit load for localized imperfections when $q_s < 0$ was $P_l/P^C = 0.917$, the corresponding limit load level when $q_s > 0$ being $P_l/P^C = 0.765$, a drop in

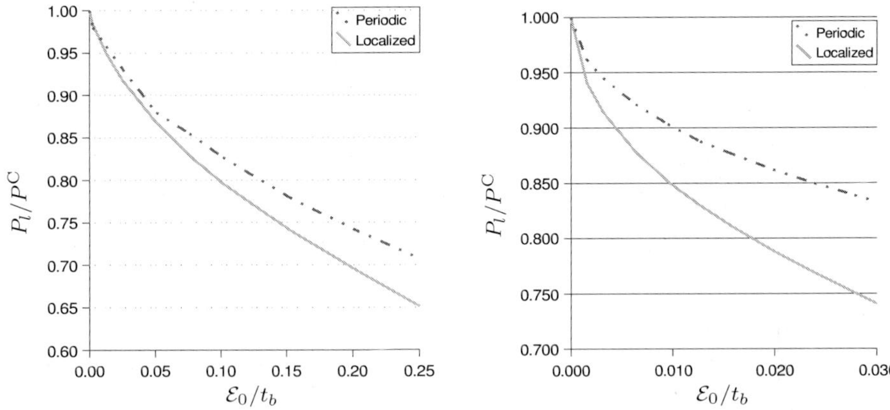

(a) Imperfect thicker face ($q_s < 0$)

(b) Imperfect thinner face ($q_s > 0$)

Fig. 10.26. Imperfection sensitivity curves for periodic and localized geometric imperfections for the monosymmetric strut with face-plate thicknesses 1.0 mm and 0.8 mm; in this case the thinner face is much more sensitive.

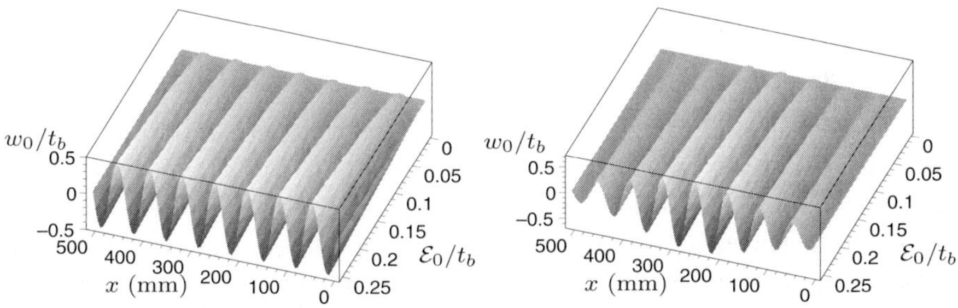

(a) Periodic imperfection

(b) Localized imperfection

Fig. 10.27. Initial imperfection profiles for periodic and worst case localized geometric imperfections for the monosymmetric strut with $t_t = 0.8$ mm and $t_b = 1.0$ mm, *i.e.* the thicker face buckles ($q_s < 0$).

ultimate strength by 17% for an imperfection amplitude A_0 that was very much less than a third of the face-plate thickness.

10.5 Concluding remarks

A sequence of increasingly sophisticated analytical models based on variational principles for the nonlinear buckling response of sandwich panels

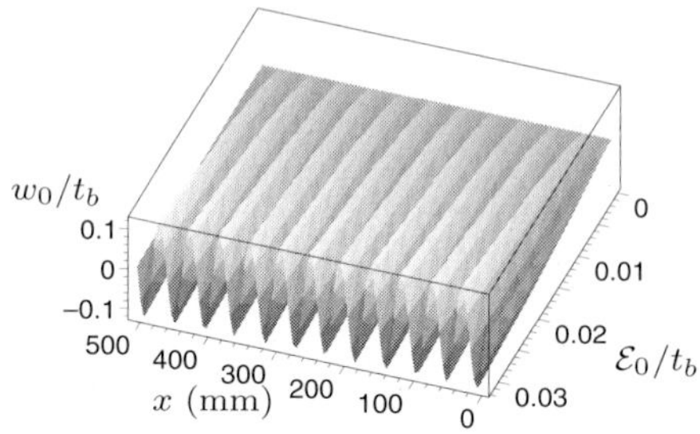

(a) Periodic imperfection

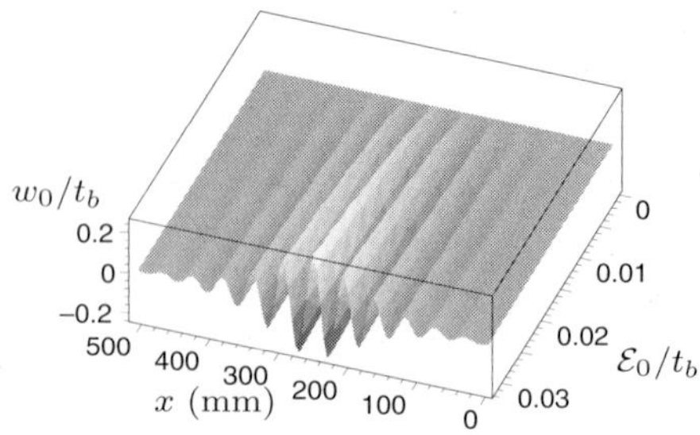

(b) Localized imperfection

Fig. 10.28. Initial imperfection profiles for periodic and worst case localized geometric imperfections for the monosymmetric strut with $t_t = 1.0$ mm and $t_b = 0.8$ mm, *i.e.* the thinner face buckles locally ($q_s > 0$).

has been presented to account for practical situations. Each enhancement has been presented alongside appropriate results that have highlighted the important interaction between overall and local modes of instability and its effect on the postbuckling response when certain parameters change, whether they are geometric or material properties. The interaction has been shown to be increased in severity by varying the shear modulus, by allowing face–core delamination and by allowing external couples to co-exist with the axial loads. Monosymmetry in the cross-section, moreover, creates asymmetry in the postbuckling response with imperfection sensitivities differing for each face-plate. The key point to note from the results is that linear eigenvalue analysis, the cornerstone to most design strength calculations in sandwich structures, does not necessarily provide safe answers for the collapse load under many circumstances when elasticity dominates the response. The mode interaction introduces a secondary instability that promotes highly unstable localized buckling which in the most severe cases, where the global and local instabilities are triggered to all intents and purposes simultaneously, exhibit snap-back responses that are unstable under dead and rigid loading. For such panels applying something akin to *knock-down factors* to linear eigenvalue results, familiar from the design of cylindrical shell structures,[40] could be the route to obtain safe design procedures for these varied and important practical structures; the level of the factors depending on the intensity of the interaction and whether snap-buckling could potentially occur.

10.6 Acknowledgements

The author would like to thank the following people for their invaluable contributions to this work: Professor Giles Hunt of the Centre for Nonlinear Mechanics at the University of Bath, Professor Luís Simões da Silva of the Department of Civil Engineering at the University of Coimbra in Portugal, Dr Avril Blackmore of the Department of Civil and Environmental Engineering at University College London and Daisuke Saito of the Department of Civil and Environmental Engineering at Imperial College London.

10.7 References

1. S. Kodiyalam, S. Nagendra, and J. DeStefano, *AIAA J.* **34**, 614 (1996).
2. E. Bannink, R. Hadcock, and H. Forsch, *J. Aircr.* **15**, 661 (1978).
3. A. C. Duthie, *Plastics and Rubber Int.* **12**, 20 (1987).
4. E. M. Knox, M. J. Cowling, and I. E. Winkle, *Mar. Struct.* **11**, 185 (1998).
5. G. W. Hunt, L. S. Da Silva, and G. M. E. Manzocchi, *Proc. R. Soc. A.* **417**, 155 (1988).
6. G. W. Hunt and M. A. Wadee, *Proc. R. Soc. A.* **454**, 1197 (1998).
7. V. Sokolinsky and Y. Frostig, *AIAA J.* **37**, 1474 (1999).
8. S. Kim and S. Sridharan, *J. Eng. Mech.-ASCE.* **131**, 1313 (2005).
9. M. A. Wadee, *J. Sandw. Struct. Mater.* **1**, 230 (1999).
10. H. G. Allen, *Analysis and design of structural sandwich panels.* (Pergamon, Oxford, 1969).
11. W. T. Koiter and M. Pignataro. In Ref. 14, p. 133.
12. J. M. T. Thompson and G. W. Hunt, *A general theory of elastic stability.* (Wiley, London, 1973).
13. G. W. Hunt, *Appl. Mech. Rev.* **39**, 1165 (1986).
14. B. Budiansky, Ed., *Buckling of Structures.* IUTAM Symposium, Cambridge, USA, 1974, (Springer, Berlin, 1976).
15. A. R. Champneys, G. W. Hunt, and J. M. T. Thompson, Eds., *Localization and solitary waves in solid mechanics.* vol. 12, *Advanced series in nonlinear dynamics,* (World Scientific, 1999).
16. A. D. Kerr, *J. Appl. Mech.-Trans. ASME.* **31**, 491 (1964).
17. C. M. Wang, J. N. Reddy, and K. H. Lee, *Shear deformable beams and plates: Relationships with classical solutions.* (Elsevier, Amsterdam, 2000).
18. M. K. Wadee, M. A. Wadee, A. P. Bassom, and A. A. Aigner, *Proc. R. Soc. A.* **462**, 817 (2006).
19. M. K. Wadee, G. W. Hunt, and A. I. M. Whiting, *Proc. R. Soc. A.* **453**, 2085 (1997).
20. M. A. Wadee, *Int. J. Solids Struct.* **37**, 1191 (2000).
21. J. M. T. Thompson and G. W. Hunt, *Elastic instability phenomena.* (Wiley, Chichester, 1984).

22. S. P. Timoshenko and J. M. Gere, *Theory of elastic stability.* (McGraw-Hill, New York, USA, 1961).
23. Y. Frostig and O. T. Thomsen, *J. Sandw. Struct. Mater.* **7**, 53 (2005).
24. R. M. Jones, *Mechanics of composite materials.* (Taylor and Francis, New York, USA, 1999), 2nd edition.
25. L. J. Gibson and M. F. Ashby, *Proc. R. Soc. A.* **382**, 43 (1982).
26. L. J. Gibson, M. F. Ashby, G. S. Schjaer, and C. I. Robertson, *Proc. R. Soc. A.* **382**, 25 (1982).
27. C. Fox, *An introduction to the calculus of variations.* (Dover, New York, USA, 1987).
28. E. J. Doedel, A. R. Champneys, T. F. Fairgrieve, Y. A. Kuznetsov, B. Sandstede, and X.-J. Wang. Technical report, Department of Computer Science, Concordia University, Montreal, Canada, (1997). (Available by FTP from `ftp.concordia.ca` in `/pub/doedel/auto`).
29. G. W. Hunt, *Nonlinear Dyn.* **43**, 29 (2006).
30. L. S. Da Silva and G. W. Hunt, *Mech. Struct. & Mach.* **18**, 353 (1990).
31. M. A. Wadee and G. W. Hunt, *J. Appl. Mech.-Trans. ASME.* **65**, 523 (1998).
32. G. A. O. Davies, P. Robinson, J. Robson, and D. Eady, *Compos. Pt. A-Appl. Sci. Manuf.* **28A**, 757 (1997).
33. L. J. Gibson and M. F. Ashby, *Cellular Solids: Structure and Properties.* Cambridge Solid State Science Series, (Cambridge University Press, Cambridge, 1999), 2nd edition.
34. M. Somers, T. Weller, and H. Abramovich, *Compos. Struct.* **17**, 295 (1991).
35. M. Somers, T. Weller, and H. Abramovich, *Compos. Struct.* **21**, 211 (1992).
36. Y. Frostig, *Compos. Struct.* **20**, 1 (1992).
37. J. W. Hutchinson, M. Y. He, and A. G. Evans, *J. Mech. Phys. Solids.* **48**, 709 (2000).
38. M. A. Wadee, *J. Mech. Phys. Solids.* **50**, 1767 (2002).
39. M. A. Wadee and A. Blackmore, *J. Mech. Phys. Solids.* **49**, 1281 (2001).
40. J. W. Hutchinson and M. Y. He, *Int. J. Solids Struct.* **37**, 6777 (2000).

CHAPTER 11

THE BOUNDARY ELEMENT METHOD FOR BUCKLING AND POSTBUCKLING ANALYSIS OF PLATES AND SHELLS

M. H. Aliabadi and P. M. Baiz

Department of Aeronautics, Imperial College London
South Kensington Campus, London, SW7 2AZ, United Kingdom
E-mail: m.h.aliabadi@imperial.ac.uk

This chapter describes recent developments on the buckling and post-buckling analysis of thin walled structures by the Boundary Elements Method (BEM). Boundary integral equations are obtained by coupling two dimensional plane stress elasticity with boundary element formulation of Reissner plate bending. The linear buckling problem is formulated as a standard eigenvalue problem, in order to obtain directly critical loads and buckling modes as part of the solution. A multi-region BEM formulation is also introduced for the analysis of local buckling in thin walled assemblies. Nonlinear equations for large deflection and post-buckling analysis are also investigated. Nonlinear system of equations is solved using an incremental solution strategy, where displacements and tractions are applied in small but finite increments. Several examples of flat and curved plates (shallow shells) with different dimensions and boundary conditions are analyzed, including plate assemblies. Results are compared with other numerical and analytical solutions, and good agreement is obtained.

11.1 Introduction

Instability in structural mechanics can be seen as the condition in which a body fails to react to the bending moment generated by a compressive stress state, leading to visible large displacements. This phenomenon in thin walled structures is of major concern in areas such as aerospace, in which the design requirements of weight critical applications usually leads to thin panels with stability problems. The development of methods for the analysis of instability have been increasing in order to ensure the integrity of the structural design. Analytical solutions are very limited in scope; while the experimental approach to the problem although give reliable data (provided that is well performed), it is an expensive procedure when all possible configurations of the practical design have to be analyzed. Consequently, numerical methods must be applied.

The Boundary Element Method (BEM)[2] like FEM is a numerical method of solving partial differential equations which have been transformed into integral equations. One of the main differences between BEM and FEM relates to the weighting functions used to transform the partial differential equations into the equivalent "weak form". FEM use simple arbitrary functions (usually polynomials), while in BEM, analytical solutions representing displacement and tractions at a point of an infinite domain due to the application of point forces at another point are used (Fundamental Solutions). In fact, the use of these analytical solutions is the reason for increased accuracy of BEM solutions over FEM ones, in particular at regions of higher stress gradients.

The application of the Boundary Element Method to linear stability problems for plate structures have been investigated since the 80's. Costa and Brebbia in 1985[12] presented a general direct formulation of the problem. The numerical implementation required domain discretization with three curvature coefficients as unknowns at every domain node. Singellakis and Kang[47] eliminated the curvatures from the domain and presented a solution which required the modelling in the domain of the deflections only. At the same time Manolis, Beskos and Pineros[34] developed a direct boundary element formulation dealing with linear elastic stability analysis of bars and plates. Tanaka and Miyazaki[53] presented elastic buckling of assembled plate structures. More recently, Syngellakis and Elzein[48] and Syngellakis[49] presented extended boundary element formulations to incorporate any combination of loading and support conditions; Nerantzaki and Katsikadelis[36] presented a boundary element formulation for buckling of plates with variable thickness; and Lin, Duffield and Shih[30] described a general boundary element formulation for different boundary conditions and arbitrary planar shapes. In all these cases classical or Kirchhoff-Love theory were assumed.

Boundary element formulations with shear deformable theory for the linear buckling problem of thin walled structures are scarce. Purbolaksono and Aliabadi[40] presented a formulation for shear deformable plates with general boundary conditions and arbitrary planar shapes. For the case of shallow shells, Baiz and Aliabadi[7,8] using a shear deformable formulation introduced the first boundary element approaches. No other studies on the linear elastic buckling analysis of shallow shell structures by BEM has been reported, for classical (Kirchhoff-Love) or shear deformable (Reissner or Mindlin) theories.

Applications of the boundary element method to nonlinear plate theory was presented by Kamiya and Sawaki,[23] with a formulation for the large deflection analysis of elastic plates based on the Berger equation. Tanaka[51] based his formulation on the von Karman's equation, obtaining boundary and domain integrals in terms of stress and displacement. Kamiya, Sawaki and Nakamura,[25] extended the formulation in[23] to include postbuckling behaviour under thermal loading. Ye and Liu[66] analyzed the finite deflection of von Karman's plates using a relaxation iterative approach. O'Donoghue and Atluri[37] presented von Karman's equations for static, dynamic and buckling applications. Other works dealing with large deflection and postbuckling of Kirchhoff's plates are those of Kawabe,[26] Qin and Huang,[41] Atluri and Pipkins,[4] Tanaka, Matsumoto and Zheng,[54,55] Elzein and Singellakis,[20] Wang, Ji and Tanaka.[60]

Geometrically nonlinear analysis of a Reissner type plate by the boundary element method was first proposed by Lei, Huang and Wang.[31,32] Later, Sun, He and Qin[50] derived exact boundary integral equations for the analysis of nonlinear Reissner plate based on a variational principle. More recent works on shear deformable plates, are those presented by Wen, Aliabadi and Young describing large deflection[63] and postbuckling[64] analysis; and Supriyono and Aliabadi[46] for combined large deflection and plasticity.

In the case of large deflection and postbuckling of shallow shells, Kamiya and Sawaki[24] presented a boundary element approach to geometrically nonlinear shallow shells with arbitrary contours. Tosaka and Miyake[58] developed a theoretical foundation for the large deflection analysis of shells, proposing three types of integral equations. Zhang and Atluri[68] presented a non linear field boundary element formulation for snap-through phenomena in thin shallow shells using the arc-length method. Ye[67] developed a mixed boundary element and finite difference method for the nonlinear analysis of spherical annular shells under axisymmetric loads. Lin and Long[29] presented a geometrically nonlinear formulation for shallow shell analysis, based on displacements rather than stress functions. In terms of shear deformable theory for large deflection of shallow shells, the only work available has been presented very recently by Dirgantara and Aliabadi[18]

The present chapter reports on the investigation of recent boundary element formulations for the buckling and postbuckling analysis of shear deformable plates and shallow shells. Initially, basic concepts of shear deformable plate theory, and boundary integral equations are described. The linear buckling problem is formulated as a standard eigenvalue problem, to provide direct evaluation of critical load factors and buckling modes. Nonlinear boundary integral equations are solved using an incremental approach, where loads and prescribed boundary conditions are applied in small but finite increments. The resulting system of equations are solved using a purely incremental technique and the Newton-Raphson technique. In the case of snap buckling a special numerical technique known as the Arc-length method is also implemented in order to handle difficulties at limit points. Several examples in which results from the proposed BEM formulations are compared with other numerical and analytical solutions, and good agreement is obtained.

11.2 Basic definitions of shear deformable plates and shallow shells

Consider a shallow shell of an isotropic linear elastic material, with uniform thickness h, Young's modulus E, Poisson ratio v and shear modulus $G = E/2(1+v)$, with a quadratic middle surface defined by R_1 and R_2, which are principal curvatures of the shell in the x_1- and x_2- directions, respectively. The indicial notation used throughout this chapter is as follows: the Greek indices (α, β, γ) will vary from 1 to 2 and Roman indices (i, j, k) from 1 to 3.

Generalized displacements are represented as w_i and u_α, where w_α denotes rotations of the middle surface (w_1 and w_2), w_3 denotes the out-of-plane displacement, and u_α denotes in-plane displacements (u_1 and u_2). The generalized tractions are denoted as p_i and t_α, where p_α denotes tractions due to the stress couples (p_1 and p_2), p_3 denotes the traction due to shear stress resultant (p_3) and t_α denotes tractions due to membrane stress resultants (t_1 and t_2) as shown in Fig. 11.1.

11.2.1 *Kinematic equations*

Strain tensors in shear deformable linear elastic theories can be derived from the deformation pattern of a differential element. The membrane strain resultant tensor can be expressed as follows:

$$\varepsilon_{\alpha\beta}^{(l)} = \frac{1}{2}\left(u_{\alpha,\beta} + u_{\beta,\alpha}\right) + w_3 k_{\alpha\beta} \tag{11.1}$$

where $(\ )_{,\beta} = \partial(\ )/\partial x_\beta$, the superscript (l) refers to the linear part, and $k_{11} = 1/R_1$, $k_{22} = 1/R_2$, $k_{12} = k_{21} = 0$. Transverse shear strain resultant

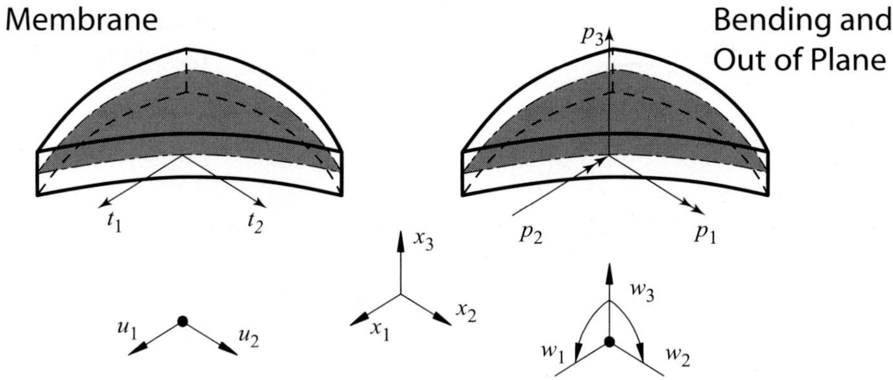

Fig. 11.1. Sign convection for generalized displacement and tractions.

can be expressed as:

$$\gamma_{\alpha 3} = w_\alpha + w_{3,\alpha} \tag{11.2}$$

And finally, the flexural strain resultant can be written as:

$$\kappa_{\alpha\beta} = 2\chi_{\alpha\beta} = w_{\alpha,\beta} + w_{\beta,\alpha} \tag{11.3}$$

It follows that shallow shells bear the same relation to plates as curved beams to straight beams. Therefore, in the literature, including the present work, shallow shells are usually referred to as curved plates. In fact, plate equations can be derived from shallow shell equations in a very straight forward manner by just considering an infinite radius ($k_{\alpha\beta} = 0$).

11.2.2 *Equilibrium equations*

Equilibrium equations for shear deformable plate bending (Reissner-Mindlin) and 2D elasticity can be written in indicial notation as follows[2] (see Fig. 11.2),

$$M_{\alpha\beta,\beta} - Q_\alpha = 0; \tag{11.4}$$

$$Q_{\alpha,\alpha} + f_3^b = 0; \tag{11.5}$$

$$N_{\alpha\beta,\beta}^{(i)} + f_\alpha^m = 0 \tag{11.6}$$

where $N_{\alpha\beta}$ denote membrane stresses, $M_{\alpha\beta}$ represent bending moments, Q_α is the shear forces for plate bending and f_3^b, f_α^m are body forces that can expressed as,

$$f_3^b = q_3 - k_{\alpha\beta} N_{\alpha\beta} \tag{11.7}$$

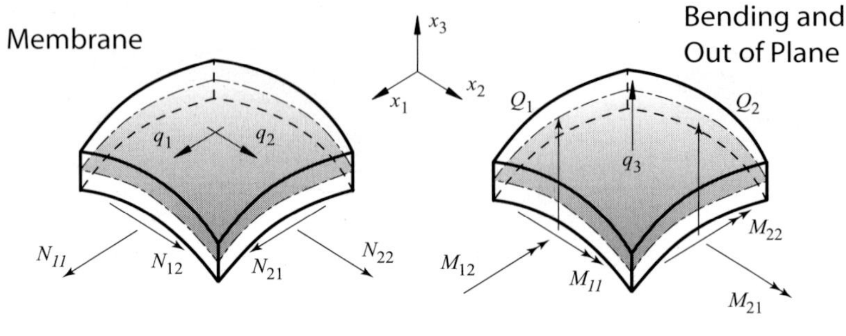

Fig. 11.2. Stress resultants.

$$f_\alpha^m = q_\alpha + N_{\alpha\beta,\beta}^{(ii)} \tag{11.8}$$

where q_i are applied body forces, and $k_{\alpha\beta}N_{\alpha\beta}$ and $N_{\alpha\beta,\beta}^{(ii)}$ are terms due to the curvature of the plate.

11.2.3 *Constitutive equations*

The relationships between stress resultants and strains were derived by Naghdi,[35] using Reissner's variational theorem of elasticity,[43] and are given by:

$$M_{\alpha\beta} = D\frac{1-\nu}{2}\left(2\chi_{\alpha\beta} + \frac{2\nu}{1-\nu}\chi_{\gamma\gamma}\delta_{\alpha\beta}\right) \tag{11.9}$$

$$Q_\alpha = C\gamma_{\alpha 3} \tag{11.10}$$

$$N_{\alpha\beta} = B\frac{1-\nu}{2}\left(2\varepsilon_{\alpha\beta} + \frac{2\nu}{1-\nu}\varepsilon_{\gamma\gamma}\delta_{\alpha\beta}\right) \tag{11.11}$$

where $\delta_{\alpha\beta}$ is the Kronecker delta function; $B(= Eh/\left(1-\nu^2\right))$ is known as the tension stiffness; $D(= Eh^3/\left[12\left(1-\nu^2\right)\right])$ is the bending stiffness; $C(=\left[D\left(1-\nu\right)\lambda_s^2\right]/2)$ is the shear stiffness and $\lambda_s = \sqrt{10}/h$ is called the shear factor.

Stress-displacement relationships can be obtained after equations (11.1)-(11.3) are substituted into equations (11.9)-(11.11):

$$M_{\alpha\beta} = D\frac{1-\nu}{2}\left(w_{\alpha,\beta} + w_{\beta,\alpha} + \frac{2\nu}{1-\nu}w_{\gamma,\gamma}\delta_{\alpha\beta}\right) \tag{11.12}$$

$$Q_\alpha = C(w_\alpha + w_{3,\alpha}) \tag{11.13}$$

and

$$N_{\alpha\beta}^{(i)} = B\frac{1-\nu}{2}\left(u_{\alpha,\beta} + u_{\beta,\alpha} + \frac{2\nu}{1-\nu}u_{\gamma,\gamma}\delta_{\alpha\beta}\right) \tag{11.14}$$

$$N_{\alpha\beta}^{(ii)} = B\left[(1-\nu)k_{\alpha\beta} + \nu\delta_{\alpha\beta}k_{\phi\phi}\right]w_3$$

$$N_{\alpha\beta}^{(l)} = N_{\alpha\beta}^{(i)} + N_{\alpha\beta}^{(ii)}$$

To make the representation more convenient, the linear membrane stress $N_{\alpha\beta}^{(l)}$ is separated into $N_{\alpha\beta}^{(i)}$ which is due to in-plane displacements and $N_{\alpha\beta}^{(ii)}$ which is due to curvature and out-of-plane displacement. Also this representation facilitates the derivation of the membrane boundary integral representations.

11.2.4 *Large deflection theory*

The behaviour of a structure is considered to be non-linear if, under static condition, the deflections of any point of the body are not proportional to the magnitude of the applied load. In the case of geometrically non-linear theory (large deflection) the membrane and bending deflections are coupled in the membrane strain equation (out-of-plane displacement contributions are included in strain-displacement relationships). The new basic assumption can be stated as follows: "Deflections (w_3) are large (several times greater than the thickness, h); but strains are still small compared to unity".

Based on this, the strain tensors can be obtained by considering the geometrical changes as the element is stretched and deformed,[57]

$$\varepsilon_{\alpha\beta} = \frac{1}{2}\left(u_{\alpha,\beta} + u_{\beta,\alpha}\right) + w_3 k_{\alpha\beta} + \left[\frac{1}{2}w_{3,\alpha}w_{3,\beta}\right] \tag{11.15}$$

from where the following expressions can be written, see equation (11.1):

$$\varepsilon_{\alpha\beta}^{(l)} = \frac{1}{2}\left(u_{\alpha,\beta} + u_{\beta,\alpha}\right) + w_3 k_{\alpha\beta}; \qquad \varepsilon_{\alpha\beta}^{(n)} = \frac{1}{2}w_{3,\alpha}w_{3,\beta}$$

Equilibrium equations are also affected by the membrane and bending coupling, resulting in the following pseudo body forces,

$$f_3^b = q_3 - k_{\alpha\beta}N_{\alpha\beta} + \left[\left(N_{\alpha\beta}w_{3,\beta}\right)_{,\alpha}\right] \tag{11.16}$$

and

$$f_\alpha^m = N_{\alpha\beta,\beta}^{(ii)} + \left[N_{\alpha\beta,\beta}^{(n)}\right] + q_\alpha \tag{11.17}$$

It is clear that nonlinear terms in the kinematic and equilibrium equations are given by the terms in square brackets. Constitutive equations remain the same in a pure geometrically nonlinear analysis, but due to the nonlinear terms in the kinematic equations the new stress-displacement relation for $N_{\alpha\beta}$ can be obtained after equation (11.15) is substituted into equation (11.11):

$$N_{\alpha\beta} = N^{(i)}_{\alpha\beta} + N^{(ii)}_{\alpha\beta} + N^{(n)}_{\alpha\beta} \tag{11.18}$$

where again, in order to make the representation more convenient, the term $N_{\alpha\beta}$ is separated into the linear $N^{(l)}_{\alpha\beta}$ $(= N^{(i)}_{\alpha\beta} + N^{(ii)}_{\alpha\beta})$ and nonlinear parts $N^{(n)}_{\alpha\beta}$:

$$N^{(i)}_{\alpha\beta} = B\frac{1-\nu}{2}\left(u_{\alpha,\beta} + u_{\beta,\alpha} + \frac{2\nu}{1-\nu}u_{\gamma,\gamma}\delta_{\alpha\beta}\right)$$

$$N^{(ii)}_{\alpha\beta} = B\left[(1-\nu)k_{\alpha\beta} + \nu\delta_{\alpha\beta}k_{\phi\phi}\right]w_3$$

$$N^{(n)}_{\alpha\beta} = \left[B\frac{1-\nu}{2}\left(w_{3,\alpha}w_{3,\beta} + \frac{\nu}{1-\nu}w_{3,\gamma}w_{3,\gamma}\delta_{\alpha\beta}\right)\right]$$

11.3 Boundary Element Method for shear deformable plates and shallow shells

Integral equations for shear deformable plate and shallow shell problems can be derived by considering the integral representations of the governing equations (11.4)-(11.6) from the following integral identities:

$$\int_\Omega \left[(M_{\alpha\beta,\beta} - Q_\alpha)\, W^*_\alpha + (Q_{\alpha,\alpha} + f^b_3)\, W^*_3\right] d\Omega = 0 \tag{11.19}$$

and

$$\int_\Omega (N^{(i)}_{\alpha\beta,\beta} + f^m_\alpha)\, U^*_\alpha d\Omega = 0 \tag{11.20}$$

where U^*_α and $W^*_i (i = \alpha, 3)$ are weighting functions and Ω is the projected domain of a shell on $x_1 - x_2$ plane, bounded by boundary Γ (see Fig. 11.3). Equation (11.19) is an integral representation related to the governing equations for bending and transverse shear stress resultants, while equation (11.20) is an integral representation related to the governing equations for membrane stress resultants. It can be seen that equilibrium equations in the integral representations (11.19)-(11.20) have the form of linear shear deformable plate bending and two dimensional plane stress (when $f^b_3 = q_3$

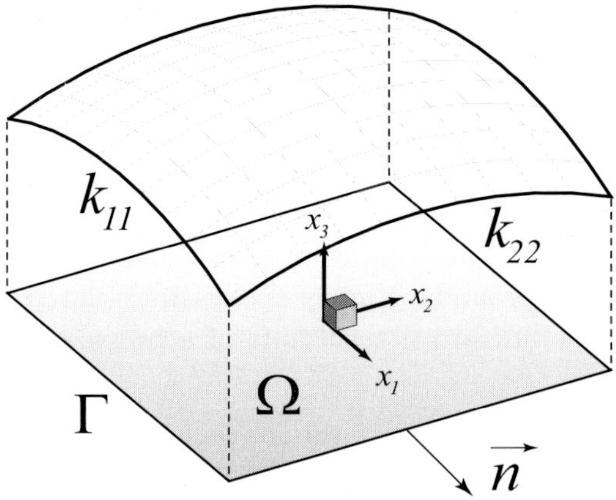

Fig. 11.3. Quadratic shallow shell and its projection on x_1-x_2 plane.

and $f_\alpha^m = q_\alpha$), for which boundary integral formulation procedures and fundamental solutions have already been developed.

11.3.1 *Rotations and out of plane integral equations*

The boundary integral representation related to the governing equations for bending and transverse shear stress resultants of a boundary source point are derived by using the weighted residual method as shown in Dirgantara.[15,17] After taking into account all the limits and the jump terms,

$$c_{ij}w_j + \fint_\Gamma P_{ij}^* w_j d\Gamma - \int_\Gamma W_{ij}^* p_j d\Gamma - \int_\Omega W_{i3}^* f_3^b d\Omega = 0 \tag{11.21}$$

After introducing f_3^b from equation (11.7) in equation (11.21) and expanding $N_{\alpha\beta}^{(l)}$, the following expression can be obtained:

$$\begin{aligned} c_{ij}w_j + \fint_\Gamma P_{ij}^* w_j d\Gamma &= \int_\Gamma W_{ij}^* p_j d\Gamma \\ - \int_\Omega k_{\alpha\beta} B \frac{1-\nu}{2} &\left(u_{\alpha,\beta} + u_{\beta,\alpha} + \frac{2\nu}{1-\nu} u_{\phi,\phi}\delta_{\alpha\beta} \right) W_{i3}^* d\Omega \\ - \int_\Omega k_{\alpha\beta} B \left[(1-\nu)k_{\alpha\beta} + \nu\delta_{\alpha\beta}k_{\gamma\gamma}\right] &w_3 W_{i3}^* d\Omega + \int_\Omega W_{i3}^* q_3 d\Omega \end{aligned} \tag{11.22}$$

where $\fint_\Gamma$ denotes a Cauchy principal value integral, c_{ij} are the jump terms, n_β are the components of the outward normal vector to the boundary Γ (see

Fig. 11.3). The value of c_{ij} is equal to $\frac{1}{2}\delta_{ij}$ when the source point is located on a smooth boundary and equal to δ_{ij} when is at the domain. W^*_{ij} and P^*_{ij} are the displacement and traction fundamental solutions respectively, derived by Vander Weeën[59] and are given in Appendix A.

11.3.2 *In plane displacement integral equations*

In the same way, the boundary integral equation related to the governing equations for membrane stress resultants of a boundary source point can be written as:[15,17]

$$c_{\theta\alpha}u_\alpha + \oint_\Gamma T^{*(i)}_{\theta\alpha}u_\alpha d\Gamma = \int_\Gamma U^*_{\theta\alpha}t^{(i)}_\alpha d\Gamma + \int_\Omega U^*_{\theta\alpha}f^m_\alpha d\Omega \tag{11.23}$$

As it was explained with $N^{(i)}_{\alpha\beta}$, the upper index (i) refers to the in-plane displacement. After introducing f^m_α from equation (11.8), and using the following relation:

$$t_\alpha = t^{(i)}_\alpha + t^{(ii)}_\alpha = N^{(i)}_{\alpha\beta}n_\beta + N^{(ii)}_{\alpha\beta}n_\beta$$

equation (11.23) can be written as follows:

$$\begin{aligned} c_{\theta\alpha}u_\alpha &+ \oint_\Gamma T^{*(i)}_{\theta\alpha}u_\alpha d\Gamma + \int_\Gamma U^*_{\theta\alpha}B\left[k_{\alpha\beta}(1-\nu) + \nu\delta_{\alpha\beta}k_{\phi\phi}\right]w_3 n_\beta d\Gamma \\ &- \int_\Omega U^*_{\theta\alpha}B\left[k_{\alpha\beta}(1-\nu) + \nu\delta_{\alpha\beta}k_{\phi\phi}\right]w_{3,\beta}d\Omega \\ &= \int_\Gamma U^*_{\theta\alpha}t_\alpha d\Gamma + \int_\Omega U^*_{\theta\alpha}q_\alpha d\Omega \end{aligned} \tag{11.24}$$

where $U^*_{\theta\alpha}$ and $T^{(i)*}_{\theta\alpha}$ are the well known fundamental solutions for in-plane displacements and membrane tractions respectively. These kernels are also given in Appendix A.

Equations (11.22) and (11.24) represent a set of five boundary integral equations. Equations in (11.22) are for rotations and out-of-plane displacement while equations in (11.24) are for in-plane displacements.

11.4 Governing integral equations for linear buckling

Eigenvalue solutions are generally used to estimate the critical buckling loads of stiff structures which carry their design loads primarily by axial or membrane action, rather than by bending action. Their response usually involves very little deformation prior to buckling. Also a linear eigenvalue analysis usually represents the first step in the stability analysis, because the eigenmodes obtained from this procedure can be used in the investigation of sensitivity of the structure to imperfections. To reduce the linear buckling problem of plates and shallow shells to an eigenvalue formulation, the distribution of membrane stresses $N_{\alpha\beta}^{(l)}$ should be known when the structure is subjected to the applied compressive load. Later, it is assumed that membrane stresses in the buckling state are λ times the value under the initial reference state.

11.4.1 *Integral equations for in plane stresses*

Linear membrane stress resultants at domain points can be evaluated from the derivative of equation (11.24) and by using the relationship in equation (11.14),

$$\begin{aligned} N_{\alpha\beta}^{(l)} &= \int_\Gamma U^*_{\alpha\beta\gamma} t_\gamma d\Gamma - \int_\Gamma T^{(i)*}_{\alpha\beta\gamma} u_\gamma d\Gamma \\ &\quad - \int_\Gamma U^*_{\alpha\beta\gamma} B\left[k_{\gamma\theta}\left(1-\nu\right) + \nu\delta_{\gamma\theta}k_{\phi\phi}\right] w_3 n_\theta d\Gamma \\ &\quad + \int_\Omega U^*_{\alpha\beta\gamma} B\left[k_{\gamma\theta}\left(1-\nu\right) + \nu\delta_{\gamma\theta}k_{\phi\phi}\right] w_{3,\theta} d\Omega \\ &\quad + \int_\Omega U^*_{\alpha\beta\gamma} q_\gamma d\Omega + B\left[(1-\nu)k_{\alpha\beta} + \nu\delta_{\alpha\beta}k_{\phi\phi}\right] w_3 \end{aligned} \tag{11.25}$$

The kernels $U^*_{\alpha\beta\gamma}$ and $T^*_{\alpha\beta\gamma}$ are linear combination of the first derivatives of $U^*_{\alpha\beta}$ and $T^*_{\alpha\beta}$ with respect to the source point and can be found in[15,17] and are also listed in Appendix A.

Another approach that could be used for the evaluation of stresses at internal points consist of a numerical differentiation of displacements after internal displacement have been found.[68] The boundary integral equations (11.25) although computational more time consuming and mathematically more cumbersome, gives more accurate results and therefore it was adopted in this work.

11.4.2 *Integral formulation for the linear buckling problem*

Appropriate forms of the linearized buckling problem can be derived by transforming the bending integral equations (11.22) into an equivalent buckling formulation. This is achieved by introducing a pseudo transversal body force and a critical load factor λ, resulting in a group of equations in terms of the prebuckling membrane stresses and the buckled displacements, as follows

$$c_{ij}w_j + \int\!\!\!\!\!\!-_{\Gamma} P^*_{ij}w_j d\Gamma = \int_{\Gamma} W^*_{ij}p_j d\Gamma$$

$$-\int_{\Omega} k_{\alpha\beta}B\frac{1-\nu}{2}\left(u_{\alpha,\beta}+u_{\beta,\alpha}+\frac{2\nu}{1-\nu}u_{\phi,\phi}\delta_{\alpha\beta}\right)W^*_{i3}d\Omega$$

$$-\int_{\Omega} k_{\alpha\beta}B\left[(1-\nu)k_{\alpha\beta}+\nu\delta_{\alpha\beta}k_{\gamma\gamma}\right]w_3W^*_{i3}d\Omega$$

$$+\lambda\int_{\Omega} W^*_{i3}q_3d\Omega + \lambda\int_{\Omega} W^*_{i3}(N^{(l)}_{\alpha\beta}w_{3,\beta})_{,\alpha}d\Omega \tag{11.26}$$

where $(N^{(l)}_{\alpha\beta}w_{3,\beta})_{,\alpha}$ is a body term due to the large deflection of w_3. This term is similar to the common extra term that appears in the nonlinear equilibrium equations of plates and shells, see equation (11.16). It is important to notice that because of the presence of the curvature terms in equation (11.26), this equation have to be solve simultaneously with equation (11.24), respectively. In the case of flat plates ($k_{\alpha\beta}=0$), only equation (11.26) is required, as presented by Purbolaksono and Aliabadi.[40]

The deflection integral equation w_3 at domain points is required as the additional equation in order to arrange an eigenvalue problem, and additionally the derivatives $w_{3,\beta}$ and $w_{3,\alpha\beta}$ have to be expressed in terms of w_3, see Appendix B. This additional equations can be written as:

$$w_3 = \int_{\Gamma} W^*_{3j}p_j d\Gamma - \int_{\Gamma} P^*_{3j}w_j d\Gamma$$

$$-\int_{\Omega} k_{\alpha\beta}B\frac{1-\nu}{2}\left(u_{\alpha,\beta}+u_{\beta,\alpha}+\frac{2\nu}{1-\nu}u_{\phi,\phi}\delta_{\alpha\beta}\right)W^*_{33}d\Omega$$

$$-\int_{\Omega} k_{\alpha\beta}B\left[(1-\nu)k_{\alpha\beta}+\nu\delta_{\alpha\beta}k_{\gamma\gamma}\right]w_3W^*_{33}d\Omega$$

$$+\lambda\int_{\Omega} W^*_{33}f_b d\Omega \tag{11.27}$$

where:

$$f_b = q_3 + N^{(l)}_{\alpha\beta,\alpha}\mathbf{f}(r)_{,\beta}\mathbf{F}^{-1}\mathbf{w}_3 + N^{(l)}_{\alpha\beta}\mathbf{f}(r)_{,\alpha}\mathbf{f}(r)_{,\beta}\mathbf{F}^{-1}\mathbf{w}_3 \tag{11.28}$$

The last two integrals in equations (11.26) have also to be expressed in terms of equation (11.28), but for the sake of brevity it will not be shown here.

This linear buckling formulation is particularly suited for the application of the Dual Reciprocity Method (DRM).[8] A similar formulation which is based on domain discretization (domain cells) instead of the DRM is given in.[7]

11.5 Multi region formulation

Let consider M assembled curved or flat plates joined at J_n as shown in Fig. 11.4(a). The global coordinate system is given by -x_1-x_2-x_3, and the local coordinate systems for each plate by -x_1^m-x_2^m-x_3^m $(m = 1, M)$.

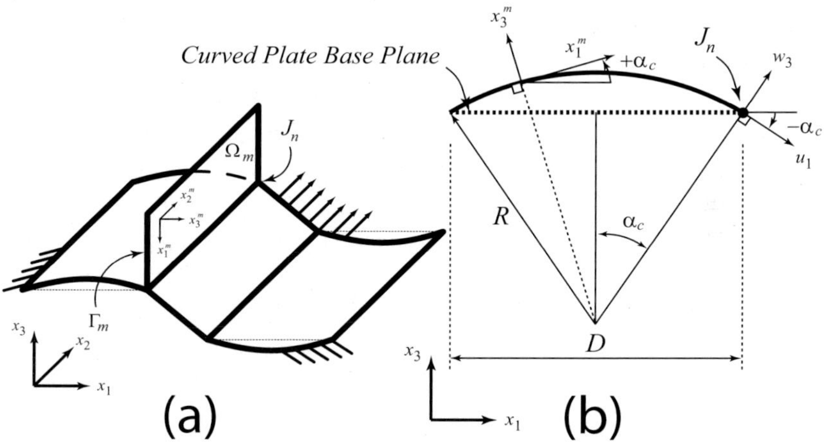

Fig. 11.4. Assembled structures and local coordinate system of a shallow shell.

In the simple case of two curved plates with the same axis orientation at the junction line (see Fig. 11.5), the continuity and equilibrium equations along the joint can be written as follows:

$$\begin{aligned} u_\alpha^m &= u_\alpha^{m+1} \\ w_i^m &= w_i^{m+1} \end{aligned} \tag{11.29}$$

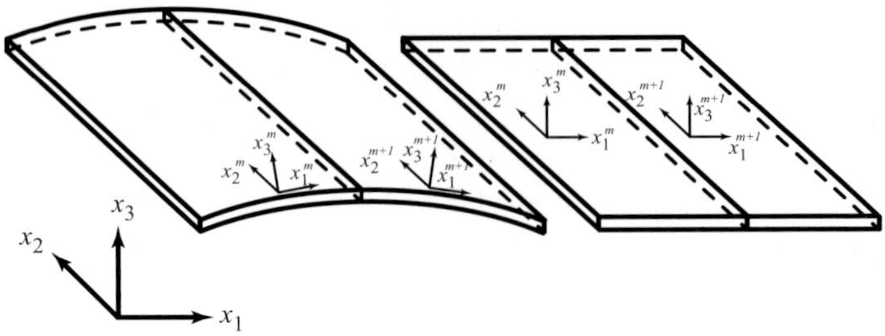

Fig. 11.5. Curved and flat plates local axis orientation.

for displacements and rotations, and

$$\sum_{m=1}^{M} t_{\alpha}^{m} = 0 \qquad (11.30)$$
$$\sum_{m=1}^{M} p_{j}^{m} = 0$$

for tractions and moments, respectively. Curved plates are defined using a curvilinear coordinate system. This means that contrary to flat plates which have a fix normal (local coordinate system) through the whole plate, the local coordinate system of a curved plate changes with the curvature (see Fig. 11.4(b)).

Because two or more angled curved and flat plates joined together are considered, an approach similar to that proposed in[69] is developed. To simplify this approach, the local coordinate systems of each region is assumed to be defined such that the x_2^m directions are all aligned with the global direction x_2, following the implementation for plate assemblies presented by Di Pisa.[14]

Based on the above simplification, w_3 and u_1 displacements for any given curved plates at a junction line (J_n) can be presented as shown in Fig. 11.4(b). Therefore, compatibility equations for each pair of adjacent curved plates (e.g. $m = 1$ and $m = 2$) could be written as follows:

$$u_1^1(n_{11}^{1c}n_{11}^1 + n_{31}^{1c}n_{13}^1) + w_3^1(n_{13}^{1c}n_{11}^1 + n_{33}^{1c}n_{13}^1) =$$
$$u_1^2(n_{11}^{2c}n_{11}^2 + n_{31}^{2c}n_{13}^2) + w_3^2(n_{13}^{2c}n_{11}^2 + n_{33}^{2c}n_{13}^2)$$
$$u_1^1(n_{11}^{1c}n_{31}^1 + n_{31}^{1c}n_{33}^1) + w_3^1(n_{13}^{1c}n_{31}^1 + n_{33}^{1c}n_{33}^1) =$$
$$u_1^2(n_{11}^{2c}n_{31}^2 + n_{31}^{2c}n_{33}^2) + w_3^2(n_{13}^{2c}n_{31}^2 + n_{33}^{2c}n_{33}^2)$$

$$\begin{aligned} u_2^1 &= u_2^2 \\ w_1^1 &= w_1^2 \\ w_2^1 &= 0 \\ w_2^2 &= 0 \end{aligned} \tag{11.31}$$

where n_{ik}^m are the components of the rotation matrix of the curved plate base plane m from local to global coordinates,[69] and n_{ik}^{mc} are the components of the rotation matrix from the curvilinear coordinate system to the curved plate base plane m.

The components of n_{ik}^{mc} are given by:

$$\begin{aligned} n_{11}^{mc} &= \cos(\alpha_c^m); & n_{12}^{mc} &= 0; & n_{13}^{mc} &= \cos(90+\alpha_c^m) \\ n_{21}^{mc} &= 0; & n_{22}^{mc} &= 1; & n_{23}^{mc} &= 0 \\ n_{31}^{mc} &= \cos(90-\alpha_c^m); & n_{32}^{mc} &= 0; & n_{33}^{mc} &= \cos(\alpha_c^m) \end{aligned} \tag{11.32}$$

where α_c^m is measured with respect to the curved plate base plane, as shown in Fig. 11.4(b). Equations in (11.31) result in a system of $5M-4$ compatibility conditions, and have to be supplemented with 4 equilibrium conditions as follows:

$$\begin{aligned} \sum_{m=1}^{M} [t_1^m (n_{11}^{mc} n_{11}^m + n_{31}^{mc} n_{13}^m) + p_3^m (n_{13}^{mc} n_{11}^m + n_{33}^{mc} n_{13}^m)] &= 0 \\ \sum_{m=1}^{M} [t_1^m (n_{11}^{mc} n_{31}^m + n_{31}^{mc} n_{33}^m) + p_3^m (n_{13}^{mc} n_{31}^m + n_{33}^{mc} n_{33}^m)] &= 0 \\ \sum_{m=1}^{M} t_2^m &= 0 \\ \sum_{m=1}^{M} p_1^m &= 0 \end{aligned} \tag{11.33}$$

to produce the required $5M$ equations. This approach relies on the assumption that the curved or flat plate flexural rigidity in its own plane is so large that it is possible to ignore its associated deformation, in another words, there is no drilling rotation.

11.6 Governing integral equations for postbuckling

The linear bifurcation analysis presented previously is a mathematical idealization that can be used to find the load that causes instability (the bifurcation load), without solving the more complicated nonlinear problem.

The relatively simple procedure of an eigenvalue problem can be effectively used to approximate the critical loads only in certain cases. The practical importance of nonlinear theory was recognized in the 1930's and the general postbuckling theory of elastic structures was presented by Koiter's thesis[27] in 1945. With potential energy as the fundamental tool, he established that imperfections may give rise to significant reductions of the critical load, which explained the discrepancy between linear theories and experiments of shell buckling.

11.6.1 *Nonlinear rotations and out-of-plane integral equations*

After introducing f_3^b from equation (11.16) in equation (11.21), and using the expression for membrane stresses given by equation (11.18), the following integral equation can be obtained:[18]

$$c_{ij}w_j + \int_\Gamma P_{ij}^* w_j d\Gamma = \int_\Gamma W_{ij}^* p_j d\Gamma$$

$$- \int_\Omega W_{i3}^* k_{\alpha\beta} B \frac{1-\nu}{2} \left[u_{\alpha,\beta} + u_{\beta,\alpha} + \frac{2\nu}{1-\nu} u_{\phi,\phi}\delta_{\alpha\beta} \right] d\Omega$$

$$- \int_\Omega W_{i3}^* k_{\alpha\beta} B \left((1-\nu) k_{\alpha\beta} + \nu \delta_{\alpha\beta} k_{\phi\phi} \right) w_3 d\Omega$$

$$- \int_\Omega W_{i3}^* k_{\alpha\beta} N_{\alpha\beta}^{(n)} d\Omega + \int_\Omega W_{i3}^* \left(N_{\alpha\beta} w_{3,\beta} \right)_{,\alpha} d\Omega + \int_\Omega W_{i3}^* q_3 d\Omega \qquad (11.34)$$

11.6.2 *Nonlinear in-plane integral equations*

Now, after introducing f_α^m from equation (11.17) into equation (11.23), and using the following relation:

$$t_\alpha = t_\alpha^{(i)} + t_\alpha^{(ii)} + t_\alpha^{(n)} = N_{\alpha\beta}^{(i)} n_\beta + N_{\alpha\beta}^{(ii)} n_\beta + N_{\alpha\beta}^{(n)} n_\beta$$

the following integral equation can be obtained:[18]

$$c_{\theta\alpha}u_\alpha + \int_\Gamma T_{\theta\alpha}^{*(i)} u_\alpha d\Gamma + \int_\Gamma U_{\theta\alpha}^* B \left[k_{\alpha\beta}(1-\nu) + \nu\delta_{\alpha\beta} k_{\phi\phi} \right] w_3 n_\beta d\Gamma$$

$$- \int_\Omega U_{\theta\alpha}^* N_{\alpha\beta,\beta}^{(n)} d\Omega - \int_\Omega U_{\theta\alpha}^* B \left[k_{\alpha\beta}(1-\nu) + \nu\delta_{\alpha\beta} k_{\phi\phi} \right] w_{3,\beta} d\Omega$$

$$+ \int_\Gamma U_{\theta\alpha}^* N_{\alpha\beta}^{(n)} n_\beta d\Gamma = \int_\Gamma U_{\theta\alpha}^* t_\alpha d\Gamma + \int_\Omega U_{\theta\alpha}^* q_\alpha d\Omega \qquad (11.35)$$

11.6.3 *Domain nonlinear terms*

Nonlinear terms in equations (11.34) and (11.35) include primarily derivatives of the out-of-plane displacement $w_{3,\beta}$ and total membrane stress resultants $N_{\alpha\beta}$ at domain points. These nonlinear terms are needed in order to solve the nonlinear system of equations. These terms can be derived from the displacement integral equations (11.34) and (11.35) as follows:

$$w_{3,\theta} + \int_\Gamma P^*_{3j,\theta} w_j d\Gamma = \int_\Gamma W^*_{3j,\theta} p_j d\Gamma$$

$$- \int_\Omega W^*_{33,\theta} k_{\alpha\beta} B \frac{1-\nu}{2} \left[u_{\alpha,\beta} + u_{\beta,\alpha} + \frac{2\nu}{1-\nu} u_{\phi,\phi} \delta_{\alpha\beta} \right] d\Omega$$

$$- \int_\Omega W^*_{33,\theta} k_{\alpha\beta} B \left((1-\nu) k_{\alpha\beta} + \nu \delta_{\alpha\beta} k_{\phi\phi} \right) w_3 d\Omega$$

$$- \int_\Omega W^*_{33,\theta} k_{\alpha\beta} N^{(n)}_{\alpha\beta} d\Omega + \int_\Omega W^*_{33,\theta} \left(N_{\alpha\beta} w_{3,\beta} \right)_{,\alpha} d\Omega + \int_\Omega W^*_{33,\theta} q_3 d\Omega \quad (11.36)$$

In the same way, linear membrane stresses $N^{(l)}_{\alpha\beta}$ can be obtained after the derivative of equation (11.35) with respect to the source point, is introduced in equation (11.18), resulting in the following integral equation:

$$N^{(l)}_{\alpha\beta} + \int_\Gamma T^{(i)*}_{\alpha\beta\gamma} u_\gamma d\Gamma = \int_\Gamma U^*_{\alpha\beta\gamma} t_\gamma d\Gamma$$

$$- \int_\Gamma U^*_{\alpha\beta\gamma} B \left[k_{\gamma\theta} (1-\nu) + \nu \delta_{\gamma\theta} k_{\phi\phi} \right] w_3 n_\theta d\Gamma$$

$$+ \int_\Omega U^*_{\alpha\beta\gamma} B \left[k_{\gamma\theta} (1-\nu) + \nu \delta_{\gamma\theta} k_{\phi\phi} \right] w_{3,\theta} d\Omega - \int_\Gamma U^*_{\alpha\beta\gamma} N^{(n)}_{\gamma\theta} n_\theta d\Gamma$$

$$+ \int_\Omega U^*_{\alpha\beta\gamma} N^{(n)}_{\gamma\theta,\theta} d\Omega + \int_\Omega U^*_{\alpha\beta\gamma} q_\gamma d\Omega + B \left[(1-\nu) k_{\alpha\beta} + \nu \delta_{\alpha\beta} k_{\phi\phi} \right] w_3 \quad (11.37)$$

It is important to notice the difference between the linear membrane stress equations (11.25) in the linear buckling problem and the present linear membrane stress equations (11.37) for the postbuckling problem.

Due to the way in which the nonlinear integral equations are solved in the present work (incremental solution strategy), equations (11.36) and (11.37) will add 5 extra degrees of freedom (3 linear membrane stresses

$N_{\alpha\beta}^{(l)}$ and 2 out-of-plane derivatives $w_{3,\beta}$) at every domain point. Therefore, in order to reduce this large system of equations, $N_{\alpha\beta}$ and $w_{3,\beta}$ are also expressed in terms of basic displacements fields (u_α and w_3) by using radial basis functions (see Appendix B).

Using radial basis functions, derivatives of the out-of-plane displacement and membrane stresses can be represented as:

$$w_{3,\beta} = \mathbf{f}(r)_{,\beta}\mathbf{F}^{-1}w_3 \tag{11.38}$$

and,

$$N_{\alpha\beta} = N_{\alpha\beta}^{(i)} + N_{\alpha\beta}^{(ii)} + N_{\alpha\beta}^{(n)}$$

$$\begin{aligned} N_{\alpha\beta}^{(i)} = B\frac{1-\nu}{2}\Big[\mathbf{f}(r)_{,\beta}\mathbf{F}^{-1}u_\alpha + \mathbf{f}(r)_{,\alpha}\mathbf{F}^{-1}u_\beta + \\ +\frac{2\nu}{1-\nu}\mathbf{f}(r)_{,\gamma}\mathbf{F}^{-1}u_\gamma\delta_{\alpha\beta}\Big] \end{aligned} \tag{11.39}$$

$$N_{\alpha\beta}^{(ii)} = B\left[(1-\nu)k_{\alpha\beta} + \nu\delta_{\alpha\beta}k_{\phi\phi}\right]w_3 \tag{11.40}$$

$$\begin{aligned} N_{\alpha\beta}^{(n)} = B\frac{1-\nu}{2}\Big[\mathbf{f}(r)_{,\alpha}\mathbf{F}^{-1}w_3\mathbf{f}(r)_{,\beta}\mathbf{F}^{-1}w_3 + \\ \frac{\nu}{1-\nu}\mathbf{f}(r)_{,\gamma}\mathbf{F}^{-1}w_3\mathbf{f}(r)_{,\gamma}\mathbf{F}^{-1}w_3\delta_{\alpha\beta}\Big] \end{aligned} \tag{11.41}$$

11.7 Numerical implementation

In order to solve the boundary integral formulations presented in the previous sections, the boundary Γ and the domain Ω must be discretized. In the present work, quadratic isoparametric boundary elements are used to describe the boundary; while for the domain, several uniformly distributed domain points are used for the implementation of the dual reciprocity method in the domain integrals of equations (11.22), 11.24), (11.25), (11.26) and (11.27). In the case of the nonlinear integral equations (11.34), (11.35), (11.36) and (11.37), domain integrals containing nonlinear terms are treated using quadratic quadrilateral isoparametric cells while the rest is treated with the dual reciprocity method.

11.7.1 *Discretization*

In order to show the numerical implementation taking place, equation (11.22) can be written in a discretized form as

$$c_{ij}w_j + \sum_{n=1}^{N_e}\sum_{l=1}^{3} w_j^{nl} \int_{\xi=-1}^{\xi=+1} P_{ij}^* \Phi^l(\xi) J_n(\xi) d\xi$$

$$= \sum_{n=1}^{N_e}\sum_{l=1}^{3} p_j^{nl} \int_{\xi=-1}^{\xi=+1} W_{ij}^* \Phi^l(\xi) J_n(\xi) d\xi - \sum_{k=1}^{M_T} B(k_{11} + \nu k_{22}) I_{2i}^{Dk}$$

$$- \sum_{k=1}^{M_T} B(\nu k_{11} + k_{22}) I_{3i}^{Dk} - \sum_{k=1}^{M_T} k_{\alpha\beta} B\left[(1-\nu)k_{\alpha\beta} + \nu\delta_{\alpha\beta}k_{\gamma\gamma}\right] I_{1i}^{Dk} + \sum_{k=1}^{M} I_{4i}^{Dk} \tag{11.42}$$

where N_e is the number of boundary elements; Φ^l is the shape function; ξ is the local coordinate; J_n is the Jacobian of transformation. M are number of DRM domain points and M_T are total number of boundary and domain points. I_{1i}^{Dm}, I_{2i}^{Dm}, I_{3i}^{Dm}, I_{4i}^{Dm} are domain integrals transformed into boundary integrals by the Dual Reciprocity Method (see next subsection). A complete description of the boundary elements and domain cells used in this work is given in Appendix C.

All the integral equations presented in previous sections have to be discretized in a similar way as equation (11.42), but for the sake of brevity the complete expressions will not be shown here.

11.7.2 *Dual Reciprocity Method (DRM)*

There are several techniques for the treatment of domain integrals in the boundary element method, some of the most important are: Analytical Integration of the domain integrals, The Multiple Reciprocity Method and The Dual Reciprocity Method. From all of these methods, the Dual Reciprocity Method (DRM) constitutes the most general technique other than domain cell integration.

The following domain integrals are transferred to the boundary by DRM (in the absence of membrane body forces, $q_\alpha = 0$):

$$I_{1i}^{D} = \int_\Omega W_{i3}^* w_3 d\Omega, \qquad I_{2i}^{D} = \int_\Omega W_{i3}^* \frac{\partial u_1}{\partial x_1} d\Omega, \tag{11.43}$$

$$I^D_{3i} = \int_\Omega W^*_{i3} \frac{\partial u_2}{\partial x_2} d\Omega, \qquad I^D_{4i} = \int_\Omega W^*_{i3} f_b d\Omega, \tag{11.44}$$

$$I^D_{5\alpha} = \int_\Omega U^*_{\alpha 1} \frac{\partial w_3}{\partial x_1} d\Omega, \qquad I^D_{6\alpha} = \int_\Omega U^*_{\alpha 2} \frac{\partial w_3}{\partial x_2} d\Omega, \tag{11.45}$$

$$I^D_{7i\beta} = \int_\Omega W^*_{i\beta 3} w_3 d\Omega, \qquad I^D_{8i\beta} = \int_\Omega W^*_{i\beta 3} \frac{\partial u_1}{\partial x_1} d\Omega, \tag{11.46}$$

$$I^D_{9i\beta} = \int_\Omega W^*_{i\beta 3} \frac{\partial u_2}{\partial x_2} d\Omega, \qquad I^D_{10i\beta} = \int_\Omega W^*_{i\beta 3} q_3 d\Omega, \tag{11.47}$$

$$I^D_{11\alpha\beta} = \int_\Omega U^*_{\alpha\beta 1} \frac{\partial w_3}{\partial x_1} d\Omega, \qquad I^D_{12\alpha\beta} = \int_\Omega U^*_{\alpha\beta 2} \frac{\partial w_3}{\partial x_2} d\Omega, \tag{11.48}$$

$$I^D_{13\theta} = \int_\Omega W^*_{33,\theta} w_3 d\Omega, \qquad I^D_{14\theta} = \int_\Omega W^*_{33,\theta} \frac{\partial u_1}{\partial x_1} d\Omega, \tag{11.49}$$

$$I^D_{15\theta} = \int_\Omega W^*_{33,\theta} \frac{\partial u_2}{\partial x_2} d\Omega, \qquad I^D_{16\theta} = \int_\Omega W^*_{33,\theta} q_3 d\Omega \tag{11.50}$$

The transformation of these domain integrals, follows the same procedure as explained in Wen, Aliabadi and Young[62] and Dirgantara.[17]

11.7.3 *Treatment of the integrals*

One of the well known difficulties in the numerical implementation of BEM formulations is regarding the presence of singular integrals. Depending on the location of source and field points, integrals can be classified as: *Regular*, in which case they can be evaluated using the standard gauss quadrature rule or; *Singular*, when the source point belongs to the element over which the integration is performed, in this case special techniques must be used. All the singular integrals appearing in the present BEM formulations are dealt with by using well established techniques and are treated separately based on their order of singularity. On the boundary, near singular integrals (when the collocation node is close to the integration element) are treated with the element subdivision technique.[2] Weakly singular integrals $O(\ln r)$ are treated using a nonlinear coordinate transformation as reported by Telles.[56] Strong singular integrals $O(1/r)$ are computed indirectly by considering the generalized rigid body motion, as explained in.[17] Domain singular integrals can also be separated in weakly

$O(1/r)$ and strong $O(1/r^2)$. Weak singular integrals are treated by a simple technique such as polar coordinate transformation, followed by a regular procedure.[2] Strong singular integrals require special techniques such as the ones described by Leitao.[33]

11.8 Numerical procedure

In this section the numerical procedure to solve the linear buckling and postbuckling problems are explained. Before solving the buckling problems, the linear problem can be solved after point collocation is performed with the discretized forms of equations (11.22),(11.24); and the known and unknown quantities are arranged as a set of linear algebraic equations:

$$[\mathbf{A}]_{5(N+L)\times 5(N+L)}\{\mathbf{X}\}_{5(N+L)} = \{\mathbf{F}\}_{5(N+L)} \tag{11.51}$$

where N and L are number of boundary nodes and domain points, respectively. $[\mathbf{A}]$ is the system matrix, $\{\mathbf{X}\}$ contains the unknowns displacements and tractions on the boundary N, as well as all the displacement in the domain L. The vector $\{\mathbf{F}\}$ is obtained by multiplying the related matrices by the known values of w_i, u_α or p_i, t_α and from the contribution of the body forces q_i.

11.8.1 *Linear buckling (eigenvalue)*

The first step towards the solution of the linear buckling problem is given by the solution of equation (11.51) after which membrane stresses at domain nodes $N_{\alpha\beta}^{(l)}$ can be obtained from equation (11.25). This is followed by determination of approximated derivatives of membrane stresses and out of plane displacement, as explained in Appendix B. Finally, the boundary integral equations of the buckling problem are assembled and solved as follows:

Equations (11.26) and (11.24) can be written in a similar way as equation (11.51),

$$[\mathbf{B}]_{NT\times NT}\{\mathbf{Y}\}_{NT} = \lambda\,[\mathbf{K}]_{NT\times LT}\{\mathbf{w}_3\}_{LT} \tag{11.52}$$

In order to arrange an eigenvalue formulation, equation (11.27) can also be written in matrix form, similar to equation (11.52):

$$[\mathbf{I}]\{\mathbf{w}_3\}_{LT} = [\mathbf{BB}]_{LT\times NT}\{\mathbf{Y}\}_{NT} + \lambda\,[\mathbf{KK}]_{LT\times LT}\{\mathbf{w}_3\}_{LT} \tag{11.53}$$

The total number of degrees of freedom is denoted by NT and the total number of domain degrees of freedom by LT, and are defined as follows

$$NT = 5 * N + 3 * L; \qquad\qquad LT = L$$

Matrices $[\mathbf{B}]$ and $[\mathbf{BB}]$ contain coefficient matrices related to the fundamental solutions. Matrix $[\mathbf{I}]$ is the identity matrix. Vector $\{\mathbf{Y}\}$ represents the unknown boundary conditions (w_i, u_α or p_i, t_α) and the unknown domain displacements (u_α, w_3). Vector $\{\mathbf{w}_3\}$ contains the unknown out of plane displacement w_3. Matrices $[\mathbf{K}]$ and $[\mathbf{KK}]$ are obtained by multiplication of the fundamental solutions with the prebuckling in plane stresses $N_{\alpha\beta}^{(l)}$ and approximation functions $f(r)$.

As it can be seen in equation (11.52) the only load considered in this transformed linearized buckling equation is the transverse body load $((N_{\alpha\beta}^{(l)} w_{3,\beta})_{,\alpha})$ multiplied by the critical load factor λ, implying that all the known values of w_i, u_α or p_i, t_α (boundary conditions), are set to zero. For this reason, there is no vector such as $\{\mathbf{F}\}$ (from equation (11.51)) in equations (11.52) and (11.53).

Equation (11.52) can be rearranged in term of the unknown vector $\{\mathbf{Y}\}$,

$$\{\mathbf{Y}\}_{NT} = \lambda\, [\mathbf{B}]^{-1}_{NT\times NT}\, [\mathbf{K}]_{NT\times LT}\, \{\mathbf{w}_3\}_{LT} \tag{11.54}$$

where matrix $[\mathbf{B}]^{-1}$ is the inverse of matrix $[\mathbf{B}]$.

The substitution of equation (11.54) into equation (11.53) yields:

$$\begin{aligned}[\mathbf{I}]_{LT\times LT}\, \{\mathbf{w}_3\}_{LT} = {} & \lambda\, [\mathbf{BB}]_{LT\times NT}\, [\mathbf{B}]^{-1}_{NT\times NT}\, [\mathbf{K}]_{NT\times LT}\, \{\mathbf{w}_3\}_{LT} \\ & + \lambda\, [\mathbf{KK}]_{LT\times LT}\, \{\mathbf{w}_3\}_{LT}\end{aligned} \tag{11.55}$$

Equation (11.55) can be written as a standard eigenvalue problem equation as follows:

$$\left([\psi] - \frac{1}{\lambda}\, [\mathbf{I}]\right) \{\mathbf{w}_3\}_{LT} = 0 \tag{11.56}$$

Buckling analysis of shear deformable shallow shell has been presented as a standard eigenvalue problem; buckling modes $\{\mathbf{w}_3\}$ and buckling load factors λ can be obtained by solving equation (11.56). This standard eigenvalue problem was solved with LAPACK[5] which is freely available on the internet.

In the case of plate and shallow shell assemblies, the same procedure is followed with the exception that NT and LT, are given by:

$$NT = \sum_{j=1}^{MT} (5 * N_j + 3 * L_j + 5 * NJ_j); \qquad\qquad LT = \sum_{j=1}^{MT} L_j$$

where MT is the total number of plates in the assembly and NJ_j is the number of junction nodes of plate j.

11.8.2 *Postbuckling*

Before nonlinear equations (11.34), (11.35), (11.36) and (11.37) are discretized, all variables should be expressed in terms of incremental quantities. In this incremental solution strategy, loads and prescribed displacements are applied in small but finite increments as follows,

$$w_i^{k+1} = w_i^k + \Delta w_i; \qquad u_\alpha^{k+1} = u_\alpha^k + \Delta u_\alpha; \qquad etc. \tag{11.57}$$

where increments are denoted by Δ, and known values of the kth step are denoted by the superscript 'k'.

The bending and out-of-plane displacement integral equation (11.34) can be written in an incremental form as follows:

$$c_{ij}\left(w_j^k + \Delta w_j\right) + \int\!\!\!\!\!-_{\Gamma} P_{ij}^*\left(w_j^k + \Delta w_j\right) d\Gamma$$

$$= \int_\Gamma W_{ij}^*\left(p_j^{*k} + \Delta p_j^*\right) d\Gamma - \int_\Gamma W_{i3}^* k_{\alpha\beta} B \frac{1-\nu}{2}$$

$$\left[\left(u_\alpha^k + \Delta u_\alpha\right)_{,\beta} + \left(u_\beta^k + \Delta u_\beta\right)_{,\alpha} + \frac{2\nu}{1-\nu}\left(u_\gamma^k + \Delta u_\gamma\right)_{,\gamma}\delta_{\alpha\beta}\right] d\Omega$$

$$-\int_\Omega W_{i3}^* k_{\alpha\beta}\left(N_{\alpha\beta}^{(n)k} + \Delta N_{\alpha\beta}^{(n)}\right) d\Omega$$

$$-\int_\Omega W_{i3}^* k_{\alpha\beta} B\left[(1-\nu)k_{\alpha\beta} + \nu\delta_{\alpha\beta}k_{\gamma\gamma}\right]\left(w_3^k + \Delta w_3\right) d\Omega$$

$$-\int_\Omega W_{i3}^*\left(\left(N_{\alpha\beta}^k + \Delta N_{\alpha\beta}\right)\left(w_3^k + \Delta w_3\right)_{,\beta}\right)_{,\alpha} d\Omega$$

$$+\int_\Omega W_{i3}^*\left(q_3^k + \Delta q_3\right) d\Omega \tag{11.58}$$

Membrane stresses can be expanded as follows:

$$
\begin{aligned}
N^k_{\alpha\beta} &= N^{(i)k}_{\alpha\beta} + N^{(ii)k}_{\alpha\beta} + N^{(n)k}_{\alpha\beta} \\
\Delta N_{\alpha\beta} &= \Delta N^{(i)}_{\alpha\beta} + \Delta N^{(ii)}_{\alpha\beta} + \Delta N^{(n)}_{\alpha\beta} \\
N^{(l)k}_{\alpha\beta} &= N^{(i)k}_{\alpha\beta} + N^{(ii)k}_{\alpha\beta} \\
\Delta N^{(l)}_{\alpha\beta} &= \Delta N^{(i)}_{\alpha\beta} + \Delta N^{(ii)}_{\alpha\beta} \\
\Delta N^{(i)}_{\alpha\beta} &= B\frac{1-\nu}{2}\left(\Delta u_{\alpha,\beta} + \Delta u_{\beta,\alpha} + \frac{2\nu}{1-\nu}\Delta u_{\gamma,\gamma}\delta_{\alpha\beta}\right) \\
N^{(i)k}_{\alpha\beta} &= B\frac{1-\nu}{2}\left(u^k_{\alpha,\beta} + u^k_{\beta,\alpha} + \frac{2\nu}{1-\nu}u^k_{\gamma,\gamma}\delta_{\alpha\beta}\right) \\
\Delta N^{(ii)}_{\alpha\beta} &= B\left[(1-\nu)k_{\alpha\beta} + \nu\delta_{\alpha\beta}k_{\phi\phi}\right]\Delta w_3 \\
N^{(ii)k}_{\alpha\beta} &= B\left[(1-\nu)k_{\alpha\beta} + \nu\delta_{\alpha\beta}k_{\phi\phi}\right] w^k_3 \\
\Delta N^{(n)}_{\alpha\beta} &= B\frac{1-\nu}{2}\left(w^k_{3,\alpha}\Delta w_{3,\beta} + \Delta w_{3,\alpha}w^k_{3,\beta}\right. \\
&\quad \left. + \frac{2\nu}{1-\nu}w^k_{3,\gamma}\Delta w_{3,\gamma}\delta_{\alpha\beta} + \text{higher order}\right) \qquad (11.59)
\end{aligned}
$$

$$
N^{(n)k}_{\alpha\beta} = B\frac{1-\nu}{2}\left(w^k_{3,\alpha}w^k_{3,\beta} + \frac{\nu}{1-\nu}w^k_{3,\gamma}w^k_{3,\gamma}\delta_{\alpha\beta}\right) \qquad (11.60)
$$

Also the nonlinear term $\left(\left(N^k_{\alpha\beta} + \Delta N_{\alpha\beta}\right)\left(w^k_3 + \Delta w_3\right)_{,\beta}\right)_{,\alpha}$ can be written as:

$$
\begin{aligned}
\left(\left(N^k_{\alpha\beta} + \Delta N_{\alpha\beta}\right)\left(w^k_3 + \Delta w_3\right)_{,\beta}\right)_{,\alpha} = {} & N^k_{\alpha\beta,\alpha}w^k_{3,\beta} + N^k_{\alpha\beta}w^k_{3,\beta\alpha} + \\
& N^k_{\alpha\beta,\alpha}\Delta w_{3,\beta} + N^k_{\alpha\beta}\Delta w_{3,\beta\alpha} + \\
& \Delta N_{\alpha\beta,\alpha}w^k_{3,\beta} + \Delta N_{\alpha\beta}w^k_{3,\beta\alpha} + \\
& \text{higher order terms} \qquad (11.61)
\end{aligned}
$$

Higher order terms involved the product of the incremental quantities ($\Delta N_{\alpha\beta}\Delta w_{3,\beta}$; $\Delta w_{3,\alpha}\Delta w_{3,\beta}$; *etc*), and because only small increments of loads and displacements are applied to the structure, these higher order terms can be ignored.

Due to the fact that loads and prescribed displacements are applied in small increments and higher order terms of the incremental quantities ($\Delta N_{\alpha\beta}\Delta w_{3,\beta}$; $\Delta w_{3,\alpha}\Delta w_{3,\beta}$; *etc*) were ignored, a purely incremental solution strategy in which increments are obtained and added to the kth state without the need of an iterative process can be used. The steps towards the nonlinear solution can be summarized as follows:

- Step1: First increment let $k = 0$, therefore nonlinear terms from the previous increment are all zero ($w_i^k = u_\alpha^k = etc = 0$), resulting in a linear system matrix [**A**], similar to equation (11.51).
- Step2: Solving system of equation, obtaining incremental quantities ($\Delta w_i = \Delta u_\alpha = etc = 0$).
- Step3: Update variables with equation (11.57), and print results.
- Step4: Obtain nonlinear terms for next load increment ($N_{\alpha\beta}^{(n)k}$, $N_{\alpha\beta}^k w_{3,\beta}^k$, etc.).
- Step5: Update new system matrix [**A**] and load vector $\{\Delta \mathbf{F}\}$.
- Step6: If final load increment was applied ($k = final$) go to Step7, otherwise go to Step2.
- Step7: Finish.

11.8.2.1 *Arc-length method*

Various forms of "arc-length methods" have stemmed from the original works of Riks[44] and Wemper.[61] The arc-length method, also known as Riks method, is a numerical technique which solves simultaneously for loads and displacements using a load parameter as an additional unknown. The basic idea is to use a scalar "*arc-length*" Δl along the static equilibrium path in the load-displacement space to control the progress of the solution, thus, unloading is automatically accommodated in tracing the equilibrium path. This means that solutions can be provided regardless of whether the response is stable or unstable. In the present work, the modified Riks method[13] and the standard "displacement control" algorithm were implemented in order to deal with unstable equilibrium paths.

Complete description of this and other postbuckling boundary element formulations for shear deformable shallow shells can be found in[9] and will be also available on future publications (currently under preparation) by the present authors.

11.9 Numerical examples

The proposed formulations are applied to several benchmark problems in order to assess its accuracy and efficiency. Rectangular cylindrical shallow shells with different curvature parameters, aspect ratios and boundary conditions will be presented. The simply supported condition refers to zero out of plane displacement ($w_3 = 0$) while the clamped condition is based on zero out of plane displacement and zero rotations ($w_i = 0$). The in plane displacements in both cases were set free ($u_\alpha \neq 0$). These boundary conditions are the same through all the boundary unless otherwise indicated.

11.9.1 *Linear buckling of curved plates*

In this example curved rectangular plates under uniform axial compression are studied. In order to facilitate the interpretation of results and establish comparison with the analytical solutions, the following non-dimensional buckling coefficient (K) and curvature parameter (Z) are defined,

$$K = \frac{N_{cr} \cdot 12(1 - v^2) \cdot b^2}{\pi^2 \cdot E \cdot h^2}; \qquad Z = \frac{b^2 \cdot \sqrt{1 - v^2}}{R \cdot h} \tag{11.62}$$

where N_{cr} represent the critical in plane stress, obtained from the multiplication of the buckling factor λ with the actual applied stress, and b denotes the length of the curved side.

The curved plates have a constant aspect ratio $a/b = 2$ and thickness $h/b = 0.015$, Young's modulus E and Poisson ratio $v = 0.3$. The radii considered are $r/b = 1.5$; 2.25; 3.25; 5; 7.5; 11.5; 17.5; 27.5; 63 and *plate*. The results are plotted in Fig. 11.6, together with FEM solutions obtained with ANSYS.[6] The FEM mesh has 72 quadratic quadrilateral elements (253 nodes). The present BEM(DRM) model has 24 boundary elements and 7X14 domain points (98 domain nodes and 48 boundary nodes). Also for comparison, results from the BEM(Cell) model in[7] with 18 boundary elements and 18 domain cells (91 internal nodes and 36 boundary nodes) are included in Fig. 11.6. As expected, an increase in the curvature also increase the linear buckling coefficients, while the decrease of the curvature converge to the flat plate solutions.

11.9.2 *Linear buckling of channel sections*

In this example the local buckling coefficients for channel sections will be compared with the analytical solution given by Allen and Bulson[3] and FEM solutions obtained with ABAQUS,[1] see Fig. 11.7. The edges are simple supported and uniform axial compressive is applied at the ends.

In this example the critical buckling coefficient (K) for the complete section is given in terms of the dimension of the central plate b_1. The thickness is the same for all the plates and l represent the length of the assembly. As it can be seen from Fig. 11.7, the solutions from all the three

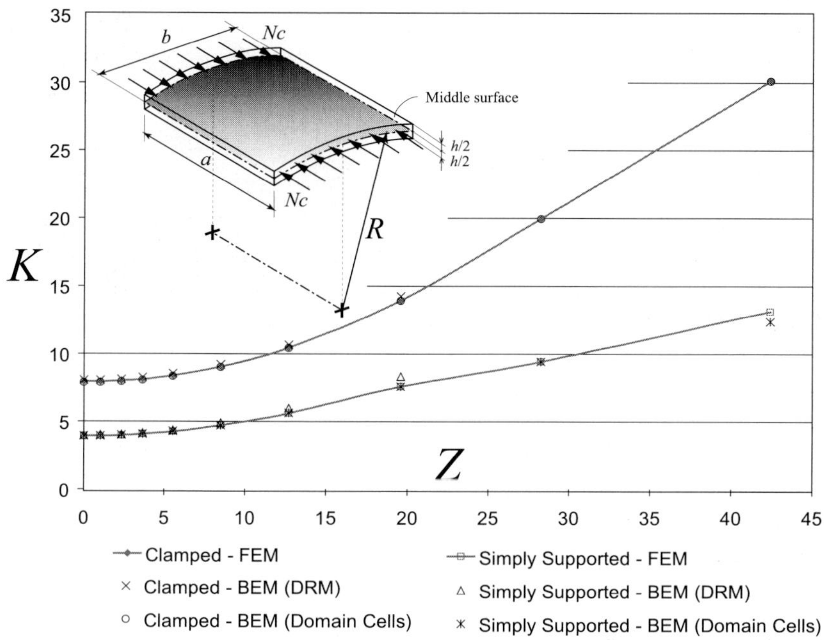

Fig. 11.6. Linear buckling coefficients for simple supported and clamped cylindrical curved plates.

different techniques agree well for channel sections of different dimensions ($b_2/b_1 = 1, 0.5, 0.2$ and $b_1/l = 1, 2, 3, 4, 5$). Bucking modes for configuration with $b_1/b_2 = 1$ and $b_1/l = 4$ are also given in Fig. 11.7. The plotted geometries correspond to a FEM model with 5652 nodes and 11022 linear triangular general purpose shell elements, and a BEM model with 168 boundary nodes (84 quadratic elements) and 180 domain nodes. As it is expected for sections buckling locally into several half waves, it can be seen from Fig. 11.7 that the critical buckling coefficients tend to be virtually independent of length (except for short length members, when $l/b \lessapprox 1$).

11.9.3 *Point load at the crown of a cylindrical shallow shell*

This example presents a hinged free point loaded cylindrical panel shown in Fig. 11.8. The results by Palazotto and Dennis[38] and Sabir and Lock[45] are used for comparison. General shell dimension are also given in Fig. 11.8

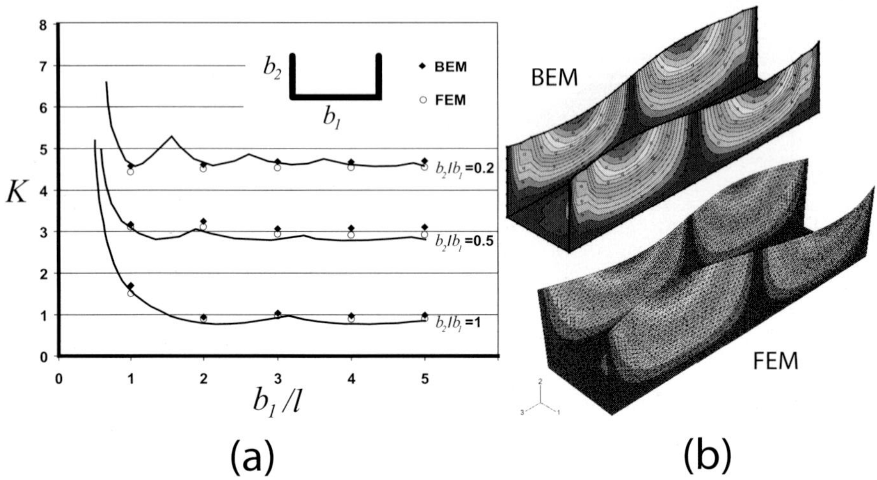

Fig. 11.7. Buckling coefficients for channel sections and buckling modes for BEM and FEM models.

and material properties are given by, E and Poisson ratio $v = 0.3$. The boundary conditions are hinge ($w_3 = u_\alpha = 0$) for the flat sides of the panel and totally free ($w_i \neq 0$ and $u_\alpha \neq 0$) for the curved sides. The mesh used by Sabir and Lock[45] contains 4x4 quadratic quadrilateral isoparametric elements for just one quarter of the geometry and the mesh used in[38] contains 4x6 quadratic quadrilateral isoparametric elements also for just one quarter of the shallow shell. The BEM mesh use 18 domain cells and 18 boundary elements to model half of the geometry. In Fig. 11.8, Large System refers to the nonlinear system that uses equations (11.36) and (11.37) while Small System uses equations (11.38)-(11.41).

In this example because of expected snap though (limit points in the load displacement response, point A in Fig. 11.8), not only applied concentrated force is considered in the analysis, but also a central displacement incrementation. Figure 11.8 shows how BEM solutions with applied purely incremental displacement can overcome this numerical difficulty. Also in Fig. 11.8 the arc-length technique is used with applied incremental load, in fact it is clear how this solution overlaps the nonlinear response obtained with the purely incremental displacement technique; with the important difference that the CPU time in the arc-length solution was more than 50 times larger than the purely incremental one.

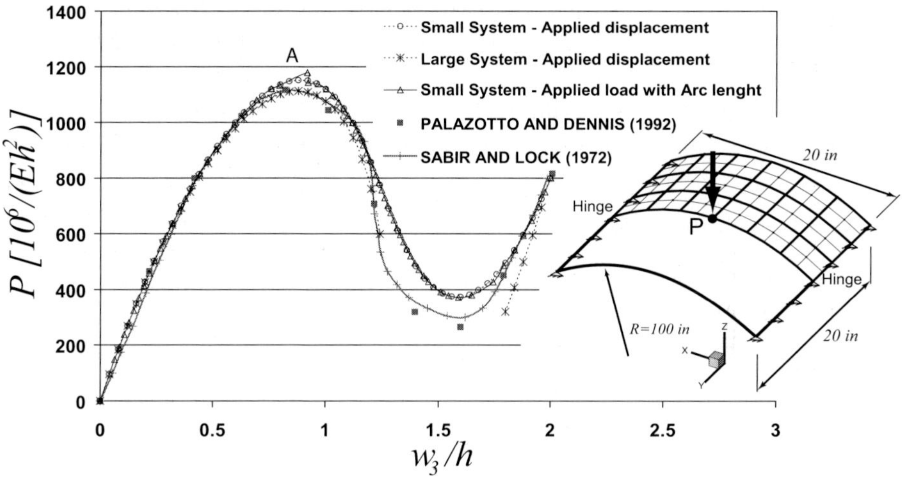

Fig. 11.8. Point loaded hinge-free cylindrical shallow shell.

11.10 Conclusions

In this chapter, boundary element formulations for the solution of buckling and postbuckling problems of shear deformable thin walled structures were presented. The linear buckling problem was presented as a standard eigenvalue problem, making possible to obtaining critical load factors and buckling modes in a single procedure. The linear buckling formulation was combined with a multi region BEM formulation for assembled plate and shallow shell structures, making possible the analysis of local buckling of thin walled structures.

Two different sets of nonlinear boundary integral equations were studied. The difference between sets is based on the determination of nonlinear domain terms: one with five extra boundary integral equations (11.36) and (11.37), and the other which uses radial basis functions, equations (11.38)-(11.41). The dual reciprocity technique was used to evaluate some of the domain integrals (containing linear terms), while the remaining domain integrals (containing nonlinear terms) were evaluated using domain discretization. Both nonlinear systems of equations were solved using an incremental solution strategy, where loads and displacements are applied in small but finite increments. Higher order terms were ignored, resulting in linearized system of equations. Because of this linearization a purely incremental solution algorithm was used. Due to the presence of limit points in some nonlinear responses (unstable postbuckling), the arc-length method was implemented with the full Newton-Raphson technique. The purely incremental solution algorithm was more efficient than the iterative approach, but it has a limitation when limit points are considered.

11.11 Acknowledgments

This work was supported by Queen Mary and Imperial College Scholarships.

Appendix A. Fundamental Solutions

The expressions for the kernels W^*_{ij} and P^*_{ij} are given by Vander Weeën[59] as follows:

$$\begin{aligned} W^*_{\alpha\beta} &= \frac{1}{8\pi D(1-\nu)}\{[8B(z)-(1-\nu)(2\ln z-1)]\delta_{\alpha\beta} - \\ &\quad [8A(z)+2(1-\nu)]r_{,\alpha}r_{,\beta}\} \end{aligned} \tag{A.1}$$

$$\begin{aligned} W^*_{\alpha 3} &= -W^*_{3\alpha} = \frac{1}{8\pi D}(2\ln z-1)rr_{,\alpha} \\ W^*_{33} &= \frac{1}{8\pi D(1-\nu)\lambda^2}[(1-\nu)z^2(\ln z-1)-8\ln z] \end{aligned} \tag{A.2}$$

and

$$\begin{aligned} P^*_{\gamma\alpha} &= \frac{-1}{4\pi r}[(4A(z)+2zK_1(z)+1-\nu)(\delta_{\alpha\gamma}r_{,n}+r_{,\alpha}n_\gamma) \\ &\quad +(4A(z)+1+\nu)r_{,\gamma}n_\alpha - 2(8A(z)+2zK_1(z)+1-\nu)r_{,\alpha}r_{,\gamma}r_{,n}] \\ P^*_{\gamma 3} &= \frac{\lambda^2}{2\pi}[B(z)n_\gamma - A(z)r_{,\gamma}r_{,n}] \\ P^*_{3\alpha} &= \frac{-(1-\nu)}{8\pi}\left[\left(2\frac{(1+\nu)}{(1-\nu)}\ln z - 1\right)n_\alpha + 2r_{,\alpha}r_{,n}\right] \\ P^*_{33} &= \frac{-1}{2\pi r}r_{,n} \end{aligned} \tag{A.3}$$

where

$$\begin{aligned} A(z) &= K_0(z) + \frac{2}{z}\left[K_1(z) - \frac{1}{z}\right] \\ B(z) &= K_0(z) + \frac{1}{z}\left[K_1(z) - \frac{1}{z}\right] \end{aligned} \tag{A.4}$$

in which $K_0(z)$ and $K_1(z)$ are modified Bessel functions of the second kind, $z = \lambda r$, r is the absolute distance between the source and the field points, $r_{,\alpha} = r_\alpha / r$, where $r_\alpha = x_\alpha(\mathbf{x}) - x_\alpha(\mathbf{x}')$ and $r_{,n} = r_{,\alpha} n_\alpha$.As it can be seen, $A(z)$ is a smooth function, whereas, $B(z)$ is a weakly singular $O(\ln r)$. Therefore W^*_{ij} is weakly singular and P^*_{ij} has a strong (Cauchy principal value) singularity $O(1/r)$.

The expressions for the kernels $U^*_{\theta\alpha}$ and $T^*_{\theta\alpha}$ are the well known (Kelvin solution) for two-dimensional plane stress problems, and are given as[17] :

$$U^*_{\theta\alpha} = \frac{1}{4\pi B(1-\nu)} \left[(3-\nu) \ln\left(\frac{1}{r}\right) \delta_{\theta\alpha} + (1+\nu) r_{,\theta} r_{,\alpha} \right] \tag{A.5}$$

$$T^{(i)*}_{\theta\alpha} = -\frac{1}{4\pi r} \left\{ r_{,n} \left[(1-\nu) \delta_{\theta\alpha} + 2(1+\nu) r_{,\theta} r_{,\alpha} \right] + (1-\nu) \left[n_\theta r_{,\alpha} - n_\alpha r_{,\theta} \right] \right\} \tag{A.6}$$

where $U^*_{\theta\alpha}$ are weakly singular kernels of order $O\left(\ln \frac{1}{r}\right)$ and $T^*_{\theta\alpha}$ are strongly singular of order $O(1/r)$.

Derivatives of the displacement fundamental solutions with respect to the field point are given as follows:

$$W^*_{\gamma 3,\alpha} = \frac{1}{8\pi D} [(2 \ln z - 1)\delta_{\alpha\gamma} + 2 r_{,\gamma} r_{,\alpha}] \tag{A.7}$$

$$W^*_{33,\alpha} = \frac{r_{,\alpha}}{8\pi D(1-\nu)\lambda} [z(1-\nu)(2\ln z - 1) - \frac{8}{z}] \tag{A.8}$$

$$U^*_{\alpha\beta,\gamma} = \frac{1+v}{4\pi B(1-\nu) r} \left[-\frac{(3-\nu)}{(1+v)} r_{,\gamma} \delta_{\alpha\beta} + \delta_{\alpha\gamma} r_{,\beta} + \delta_{\beta\gamma} r_{,\alpha} - 2 r_{,\beta} r_{,\gamma} r_{,\alpha} \right] \tag{A.9}$$

The kernel $W^*_{\gamma 3,\alpha}$ is regular, while $W^*_{33,\alpha}$ and $U^*_{\alpha\beta,\gamma}$ are weakly singular in the domain, singularity $O(1/r)$.

The expressions for the kernels $U^*_{\alpha\beta\gamma}$ and $T^*_{\alpha\beta\gamma}$ are:[17]

$$U^*_{\alpha\beta\gamma} = \frac{1}{4\pi r} \left[(1-\nu) \left(\delta_{\gamma\alpha} r_{,\beta} + \delta_{\gamma\beta} r_{,\alpha} - \delta_{\alpha\beta} r_{,\gamma} \right) + 2(1+\nu) r_{,\alpha} r_{,\beta} r_{,\gamma} \right] \tag{A.10}$$

$$\begin{aligned} T^{(i)*}_{\alpha\beta\gamma} = \frac{B(1-\nu)}{4\pi r^2} & \{ 2 r_{,n} \left[(1-\nu) \delta_{\alpha\beta} r_{,\gamma} + \nu \left(\delta_{\gamma\alpha} r_{,\beta} + \delta_{\gamma\beta} r_{,\alpha} \right) \right. \\ & \left. -4(1+\nu) r_{,\alpha} r_{,\beta} r_{,\gamma} \right] + 2\nu \left(n_\alpha r_{,\beta} r_{,\gamma} + n_\beta r_{,\alpha} r_{,\gamma} \right) \\ & + (1-\nu) \left(2 n_\gamma r_{,\alpha} r_{,\beta} + n_\beta \delta_{\alpha\gamma} + n_\alpha \delta_{\beta\gamma} \right) && \text{(A.11)} \\ & - (1-3\nu) n_\gamma \delta_{\alpha\beta} \} && \text{(A.12)} \end{aligned}$$

Appendix B. Radial Basis Functions

A generic function $g(x, y)$ (e.g. w_3, or $N_{\alpha\beta}$) can be approximated as follows:[63]

$$g(x_1, x_2) = \sum_{m=1}^{M} f^m(r)\Psi^m \tag{B.1}$$

where: $r = \sqrt{(x_1 - x_1^m)^2 + (x_2 - x_2^m)^2}$, Ψ^m are the interpolating coefficients, $f(r)^m$ is the interpolating function (radial basis function) and M is the total number of interpolating points. These interpolating points in the present study, are the nodes of the domain cells used to evaluate the domain integrals or the domain points used for the applications of the Dual Reciprocity Method (DRM points).

In matrix form, equation B.1 can be rewritten as follows:

$$\mathbf{g} = \mathbf{F}\{\Psi\} \tag{B.2}$$

where: $\mathbf{g}$ is a vector containing the value of $g(x_1, x_2)$ at all the M points, $\{\Psi\}$ is an unknown vector of coefficients; $\mathbf{F}$ is the matrix obtained from evaluating $f(r)$ at the M points.

The coefficients $\{\Psi\}$ can be computed as follows:

$$\{\Psi\} = \mathbf{F}^{-1}\mathbf{g} \tag{B.3}$$

Differentiation of $g(x_1, x_2)$ can be obtained as follows:

$$\frac{\partial g(x_1, x_2)}{\partial x_\alpha} = \sum_{m=1}^{M} \frac{f(r)^m}{\partial x_\alpha}\Psi^m \tag{B.4}$$

Using equation (B.3), the first derivative of $g(x_1, x_2)$ given in equation (B.4) can be expressed in matrix notation, as follows:

$$g_{,\alpha}(x_1, x_2) = \mathbf{f}(r)_{,\alpha}\mathbf{F}^{-1}\mathbf{g} \tag{B.5}$$

The second derivative of $g(x_1, x_2)$ can also be obtained using the same procedure:

$$g_{,\alpha\beta}(x_1, x_2) = \mathbf{f}(r)_{,\beta}\mathbf{f}(r)_{,\alpha}\mathbf{F}^{-1}\mathbf{g} \tag{B.6}$$

Several Radial Basis Functions have been tested by Aliabadi[2] and co-workers, and one of the most satisfactory functions is the following:

$$f(r) = \sqrt{c^2 + r^2} \tag{B.7}$$

the derivative of this function is given by:

$$\frac{\partial f(r)}{\partial x_\alpha} = \frac{(x_\alpha - x_\alpha^m)}{\sqrt{c^2 + r^2}} \tag{B.8}$$

where c is a constant, which value strongly depends on dimensions of the BEM model.

Appendix C. Elements

The quadratic continuous shape functions for the boundary are defined as:

$$\begin{aligned}\Phi^1(\xi) &= \frac{1}{2}\xi(\xi-1)\\ \Phi^2(\xi) &= (1-\xi)(1+\xi)\\ \Phi^3(\xi) &= \frac{1}{2}\xi(\xi+1)\end{aligned} \tag{C.1}$$

For the case of semi-discontinuous boundary elements:

$$\begin{aligned}\Phi_{S1}^1(\xi) &= \frac{9}{10}\xi(\xi-1) &;\quad \Phi_{S3}^1(\xi) &= \frac{6}{10}\xi\left(\xi-\frac{2}{3}\right)\\ \Phi_{S1}^2(\xi) &= -\frac{3}{2}(\xi-1)\left(\xi+\frac{2}{3}\right) &;\quad \Phi_{S3}^2(\xi) &= -\frac{3}{2}(\xi+1)\left(\xi-\frac{2}{3}\right)\\ \Phi_{S1}^3(\xi) &= \frac{6}{10}\xi\left(\xi+\frac{2}{3}\right) &;\quad \Phi_{S3}^3(\xi) &= \frac{9}{10}\xi(\xi+1)\end{aligned} \tag{C.2}$$

where Φ_{S1}^m correspond to nodes placed at $\xi = -\frac{2}{3}, 0, +1$, while Φ_{S3}^m is for nodes placed at $\xi = -1, 0, +\frac{2}{3}$. See Fig. C.1. The position of the internal node in semi-discontinuous element is chosen arbitrarily at $-\frac{2}{3}$ or $+\frac{2}{3}$, not very close to the element end point to avoid near singularity problems.

J_n is the Jacobian of transformation for boundary elements, and is defined as:

$$J_n(\xi) = \sqrt{\frac{\partial x_\theta(\xi)}{\partial \xi}\frac{\partial x_\theta(\xi)}{\partial \xi}} \tag{C.3}$$

where $\frac{\partial x_\theta(\xi)}{\partial \xi}$ is the derivative of the global coordinates x_θ with respect to the local coordinate ξ.

The quadratic quadrilateral shape functions for the domain cells are given as:

$$\begin{aligned}
\Psi^1(\xi,\eta) &= \xi\eta\frac{(-a_2+\xi)(-a_3+\eta)}{a_4a_5a_1a_6}\\
\Psi^2(\xi,\eta) &= \xi\eta\frac{(a_4+\xi)(-a_3+\eta)}{a_2a_5a_1a_6}\\
\Psi^3(\xi,\eta) &= \xi\eta\frac{(a_4+\xi)(a_1+\eta)}{a_2a_5a_3a_6}\\
\Psi^4(\xi,\eta) &= \xi\eta\frac{(-a_2+\xi)(a_1+\eta)}{a_4a_5a_3a_6}\\
\Psi^5(\xi,\eta) &= \eta\left((a_2-\xi)(a_4+\xi)\right)\frac{(-a_3+\eta)}{a_2a_4a_1a_6}\\
\Psi^6(\xi,\eta) &= \xi\left((a_3-\eta)(a_1+\eta)\right)\frac{(a_4+\xi)}{a_1a_3a_2a_5}\\
\Psi^7(\xi,\eta) &= \eta\left((a_2-\xi)(a_4+\xi)\right)\frac{(a_1+\eta)}{a_2a_4a_3a_6}\\
\Psi^8(\xi,\eta) &= \xi\left((a_3-\eta)(a_1+\eta)\right)\frac{(-a_2+\xi)}{a_1a_3a_4a_5}\\
\Psi^9(\xi,\eta) &= \left((a_2-\xi)(a_4+\xi)\right)\frac{((a_3-\eta)(a_1+\eta))}{a_2a_4a_1a_3}
\end{aligned} \tag{C.4}$$

The parameters a in equation C.4 are defined in Fig. C.1. The continuous case is obtained when, $a_1 = a_2 = a_3 = a_4 = 1.0$. The discontinuous case requires values in the range 0.0 to 1.0, to be chosen for a_1, a_2, a_3, a_4. In this work the following parameter were chosen for the totally discontinuous case: $a_1 = a_2 = a_3 = a_4 = 2/3$. A whole range of transitional cells can be generated by varying one or more of these parameters (semi discontinuous cells).

The Jacobian of transformation for cell elements is defined as:

$$J_k(\xi,\eta) = \sqrt{(N_{31}^2 + N_{32}^2 + N_{33}^2)} \tag{C.5}$$

where N_{ij} is a minor of

$$\begin{bmatrix}
\dfrac{\partial x_1(\xi,\eta)}{\partial\xi} & \dfrac{\partial x_2(\xi,\eta)}{\partial\xi} & \dfrac{\partial x_3(\xi,\eta)}{\partial\xi}\\[2ex]
\dfrac{\partial x_1(\eta)}{\partial\eta} & \dfrac{\partial x_2(\xi,\eta)}{\partial\eta} & \dfrac{\partial x_3(\xi,\eta)}{\partial\eta}\\[2ex]
1 & 1 & 1
\end{bmatrix} \tag{C.6}$$

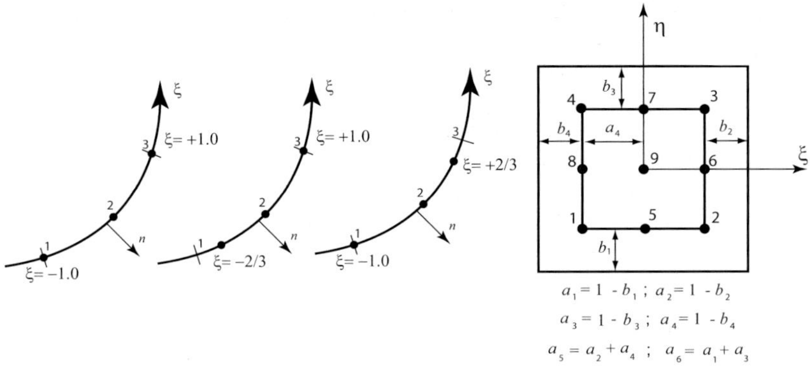

Fig. C.1. Types of elements used.

11.12 References

1. ABAQUS, Inc., Documentation, version 6.5 (2006).
2. Aliabadi, M.H., *The Boundary Element Method, vol II: application to solids and structures*, Chichester, Wiley (2001).
3. Allen, H.G., Bulson, P.S., *Background to Buckling*, (Mc Graw Hill, UK, 1980).
4. Atluri, S.N., Pipkins, D.S., *Boundary Element Analysis of Plate and Shells*, D.E. Beskos (Ed.), Spinger Verlag, Berlin, 141-166 (1991).
5. Anderson, E., Bai, Z., Bischof, C., Blackford, S., Demmel, J., Dongarra, J., Du Croz, J., Greenbaum, A., Hammarling, S., McKenney, A., Sorensen, D., *LAPACK Users' Guide*, third edition, Society for Industrial and Applied Mathematics (1999).
6. ANSYS, Inc. is a UL registered ISO 9001: Company. Version 9.0. (1994).
7. Baiz, P.M., Aliabadi, M.H., *Computer Modelling in Engineering & Sciences*, **13**(1): 19 (2006).
8. Baiz, P.M., Aliabadi, M.H., *Engineering Analysis with Boundary Elements*, **31**(4): 361 (2007).
9. Baiz, P.M., *Boundary Element Formulations for Instability and Fracture Mechanics Analysis of Plates and Shells*, (PhD Thesis, Imperial College London, 2006).
10. Barcellos, C. A., Silva, L. H. M., *Boundary Element Technology*, C. A. Brebbia and W. S. Venturini (Eds.), Computational Mechanics Publications, 123-130 (1989).
11. Brush, D.O., Almorth, B.O., *Buckling of Bars, Plates and Shells*, (McGraw-Hill, New York, 1975).
12. Costa, J.A., Brebbia, C.A., *Boundary Element VII*, C.A. Brebbia (Ed.), Springer-Verlag, Berlin, 429-442 (1985).
13. Crisfield, M.A., *Computers and Structures*, **13**: 55 (1981).

14. Di Pisa, C., *Boundary Element Analysis of Multi-layered Panels and Structures*, (PhD Thesis, Queen Mary University of London, 2005).
15. Dirgantara, T., Aliabadi, M.H., *International Journal for Numerical Methods in Engineering*, **45**: 1257 (1999).
16. Dirgantara, T., Aliabadi, M.H., *Communications in Numerical Methods in Engineering*, **17**: 749 (2001).
17. Dirgantara, T., *Boundary Element Analysis of Crack in Shear Deformable Plates and Shells (Topics in Engineering vol 43)*, (WIT Press, Computational Mechanics Publications, U.K., 2002).
18. Dirgantara, T., Aliabadi, M. H., *Computer Methods in Applied Mechanics and Engineering*, **195**: 4635 (2006).
19. Donnell, L.H., *NACA Report 479* (1933).
20. Elzein, A., Singellakis, S., *Advances in Boundary Elements XI, vol 3*, Ed. Brebbia C.A. Connor J.J., Springer-Verlag 269-284 (1989).
21. Flügge, W., *Stresses in Shells*, (2nd edition, Springer-Verlag, Berlin, 1973).
22. Gerard, G., Becker, H., *Handbook of Structural Stability Part III - Buckling of Curved Plates and Shells*, (NACA TN 3783, Washington D.C., 1957).
23. Kamiya, N., Sawaki, Y., *International Journal for Non-linear Mechanics*, **17**: 187 (1982).
24. Kamiya, N., Sawaki, Y., *Methodes Numeriques de l'Ingenieur, Comptes Rendus du Troisieme Congres International*, 63 (1983).
25. Kamiya, N., Sawaki, Y., Nakamura, Y., *Engineering Analysis*, **1**: 40 (1984).
26. Kawabe, H., *Theory and Application of Boundary Element Methods*, (Ed.) M. Tanaka and Q. Du, Pergamon, Oxford, 367-374, (1987).
27. Koiter, W.T., *Over de stabiliteit van het elastisch evenwicht*, (PhD Thesis, Delft, Amsterdam, 1945).
28. Libai, A., Simmonds, J.G., *The Nonlinear Theory of Elastic Shells*, (Academic Press Inc., London, 1988).
29. Lin, J., Long S., *Engineering Analysis with Boundary Elements*, **18**: 63 (1996).
30. Lin, J., Duffield, R.C., Shih, H.R., *Engineering Analysis with Boundary Elements*, **23**: 131 (1999).
31. Lei, X.Y., Huang, M.K., Wang, X., *Computers and Structures*, **37**: 911 (1990).
32. Lei, X. Y., Huang M.K., Wang, X., *Applied Mathematical Modelling*, **19**: 194 (1995).
33. Leitao, V.M.A., *Boundary Element in Nonlinear Fracture Mechanics (Topics in Engineering Vol 21)*, (Computational Mechanics Publications, 1994).
34. Manolis, G.D., Beskos, D.E., Pineros, M.F., *Computers and Structures*, **22**: 917 (1986).
35. Naghdi, P.M., *Quarterly of Applied Mathematics*, **14**: 369 (1956).
36. Nerantzaki, M.S., Katsikadelis, J.T., *Engineering Analysis with Boundary Elements*, **18**: 149 (1996).
37. O'Donoghue, P.E., Atluri, S.N., *Computers and Structures*, **27**: 427 (1987).
38. Palazotto, A., Dennis, S., *Nonlinear Analysis of Shells Structures*, (AIAA Education Series, Inc. Washington D.C., 1992).

39. Purbolaksono, J., Aliabadi, M.H., *Journal of Engineering Mathematics*, **51**: 211 (2005).
40. Purbolaksono, J., Aliabadi, M.H., *International Journal for Numerical Methods in Engineering*, **62**: 537 (2005).
41. Qin, Q., Huang Y., *Applied Mathematical Modelling*, **14**: 544 (1990).
42. Reissner, E., *Quarterly of Applied Mathematics*, **5**: 55 (1947).
43. Reissner, E., *Journal of Mathematics and Physics*, **29**: 90 (1950).
44. Riks, E., *International Journal of Solids and Structures*, **15**: 529 (1979).
45. Sabir, A.B., and Lock, A.C., *Variational Methods in Engineering*, C.A. Brebbia and H. Tottenham, (eds.), Southampton Press, UK (1972).
46. Supriyono, Aliabadi, M.H., *Engineering Analysis with Boundary Elements*, **30**: 31 (2006).
47. Syngellakis, S., Kang M., *Engineering Analysis*, **4**: 75 (1987).
48. Syngellakis, S., Elzein, A., *International Journal Numerical Method in Engineering*, **37**: 1763 (1994).
49. Syngellakis, S., *Advances in Boundary Element Series*, (Computational Mechanics Publications, Southampton UK, 1998).
50. Sun, Y.B., He, X.Q., Qin, Q.H., *Computer and Structures*, **53**: 649 (1994).
51. Tanaka, M., *Developments in Boundary Element Methods*, C.A. Brebbia (Ed.), Springer-Verlag, Berlin, 115-136 (1984).
52. Tanaka, M., Miyazaki, K., *Boundary Elements VII*, C.A. Brebbia (Ed.), Springer-Verlag, Berlin, 4-3/4-16, (1985).
53. Tanaka, M., Miyazaki, K., *Boundary Elements VIII*, C.A. Brebbia (Ed.), Springer-Verlag, Berlin, 537-546 (1986).
54. Tanaka, M., Matsumoto, T., Zheng, Z., *Engineering Analysis with Boundary Elements*, **17**: 123 (1996).
55. Tanaka, M., Matsumoto, T., Zheng, Z., *Engineering Analysis with Boundary Elements*, **23**: 399 (1999).
56. Telles, J. C. F., *International Journal for Numerical Methods in Engineering*, **24**: 959 (1987).
57. Timoshenko, S., Gere, J.M., *Theory of Elastic Stability*, (McGraw-Hill, New York, 1961).
58. Tosaka, N., Miyake, S., *Boundary Elements VII*, C. A. Brebbia (Ed.), Springer-Verlag, Berlin, 4-43/4-52 (1985).
59. Van der Weeën, F., *International Journal for Numerical Methods in Engineering*, **18**: 1 (1982).
60. Wang, W., Ji, X., Tanaka, M., *Computational Mechanics*, **26**: 58 (2000).
61. Wempner, G.A., *International Journal of Solids and Structures*, **7**: 1581 (1971).
62. Wen, P.H., Aliabadi, M.H., Young, A., *Engineering Analysis with Boundary Element*, **24**: 583 (2000).
63. Wen, P.H., Aliabadi, M.H., Young, A., *Computer and Structures*, **83**: 870 (2005).
64. Wen, P.H., Aliabadi, M.H., Young, A., *Journal of Strain Analysis for Engineering Design*, **41**: 239 (2006).

65. Wilde, A.J., Aliabadi, M.H., *Communications in Numerical Methods in Engineering*, **14**: 505 (1998).
66. Ye, T.Q., Liu, Y.J., *Applied Mathematical Modelling*, **9**: 183 (1985).
67. Ye, J., *Engineering Analysis with Boundary Elements*, **9**: 283 (1992).
68. Zhang, J.D., Atluri, S.N., *International Journal for Numerical Methods in Engineering*, **26**: 571 (1988).
69. Zienkiewicz, O.C., Taylor, R.,*The Finite Element Method, vol 2: Solid Mechanics*, (B-H, Oxford, 2000).

CHAPTER 12

PROGRESSIVE FAILURE IN COMPRESSIVELY LOADED COMPOSITE LAMINATED PANELS: ANALYTICAL, EXPERIMENTAL AND NUMERICAL STUDIES

Shiladitya Basu

Technical Professional – Marine, Granherne, Inc, Houston, TX

Anthony M. Waas*

Professor, Aerospace Engineering Department, University of Michigan

Damodar R. Ambur

Chief, Structures Division, NASA Glenn Research Center, Cleveland, OH 44135

A mechanism-based progressive failure analyses approach is presented for fiber reinforced composite laminates that are loaded predominantly in compression. Each ply of the laminate is modeled as a nonlinear elastic degrading lamina (characterized through laboratory scale experiments, in shear and in transverse directions) in a state of plane stress. In the fiber direction, elastic behavior prevails, however, in the present work, the phenomenon of fiber microbuckling, is explicitly accounted for by allowing the *in situ* fiber rotation to be a variable in the problem. The present model disbands the notion of a fixed compressive strength, and instead uses the mechanics of the failure process to provide the *in situ* compression strength in a lamina, dictated by the current local stress state, the current state of the degrading lamina and the local fiber rotation. These features are built into a user defined material subroutine that is implemented through the ABAQUS† commercial code in conjunction with Classical Lamination Theory. The inputs to the present work are data for lamina transverse property degradation, the elastic lamina orthotropic properties, the ultimate tensile strength of the lamina in the fiber direction, the laminate stacking sequence and the structural panel

*Author to whom all correspondence should be addressed (dcw@umich.edu, Tel:734-764-8227, Fax:734-763-0578).
†Registered trademark of ABAQUS, Inc. currently, Simulia.

geometry. The validity of the approach is demonstrated through numerical simulations of the response of a double notched flat unstiffened panel subjected to compression. The predictions of the simulations are compared against experimental data and good agreement is reported.

12.1 Introduction

A mechanism-based progressive failure analyses (PFA) approach is presented for fiber reinforced composite laminates that are loaded predominantly in compression. Each ply of the laminate is modeled as a nonlinear elastic degrading lamina in a state of plane stress according to Schapery theory (ST). In this theory, each lamina degrades as characterized through appropriate laboratory scale experiments. In the fiber direction, elastic behavior prevails, however, in the present work, the phenomenon of fiber microbuckling, which is responsible for the sudden degradation of the axial lamina properties under compression, is explicitly accounted for by allowing the fiber rotation at a material point to be a variable in the problem. The latter is motivated by experimental and numerical simulations that show that local fiber rotations in conjunction with a continuously degrading matrix are responsible for the onset of fiber microbuckling leading to kink banding. The present model, thus, disbands the notion of a fixed compressive strength, and instead uses the mechanics of the failure process to provide the in-situ compression strength of a material point in a lamina, the latter being dictated strongly by the current local stress state, the current state of the degrading lamina transverse material properties and the local fiber rotation. These features are built into a user defined material subroutine that is implemented through the commercial finite element (FE) software ABAQUS in conjunction with Classical Lamination Theory (CLT). CLT considers a laminate to be modeled as a collection of perfectly bonded lamina. The inputs to the present work are laboratory scale, coupon level test data that provide information on the lamina transverse property degradation (i.e. appropriate, measured, strain-stress relations of the lamina transverse properties), the elastic lamina orthotropic properties, the ultimate tensile strength of the lamina in the fiber direction, the stacking sequence of the laminate and the geometry of the structural panel. The validity of the approach advocated is demonstrated through numerical simulations of the response of various composite structural panels that are loaded to complete failure. A flat, unstiffened panel with a cutout subjected to in-plane shear loading, a set of centrally notched unstiffened stitched panels subjected to axial compression and a double notched unstiffened stitched panel subjected to axial compression are selected for study. The predictions of the simulations are compared against experimental data and good agreement is reported.

A large body of literature devoted to PFA of composite laminated structures is now present. Many of the PFA schemes introduced and available today have relied on the phenomenological approach of defining strength criteria for a single lamina when subjected to different single component stress states. These methods define the onset of failure through specific indices that are expressed as functions of the current stress state. When any of these indices exceeds a predefined critical value, the material at that point is said to have failed.[1,2] When a material point has failed, for subsequent loading, it is assumed to have a reduced stiffness that is predetermined in an empirical manner. Depending on the type of failure (for instance, fiber breaking and/or matrix cracking due to tension along the fibers, fiber kink-banding due to compression along the fibers, fiber/matrix debonding due to in-plane shear), different elastic moduli are set to zero. In addition, linear elastic material behavior is assumed throughout the analysis. In a laminated composite plate, the stiffness at a material point is determined by the current local stress state and the local 'state' of the material. The local stress state, in general, is multi-axial and a material point at the current state may have accumulated damage, dictated by the past loading history. Thus, 'strength' at a material point is influenced to a great extent by the current stress and strain state and further, is predicated on the mechanism of failure. PFA methodologies that abruptly change material properties based on strength 'indices' and rely on a linear elastic analysis may be unable to realistically account for features associated with the mechanisms of failure. Recently, an improved PFA scheme that still relies on strength indices but indirectly accounts for fiber misalignment has been introduced by Pinho et al.[3] This PFA methodology is an improvement over the previous LaC02 criteria developed by Dávila et al.[4] For progressive failure modeling, a framework that accounts for the continued degradation of the material is needed.

Schapery[5] introduced a thermodynamically based theory (referred to as ST) that uses internal state variables (ISVs) to analyze damage evolution in composite laminates. These ISVs are related to mechanical aspects of damage mechanisms. The ISVs are related to the energy required for the evolution of the damaged states. Different damage mechanisms can have distinct ISVs to track the damage evolution. For instance, matrix microcracking can be expressed via one ISV, while transverse cracking can be represented by another ISV.[6] The evolution of these ISVs with global loading is determined at each material point and as functions of load history through the satisfaction of a thermodynamic criterion (an evolution equation), throughout the loading history. Therefore, the 'state' of the material is continually updated according to the stress and strain history that a material point experiences throughout the loading history. The greatest advantage of ST is its ability to relate material degradation data obtained from coupon level laboratory specimens to the ISVs that enter the constitutive relations. This feature provides great utility for a PFA methodology for strength prediction.

Various experimental[7–9] and analytical[10] studies show that the maximum load sustained by a lamina in the axial direction (σ_{11}^{cr}), in compression, depends greatly on the presence of the other in-plane stress components such as σ_{22} and τ_{12}. Experimental, analytical and micromechanical studies show that the main physical event occurring during kink banding is the rotation of fibers in a band within a degrading matrix. The rotation of fibers gives rise to high localized shear strains that drive the shear degradation of the local matrix. The shear degradation in turn drives the rotation of the fibers creating a positive feedback loop. The simultaneous presence of these phenomena results in a limit load situation for a lamina under axial compression. This limit load, which is dictated by the local stress state and the state of the transverse lamina properties, can be relieved (increased) by the presence of other stress components or, in certain cases, can be elevated (reduced). The present approach captures these two phenomena in a macroscopic (lamina level) numerical setting and is able to reproduce the fine scale (individual fiber-matrix level) micromechanical observations seen in experiments and in associated numerical models.[8,11–24] It is to be noted that Merodio and Pence[25] have presented a comprehensive analysis of the formation of kinks, defined as surfaces across which the deformation gradient and hydrostatic pressure suffer jumps, in directionally reinforced neo-Hookean materials. Their results, which are carried out for purely elastic materials, bear some resemblance to the results reported in the literature for fiber reinforced materials with a polymer matrix.[8,9] This work has recently been further extended by Fu and Zhang,[26] who have considered the rotation of the kink band with continued loading.

The present chapter is structured as follows. In Section 12.2, various aspects of the proposed PFA are introduced. Theoretical modeling details and the numerical inputs are discussed. An initial validation is carried out by predicting the behavior of a unidirectional lamina under axial compression, for which a set of rigorous micromechanical computation results exists. In Section 12.3, the salient experimental details of the various structural panel configurations are described. Section 12.4 reports the input data used for the PFA of these panel configurations and discusses the results of the analyses.

12.2 Macroscopic Model for Kink Banding Instabilities in Fiber Composites

In this section, a mechanism-based lamina level modeling approach is used as the basis to develop a macroscopic (lamina level) model to capture the mechanisms of kink banding. Laminae are modeled as inelastic degrading *homogenized* layers in a state of plane stress according to ST. However, the principal orthotropic material axes are allowed to rotate as a function of deformation. In ST, each lamina degrades as characterized through laboratory scale experiments. In the fiber direction, elastic behavior prevails, however, in the present work, the phenomenon of fiber microbuckling leading to kink banding, which is responsible for the sudden degradation of the axial lamina properties under compression is explicitly accounted for by allowing the fiber rotation at a material point to be a variable in the problem. These features are built into a user defined material subroutine that is implemented through the commercial finite element (FE) software ABAQUS. The notion of a fixed compressive strength of a lamina is disbanded and instead the mechanics of the failure process is used to provide the *in situ* compression strength of a material point in a lamina. This strength is dictated strongly by the current local stress state, the current state of the lamina transverse material properties and the local fiber rotation. The inputs to the present model are laboratory scale, coupon level test data (at the lamina level) that provide information on the lamina transverse property degradation (i.e. appropriate, measured, strain-stress relations of the lamina transverse properties), the elastic lamina orthotropic properties and the geometry of the lamina. The validity of the approach advocated is demonstrated through numerical simulations of unidirectional lamina with initial fiber imperfections. The predictions of the simulations reported in this section are compared against previously reported results from micromechanical analyses.[8,20,21] Good agreement between the present macroscopic modeling approach and the previous micromechanical observations are reported.

12.2.1 *Progressive Failure Analysis using Schapery Theory*

Elements of the PFA approach

Non-linear Constitutive Formulation

Schapery[27] developed nonlinear elastic constitutive relations for an orthotropic lamina using a work potential approach which accounted for the effect of microdamage. A work potential approach can be justified if a lamina shows path independence with respect to loading, at least in a limited sense. In the principal material coordinate system of a lamina, with 1 being the fiber direction, the lamina stress-strain relations are,[28]

$$\begin{aligned} \sigma_{11} &= Q_{11}\epsilon_{11} + Q_{12}\epsilon_{22} \\ \sigma_{22} &= Q_{12}\epsilon_{11} + Q_{22}\epsilon_{22} \\ \tau_{12} &= Q_{66}\gamma_{12} \end{aligned} \tag{12.1}$$

where,

$$\begin{aligned} Q_{11} &= \frac{E_{11}}{1-\nu_{12}\nu_{21}} \;;\; Q_{22} = \frac{E_{22}}{1-\nu_{12}\nu_{21}} \\ Q_{12} &= \nu_{12}Q_{22} \;;\; Q_{66} = G_{12} \\ \nu_{21} &= \frac{\nu_{12}E_{22}}{E_{11}} \end{aligned} \tag{12.2}$$

The material moduli given by E_{11}, E_{22} and G_{12} are the secant moduli of the lamina.

In the existing literature on damage mechanics as applied to continuous fiber laminated composite materials, the effect of damage is incorporated through the change in transverse Young's modulus E_{22} and in plane shear modulus G_{12}. For instance, Sun and Chen[29] proposed a one parameter plastic potential in conjunction with orthotropic incremental plasticity theory to study the evolution of E_{22} and G_{12} in tension. Schapery and Sicking,[6] used ST to study the evolution of E_{22} and G_{12}. These previous studies were not concerned with the state of the lamina beyond first failure in the fiber direction. Yet, it is recognized[4,30] that such damage is dominant for compression loaded composite structures. Lamina level coupon tests in tension have shown that fiber direction modulus, E_{11} and Poisson's ratio, ν_{12} can be assumed to be independent of microdamage that influence E_{22} and G_{12}.[31] This situation is also true for compression until the onset of kinking (the axial compression load reaches a maximum limit load at the point in which a kink band starts to form, for example, as shown in Lee and Waas[8]). During kink band formation and propagation, it is likely that microdamage mechanisms do influence E_{11}, ν_{12}, E_{22} and G_{12}. Subsequent to kink banding, other failure mechanisms such as delamination can take place.[11] Delamination, which can be a dominant failure mechanism,[32,33] is not accounted for here, but has received attention recently in the context of PFA.[34]

Elements of Schapery Theory

In Schapery and Sicking,[6] ISVs are used to incorporate *inelastic* behavior in the material response. Earlier, Schapery,[5] introduced a more general thermodynamic framework to study materials that undergo damage. In these developments, the total work done, W_T, in a mechanical process is composed of the inelastic work, W_s and the work of deformation W.

$$W_T = W + W_s \tag{12.3}$$

The irrecoverable portion of total energy (W_s) can be determined from the material stress-strain response as shown in Fig. 12.1. ISVs are described through S_i's. Each S_i is associated with a particular damage mechanism. To satisfy the path independence of total work, these ISVs have to satisfy the following relation,

$$f_i = \frac{\partial W_s}{\partial S_i} \tag{12.4}$$

The left hand side of Eq. (12.4) is referred to as the *thermodynamic force* related to the ith ISV. This represents the amount of *force* required for the ith damage mode to progress/accumulate. The available force or the ith *driving force* is given by the right hand side of Eq. (12.5). When this available force exceeds the required force $\partial W_s/\partial S_i$, then the material undergoes a structural change that is associated with S_i.

$$f_i \equiv -\frac{\partial W}{\partial S_i} \tag{12.5}$$

Stated another way,

$$\dot{S}_i > 0 \tag{12.6}$$

where, the overdot represents a temporal derivative. However, as pointed out by Schapery,[5] it is to be noted that if the available thermodynamic force, $(-\partial W/\partial S_i)$ is less than the required thermodynamic force $(\partial W_s/\partial S_i)$ then,

$$\dot{S}_i = 0 \tag{12.7}$$

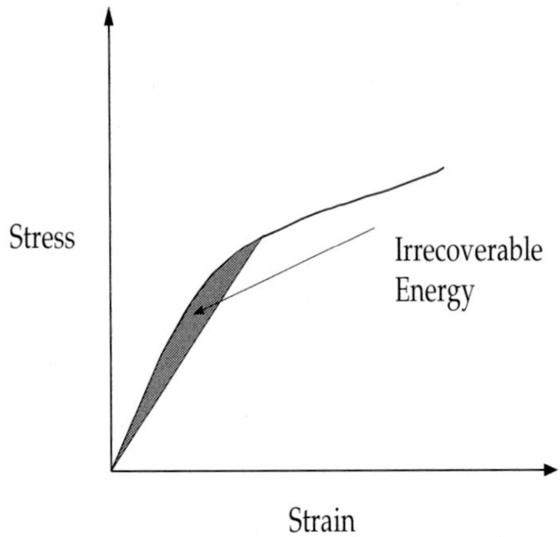

Fig. 12.1. Definition of irrecoverable energy using a generic stress-strain curve.

Furthermore, S_is need not change continuously with the loading. Schapery considered two ISVs. They were the energies associated with matrix micro-cracks (S) and of the transverse intra-ply cracks (S_c), respectively. Inelastic work is described as,

$$W_s = S + S_c \tag{12.8}$$

In the present work, degradation of the transverse ply properties is captured through one ISV, denoted by S. This ISV affects the moduli E_{22} and G_{12} through Eq. (12.9)

$$\begin{aligned} E_{22} &= E_{22_0} e_s(S) \\ G_{12} &= G_{12_0} g_s(S) \end{aligned} \tag{12.9}$$

Here, E_{22_0} and G_{12_0} are transverse and shear moduli of the virgin material, i.e., at zero strain and zero damage; $e_s(S)$ and $g_s(S)$ are functions relating these two moduli to the S. The functions e_s and g_s are expressed as polynomial relations in S. These polynomial relations are obtained by analysing coupon level test results.[6]

The strain energy density (or work of deformation) can be written as

$$W = \frac{1}{2}\left(Q_{11}\epsilon_{11}^2 + Q_{22}\epsilon_{22}^2\right) + Q_{12}\epsilon_{11}\epsilon_{22} + \frac{1}{2}Q_{66}\gamma_{12}^2 \tag{12.10}$$

To incorporate geometric nonlinearities, Green's strains and the second Piola Kirchoff stresses need to be used in the expression for W. For small strains, Eq. (12.10), would contain only the first order terms in the strain-displacement relations. It has been shown[6] that material non-linearities as incorporated in Eq. (12.10) are still significant for fiber reinforced composites even when inclusion of geometric nonlinearities are not called for.[35] Using Eqs (12.1)-(12.5), the evolution equations for S is as follows,

$$\frac{\epsilon_{11}^2}{2}\frac{\partial Q_{11}}{\partial S} + \frac{\epsilon_{22}^2}{2}\frac{\partial Q_{22}}{\partial S} + \epsilon_{11}\epsilon_{22}\frac{\partial Q_{12}}{\partial S} + \frac{\gamma_{12}^2}{2}\frac{\partial G_{12}}{\partial S} = -1 \qquad (12.11)$$

For most materials $\nu_{12}\nu_{21} \ll 1$ which implies that, the Q_{ij}'s can be replaced by E_{ij}'s. In the present work, it is assumed that the fiber direction stiffnesses are unaffected by S. Thus, the term containing Q_{11} in Eq. (12.11) can be neglected. It is also assumed that ν_{12} changes with S such that $\nu_{12}E_{22}$ remains constant during loading. The implications of this assumption are two fold. Firstly, it allows for a change in ν_{12} with damage accumulation in the lamina. Secondly, it makes it possible to uncouple ϵ_{11} from the damage accumulation. If ϵ_{11} is present in the evolution equation (Eq. (12.11)), then damage becomes a direct function of the longitudinal strain, which is an improper representation of the underlying mechanism. Equation (12.11) can be reduced to,

$$\frac{\epsilon_{22}^2}{2}\frac{\partial E_{22}}{\partial S} + \frac{\gamma_{12}^2}{2}\frac{\partial G_{12}}{\partial S} = -1 \qquad (12.12)$$

For an inelastic process, the entropy production rate is non-negative. Hence,

$$\dot{S} \geq 0 \qquad (12.13)$$

Physically, $\dot{S}$ is non-negative because healing (or reversible damage) is not allowed for in the damage mechanism considered.

From experiments,[6] it has been observed that for small strains, S behaves as ϵ^3. This is based on the fact that the moduli are constant for small strains. Thus to express the moduli, E_{22} and G_{12} in terms of a polynomial of S, a reduced variable S_r can be used,

$$S_r \equiv S^{1/3} \qquad (12.14)$$

Replacing S in Eq. (12.12) by S_r (from Eq. (12.14)) and using the chain-rule of differentiation, the evolution equation for S_r is,

$$\frac{\epsilon_{22}^2}{2}\frac{\partial E_{22}}{\partial S_r} + \frac{\gamma_{12}^2}{2}\frac{\partial G_{12}}{\partial S_r} = -3S_r^2 \qquad (12.15)$$

During laboratory experiments damage in a lamina reaches a saturation limit and the specimen then fails in a catastrophic manner. This results in a termination of the experimental data and a total loss of lamina stiffness and strength. The corresponding value of S_r, obtained from the resultant stress-strain plot, is denoted by S_r^*. The constitutive model described thus far holds true for $S_r \leq S_r^*$. For $S_r > S_r^*$, numerical stability considerations are important for a stable PFA method. The method used in the present section is detailed in Section 12.2.2.

Fiber rotation under axial compression

Results from previous experimental, analytical and numerical studies have shown that the main feature of laminate failure under predominantly compression loading is localized fiber bending leading to kink band formation. The presence of initial fiber misalignment in a lamina that experiences compression along the fibers gives rise to local shear strain in the matrix adjacent to the misaligned fibers. These shear strains accelerate degradation of the local matrix shear stiffness. Imperfect fiber systems are prone to rotate under axial compression. That is, the fibers within a lamina that are subjected to axial compression have the propensity to *change* their alignment. This change is dictated by the local multiaxial stress state and the local shear stiffness of the matrix. As the local shear strains degrade the matrix shear stiffness, the resistance to localized fiber bending diminishes and fiber rotation gradually begins to build up which in turn creates more local imperfection and local shear strain. Thus a positive feedback loop is established between these two competing events. A point is reached when the in-situ shear stiffness is not sufficient to prevent the additional fiber rotation. This point usually coincides with the peak load in an axial material response curve. Beyond this point, the matrix is unable to resist any fiber rotation and the fibers rotate aligning themselves in localized deformed bands, commonly described as 'kink bands'.

Consider a fiber reinforced lamina under a generalized load state as indicated in Fig. 12.2(a). A band of misaligned fibers is shown sandwiched between two regions where the fibers are nominally straight and aligned. This equilibrium configuration is conceptualized from images of kink bands captured experimentally[36] (Fig. 12.2(b)). In these real kink bands, there is a conical region where fiber bending is dominant and this region is sandwiched between the regions where there is dominant shearing (inside the band) and the far-field aligned region where there is uniform deformation (Fig. 12.2(c)). Owing to the fiber bending dominant region, the kink band boundary inclination ($\beta + \beta_0$) has the flexibility of changing during the loading, i.e., there is continuous exchange of material between the fiber bending dominant region and the shearing dominant region. In the idealization presented here, the fiber bending dominant region is disregarded and the shearing dominant region is assumed to be held in equilibrium between the two nominally aligned regions. A reference frame $x - y$ can be defined

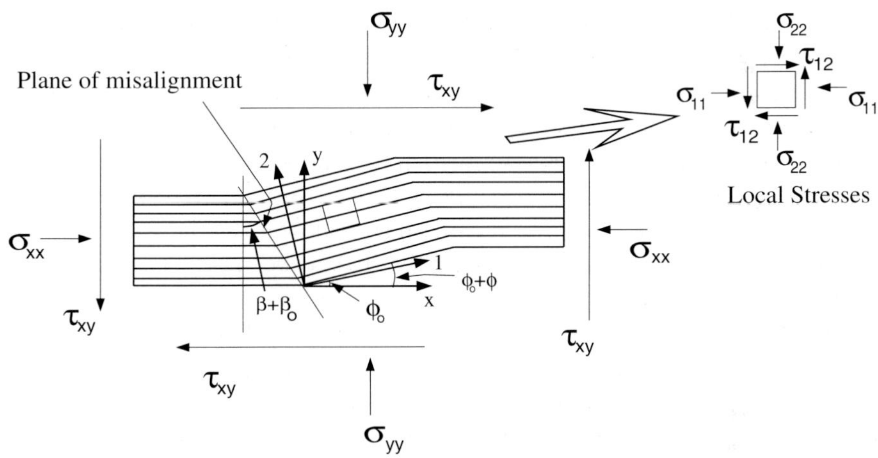

Fig. 12.2. (a) A unidirectional lamina with a band of misaligned fibers in equilibrium under a multiaxial stress state. Inset shows the stress state inside the misaligned band.

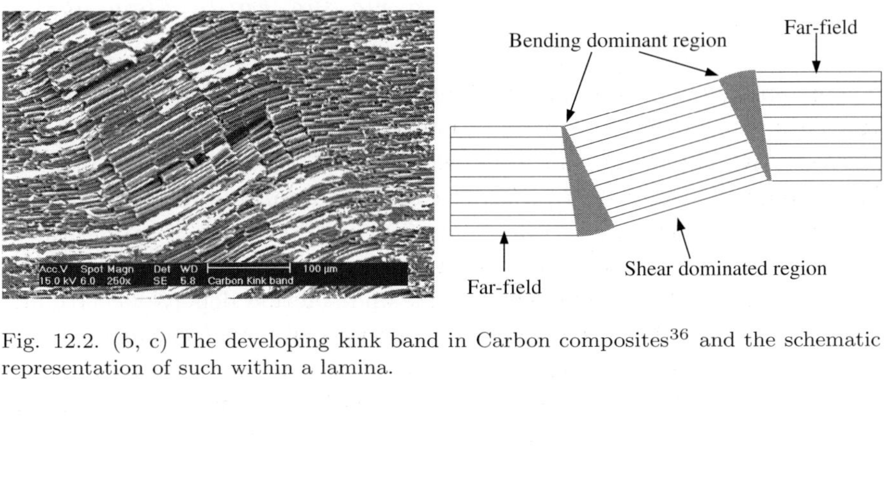

Fig. 12.2. (b, c) The developing kink band in Carbon composites[36] and the schematic representation of such within a lamina.

such that the $x-$direction is parallel to the nominal fiber direction in the lamina and the $y-$direction is normal to it. In subsequent discussions, the $x-y$ reference frame will be termed as the 'global' frame. The misaligned band of fibers, in the current configuration, is defined through two angles, $\hat{\phi} = \phi + \phi_0$ and $\hat{\beta} = \beta + \beta_0$. The angles (ϕ_0, β_0) are constants and the angles ϕ and β are variables that may change as a function of current far-field stress state. The reference frame, 1-2, is defined in the unstressed initial configuration of the misaligned lamina. Fibers inside the misaligned band are parallel to the '1' direction in the initial state. A current reference frame $1^{'}$-$2^{'}$ is defined where '$1^{'}$' is always parallel to the current fiber direction inside the band , thus the 1-axis rotates to $1^{'}$-axis during loading. The '2' and '$2^{'}$' directions are always orthogonal to the '1' and '$1^{'}$' directions respectively. The 1-2 frame will be defined as the 'local' frame and the $1^{'}$-$2^{'}$ system will be termed the instantaneous frame. Initially, when the matrix retains most of its in-situ shear stiffness, the 1-2 and $1^{'}$-$2^{'}$ axes systems will be nearly co-incident. As the matrix shear stiffness degrades, local shear strain, γ_{12}, will start to increase rapidly. It can be shown that the angle between the 1-axis and the $1^{'}$-axis, given by ϕ and γ_{12}, for small strains, are related by

$$\gamma_{12} = \gamma_{12}^{\infty} + \phi - \beta \tag{12.16}$$

Here, γ_{12}^{∞} is the contribution from the externally applied shear loading, if any. For a uniaxial compression loading and assuming $\beta = \beta_0 = 0$, we arrive at,

$$\gamma_{12} \approx \phi$$

12.2.2 *Numerical Implementation via the Finite Element (FE) Method*

The material behavior outlined earlier is modeled in the numerical domain using the commercially available FE package *ABAQUS*. *ABAQUS* has the capability of integrating user defined material behavior with its existing element library through user defined material subroutine, UMAT.[37] This subroutine is called at each material point for which the constitutive law is defined through the user defined option. A UMAT receives from the solver in ABAQUS, stresses, strains and solution dependent variables (SDVs) from the previous equilibrium step ($(n-1)^{th}$ step), the increment of strain in the current loading step (n^{th} step), and various other parameters. After calculations, the UMAT returns to the solver the updated stresses and SDVs, if any, and the incremental tangent stiffness matrix, $\partial\sigma_{ij}/\partial\epsilon_{ij}$ in the current loading step.

When the increment of shear strain is small, then the instantaneous fiber rotation can be equated to the change in shear strain[31] $\mathrm{d}\gamma^n_{1'2'}$. From the constitutive relation one can also write,

$$\gamma^n_{1'2'} = S^n_{66}\tau^n_{1'2'} \tag{12.17}$$

where, $S_{66} = 1/G_{12}$. Taking differentials on both sides of Eq. (12.17),

$$d\gamma^n_{1'2'} = S^n_{66} d\tau^n_{1'2'} + dS^n_{66}\tau^n_{1'2'} \tag{12.18}$$

Equation (12.18) provides an expression for the change in angle $\mathrm{d}\phi^n$. This change is added to the fiber angle value of the previous step to obtain the current fiber angle ϕ^n.

$$\phi^n = \phi^{n-1} + d\phi^n \tag{12.19}$$

This updated angle is used in the current increment, to transform the stresses and the material incremental constitutive matrix computed in the instantaneous $(1' - 2')$ frame to the local (1-2) frame, to return to the solver in ABAQUS. In the absence of damage (or when the damage is small) the angle ϕ^n will be small. But with the accumulation of damage, ϕ^n starts to increase, leading to local fiber direction instability. It should be noted that the definition of fiber rotation via Eq. (12.18) allows the possibility of elastic rotation recovery (if a material point unloads) that is instrumental in deformation localization during kink banding.

12.2.3 *Numerical Predictions*

The capabilities of the present lamina level modeling scheme are assessed by simulating the rigorous micromechanical analyses performed earlier on unidirectional axially loaded composite lamina.[21] This earlier study considered the composite lamina to consist of discretely layered fiber and matrix phases. Fiber elements were assigned linear elastic properties and the matrix elements were modeled as elastic-plastic material obeying a J2 incremental flow theory of plasticity with a Mises yield criterion and isotropic hardening. Similar FE studies that were performed earlier[8,17,20,38] considered non-uniform fiber spacing effects, 3D effects (variation in fiber packing) and in a limited manner, multi-axial loading. These studies established that a 2D representation of the composite with uniform packing sufficed to capture the important aspects of kink banding provided that the fiber volume fraction was at least 50%. For the present study, the laminae are assumed

Table 12.1. Mesh dimensions in the validation analyses.

	Length (mm)	Width (mm)	Aspect Ratio	Elements	Nodes	d.o.f.
Mesh A	1.00	1.00	1.0	1	8	16
Mesh B	3.00	1.00	3.0	3	18	36
Mesh C	3.00	1.00	3.0	3600	11041	22082

to be degrading homogenized orthotropic media, with fiber volume fraction of $v_f = 50\%$. Two lamina geometries with aspect ratio AR = 1 (1 mm × 1 mm) and AR = 3 (1 mm × 3 mm) are considered (data in Table 12.1). These laminae are discretized using quadratic plane stress continuum elements CPS8 available in the commercial finite element software package ABAQUS. The $AR1$ lamina is discretized using a single CPS8 element. The $AR3$ lamina is meshed using two different finite element discretization schemes which are shown in Fig. 12.3). In generating mesh C for $AR3$ a gradation of element sizes is used, with a finer discretization near the free boundaries. This was done to preclude any artificial stiffening that can be introduced due to edge effects if a uniform mesh is used in conjunction with slight misalignment of the principal material axis with respect to the loading direction. For all meshes, static analyses are performed using displacement control loading. Linear eigenvalue buckling analyses are also performed to generate perturbations in the form of the linear eigen modes to be used in subsequent response analyses. Geometric nonlinearity is included in the response analysis through the NLGEOM option in *ABAQUS*.

Elastic material properties for the AS4/3501-6[39] material system are described in Table 12.2 which are consistent with the lamina properties derived from the fiber and matrix properties described in Basu.[21] The nonlinear behavior in shear and in the transverse direction are provided as input, as normalized moduli variation against S_r as shown in Fig. 12.4. A material point is denoted as 'damaged' when S_r at that location reaches S_r^* (or the end of the input stress-strain curve). For the present analysis, this refers to a 55% degradation in the *in situ* secant shear modulus, G_{12}. The complete material constitutive behavior is modeled via UMAT. Unit thickness in the Z-direction of the lamina is assumed for the plane stress element section definition. Boundary conditions for the laminae are applied at nodal positions and are similar to the micromechanical analyses performed earlier as indicated in Fig. 12.5. The edge AB of the mesh is constrained from moving in the axial or X- direction, but allowed to

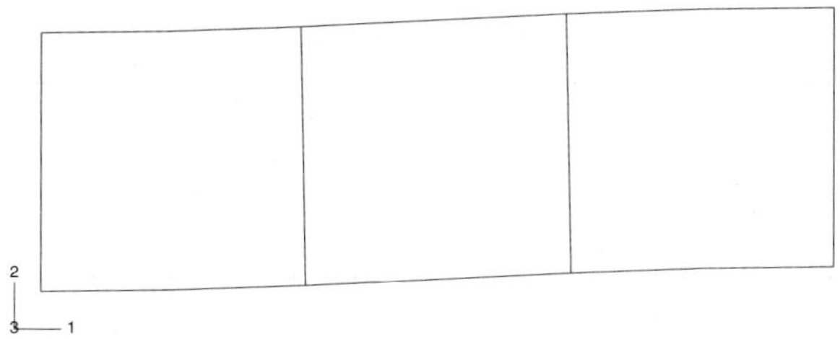

(c) Mesh B

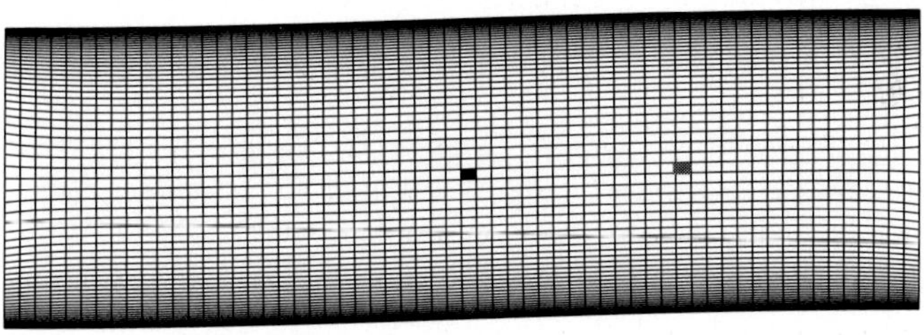

(d) Mesh C

Fig. 12.3. Finite element meshes used in the present analyses for the AR3 lamina. The element darkened at the centre is element number 330. The darkened element to the right is element 316. Quantities extracted from these two elements are presented later.

Table 12.2. Material constants.

	AS4/3501-6 (GPa)
E_{11}	116.00
E_{22}	12.74
G_{12}	4.78
ν	0.27

'breath' in the transverse or Y- direction. The centre point of this edge is prevented from moving at all, eliminating any possible rigid body motion. Edge CD is used to provide the displacement control loading simulating an axial compressive load. Edges AD and BC are allowed to deform as per the equilibrium requirements of the boundary value problem, i.e., on these edges $\sigma_{yy} = 0$ and $\tau_{xy} = 0$. Geometric and material perturbations are both used in the present analyses. Geometric perturbation is provided using the lowermost linear eigenvalue buckling mode such that the perturbed centre-

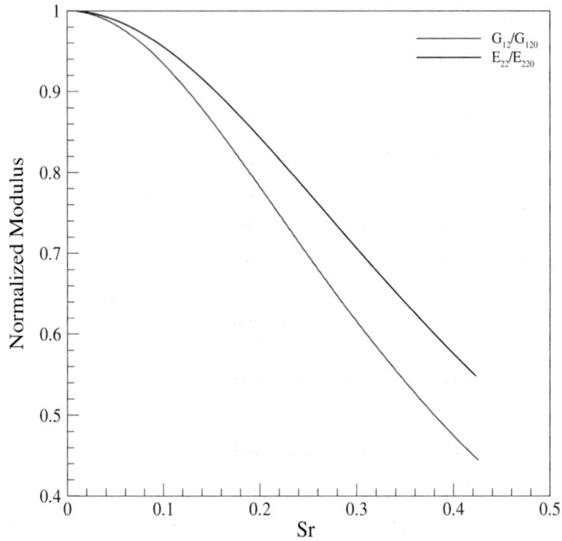

Fig. 12.4. Normalized moduli plotted against the damage parameter S_r.

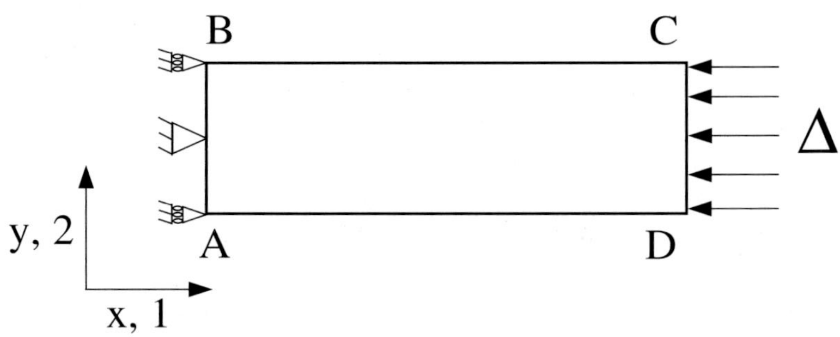

Fig. 12.5. Boundary conditions and loading used in the present analyses.

line of the mesh creates an angle of 2^0 with the horizontal at the centre of the mesh. In addition, a 2^0 initial fiber angle value (Eq. (12.19)) is used as an initial misalignment. These imperfection magnitudes are similar to the ones used in Basu[21] and are commonly observed as typical misalignment angles for pre-preg based laminates.

12.2.4 *Results and Discussion*

Results from the FE study outlined in the previous section are presented here. For each mesh, the load-end shortening data are normalized as described in Eq. (12.20).

$$\sigma = \frac{P}{A} \; ; \; \epsilon = \frac{\Delta}{L} \tag{12.20}$$

Here, A and L denote the initial cross-sectional area and the initial axial length respectively. Deformed shape of the finite element mesh at various load levels are presented. Contour plots showing the evolution of the instantaneous fiber rotation angle, ϕ, and deformation magnitude are also presented.

AR1 lamina results

Normalized load - end shortening response (σ -ϵ) for the $AR1$ lamina is shown in Fig. 12.6 with the corresponding result from the micromechanical study[21] presented for comparison. The macro-level response for this aspect ratio matches quite well with the micromechanical prediction. As the mesh A has only 16 degrees of freedom compared to the nearly 128,000 for the micromechanical analysis, it behaves stiffer than the micromechanical analysis mesh. This is observable in the slight deviation near the peak of the $\bar{\sigma}$ - $\bar{\epsilon}$ response of mesh A. On the other hand, such a low resolution of the current mesh, does not allow it to capture the gradients of stresses, strains or the deformation field accurately. This actually magnifies the effect of the material imperfection used in the present analysis and we obtain a lower peak load for the $AR1$ lamina compared to the corresponding micromechanical analysis. The post-peak response from the micromechanical analysis shows a 'snapback' where the stress and the strain both decrease. Subsequently, the response shows a softening behavior where the stress continues to drop with increasing strain.

AR3 lamina results

Normalized load - end shortening responses (σ-ϵ) of the $AR3$ lamina are presented in Fig. 12.7. Both mesh B and mesh C are able to capture the overall global behavior though they use widely different finite element discretizations. It can be seen that the macro-level responses are stiffer than the micromechanical response which is due to the fact that both these meshes have significantly lower number of degrees of freedom compared to

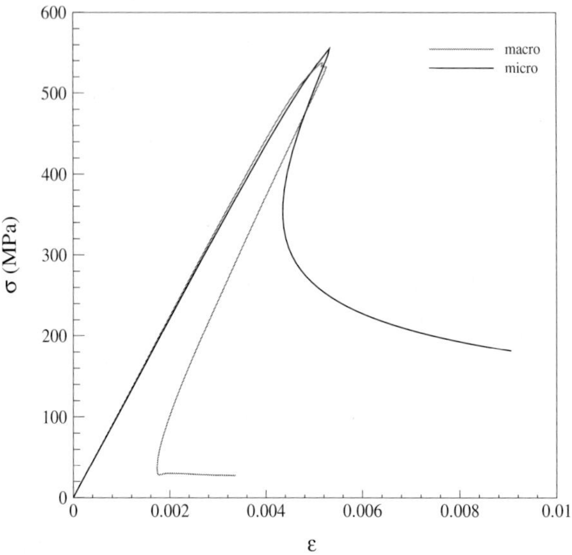

Fig. 12.6. Normalized stress-strain response for $AR1$ lamina.

the micromechanical mesh (less than 10%). Mesh B has the lowest possible resolution for this lamina and hence behaves similar to mesh A. For mesh C, the material imperfection is provided by normalizing the shear strain distribution. Thus an imperfection magnitude of 2^0 is effectively lower than an uniform 2^0 imperfection which appears for mesh B. This is apparent from the peak load predictions of mesh B and mesh C. When the imperfection magnitude of mesh C is increased (effectively bringing the imperfection magnitude closer to the imperfection magnitude of the micromechanical mesh) the peak load matches closer with the micromechanical prediction as shown in Fig. 12.7(b). Plots showing the deformation shapes at various load levels are shown in Fig. 12.8. From these plots it can be observed that a band of elements (four element length wide) at mthe centre of the lamina undergo severe deformation during the global compression. Using this band as the kink band, the ratio of the kinked and unkinked volumes for this homogenized (or smeared) system is 0.07 compared to the corresponding ratio of 0.08 obtained from the micromechanical analysis. Contour plots (in gray-scale) showing the distribution of local fiber rotation magnitudes are presented in Fig. 12.9. Regions of higher fiber rotation are shaded

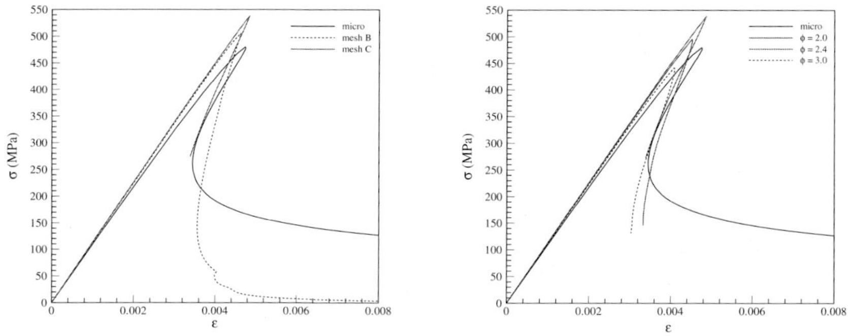

(a) Different Discretizations (b) Different Imperfections

Fig. 12.7. Normalized responses for $AR3$ lamina.

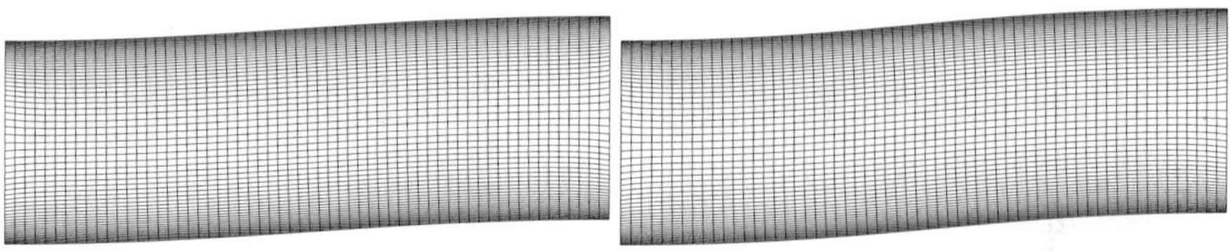

(a) 411 MPa (b) 495 MPa

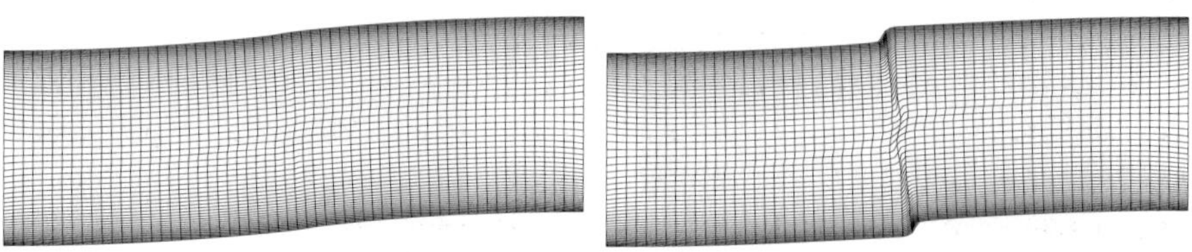

(c) 420 MPa (d) 146 MPa

Fig. 12.8. Plots of deformed shape for mesh C at different load levels.

in lighter shades of gray. In the pre-peak region, the complete geometry deforms. As the loading progresses, the fibers rotate locally based on the initial imperfection present in the system. This allows some areas to rotate faster than the rest of the lamina. Locations where fiber rotation is lower store significant amount of elastic energy. Beyond the peak load, this energy is released and redistributed in the lamina. This results in further rotation of the higher rotation locations leading to a deformation localization which is manifested as a kinked band of fibers. The *in situ* fiber rotation ϕ is plotted for two different locations on mesh C in Fig. 12.10. This plot reveals that beyond the peak-lead, some portion of the lamina continues straining monotonically whereas some portions release the elastic straining they have undergone. This shows that the elastic strain redistribution which is connected to the stored elastic energy causes deformation localization ultimately leading to kink banding. The mechanism of kink banding observed from this macro-level smeared approach is similar to the mechanisms observed in the micromechanical analyses, which validates the analysis approach used here. Thus, it appears that the smeared system, designed to capture the salient features associated with the kink band formation, performs as intended.

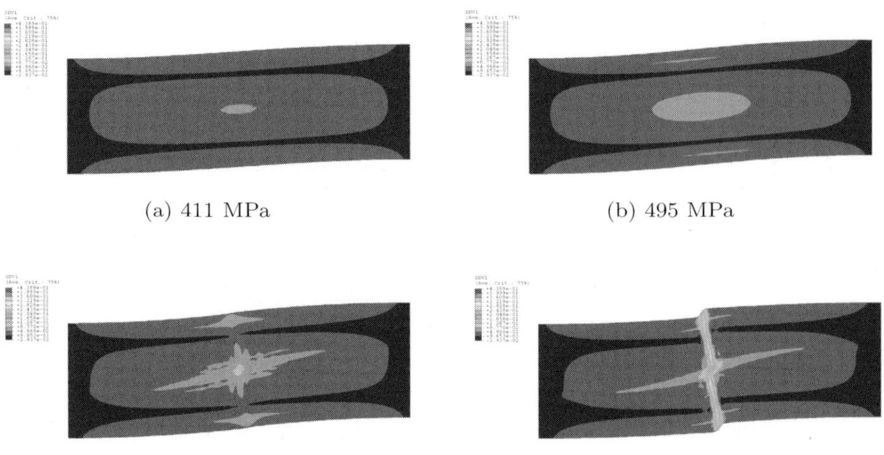

(a) 411 MPa (b) 495 MPa

(c) 420 MPa (d) 146 MPa

Fig. 12.9. Contours plots showing the in-situ fiber angle ϕ at various load levels for mesh C.

In this section, a mechanism based lamina level modeling approach is developed to study fiber kinking and validated against rigorous micromechanical analyses of unidirectional laminae. This modeling approach uses

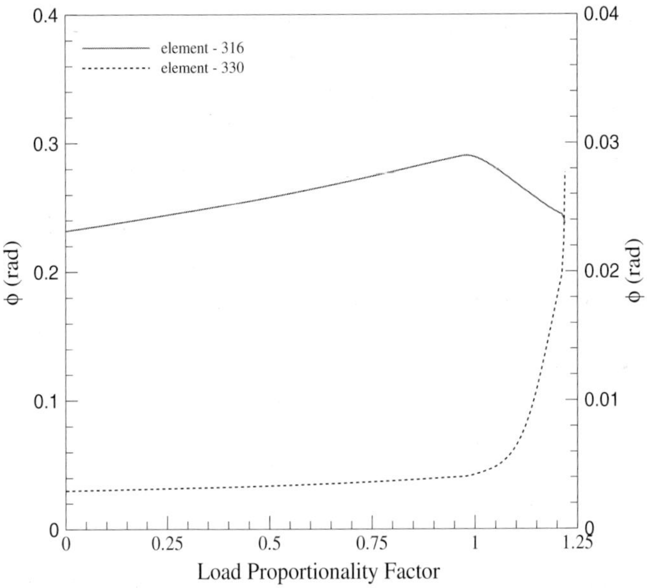

Fig. 12.10. Plot of in-situ fiber rotations from two different locations from the mesh C analysis. Element 316 result is plotted on the right hand side vertical axis. Element 330 result is plotted on the left hand side vertical axis.

the complete non-linear stress-strain relations for the lamina in shear and in transverse tension/compression as input, along with readily available lamina level elastic properties. Using only these (a minimum number) as inputs, compressive failure due to fiber kinking is modeled for unidirectional laminae of varying geometry and varying levels of finite element discretization. The present modeling approach is able to capture the pertinent features of the micromechanical analysis namely, the peak load, the deformation localization and kink band width. Quantitatively the predictions of the present approach matches with the micromechanical analyses very well. The present approach has shown a drastic reduction in the computational costs in capturing a highly nonlinear and complex deformation localization phenomenon. In Section 12.4, the modeling approach presented here is used in conjunction with classical lamination theory to study progressive failure behavior of notched laminated panels.

12.3 Description of Experimental Studies on Composite Laminated Panels

Three structural panel configurations, for which a set of laboratory data is available, are studied using the present PFA approach. These structural panels are identified according to their configurations; FSP for flat shear panel, SNP for single notched panel and DNP for double notched panel. The DNPs are thick (70-ply) unstiffened stitched double notched carbon fiber reinforced laminated panel and are subjected to axial compression loading. The SNPs are also thick unstiffened stitched carbon fiber reinforced laminated panels with a centrally located notch. These are also subjected to global axial compression loading. The third configuration FSP, is a relatively thin (24-ply) flat unstiffened laminated shear panel with a centrally located circular cutout and was loaded deep into the postbuckling regime. In the present section, experimental details for the DNPs will be provided. Results for the SNP and FSP configurations are not presented here for brevity and are discussed elsewhere.[21,40] Double notched panels, with notch tip radii equal to the laminate thickness are cut from a master laminate (Fig. 12.11). The notches act as stress concentrators and create a high stress zone, making it easier to observe damage initiation and progression. The use of a stitched laminate eliminates delamination among adjacent plies of a multidirectional laminate making it possible to isolate progressive failure occurring within lamina without the additional complication of ensuing delamination buckling. However, stitching can induce ply-level mesobuckling depending on the stitch tension and stitch spacing.[41]

12.3.1 *Experimental Details of Stitched Double Notched Panels (DNPs)*

Multidirectional composite laminated specimens used in the DNP study are cut from a master laminate which was obtained from NASA Langley Research Center and is similar to the panels used by McGowan *et al*[42] for SNP studies. The master laminate is fabricated from Hercules Inc. AS4/3501-6 graphite-epoxy materials. It is stitched through the thickness and parallel to the principal fiber direction using E.I. DuPont de Nemours Inc. *Kevlar*$^{(R)}$ thread. The master laminate is made of ten prefabricated 'stacks' of dry preknitted lamina each of which is 1.4 mm thick and has seven layers of unidirectional plies with a stacking sequence of $[\pm 45/0/90/0/\mp 45]$. The 0^0 layers are parallel to the stitching direction. The resulting laminate

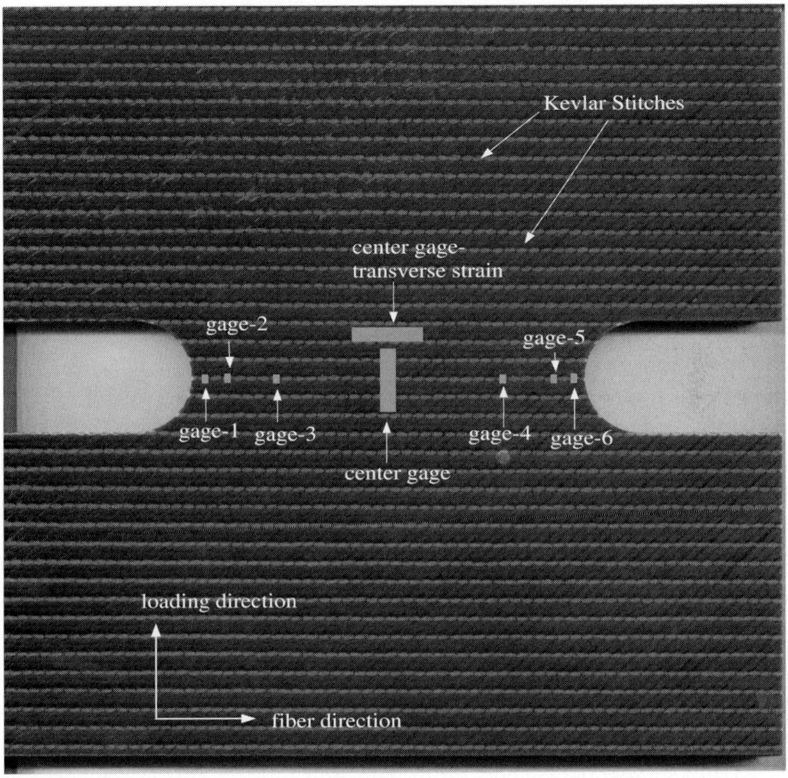

Fig. 12.11. Top surface of a DNP specimen. Strain gage locations are shown with white boxes.

is then infused with resin by using resin film infusion technique. The cured thicknesses of individual ±45, 0, and 90-deg layers are given in Table 12.3.

Schematics of an experimental DNP specimen is shown in Fig. 12.12. Each specimen has overall dimensions of 210 mm x 210 mm x 15 mm. Two 52.5 mm deep notches with a notch tip radius of 15 mm are cut on opposite side of each specimen. These notches create a localized high stress region which induces failure initiation and provides a predefined location for probing and obtaining diagnostics of damage initiation. Multiple DNPs have been studied. For convenience, the specimen face with thicker visible stitches is marked as the 'top' surface and the reverse surface is labeled as the 'bottom' surface. Figure 12.11 shows the top surface of a typical specimen. A water-jet cutting procedure is used to cut the specimens from

Table 12.3. Properties of AS4/3501-6 plies.[42]

	AS4/3501-6		
	$\pm 45^0$	0^0	90^0
thickness (mm)	0.15	0.16	0.17
E_{11} (GPa)	111.0	113.0	110.0
E_{22} (GPa)	11.0	11.0	11.0
G_{12} (GPa)	5.5	5.5	5.5
G_{13} (GPa)	5.5	5.5	5.5
G_{23} (GPa)	2.8	2.8	2.8
ν_{12}	0.34	0.34	0.34

the master laminate. All four sides of each specimen are ground with a fine grit grinding wheel after cutting to ensure parallelism between opposite sides. The unnotched sides of the specimens are used as loading edges. Each loading edge is slotted in end supports and puttied using Devcon steel putty to end caps that are made from hardened steel. This way, a 'clamped' boundary condition is obtained at each loading edge. Notched sides are slotted into anti-buckling guides. A schematic of the boundary conditions is shown in Fig. 12.12. Experiments are carried out in an MTS machine with a crosshead displacement rate of 0.01 mm/s and an axial load capacity of 500 KN (110,000 lbf). Back to back strain gages are used at the center of each test specimen to check for specimen bending. Six quarter bridge strain gages are placed along the line joining the notch tips on the 'top' surface of each specimen to monitor strains during loading. These locations are shown in Fig. 12.11. A 500 KN capacity load-cell placed in the MTS actuator is used to measure the compressive load throughout the loading history.

12.4 Progressive Failure Analysis of Multidirectional Composite Laminated Panels

The main objective of the present section is to validate the new PFA approach for composite laminates that was introduced in Section 12.2 by comparing the numerical predictions against available experimental data (Section 12.3). In this section, the numerical modeling details such as the finite element discretization, material input data etc., for the PFA are described. The PFA predictions are compared against available experimental data at a global (load-vs.-load point displacement) and local level (damage contours, strain gage data). Various aspects of the PFA predictions are

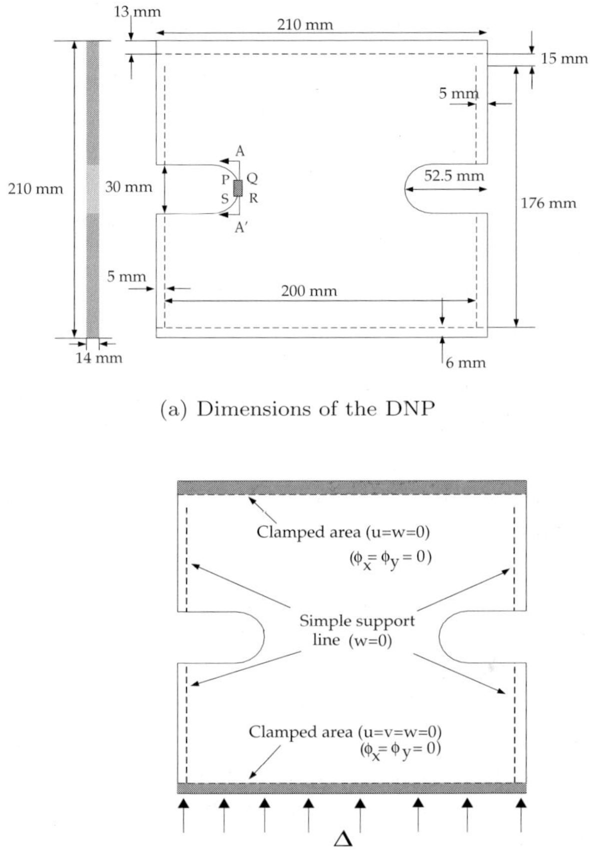

(a) Dimensions of the DNP

(b) Boundary conditions of the DNP

Fig. 12.12. Schematics of the specimens and test conditions used in the present study. Section AA' is used to cut Test-3 specimen for damage diagnostics.

reported and discussed. In this section, one of the three validation cases will be described in detail. Other two set of results are described elsewhere.[21,40]

12.4.1 *Numerical Simulations - Modeling Details*

The predictive capabilities of the present PFA methodology is assessed by simulating the experimental results of the DNP panels described in the previous section. Test sections of the DNP experimental geometry were discretized using the shear deformable three noded (S3) and four noded

(S4/S4R) shell elements available in *ABAQUS*. Figure 12.13 shows the DNP mesh. A summary of the nodal and elemental data are presented in Table 12.4. Static analyses were performed using displacement control loading. An arc length solution method[43] is used through the RIKS option available in *ABAQUS*. Geometric nonlinearity is included in the response analysis through the NLGEOM option. As the experimental data showed that negligible amount of global out-of-plane bending was present, a 2-D analysis was performed for the DNP. Only one stack of $[\pm 45/0/90/0/\mp 45]$ (seven layers) for the DNP was modeled. The thickness of each individual ply was chosen as the total thickness of all similar orientations such that the total model thickness equals the total panel thickness. The present approach does not consider any out of plane damage mode which allows lumping layers with similar orientations in this manner. Elastic material

Table 12.4. DNP mesh data.

		DNP
Elements		11713
	S3/S4	11713
Nodes		11951
Degrees of Freedom		71706

properties for the AS4/3501-6[39,42] material system are described in Table 12.3. Nonlinear shear stress-strain curve for this material system is shown in Fig. 12.14 (darker curve, 'softer' among the two shown). A material point is denoted as 'damaged' when it reaches the end of the input stress-strain curve. For the present analysis, this refers to a 55% degradation in the *in situ* shear modulus, G_{12}. Complete material constitutive behavior is modeled via the user material subroutine option of ABAQUS. Section lay-up of the laminates are defined using the *SHELL SECTION, COMPOSITE option available for shell elements. Thickness effects are incorporated by using three integration points through the thickness for each layer. Boundary conditions for the panels are applied at nodal positions. For the DNPs, nodes corresponding to the top grip location are constrained from having any motion. Nodes occupying the bottom grip location are constrained to move only in the 1-2 plane. Rows of nodes corresponding to the knife edge support locations are constrained from moving out-of-plane. Loading applied by the MTS cross-head is simulated by specifying displacement in the 2-direction for nodes at the bottom grip location. A column of nodes near

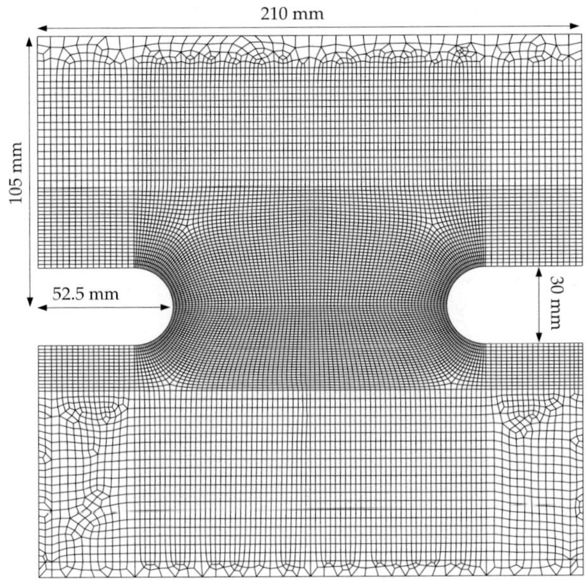

Fig. 12.13. Finite element meshes used for DNP study.

the left and right edges of the panel is put under simple-support condition to mimic the knife edge supports used in the experiments.

Depending on the process of manufacture, a composite laminate contains a material imperfection signature (MIS). For pre-preg based laminates, the MIS is characterized by how much the fibers have deviated from their intended alignment. Fiber misalignment angles of 1^0-3^0 are not uncommon. Motivated by these factors, geometric and material perturbations are both used in the DNP analyses. In-plane geometric perturbation is provided using a negative shear loading on the bottom grip such that the bottom grip is offset by a predetermined amount with respect to the vertical (Y-direction). An offset angle of 2^0 was used in the analysis. Material axis system of the axial layers (90^0 layers) are offset from the Y-direction by 2^0 in addition to the in-plane geometric perturbation. A group of elements (4 element rows wide) situated between the notch tips were also given an additional 2^0 angle for the initial value of ϕ. The geometric and the material axis perturbations are defined such that they act in harmony. It is to be noted that material imperfections (misalignment of fibers within a layer) are usually on the order of 1^0-2^0 for pre-preg based laminates.

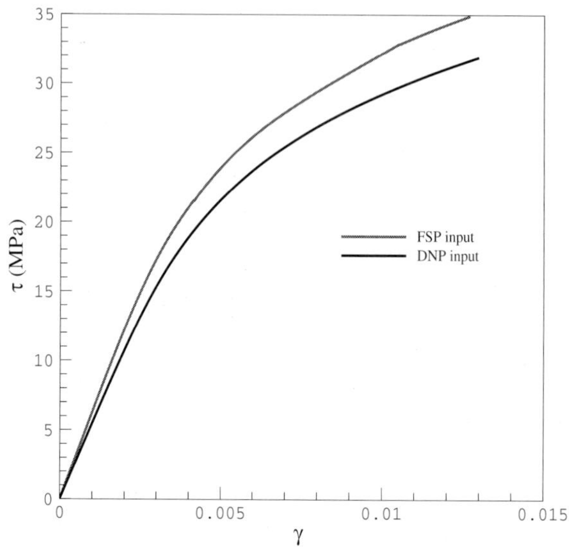

Fig. 12.14. Material inelastic input curves in shear, used in the present study.

12.4.2 *Results for the Stitched Panels – DNPs*

Results from the DNP tests and the FE simulation are presented and salient features of compressive damage diagnostics obtained from the specimens are discussed in the present section. Experimental results are discussed first followed by the comparison with the FE simulation results. Strain readings from the back to back gages at the center of the specimen showed insignificant amount of bending being present. This was true for all three specimens examined. However, it is likely that the panels had initial geometric imperfection shapes unlike that of the lowest eigen mode shapes assumed here in the analysis to follow. This is evident in the back to back strain gage readings from the right and left notch tip areas that show differences between them suggesting that the panel deformation also shows a 'skewed x-wise' bending, unlike the first elastic eigen mode shape. A linear elastic FE study predicted that the notch tip compressive strain in the loading direction will exceed 0.01 (10000 μ-strain) at a global reaction force of 220 KN (Table 12.5). This value of axial strain is typical for fiber kinking for the present class of materials.[8,38] Figure 12.15 presents the membrane strains (obtained from the back to back center gages shown in

Table 12.5. Linear elastic FE predictions for DNPs.

Event	Global Load (KN)
Notch tip axial strain, −0.010	220
Notch tip axial strain, −0.015	330
Global buckling	1300

Negative strains indicate compression.

Fig. 12.11) plotted against the reaction forces. Comparison with the finite element analysis prediction shows that the overall panel response remains largely linear during loading. Nonlinearity in the global response sets in beyond 200 KN which is near the predicted load level for fiber kinking near the notch tip. The resultant P-Δ data from the DNP tests are plotted in Fig. 12.16. The responses show a bilinear behavior with a distinct knee around a load of 220 KN. A slight 'kink' in the P-Δ responses near 340 KN is also observed. There was no audible or visible indication of damage prior to this load level and the DNP specimens failed catastrophically at an external load of 370 KN. The specimens separated in two halves during their removal from the end supports. Comparison of the catastrophic failure loads of the DNP specimens (Table 12.6) with the predicted linear buckling load (Table 12.5) shows that specimen failure did not occur due to global buckling. Instead, progressive damage accumulation in all the layers in the vicinity of each notch tip leads to a zone of 'degraded' material. These zones grow towards each other just prior to catastrophic failure and the panel undergoes an out-of-plane shear failure separating into two halves as shown in the post-experiment image of Fig. 12.17.

Table 12.6. Experimental observation of the DNP specimens.

	Ultimate Failure Load (UFL) (KN)	Axial strain at UFL (μ-strain)
Test-1	372	−6110
Test-2	375	−5943

Negative strains indicate compression.

Figure 12.17 shows the top surface of the first DNP specimen (Test-1) after it was removed from the fixture. As it can be seen, damage which initiated near the notches grew suddenly across the specimen above the center stitch. On the bottom surface, a damage zone propagated across the specimen below the center stitch (Fig. 12.17(b)). Presence of the center stitch seems to prevent damage initiation exactly at the notch tip. In view

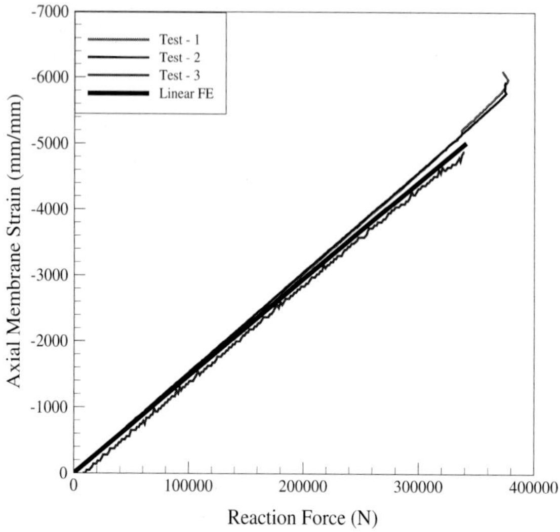

Fig. 12.15. Membrane strain plotted against reaction force from three DNP specimens and the FE analysis. Notice the repeatability of the test results and agreement with the FE prediction.

of these observations from the panels studied, a third DNP specimen was tested such that damage would initiate in the vicinity of the notch tips but would not progress towards catastrophic failure. This specimen was loaded up to 340 KN and subsequently unloaded. Strain gage readings from the gages near the notch tips showed nonlinearity at this load level indicating the presence of internal damage. The unloaded specimen was taken off the grips and cut along the center line of the left notch tip gage (section AA', Fig. 12.12). Post-experiment damage diagnostics performed using optical microscopy and Scanning Electron Microscopy (SEM) showed that fiber kinking in the vicinity of the notch tip is the main source of damage.

Load-displacement comparison

Load-vs.-load point displacement data for the DNP test specimens are compared with the FE result in Fig. 12.16. The experimental data show regions of 'settling' in the initial stages. The FE simulation results are plotted after accounting for the settling of laboratory specimens. The overall behavior of the FE simulation is similar to the experimental observations. Initial

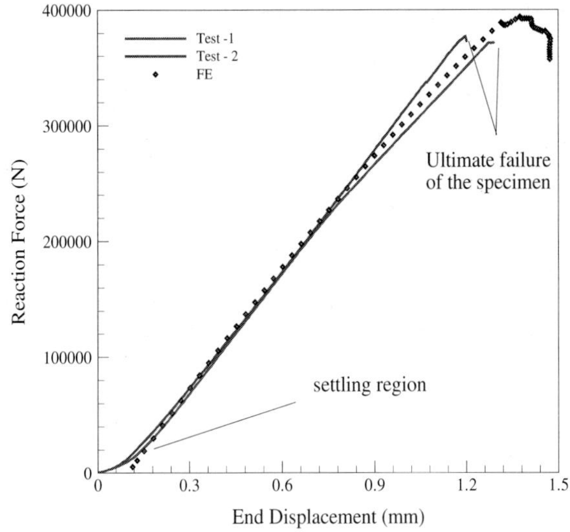

Fig. 12.16. Plot of P-Δ data comparison of the DNP experiments and the numerical simulation.

slopes of the curves presented match quite well. Beyond 220 KN, the FE result remains bounded by the two experimental responses. The FE analysis predicts a peak load of 394 KN which is within 6% of the experimentally observed maximum load.

Another mesh with 1.6 times the degrees of freedom of the initial mesh (Fig. 12.13) has been used to examine the mesh dependency of the analysis results. It is found that the slope of the P-Δ response from the finer mesh matches with the corresponding slope of the coarser mesh. The finer mesh predicts a peak load of 390 KN which is within 1% of the coarser mesh prediction. These observations, and maps of damage contours that indicate in-situ shear modulus (G_{12}) degradation to be discussed subsequently, show that the results presented here are indeed mesh independent. Consequently, results from the coarser mesh are used for further discussion.

It should be noted that prior to the peak at 394 KN, the response passes through a smaller crest at 388 KN which is followed by a softening regime (Fig. 12.18). Subsequent loading of the laminate leads to an increase in load, at a reduced slope (compared to the initial slope) until the global peak is reached at 394 KN. Though the predicted behavior in Fig. 12.18 is similar

(a) Left Notch (b) Right Notch

Fig. 12.17. (a,b) A DNP specimen is shown after being loaded beyond catastrophic failure.

to the first test result, it appears to be absent in the second test. Fiber kinking is a dynamic event resulting in a catastrophic loss of strength and stiffness of the system under loading. Therefore dynamic events that occur at very short time scales may not be appropriately captured in the global P-Δ response. Beyond this maximum load, the global reaction force drops in a stepwise fashion, indicating discrete damage progression. Ultimately at a load of 375 KN, the laminate is unable to sustain any more load and a vertical drop is observed in the P-Δ response, indicative of a catastrophic global failure event.

Strain-gage measurements

The strain readings recorded from the DNP panels are shown in Fig. 12.19. The locations of the gages are indicated in Fig. 12.11 and tabulated in Table 12.7. All strains initially are linear with the applied load, but a careful examination shows a softening behavior as loading proceeds. This nonlinearity is progressive and is indicative of damage accumulation in the panel. As expected, the gages closest to the notch tips record the highest values and the intensity decreases with distance away from the notch tips. The back to back notch tip gages show bending and also differ between the right notch and left notch (Fig. 12.19(a),(c) and Fig. 12.19(b),(d)), suggesting a width wise skewed (in the x-direction) bending during the deformation. An acceleration of the notch tip strain nonlinearity is observed

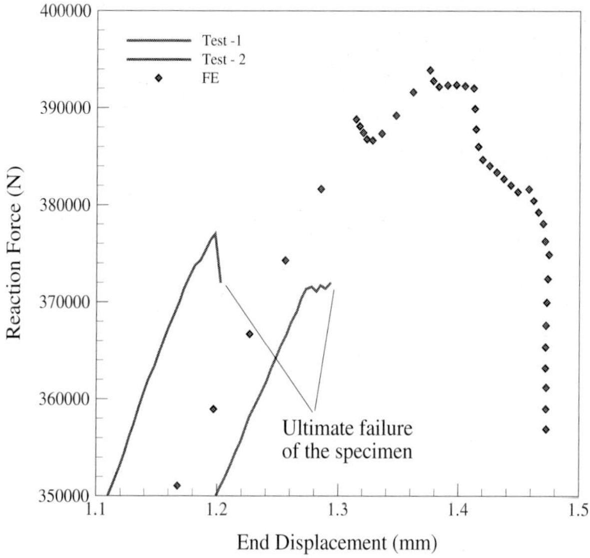

Fig. 12.18. Close up near the peak of the P-Δ data shown in Figure 12.16.

at 300 KN (gage 6 in Fig. 12.19(d)) and the gage is rendered malfunctional at about 310 KN. The FE results, based on a panel without any out-of-plane geometric imperfection, does not show any differences in the back to back notch tip readings. The FE data is also 'stiffer' as expected, indicating that the panel properties are most likely overestimated. However, the trends in all the strains captured by the FE simulation are correct with respect to softening nonlinearity and with respect to the acceleration of the nonlinearity as loading proceeds.

Indeed, all the FE strain readings become near vertical at approximately 388 KN, indicative of the softening induced localized instability experienced by the panel. An examination of the strain readings appears to indicate that the strain predictions of the FE simulations are slightly shifted. That is, if all the panel strain readings are correlated with the FE strains at locations that are further away from the notch, then the readings show a remarkable agreement. This suggests that a 'scaling' of the material properties and proper accounting for the initial geometric imperfections should bring agreement much closer. It must be noted that the FE simulations are based on classical lamination theory ideas (strictly 2D) which neglect

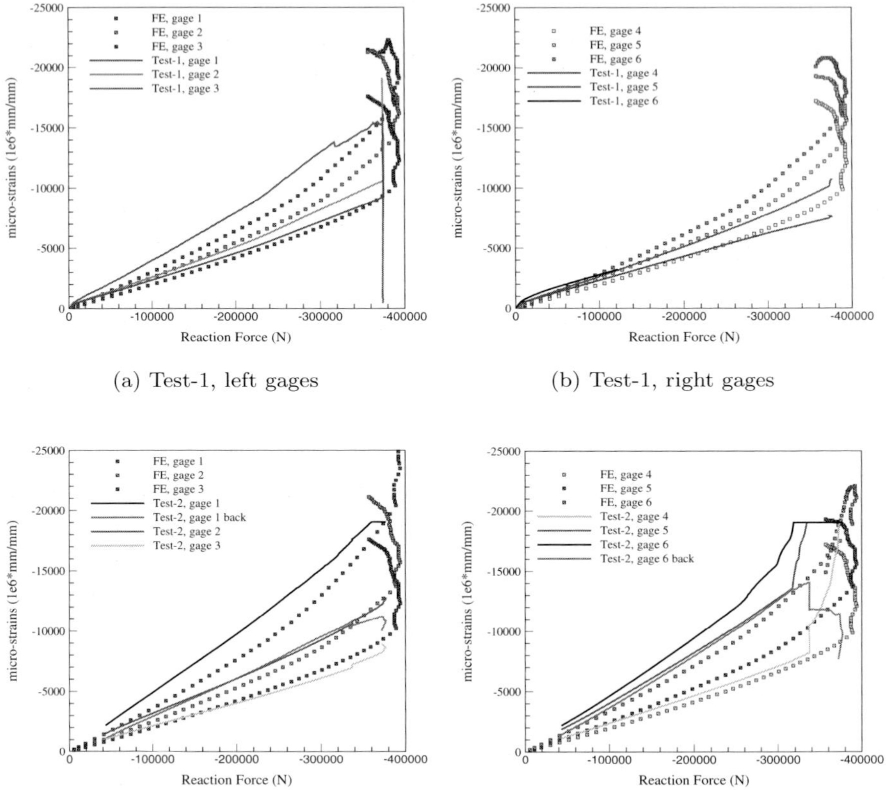

(a) Test-1, left gages

(b) Test-1, right gages

(c) Test-2, left gages

(d) Test-2, right gages

Fig. 12.19. Comparison of measured strain data from experiments with the FE results. strain gage locations are given in Table 12.7.

Table 12.7. Strain measurement stations on the DNP panel.

Distance from the left notch tip	Test-1	Test-2
Gage 1	5 mm	2 mm
Gage 1 back	-	3 mm
Gage 2	12 mm	10 mm
Gage 3	25 mm	25 mm
Distance from the right notch tip		
Gage 4	24 mm	25 mm
Gage 5	12 mm	11 mm
Gage 6	6 mm	3 mm
Gage 6 back	-	3 mm

possible 3D effects that can also contribute to the discrepancies. A study that examines the effects of material property uncertainty was carried out in a limited sense for a different configuration.[21]

Damage contours at various load levels

Contours of the damage variable S_r are presented at increasing load levels in Fig. 12.20 and Fig. 12.21. These contours represent the effect of matrix cracking as captured through S_r. Results are presented for the top four layers, $\pm 45^0$, 0^0 and 90^0 layers, where the layer orientation is given with respect to the global x-direction. The dark areas in the region between the two notches correspond to shear stiffness loss of 55% or more. Progressively lighter shades denotes lesser and lesser amount of damage accumulation. Any darker shade away from the notch tips indicate minimal or zero damage. In the following discussion, damage state or level of degradation is used interchangeably with the loss of shear stiffness. It is readily observable that different layers are at different levels of degradation at a given global reaction force. It is also noticeable that the 90^0 layer which is under global axial compression, shows the least amount of damage whereas the $\pm 45^0$ layers show the most damage. The 90^0 layer damages at a rate lower than the rate of damage in the other layers. This can be observed in Fig. 12.22. This figure also shows that when all the other layers reach S_r^*, the rate of damage accumulation increases in the 90^0 layer, indicating a load redistribution among the damaged and undamaged layers. The $\pm 45^0$ layers make up nearly 60% of the laminate, thus the global stiffness of the system is greatly affected by the degradation of these layers. In a through-the-thickness view at the notch tips, at a load of 388 KN, only 14% of the layers have substantial *in situ* shear stiffness remaining and the rest have lost 55% or more of their shear stiffnesses. Thus the laminate weakens in both in-plane and out-of-plane directions. Damage zones grow a little offset from the notch tips in the 90^0 layer and converges towards the notch tips with increasing load. In the $\pm 45^0$ layers, damage nucleates from the notch tips in a small region. Then the two separate damage zones propagate towards each other along the center line of the specimen, finally establishing a region of uniformly damaged material between the two notches. Evidence of this phenomenon is readily visible in the damage contours of the 0^0 layer. At the global peak load of 394 KN, most of the layers have lost 50% or more in *in situ* shear stiffness between the notches. The remaining layers (90^0 layers) have also established a zone extending from the left

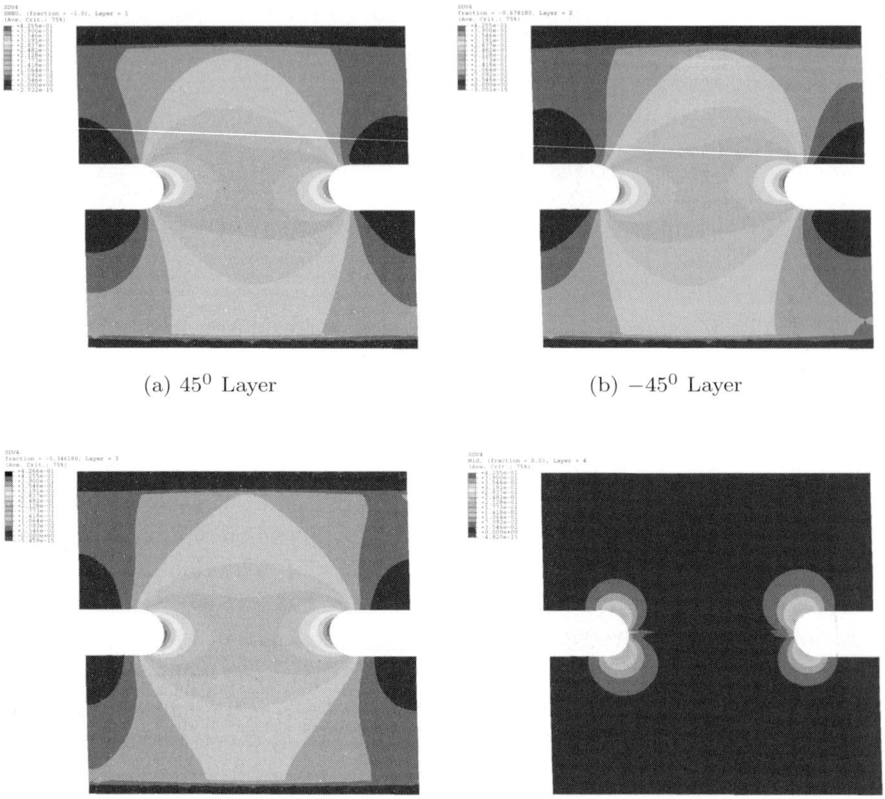

(a) 45^0 Layer

(b) -45^0 Layer

(c) 0^0 Layer

(d) 90^0 Layer

Fig. 12.20. Contours of S_r (P = 217 KN) are shown for the top four layers. The 0^0 layer is parallel to the global x-direction. Dark areas in these contours correspond to 55% or higher degradation in in-situ shear stiffness.

notch to the right notch, which has lost nearly 20% of its *in situ* shear stiffness. The only load path for this panel passes through the region between the notches. Thus when the panel looses a substantial portion of its *in situ* shear stiffness in that region, the load path is essentially severed. This leads to a staggered drop in the global load with increasing end compression as the material in the 90^0 layer looses more and more of its shear stiffness. Ultimately at 375 KN, the load path becomes unsustainable which triggers a near vertical load drop. The *in situ* G_{12} are also plotted for the outermost 45^0 layer[22] at the peak load. It showed that the damage contours

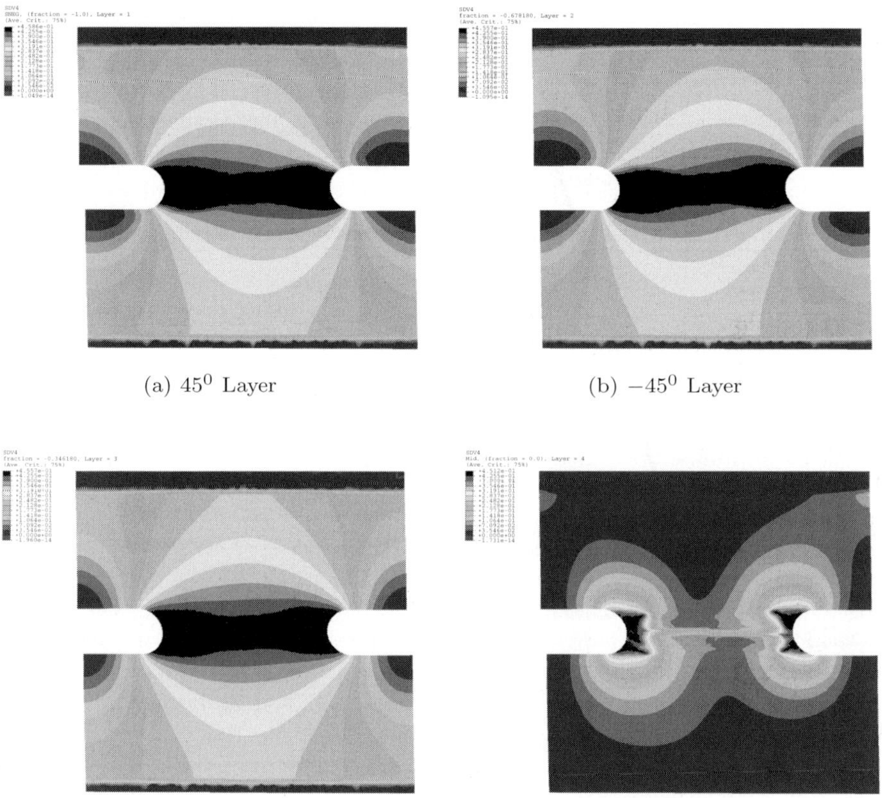

(a) 45^0 Layer

(b) -45^0 Layer

(c) 0^0 Layer

(d) 90^0 Layer

Fig. 12.21. Contours of S_r at P = 357 KN. (the final point of the FE P-Δ response curve).

(dark coloured area ahead of notch tips denotes 55% or more degradation) occupy similar sized areas in both the meshes. This comparison of a *local* measure combined with the similarity in global measures (slope of the load-displacement curve and peak load) emphasizes the mesh independent nature of the present analysis.

This breakdown of the load path and damage growth in the other layers (which may have lost nearly its entire load carrying capacity) coupled with the strain gage readings becoming near vertical (see Fig. 12.19), signals the onset of catastrophic failure.

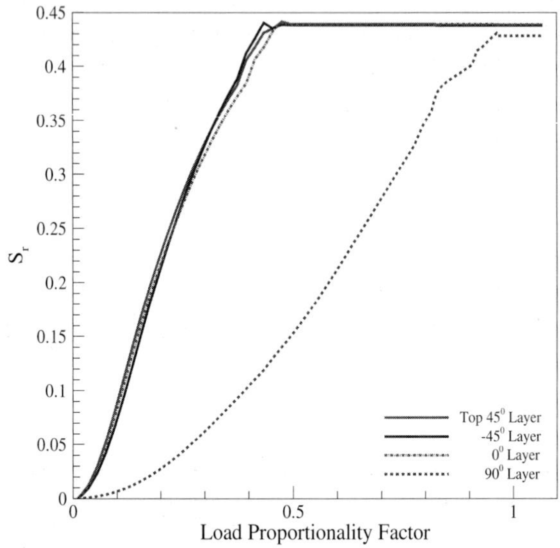

Fig. 12.22. Plot of S_r evolution with loading at the left notch tip. Different rates of damage accumulation are observed for different layers.

12.5 Concluding Remarks

A failure mechanism based progressive failure analysis methodology has been developed and validated against experimental data for three types of laminated structural panels. The methodology uses the complete non-linear stress-strain relations for a lamina in shear and in transverse tension/compression in the principal material orthotropic directions as input, along with readily available lamina level elastic properties. Using only these (a minimum number) as inputs, progressive failure and damage growth in different types of laminated structural panels, loaded under a pre-dominantly compression loading have been simulated. In particular by modeling the mechanism of kink banding formation that is responsible for limiting the lamina axial compression strength, the maximum load sustained by the panels and the subsequent damage evolution have been captured fairly accurately. The notion of a fixed compression strength has been disbanded, instead the *in situ* compression strength of a lamina is determined as a part of the solution process, and as a function of loading history. That is, kink band formation is a natural outcome of the solution process. A macroscopic (lamina level) description of kink banding is made possible by including fiber rotation at a material point as an unknown in the problem formulation. The success of the present PFA methodology is encouraging and points the way for extending the methodology to analyze other structural configurations.

12.6 References

1. Z. Hashin, *J Appl Mech - T ASME.* **47**, 329–334 (June, 1980).
2. F.-K. Chang and L. B. Lessard, *J Compos Mater.* **25**, 2–43 (January, 1991).
3. S. T. Pinho, C. G. Dávila, P. P. Camanho, L. Iannucci, and P. Robinson. NASA/TM-2005-213530 (February, 2005).
4. C. G. Dávila, N. Jaunky, and S. Goswami, 44th *SDM Conference, AIAA, Norfolk* (April. 2003).
5. R. A. Schapery, *J Mech Phys Solids.* **38**(2), 215–253 (June, 1990).
6. R. A. Schapery and D. L. Sicking, 7th *International Conference on Mechanical Behavior of Materials.* **47**, 45–76 (June, 1995).
7. T. J. Vogler and S. Kyriakides, *Int J Solids Struct.* **36**, 557–574, (1999).
8. S. H. Lee and A. M. Waas, *Int J Fracture.* **100**(3), 275–306 (December, 1999).
9. K. Oguni, C. Y. Tan, and G. Ravichandran, *J Compos Mater.* **34**(24), 2081–2097, (2000).
10. S. Basu, A. M. Waas, and D. R. Ambur, *J Mech Phys Solids.* **54**(3), 611–634 (March, 2006).
11. A. M. Waas, C. D. Babcock Jr., and W. G. Knauss, *Int J Solids Struct.* **26** (9-10), 1071–1098, (1990).
12. A. R. Khamseh and A. M. Waas, *J Eng Mater -T ASME.* **114**(3), 304–310 (July, 1992).
13. C. Soutis, P. T. Curtis, and N. A. Fleck, *P Roy Soc Lond A Mat.* **440**(1909), 242–256 (February, 1993).
14. C. T. Sun and A. W. Jun, *Compos Sci Technol.* **52**(4), 577–587, (1994).
15. I. Chung and Y. Weitsman, *Int J Solids Struct.* **32**(16), 2329–2344 (August, 1995).
16. H. M. Jensen and J. Christoffersen, *J Mech Phys Solids.* **45**(7), 1121–1136 (July, 1997).
17. T. J. Vogler and S. Kyriakides, *Acta Mater.* **45**(6), 2443–2454, (1997).
18. I. M. Daniel and H. M. Hsiao, *Int J Fracture.* **95**(1–4), 143–158, (1999).
19. Z. P. Bazant, J. J. H. Kim, and I. M. Daniel, *Int J Fracture.* **95**(1–4), 103–141, (1999).
20. C. S. Yerramalli and A. M. Waas, *Computer Modeling in Engineering and Sciences.* **6**(1), 1–16, (2004).
21. S. Basu. PhD thesis, Aerospace Engineering Department, University of Michigan, Ann Arbor, (2005).
22. S. Basu, A. M. Waas, and D. R. Ambur, *Int J Solids Struct.* **44**(9), 2648–2676 (May, 2007).

23. H. Suemasu, H. Takahashi, and T. Ishikawa, *Compos Sci Technol.* **66**(5), 634–641 (May, 2006).
24. T. Ishikawa, *Adv Compos Mater.* **15**(1), 3–37, (2006).
25. J. Merodio and T. J. Pence, *J Elasticity.* **62**, 145–170, (2001).
26. Y. B. Fu and Y. T. Zhang, *Int J Solids Struct.* **43**(11–12), 3306–3323 (June, 2005).
27. R. A. Schapery, *Mech Compos Mater St.* **AMD-100**, 1–9 (June, 1989).
28. C. T. Herakovich. (John Wiley & Sons, Inc., 1998).
29. C. T. Sun and J. L. Chen, *J Compos Mater.* **23**, 1009–1020 (October, 1989).
30. C. G. Dávila, D. R. Ambur, and D. M. McGowan, *J Aircraft.* **37**(5), 898–905 (September–October, 2000).
31. R. A. Schapery, *Int J Solids Struct.* **32**(6/7), 739–765, (1995).
32. G. A. O. Davies, P. Robinson, and J. Robson, *Compos Part A-Appl S.* **28** (8), 757–765, (1997).
33. G. A. O. Davies, D. Hitchings, and J. Ankersen, *Compos Sci Technol.* **66**(6), 2006 (May, 2006).
34. V. K. Goyal, N. Jaunky, E. R. Johnson, and D. Ambur, 43rd *SDM Conference, AIAA, Denver.* (AIAA 2002-1745), 1–12, (2002).
35. R. A. Schapery. Private communication, (2002).
36. C. S. Yerramalli and A. M. Waas, *Int J Solids Struct.* **40**(5), 1139–1164 (March, 2003).
37. A. v6.3. Hibbit, Karlson and Sorenson, Pawtucket, RI, (2003).
38. S. Kyriakides, R. Arseculeratne, E. J. Perry, and K. M. Liechti, *Int J Solids Struct.* **32**(6/7), 689–738, (1995).
39. P. Soden, M. J. Hinton, and A. S. Kaddour, *Compos Sci Technol.* **58**, 1011–1022, (1998).
40. S. Basu, A. M. Waas, and D. R. Ambur, 47th *SDM Conference, AIAA, Newport, RI* (May. 2006).
41. S. Drapier and M. R. Wisnom, *Compos Sci Technol.* **59**(8), 1287–1297, (1999).
42. D. M. McGowan, C. G. Dávila, and D. R. Ambur, 42nd *SDM Conference, AIAA, Seattle* (April. 2001).
43. E. Riks, *J Appl Mech-T ASME.* **39**(4), 1060–1065 (December, 1972).

CHAPTER 13

MICRO- AND MESO-INSTABILITIES IN STRUCTURED MATERIALS AND SANDWICH STRUCTURES

Thomas Daxner, Dieter H. Pahr, and Franz G. Rammerstorfer

Institute of Lightweight Design and Structural Biomechanics,
Vienna University of Technology,
Gusshausstrasse 27-29, A-1040 Vienna, Austria,
daxner@ilsb.tuwien.ac.at

The reason for the onset of undesirable nonlinear mechanical behaviour of structures can often be found in instability phenomena on length scales that are much smaller than those of the component. Micro-buckling in cellular materials, e.g., may lead to layer-wise collapse of the cell structure, corresponding to a characteristic plateau in the overall compressive force-displacement relationship. Correspondingly, local buckling of the face sheets or the cores of sandwich plates may mark the onset of buckling of the whole sandwich plate. To account for the importance of understanding and predicting of buckling on small length scales this paper discusses buckling phenomena in open and closed-cell cellular materials as well as in sandwich structures with unstructured and structured cores. Analytical and numerical methods for predicting corresponding critical loads are presented. Most of these methods rely on the assumption of an actual, or representative, periodic micro-structure for facilitating the analysis. The influence of boundary conditions and some localisation phenomena are also discussed.

13.1 Introduction

Considerations regarding structural stability are typically related to the question of critical loads for whole constructions or for structural components. However, structural stability is becoming more and more also a question of materials design. This is why, for instance, light-weight materials frequently exhibit special micro-structures, such as the cellular microstructure of closed or open-cell foams, as well as meso-structures, such as honeycomb cores appearing in sandwich structures with thin layers of homogeneous or composite materials.

In order to calculate the load carrying capacity of structures made of such materials, or of sandwich structures, strength and stability considerations must be taken into account not only on the structural, i.e., the macro level, but also on the meso and micro levels. For instance, stability loss occurring at length scales which are several orders of magnitude smaller than the size of the structural part can be the starting point of failure of the complete structure. Typical material instabilities, i.e., instabilities on the level of the materials' micro-structure, are the buckling of struts and cell walls in polymer or metallic foams. Such phenomena are treated in terms of micro-instabilities.

Typical kinds of meso-instabilities are localisations in the form of bands of buckled foam cells or the formation of folds in sandwich plates (or shells) leading to phenomena similar to material instabilities such as the formation of necks or shear bands by plastic localisation in polycrystalline materials.

Wrinkling of face layers of sandwich plates under macroscopic in-plane loading or bending is studied as a particular form of meso-instability of sandwich structures. Although wrinkling starts as a bifurcation process with a distinct periodic eigenfunction, the post-critical behaviour shows the formation of single localised folds being the result of crushing of the micro-structure of the core, frequently in combination with local plastic yielding of the face layers.

In sandwiches with honeycomb cores and very thin face layers a special kind of meso-instability, namely face layer dimpling, can be observed, too. It is shown by computational means that this kind of mesoscopic structural instability represents an interactive buckling of cell walls and face sheets. In corrugated paper local buckling of face layers or buckling of the paper core is often the starting point for global failure of a structure made of corrugated paper (e.g., large boxes). These phenomena are treated both analytically and numerically.

13.2 Instabilities in Micro-Structured Materials

13.2.1 *Micro-Structured Materials — Introduction*

In this section we consider instabilities in materials that feature a micro-structure which is prone to buckling. Micro-structure herein refers to a structure that becomes visible only at length scales that are several orders of magnitude smaller than the length scale of the component made of the considered materials. Buckling instabilities occur predominantly in structures that contain slender structural members or thin, shell-like components. With regard to micro-structured materials such elements can be found in cellular materials of low effective density.

Cellular materials can be of natural or of artificial, man-made origin. In both cases they offer significant advantages with respect to their high weight-specific stiffness and strength. In addition to their advantageous weight-specific mechanical properties they offer the potential of being used in a multi-functional manner, because they are either permeable for fluids in the case of open-cell foam, i.e., sponges, which makes them suitable for applications such as heat-exchangers or catalytic converters, or impermeable in the case of closed-cell foams, thereby providing insulation and buoyancy. Aspects like lightweight design, energy absorption and heat dissipation are discussed in.[1,2]

Cellular tissue can be found in most living beings, as some of the more well-known examples like wood, sponge, trabecular bone, and cork illustrate. Understanding their mechanics is not only an end in itself, but also important for designing biomaterials for bone-replacement or for use as porous scaffolds. A recent review of the subject of biomechanics of cellular solids can be found in.[3] The behaviour of wood cells is discussed, e.g., in.[4] More general reviews of the mechanics of cellular materials are given in[5] and.[6]

Cellular engineering materials comprise polymeric foams, metallic foams and ceramics. Besides the previously made distinction into open-cell and closed-cell topologies one can choose between materials with stochastic micro-structure and periodically fabricated or assembled 'materials' (in a wider sense). In the modelling and simulation of cellular materials we often find models with a periodic micro-structure as idealised representations of random micro-structures. The implications of this approach will be discussed in the following sections.

13.2.2 *Micro-Structured Materials — Methods*

Cellular materials have drawn a considerable amount of scientific interest both on the experimental level and in the field of micro-mechanical simulation, see, e.g.[7]

With respect to experimental approaches it has to be noted that due to the large differences in the length scales of the micro-structure and the whole test sample, which — in most cases — is desirable for obtaining a macroscopically homogeneous effective material behaviour, the information about specific local buckling mechanisms is lost in the noise of hundreds of such events taking place at practically the same time during a typical foam compression test. Rarely, a buckling mode can be observed on the level of the whole specimen, see, e.g.,[8] for a long wavelength buckling mode in polyester urethane foam. Instead, the collapse of individual cells or layers of cells manifests itself in a much reduced current stiffness of the sample or even a horizontal plateau in the force-displacement diagram.

If the focus of experimental interest is on the actual micro-mechanical deformation mechanisms then the size of the sample has to be reduced in order to make micro-structural features stand out. Zhu *et al.*[9] demonstrate compression tests on individual polymer foam cells. Another possible approach in this respect is the testing of small samples under surveillance by micro computed tomography (CT). We shall cite[10] as an example for the detailed investigation of the compression mechanisms in polyurethane foam. A different approach becomes available as soon as it is possible to enlarge the micro-geometry far enough to make the deformation mechanisms accessible to visual inspection. Rapid prototyping methods offer this possibility as will be discussed in Sec. 13.2.3.

With respect to the prediction of instabilities in cellular materials the methods range from simple analytical estimation formulae to complex finite element models of the actual micro-structures that are scanned with micro CT. Most numerical models of the micro-structure are based on periodic micro-structures, where a periodically repeating unit cell, which fills space both in the undeformed and the deformed configuration, is used as an idealised representation of the actual (potentially random) micro-structure. Models for different micro-structural topologies will be presented in Secs. 13.2.3 to 13.2.5.

As soon as appropriate periodic boundary conditions are imposed on unit cell models, the prediction of buckling loads and buckling modes becomes fairly straightforward, at least as long as the buckling process is an elastic one. One has, however, to be careful with respect to the size of the periodic unit cell with respect to the number of individual cells that it contains, for this number has to be large enough as to allow for the formation of the critical buckling mode. It has been demonstrated in[8,11,12] that the actual buckling mode of a periodic model for open-cell polyurethane foam can span over several base cells, see Fig. 13.1.

Determining the number of base cells (along all vectors of periodicity) necessary to approach the actual buckling eigenvalue of the infinite periodic structure can be a very expensive process in terms of model size and,

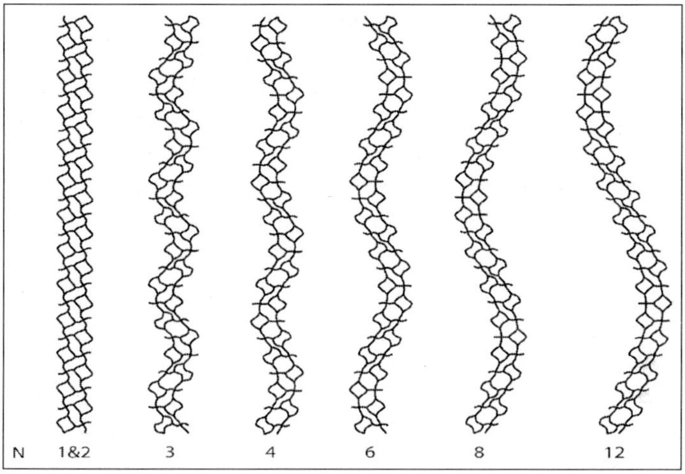

Fig. 13.1. Influence of the number of unit cells N stacked in the loading direction on the buckling mode shape of a tetrakaidecahedral open-cell micro-structure.[8] The undeformed cells are slightly elongated in the vertical direction.

consequently, computational resources. It was, however, demonstrated in[12] that the critical buckling stresses can be predicted very efficiently based on the (tangent) stiffness matrix of a *single* (deformed) periodic base cell. The corresponding numerical procedure is derived from Bloch wave theory. At its heart, this theory is based on the assumption of a bifurcation mode that takes the following form:[12]

$$\underline{\tilde{\boldsymbol{u}}}(\underline{\boldsymbol{x}}) = \underline{\boldsymbol{u}}_{\mathrm{per}}(\underline{\boldsymbol{x}}) \exp\left(\mathrm{i}\left[\frac{m_1 x_1}{l_1} + \frac{m_2 x_2}{l_2} + \frac{m_3 x_3}{l_3}\right]\right), \tag{13.1}$$

where $\underline{\tilde{\boldsymbol{u}}}$ is the displacement vector as a function of material coordinates $\underline{\boldsymbol{x}}$, $\underline{\boldsymbol{u}}_{\mathrm{per}}(\underline{\boldsymbol{x}})$ is a displacement field that has the same periodicity as the base cell, i.e.:

$$\underline{\boldsymbol{u}}_{\mathrm{per}}(x_1 + n_1 l_1,\ x_2 + n_2 l_2,\ x_3 + n_3 l_3) = \underline{\boldsymbol{u}}_{\mathrm{per}}(x_1, x_2, x_3), \tag{13.2}$$

where l_i $(i = 1, 2, 3)$ are the dimensions of the base cell, $\mathrm{i} = \sqrt{-1}$ is the imaginary number, and $m_i/l_i = \omega_i$ $(i = 1, 2, 3)$ are the wave numbers. Together with the instantaneous stiffness matrix of the base cell and the displacement field (13.1) the Bloch wave method leads to the formulation of an eigenvalue problem that allows for the calculation of long wavelength buckling modes and corresponding eigenvalues. For details the reader is referred to.[12]

Disregarding the possibility of long wavelength buckling modes can lead to an overestimation of the buckling loads. In[12] the buckling load of a single, slightly elongated tetrakaidecahedral beam element unit cell was found to be approximately 16% higher than the relevant buckling eigenvalue for a unit cell that contained a stack of 12 base cells, compare also Fig. 13.1.

For the case of bifurcation buckling with a stable postbuckling response it is possible to increase the applied load further until a structural limit load is reached. In such situations it is desirable to be able to perform a post-buckling analysis of the structure for determining the limit load. Within the capabilities of standard finite element codes this is usually achieved by imposing small geometric imperfections, which are typically chosen as being affine to the bifurcation displacement vector corresponding to the relevant buckling eigenvalue, on the initial geometry and calculating the non-linear response of the imperfect structure. In the case of an unstable buckling response the same procedure can be used for assessing the sensitivity of the structure to geometrical imperfections.

13.2.3 *Open-Cell Topologies*

The product of many natural growth processes and many technical production processes are micro-structural networks or lattices of slender, beam-like members. These topologies are often referred to as open-cell foams, or sponges. These names indicate that these cellular materials are permeable to fluids.

Examples of natural open-cell materials are spongious/trabecular bone and natural sponges.[6] Many engineering materials like polymeric foams, cellular ceramics and some metallic cellular materials also show an open-cell structure. Regular open-cell structures can be produced, e.g., by rapid prototyping.[13,14]

Open-cell structures deform primarily by bending of their struts. These struts are often the sole remainder of a previously closed-cell structure which lost its cell wall faces in the course of the production process. Figure 13.2 shows a bent strut in an open-cell metallic foam illustrating the importance of bending deformation modes for these structures. This figure also shows the characteristic triangular cross-section of the struts of many open-cell foams which can be explained with the laws of physics of wet foams. In the case of trabecular bone, the edge network is the result of a stimulus-driven adaptation process that combines bone tissue deposition and resorption to adapt the micro-geometry to imposed *in vivo* loading conditions.

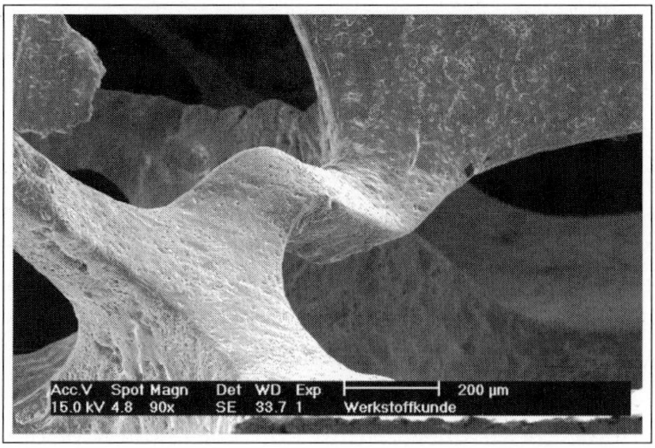

Fig. 13.2. REM image of a bent strut in an open-cell metallic foam; (image courtesy of Institute of Materials Science and Testing, Vienna University of Technology.)

Instabilities in open-cell foams typically take place in the form of progressive snap-through and/or bifurcation buckling of struts or groups of struts. The former is perhaps the more frequently observed (and predicted) mode of deformation, because bifurcation buckling in the form of elastic Euler buckling or inelastic buckling requires an orientation of the concerned strut which is nearly parallel to the loading direction in the uniaxial case.

For trabecular bone, Hayes and Carter[15] present microscopic evidence for the failure of trabeculae by shear band formation and by buckling, the buckling mode being that of a compact beam under axial compression, the rotations of which are constrained at both ends while lateral movement is possible at one end. Townsend *et al.*[16] tested single human trabeculae in wet and dry states. Their results indicate that the slenderness ratio of the trabeculae can be used for estimating the buckling stress. The dried bone material failed in a buckling mode with brittle postbuckling failure while the buckling stress of the wet specimens was apparently limited by a yield cut-off stress for low slenderness ratios (length/radius < 50), thus indicating ductile bone behaviour. Similar results were obtained in.[17]

Convincing evidence for the elastic buckling of individual struts can be found in[18] for open-cell polyester urethane foam. Certainly, the buckling of struts softens the overall response of the open-cell structure. A dominant role in the loss of overall stability of open-cell materials must, however, also be attributed to the effects of geometric nonlinearity due to the bending deformation of the struts and the corresponding distortion of the cell shape, ultimately leading to snap-through behaviour in cells, as compression tests on individual polyurethane foam cells in[9] demonstrate. Softening by geometrical nonlinearities leading to snap-through behaviour was also predicted by an analytical model of a periodic tetradecahedral lattice in[9] and an analytical analysis of a tetrahedral model structure in.[19]

More complex models of irregular and imperfect geometries show both geometrical cell distortions and isolated buckling events at the same time, compare,[20] where finite element analyses of a random Voronoi tessellation based open-cell structure is simulated under compression loading. This study did not include dedicated buckling eigenvalue analyses, making it unclear if any bifurcation points were crossed in the course of the simulated deformation. Such buckling eigenvalue analyses were performed very thoroughly on periodic tetrakaidecahedral unit cell models of open-cell foams in[12] and augmented with experimental results in.[8] Their ability to detect long wavelength buckling modes by means of Bloch wave methods was already discussed in Sec. 13.2.2. Failure envelopes of various cellular materials under multi-axial loading conditions were documented in,[21] where obvious differences in the behaviour under tensile and compressive loading conditions indicate geometric nonlinearities and local buckling instability mechanisms.

The influence of imperfections on the behaviour of open-cell microstructures can be assessed by producing and testing such structures with well-defined imperfections by means of rapid prototyping.[13,14] Such methods are demonstrated in[22–25] along with numerical experiments on finite sized cubical and cylindrical samples of perfect and imperfect simple cubic lattices, see Fig. 13.3, where the experimentally observed and the numerically predicted configurations for uniaxial compression tests are compared. The bulk material of the struts exhibited elastic-plastic behaviour and was represented accordingly in the simulations.

The simple cubic sample, that is presented in Fig. 13.3, was compressed in the direction of the diagonal of one of the faces of the cube that is spanned by three mutually perpendicular struts that emanate from the same vertex, thus forming the base cell. As it is shown in both the experiment and the simulation results this leads to the formation of a diagonal band of collapsed 'cells' in the sample which appears to be related to the overall geometry of the sample since it runs from the top/left corner of the frontal face of the sample to the bottom right corner. The simulated configuration of the geometrically undisturbed sample in Fig. 13.3 (bottom/left) shows a symmetric 'X'-shaped localisation pattern, which is not as clearly visible in the corresponding tested sample shown in the top/left image in Fig. 13.3.

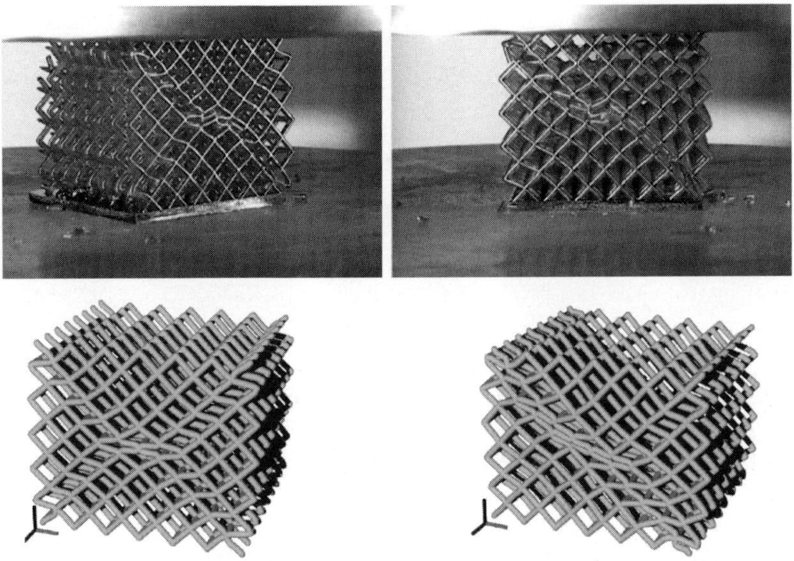

Fig. 13.3. Comparison of images from crush experiments (top row) and numerically predicted localisation patterns (bottom row) for a perfect (left column) and a slightly disturbed (right column) simple cubic lattice block (image courtesy of M. Luxner).

This already indicates that the imperfections which are pertinent to the rapid-prototyping process favour an asymmetric collapse mode. As soon as imperfections are introduced into the simulation model, the predicted localisation mode also becomes asymmetric, see Fig. 13.3 (bottom/right). Furthermore, the localisation band appears to become wider and affects more than just one band of cells.

This trend is further documented in Fig. 13.4, where deformed configurations of simple cubic lattice samples under uniaxial compression in different directions with regard to their axes of symmetry and with different degrees of geometrical imperfection are presented in their state shortly after the limit load is reached in the overall force-displacement relationship. The imposed imperfections concerned the random displacement of the lattice vertices by a fixed distance δ, which is expressed as a fraction of the edge length of a single base cell in Fig. 13.4.

Depending on the orientation of the sample and the severity of the imperfection the localisation of the deformation is more or less pronounced. The 'perfect' samples show distinct localisation bands which are one or two cell layers wide. In the corresponding overall force-displacement curves the

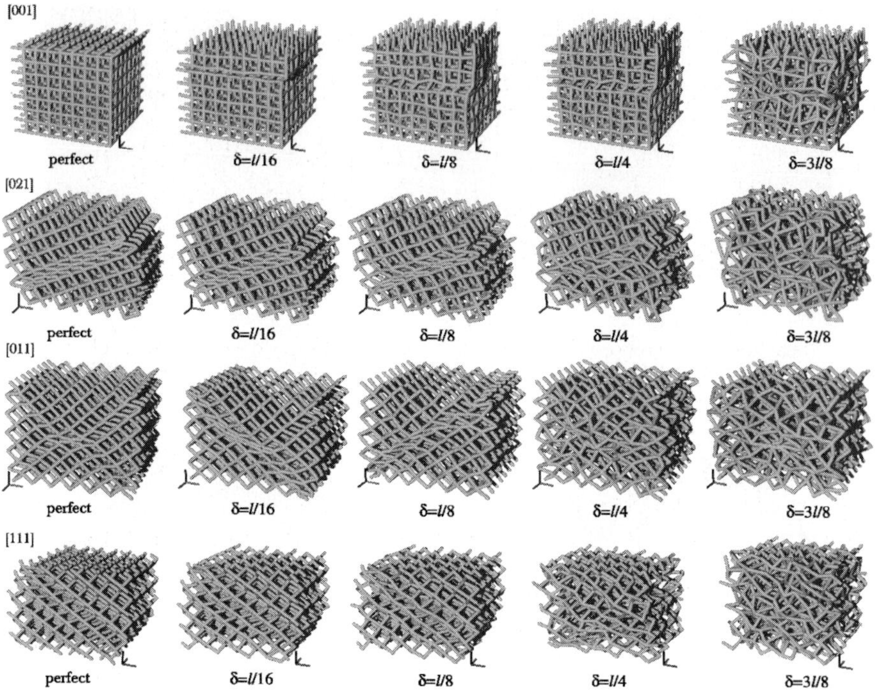

Fig. 13.4. Predicted deformed configurations for a number of finite simple cubic lattice blocks under uniaxial compression; different degrees of disorder, quantified by vertex displacements δ, being expressed as fractions of the edge length of a single base cell.[25]

formation of the localisation zones coincides with a distinct drop of the imposed load.[25] As larger vertex displacements are introduced, the overall response becomes softer and the limit load is reduced. These detrimental effects are somewhat offset by the fact that the drop of the load after the onset of collapse is much less severe than it is the case for the geometries with smaller imperfections, i.e., the imperfect structures retain most of their load-carrying capacity after they reach their respective limit loads. As far as the deformed configurations are concerned, there are no more distinct localisation zones visible in the imperfect samples, i.e., the deformation is comparatively uniform even at high overall compressive strains.

13.2.4 *Closed-Cell Foams*

Closed-cell foam emerges naturally in many production processes for cellular materials. With regard to metallic closed-cell foams a review of different

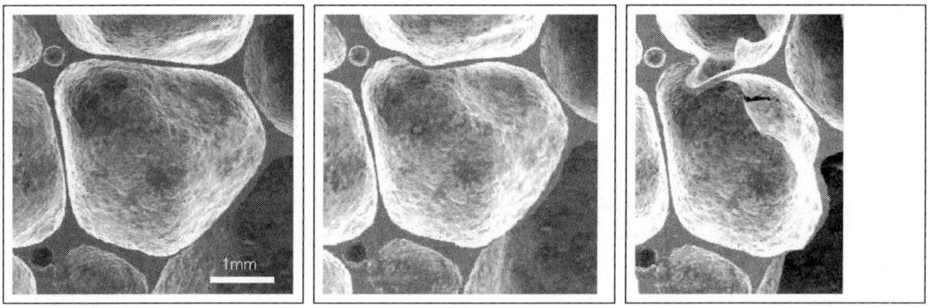

Fig. 13.5. Deformation sequence of a wrought alloy aluminium foam showing postbuckling collapse of a cell wall oriented in deformation direction; images courtesy of Institute of Materials Science and Testing, Vienna University of Technology.[27]

production methods can be found in.[26] Closed-cell foam differs fundamentally from open-cell foam or sponges in terms of deformation behaviour. While sponges deform primarily by bending of the struts that build the skeleton of the micro-structure, the dominant deformation mechanism in closed-cell foam is cell wall stretching, shearing, and buckling.[6]

Consequently, the initial mode of instability in closed-cell cellular materials is buckling of the cell walls along the direction(s) of compressive principal stress(es), see Fig. 13.5. The cell face can typically sustain tensile principal stresses beyond the onset of buckling, but this postbuckling load carrying capacity is limited by cell wall fracture.[28] Cell walls can also fracture wherever plastic hinges form during the collapse of individual cells.[29] As soon as the load-carrying capacity of the cell walls is exhausted at higher compressive strains, the edge regions start to deform in modes that are similar to those that were described in Sec. 13.2.3 for open-cell structures. Bastawros *et al.*[30] observe localised collapse of bands of cells in aluminium foam and distinguish between cell collapse mechanisms that include only a distortion of the cells and mechanisms that, additionally, cause a rotation of cells.

The onset and effects of cell wall buckling on the mechanical behaviour of closed-cell polymer foams was modelled very pragmatically by Mills and Zhu,[31] who assumed for the analytical description of the compressive behaviour of their tetrakaidecahedral foam model that elastic buckling appears from the onset of compressive deformation of the cell walls, and that the buckled cell walls completely loose their load-carrying capacity in the direction of the principal compressive strain. This assumption proved to be conservative when the calculated elastic moduli were compared to experimentally measured ones, giving moduli that amounted to only 30% of those measured in tension. It was, however, also noted that the measured compressive elastic modulus of PVC foams can be as low as 50% of the tensile elastic modulus, owing to instant buckling under compression. Coupling their model with a simple yield criterion and using the corresponding yield stress as a fitting parameter, Mills and Zhu[31] were able to predict the compressive stress-strain relationship remarkably accurately up to 70% nominal compressive strain. They remarked, that the dominant contribution to the compressive resistance of the considered low density polyethylene foam comes from the compression of the enclosed gas at high compressive strains.

A different modelling approach, which was based on analytical 'superfolding elements' was presented by Santosa and Wierzbicki[32] who proposed a truncated cube model for closed-cell aluminium foam and described the plastic collapse of this model by means of plastic yield lines and extensional zones in subsections of the model structure, namely web sections, cruciform sections and pyramidal structures. They were able to verify their analytical model with a corresponding finite element model and found it capable of fitting experimental results for the density dependence of the collapse stress of Hydro aluminium foam, provided that an appropriate yield stress was selected for the cell wall material.

Buckling of cell walls certainly reduces the load carrying capacity of the concerned walls. It has to be noted, though, that in the micro-structure of real closed-cell foams the cell walls might already be pre-deformed in a way that is similar to that of the buckling mode. The problem changes then from calculating the buckling load to assessing the sensitivity of the structure to imperfections such as curved or corrugated cell walls. A corresponding study on tetradecahedral shell models for closed-cell metallic foam was presented by Simone and Gibson.[33] They note that typical cell walls in those foams had curvatures with radii of curvature above 160% of the face edge length. For the latter radius they predicted reductions of the peak stress of 20% and 33% for relative densities of 10% and 5%, respectively, compared to the peak stress for plane faces.

As a generic example of closed-cell foam, a special representative of regular period cellular structures will be presented in the following. This closed-cell structure, which is sometimes referred to as a 'three-dimensional honeycomb' (3D-HC), is the result of a production process which involves the expansion of organic spheres that are coated with a metal powder slurry.[34,35] The expansion process merges the coatings of adjacent spheres

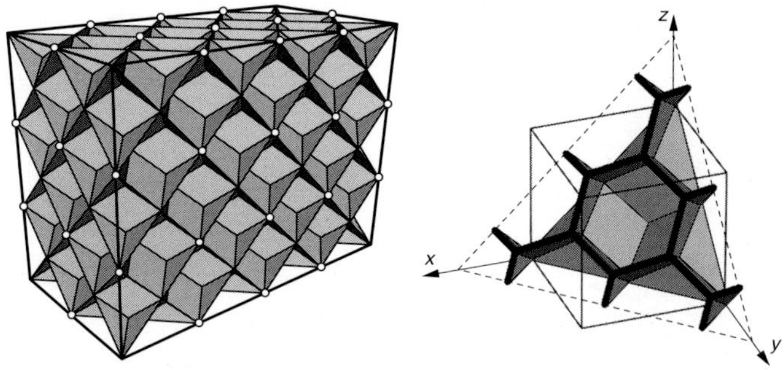

Fig. 13.6. A periodic array of rhombic dodecahedra, building up a 'three-dimensional honeycomb' (left), and a hexagonal cross-section thereof (right).

to form the cell walls of a closed-cell structure which is isolated by removing the precursor spheres and stabilised by sintering.

Depending on the effort that is put into the arrangement of the precursor spheres more or less regular cellular structures can be obtained. For obtaining suitable generic models for the cellular structure a regular face-centred cubic arrangement of the spheres is assumed corresponding to a close packing of spheres in an infinite cellular medium. Expanding the spheres results in a thin-walled structure that can be seen as a Voronoi tessellation of the face-centred cubic lattice, see Fig. 13.6 (left). The building block of such a tessellation is the rhombic dodecahedron, which is a polyhedron with 12 rhombic faces. The chosen designation 'three-dimensional honeycomb' stems from the fact that this structure can be cut by specific planes for obtaining regular hexagonal cross-sections, compare Fig. 13.6 (right).

If this polyhedral structure is subjected to macroscopic loading, the deformation of the rhombic faces can be characterised as being dominated by in-plane shear. Cell wall bending plays a subordinate role, and becomes important only at high relative densities, or thick cell walls, respectively.

Naturally, the shear-dominated deformation of the cell walls also influences the buckling pattern. Since most macroscopic stress states, including tensile ones, translate to shear deformations on the cell wall level, the respective elastic buckling modes are very similar. As an example, the buckling eigenform for a 3D-HC unit cell under uniaxial compression in the '1'-direction is given in Fig. 13.7(a) as the result of a linear initial stability analysis. A periodic unit cell model composed of linear shell elements was used for obtaining this result.

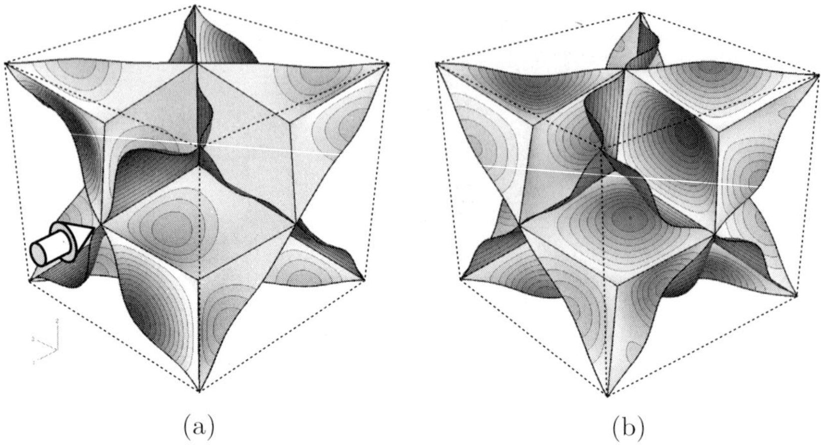

Fig. 13.7. Predicted buckling eigenmode for a 3D-honeycomb under (a) macroscopic uniaxial compression, and (b) macroscopic hydrostatic compression.

The buckling eigenform shows that the rhombic cell walls buckle like plates which are clamped on all sides. No significant macroscopic deformation is visible nor do the cell walls show any sign of interacting with each other. It may be concluded that the buckling of this specific structure can be attributed to the buckling of the individual faces of the polyhedra.

The shear deformation of the cell walls can lead to cell wall buckling even when the macroscopic stress state is a tensile one. Fig. 13.8 shows a short wavelength buckling mode on faces of the 3D-HC which are aligned with the macroscopic loading direction (indicated by arrows in the positive and negative '1' direction). This buckling mode is triggered by the compressive principal stress component that is acting perpendicularly to the macroscopic loading direction in the concerned faces. The depicted short wavelength mode appears shortly before the uniaxial, tensile macroscopic limit load of the model is reached. This can be seen from the fact that some of the faces parallel to the macroscopic loading direction, marked with the letter 'U' in Fig. 13.8, have not yet buckled. This buckling mode affects a configuration that is already severely affected by plastic yielding. The initial, elastic buckling mode exhibits individual dimples in the concerned cell walls, similar to those displayed in Fig. 13.7(a). These dimples, however, vanish as soon as the tensile principal stress along the face diagonal in the loading direction increases. The macroscopic stress has to be increased

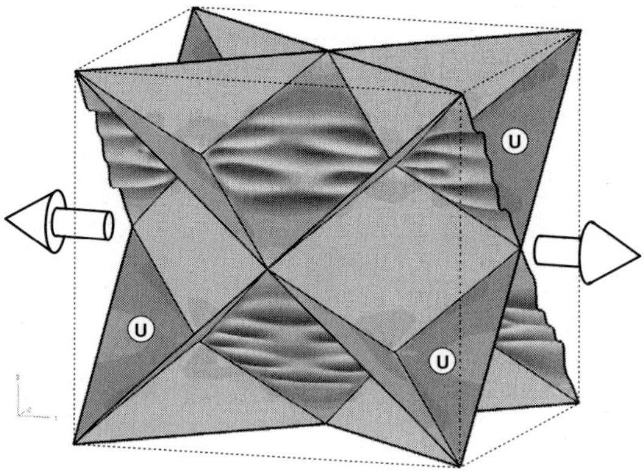

Fig. 13.8. Contour plot of the predicted maximum principal stress in a 3D honeycomb plotted on a deformed configuration (displacements magnified by a factor of 10) close to the tensile uniaxial limit load. Faces marked with (U) will be the next to buckle.

further for the short wavelength elasto-plastic buckling pattern to appear progressively on faces aligned with the macroscopic loading direction.

No shear deformation, but rather biaxial compression, is involved in the uniform hydrostatic compression of 3D-HC on the macroscopic level. Here, the predicted buckling pattern, which can be found in Fig. 13.7(b), is much more uniform and all faces seem to buckle at once, owing to the uniformity of the compressive global and local stress states.

Having presented the buckling modes under some important macroscopic load cases we now want to compare the respective macroscopic buckling stresses. Figure 13.9 shows the macroscopic elastic buckling stresses for uniaxial compressive loading along the base axes of the unit cell ([100]), along the space diagonal of the unit cell ([111]), for pure shear loading, and for loading under hydrostatic pressure. The critical stresses are normalised by the elastic modulus E_S of the bulk material and plotted over the relative density ρ^*/ρ_S. For the simplified study presented here, where no correction for taking into account the volume of the edges and the vertices was performed, the relative density was linearly proportional to the face wall thickness, which enters the bending stiffness of the cell walls with a power of three. Consequently, the normalised elastic macroscopic buckling stresses were also proportional to the third power of the relative density,

$$\frac{\sigma^*_{\mathrm{el}}}{E_S} \approx \left[\frac{\rho^*}{\rho_S}\right]^3 . \tag{13.3}$$

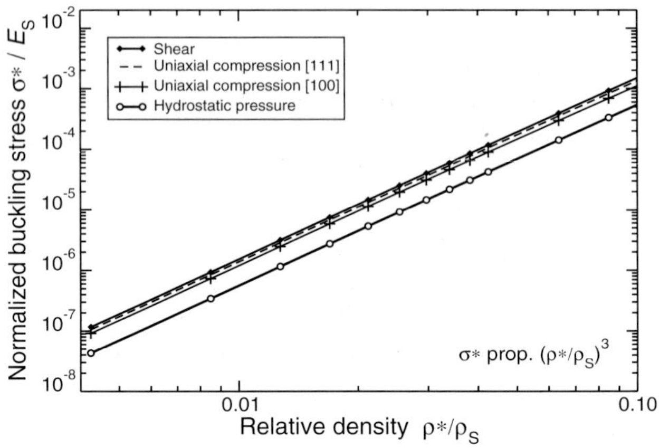

Fig. 13.9. Macroscopic elastic buckling stresses as a function of relative density of 3D-honeycombs under different macroscopic loads.

Furthermore, the relative differences between the macroscopic buckling stresses for the different loading cases are more or less constant. Taking the macroscopic buckling stress for the uniaxial compression along the unit cell axes as the reference value, the following relations hold: The macroscopic uniaxial elastic buckling stress along the space diagonal is approximately 17% higher than the reference buckling stress. The macroscopic shear buckling stress is approximately 30% higher than the macroscopic uniaxial compressive buckling stress, and, lastly, the macroscopic hydrostatic buckling stress has the lowest of all four values and is, on average, 53% lower than the reference stress.

Certainly a more detailed model could have been used for predicting the buckling behaviour of the presented model topology, especially with respect to the more accurate representation of the edge and vertex regions, but the principal features of the buckling and postbuckling of closed-cell foams could be demonstrated nonetheless for this comparatively simple structure. It was most remarkable that, at least for the demonstrated example of the 3D honeycomb, the buckling events are hardly discernible in the overall macroscopic stress-strain curve owing to the stable postbuckling behaviour of the buckled faces. Nevertheless, the progressive buckling of the cell faces led to a softening of the structure, the stability of which relies — after the rupture of the cell faces under the further growing tensile principal stresses — on the load carrying capacity of the cell edge regions.

13.2.5 *Mixed Topologies*

Most materials which exhibit a cellular micro-structure either fall in the categories of open-cell sponges or closed-cell foams. There are, however, topologies that do not fall into those two distinct categories. As an example for such a mixed topology, open-cell foam (sponges) with hollow struts will be presented in the following.

It is relatively straightforward to produce open-cell polymer foams, as is demonstrated by the extensive use of such foams in packaging, insulation and crash protection applications. These polymer sponges can be used as templates in coating processes that produce hollow metallic shells with a surface geometry which is affine to that of the precursor sponge. One of these technologies is electro-deposition of metal ions from an electrolyte onto polymeric foam, while another technology involves vapour deposition to form a metallic layer from gaseous metals or gaseous metallic compounds. Both the galvanic and the vapour deposition processes are generally followed by removing the polymer precursor structure by thermal or chemical treatment.[26]

Since the metallic coating envelops the precursor foam in the form of a layer of near constant thickness, the removal of the polymer precursors' struts leaves behind a structure that can be understood as a network of hollow metallic struts with roughly triangular enclosed cross-sections owing to the shape of the so-called Plateau borders of the polymer template sponge (see Fig. 13.10). A detailed description of the micro-geometry of open-cell nickel foams produced by the company NiTECH, including qualitative and quantitative information about the deformation of the foam under loading can be found in.[36] A similar material with the trade name Incofoam is described in,[37] compare Fig. 13.10 (left).

The fact, that the struts of these metal sponges are hollow, introduces, along with the characteristic shape of the strut cross-sections, different structural levels, on which instabilities can occur: the lower hierarchical level is the level of the strut walls, or the ribs that are formed by the junction of the strut walls, and the higher hierarchical level is the level of the strut skeleton, which behaves, effectively, like a conventional open-cell sponge, i.e., by bending of the struts. Depending on the relative density of the polymer template sponge, this distinction can become somewhat blurred.

In,[38] experimental evidence is presented for the interaction of elasto-plastic buckling on the cell wall as well as on the strut level in a nickel sponge. Two micro computed tomography images from[38] are reprinted

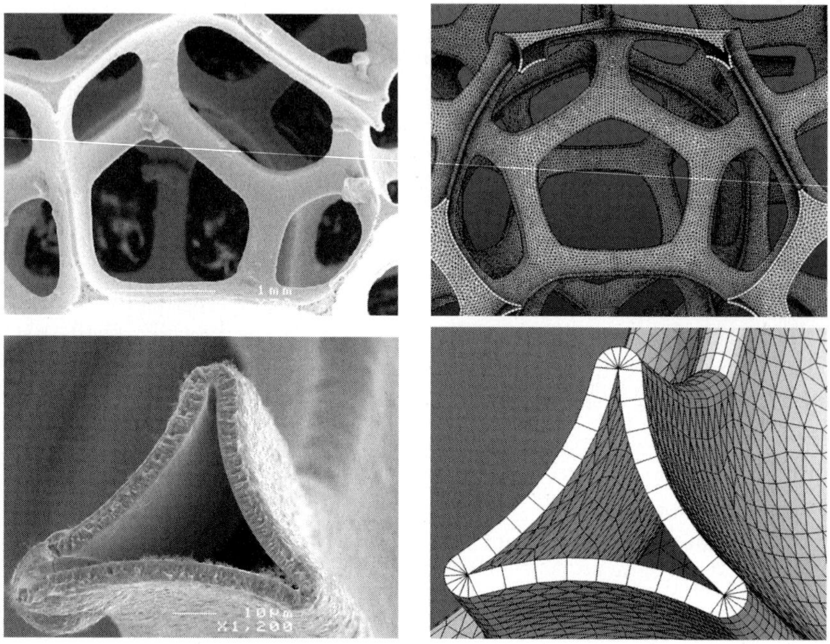

Fig. 13.10. Comparison of a commercial nickel sponge with hollow struts (left; from,[37] courtesy of V. Paserin, INCO Ltd.) with subsections of the proposed finite element unit cell model of a generic sponge with hollow struts.

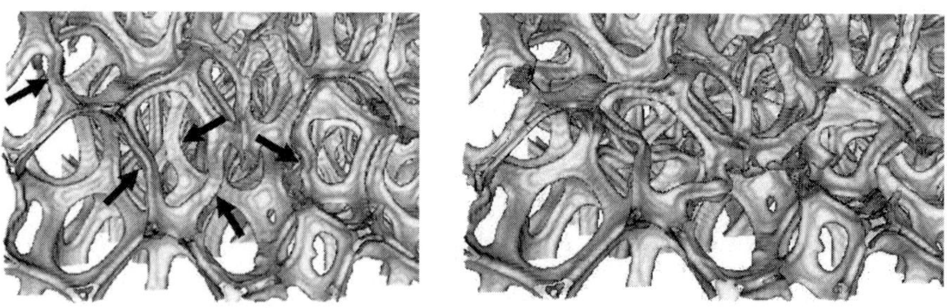

(a) (b)

Fig. 13.11. Experimental evidence for buckling of the ribs of the hollow struts in a nickel foam loaded in the vertical direction (from[38]).

in Fig. 13.11; black arrows mark struts that buckle inelastically during the deformation process under consideration. It can be seen how buckling of a strut rib can cause the formation of a 'yield joint' at the centre and the ends of struts. On the other hand, struts can buckle without signs of strong local deformation.

For further investigating the different modes of buckling in sponges with a hollow strut topology, a periodic finite element model was proposed in.[39] This FE model, which was composed of continuum shell and continuum solid finite elements, is compared to the real micro-structure in Fig. 13.10. The connectivity and geometry of the struts' mid-axes correspond to that of the cell wall edges in a Wearie-Phelan foam,[40,41] and the template geometry is based on the shape of the Plateau border network of a wet liquid foam, which was predicted with the program '*Surface Evolver*'.[42]

The finite element model allows for a free choice of template and coating parameters without technological restrictions and, therefore, a detailed investigation of the different possible modes of buckling. Figure 13.12 shows predicted buckling modes for two different hollow sponges (right) along with the corresponding undeformed configurations (left).

For the model in Fig. 13.12(a) a template sponge with a comparatively low relative density of 1% was chosen in order to set up a model of a hollow sponge with very slender struts. A hydrostatic pressure load was applied on the macroscopic level and translated to appropriate master node forces. A linear initial stability analysis was performed on this setup, and the corresponding buckling mode is shown in Fig. 13.12(b). This image shows a buckling pattern that does not show any significant cell wall deformation, but rather a deformation mode of the struts which is in accordance with classical (Bernoulli-Euler) beam theory. The periodic nature of the buckling pattern is clearly visible and reinforces the need for periodic boundary conditions on the unit cell.

A slightly different result arises when the relative density of the template sponge is set to a higher value of 4%, giving a final finite element configuration as depicted in Fig. 13.12(c). When a macroscopic pressure load is applied to this model and a linear initial stability analysis is performed, the result looks very different from the strut buckling mode observed in Fig. 13.12(b), as is documented in Fig. 13.12(d). Here, the shell of the hollow sponge suddenly forms local bumps and also distinct kinks on the ribs at the junction lines of the cell walls. It can be concluded, that the higher effective bending stiffness of the struts, which is a consequence of their higher moment of inertia owing to the fact that the compactness of the template sponge moves the coating material farther away from the axes of inertia, prevents beam-like buckling. At the same time, the average curvature of the hollow sponge's shell is lower than in the case of the slender,

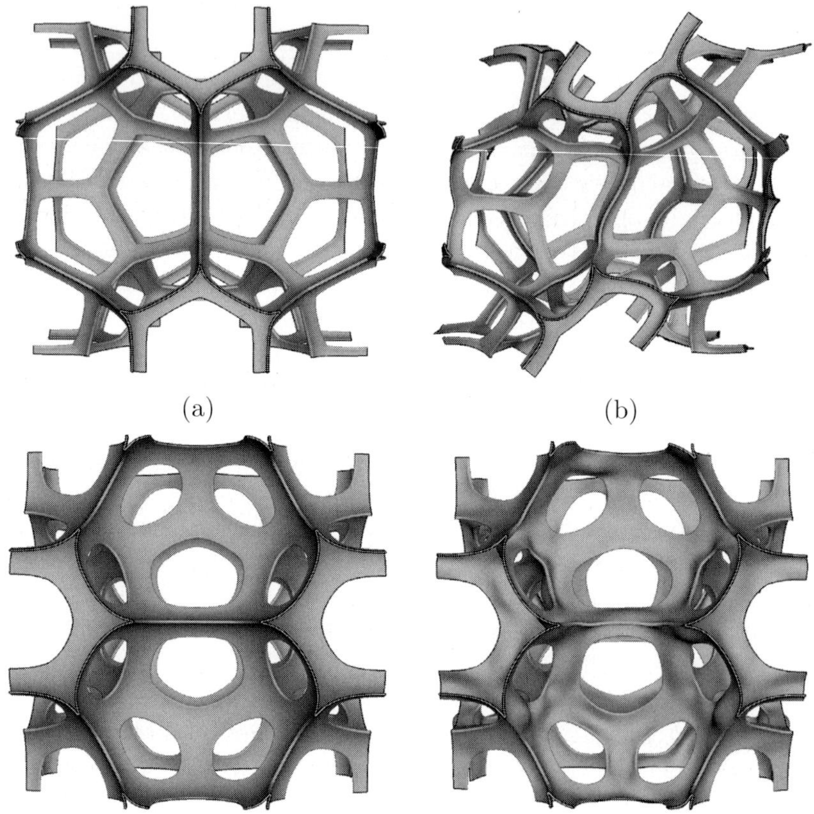

(a) (b) (c) (d)

Fig. 13.12. Predicted strut buckling (top row) versus strut wall buckling (bottom row) in a FE model of a sponge with hollow struts under hydrostatic loading.

low density template sponge. This leads to a reduced resistance of the shell to local buckling and changes the buckling mode from beam-like buckling to shell-like buckling.

The same effect can be obtained if the coating thickness is increased on a hollow sponge with slender struts. In this case, the bending stiffness of the struts prevents strut buckling while the high curvature of the coating avoids shell buckling. Buckling nevertheless occurs in the form of buckling of the ribs, which experience the highest compressive stresses due to their distance from the neutral axes of the struts. Figure 13.13 shows the undeformed and the predicted buckled configurations of a vertex in a hollow sponge. Owing

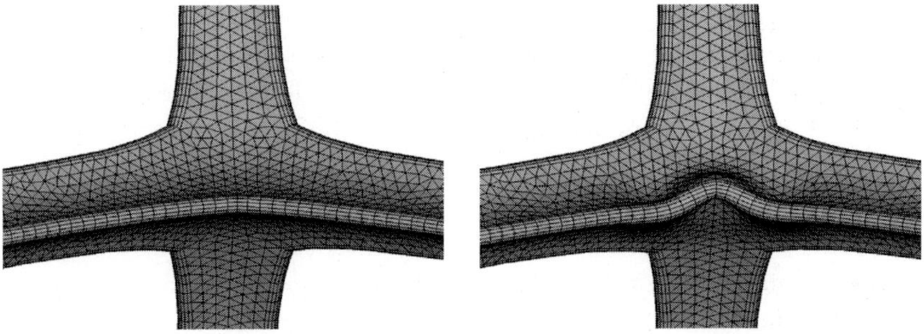

(a) (b)

Fig. 13.13. Detailed view of the predicted mode of local buckling of a rib along the struts of a hollow sponge.

to the fact that the cross-sectional bending moments of the struts reach their highest values close to the vertices, buckling has also to be expected in these regions. For the presented configuration, the buckling instability is expressed in a lateral buckling of the rib that has formed around the edges of the Plateau borders.

13.2.6 *Micro-Structured Materials — Summary*

In the previous sections the relevance of instabilities on the micro-mechanical level of cellular materials on the effective, macromechanical behaviour of such materials was discussed. It has to be noted that for materials, which fulfil the necessary conditions for being treated as homogeneous materials on the level of whole components, that is, materials the micro-structure of which is several orders of magnitude smaller than the length scales of the components they belong to, the micro-structural deformation behaviour can be replaced without loss of information by the effective linear elastic properties of the homogenised material as long as the deformation can be treated as linear elastic on all length scales.

As soon as instabilities, such as bifurcation buckling, define the onset of markedly nonlinear behaviour, the deformation mechanisms on the micro-mechanical length scale have to be investigated carefully again, as they commonly lead to a decrease of the current effective material stiffness, or — in worse cases — to the collapse of the cellular structure. Whether the postbuckling behaviour is a stable or an unstable one can be examined by means of micro-mechanical simulation methods, which were introduced in Sec. 13.2.2 and further discussed for open-cell, closed-cell, and hollow, spongy topologies in Secs. 13.2.3 to 13.2.5, respectively.

A special emphasis has to be put on the insight that relevant buckling modes can have wavelengths that are several times greater than the size of characteristic periodical base cells. Methods for detecting such long wavelength buckling modes in a computationally efficient way were presented in Sec. 13.2.2.

Little effort has been made so far to treat elasto-plastic buckling in cellular materials, or models thereof, in greater detail. Instead, nonlinear analyses of imperfect micro-geometries have been performed. As elasto-plastic buckling is most likely the dominant failure mode of metallic cellular materials, especially those of higher relative densities, perhaps more attention should be focused on this mode of buckling; this might be true both in terms of the necessity of collecting experimental evidence as well as in terms of theoretical investigations.

13.3 Instabilities in Sandwich Structures

Sandwich structures exhibit very high structural efficiencies (ratios of strength or stiffness to weight) and, therefore, are of interest in the design of aerospace structures. A sandwich consists of two thin load bearing face sheets glued on a lightweight core that prevents the face sheets from buckling individually. Quasi-homogeneous cores (e.g., made of foam) or cellular cores are typically used.

Failure modes of sandwich structures include: global buckling, shear crimping, face sheet wrinkling, face sheet dimpling, face/core yielding, and core-face debonding.[43,44]

In this section, different types of instabilities, which are characteristic for sandwich structures, are predicted by analytical and numerical means. For sandwiches with core materials that can be treated as being homogeneous, wrinkling of face layers is studied in a general setting including unsymmetric setups and orthotropic constituent materials. With respect to sandwiches with structured cores, buckling mechanisms characteristic for honeycomb cores as well as buckling phenomena in corrugated board are discussed.

13.3.1 *Sandwiches with Homogeneous or Homogenised Cores*

Semi-analytical Approach

Recently a quite general semi-analytical model for predicting the onset of wrinkling failure in sandwiches was presented.[45,46] It takes into account

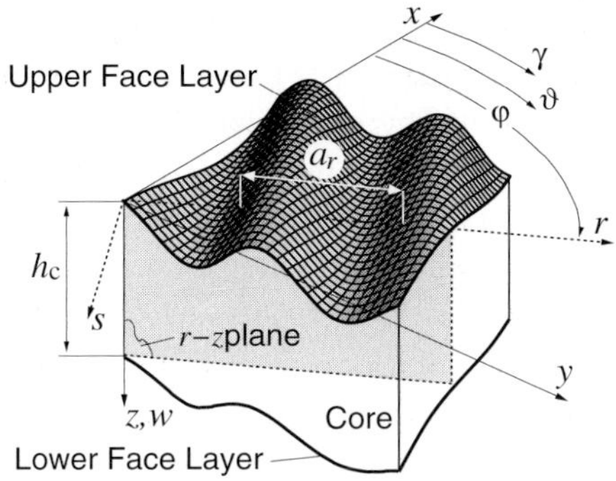

Fig. 13.14. Schematic representation of the analytical model.

unsymmetric sandwiches, orthotropic constituent materials and arbitrary loading situations. The basic setting of the wrinkling problem is shown in Fig. 13.14. A Rayleigh-Ritz method was employed (free parameters: wave length a_r and wave angle φ), leading to a numerically efficient calculation scheme although the complexity of the problem is considerable and numerical optimisation has to be used.

The semi-analytical approach treats the wrinkling phenomenon as coupled bifurcation buckling of two parallel thin plates with elastic or elastic-plastic homogenised material between them, see Fig. 13.14.

One of the faces is defined as the "upper face layer", and the core together with the other face layer (called "lower face layer") acts as its "foundation". As far as elastic buckling is considered, the upper face layer is treated as an infinite, thin plate on an elastic foundation under in-plane loading, the out-of-plane displacement w being described by

$$B_{xx}\frac{\partial^4 w}{\partial x^4} + 2B_{xy}\frac{\partial^4 w}{\partial x^2 \partial y^2} + B_{yy}\frac{\partial^4 w}{\partial y^4} + P_{xx}\frac{\partial^2 w}{\partial x^2} + P_{xy}\frac{\partial^2 w}{\partial x \partial y} + P_{yy}\frac{\partial^2 w}{\partial y^2} + w\, k_{\text{thin}} = 0 \qquad (13.4)$$

with P_{ij} being the membrane forces in the upper face layer having effective bending and twisting stiffnesses B_{ij}, which in the case of layered composite faces can be calculated by using classical lamination theory.

Since short wave buckling must be expected, the foundation, with the stiffness k_{thin} , must not be idealised by simple models such as, e.g., Winkler or Pasternak foundations. k_{thin} can be determined on the basis of continuum mechanics by using an analytical model as shown in.[46] It depends, in addition to the properties of the core (using effective core material properties) and of the lower face layer, on the combined buckling mode, i.e., $k_{\text{thin}} = k_{\text{thin}}(a_r)$. The search for the first appearance of a non-trivial solution of (13.4),

$$w = w^* \sin\left(\frac{\pi r}{a_r}\right) = w^* \sin\left(\frac{\pi(x\cos\varphi + y\sin\varphi)}{a_r}\right), \tag{13.5}$$

leads to the critical membrane force intensity.

Based on this model, eigenvalue analyses can be performed for calculating the critical load configurations for given wavelengths a_r and fold angles φ. Certainly, the physically relevant eigenvalue is the one which is minimal with respect to a_r and φ. Therefore, as described in detail in,[46] an optimisation procedure is required for finding critical winkling loads. This optimisation can hardly be done analytically, but is performed numerically. This is the reason why this approach is called a semi-analytical one. It allows for the determination of critical mesoscopic stress states in the face layers as well as the prediction of the buckling modes for very general material and load combinations, see.[45]

Application of the Unit Cell Approach

Since the deformation field (single sinusoidal mode shape) had to be chosen *a priori* for the above described Ritz ansatz (13.5), there is a need to verify this deformation field with a more general discretisation scheme, e.g., FE unit cell calculations, see.[47]

The chosen FE unit cell models represent a cutout of an infinite periodic sandwich plate, by enforcing periodic boundary conditions at the model boundaries, assuming that the wrinkling problem is periodic in the plane of the sandwich plate (x- and y-direction).

This unit cell approach incorporates the drawback that the overall model dimensions in x- and y-directions act as a frequency filter regarding the wavelengths of the periodic patterns that can possibly be found with such a model, compare.[48] If the model dimensions in x- and y-directions are denominated by l_x and l_y, respectively, the solution of the FE model has to be periodic with the wavelengths a_x and a_y as follows:

$$a_x = \frac{l_x}{n}, \qquad a_y = \frac{l_y}{m}, \qquad n, m \in [1, 2, 3, \ldots], \tag{13.6}$$

where a_x and a_y are connected to the free variables a_r and φ of the analytical model by:

$$\varphi = \arctan\left(\frac{a_x}{a_y}\right), \qquad a_r = a_x \cos(\varphi). \tag{13.7}$$

Thus, the in-plane dimensions of the model determine the wavelengths of the solution. The maximum number of waves (n and m in Eq. (13.6)) which can be calculated is naturally limited by the mesh density of the model. In order to find the lowest buckling eigenvalue and the corresponding wavelength with a certain accuracy, two different strategies were chosen, a small and a large unit cell approach.

Small FE Unit Cell Model

A small unit cell model, which captures just one full buckling wave, is used, and the model dimensions (l_x and l_y) are varied in order to minimise the calculated eigenvalue. The mesh size is chosen to be able to represent one wave across the cell with sufficient accuracy. This leads to a very small FE model requiring only a very short computing time. This approach requires a minimisation scheme, multiple FE calculations and corresponding pre- and postprocessing, all of which were realized within a special code that invokes the FE program multiple times. Since the results based on the analytical approach gave very good initial estimates for the critical wavelengths, this approach proved to be especially effective because only a small number of evaluations was necessary to reach the minimum eigenvalue.

Large FE Unit Cell Models

The small FE models enforce a total periodicity with a certain wavelength and wave angle (defined by the FE model dimensions). It is, therefore, not possible, that the deformation pattern of the upper and the lower face layer differ in terms of wavelength and wave angle. This is an assumption which has already been imposed on the analytical derivation. In order to allow for a more general deformation pattern than in the small FE unit cell models, large FE unit cell models are set up. These models are formulated exactly like the small models, but the model size and meshing allows for 10×10 waves to form in the unit cell (i.e., the size of 100 small FE unit cells).

Finite Plate Example Problem

This example problem is chosen for investigating the applicability of the main assumption of the analytical approach,[45,46] namely the deformation

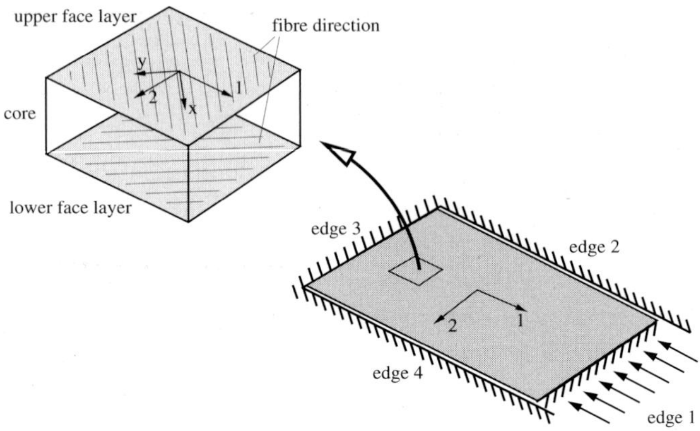

Fig. 13.15. Setup of the finite sandwich plate example.

pattern used in the Ritz ansatz (constant wave length a_r and wave angle φ throughout the thickness). The setup of the example problem is shown in Fig. 13.15.

The presented sandwich plate consists of two identical unidirectional carbon fibre reinforced face layers which are oriented at 90° with respect to each other. The mechanical properties of the face layers, expressed in the local x-y coordinate system (as defined in Fig. 13.15), are given by: $E_x = 117\,000$ N/mm^2, $E_y = 8000$ N/mm^2, $G_{xy} = 2700$ N/mm^2, $\nu = 0.4$, $t_\mathrm{f} = 1$ mm. The core is chosen to have the following effective mechanical properties: $E_z^{(\mathrm{c})} = 55.2$ N/mm^2, $E_x^{(\mathrm{c})} = E_y^{(\mathrm{c})} = 0.0552$ N/mm^2, $\nu_{xz}^{(\mathrm{c})} = \nu_{yz}^{(\mathrm{c})} = 0$, $\nu_{xy}^{(\mathrm{c})} = 0.3$, $G_{xz}^{(\mathrm{c})} = G_{yz}^{(\mathrm{c})} = 26.2$ N/mm^2, $G_{xy}^{(\mathrm{c})} = 0.0212$ N/mm^2.

The coordinate system x-y-z of the upper face layer and the sandwich core is oriented at an angle of 45° with respect to a global coordinate reference frame (1-2-3-system) used to define the fully clamped boundary conditions and the deformation based loading. The local coordinate system of the lower face layer is perpendicular to the one of the upper face layer.

The plate is loaded by a prescribed global compressive strain ε_{11} of 10^{-4}. The high stiffness of the face layers in their fibre direction leads to the fact that the layers take up load mainly in the direction of their fibres. Both faces are loaded equivalently in their own material coordinate frames. Thus, in the case of a very thick core, there are two eigenvalues at the same load level which are based on different wave directions. The problem is defined

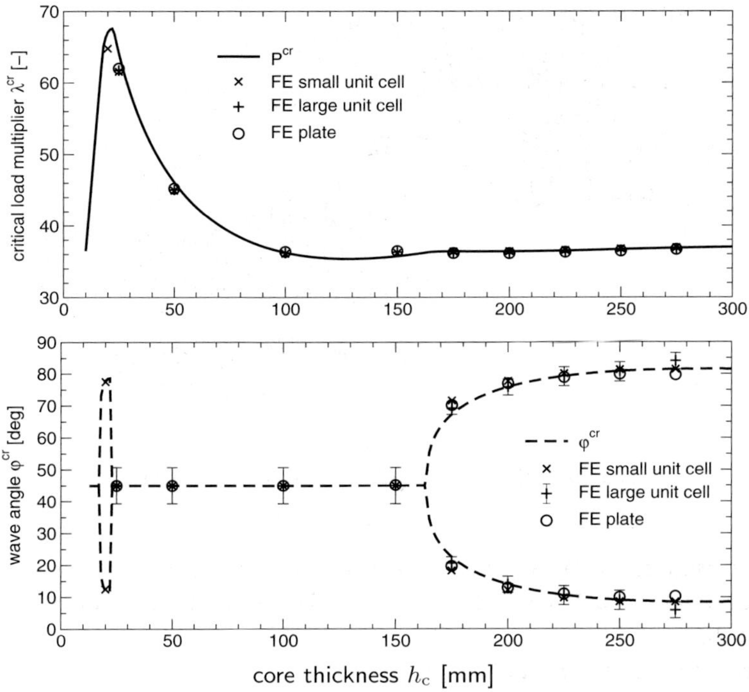

Fig. 13.16. Analytical and FE results for the example problem. The critical load multiplier λ^{cr} refers to a reference load which corresponds to $\varepsilon_{11} = 10^{-4}$.

such that these directions differ significantly (see Fig. 13.16 for thick cores, i.e., high values of h_c) in order to investigate the associated effects.

Results

The parameters eigenvalue, wave length and wave angle, respectively, were evaluated from the FE results, and Fig. 13.16 shows the corresponding results together with the results of the analytical model.

The FE results are in very good agreement with the analytical model.[45,46] The FE results show critical loads which are mostly a little bit lower as compared to the analytical approach, a fact which was also observed by Hadi[49] (concerning 2D FE wrinkling calculations). The results of the large FE unit cell models are shown with error bars indicating the range of a^{cr} and φ^{cr} which is covered by the deformation pattern. This deformation pattern can only change stepwise between discrete configurations

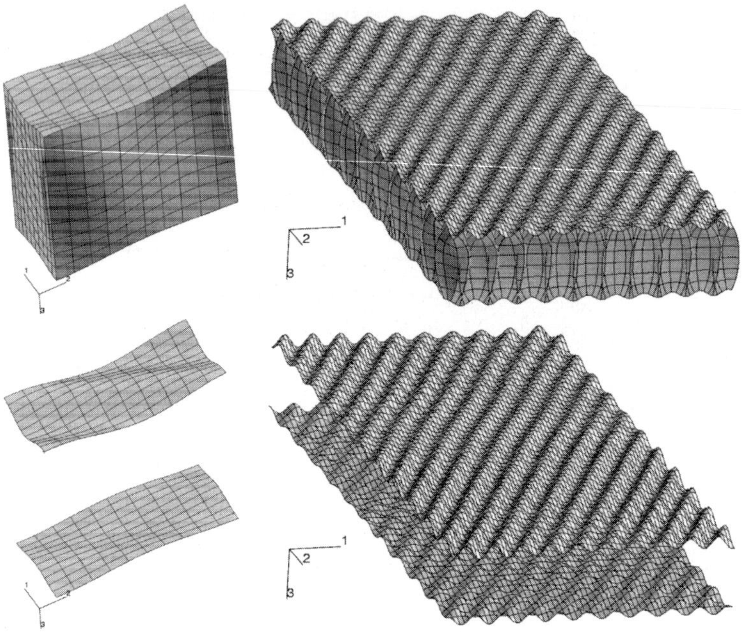

Fig. 13.17. Predicted buckling patterns for the small (left) and the large FE unit cell model (right) of a finite sandwich plate.

in these models. The size of the error bars depends on the aspect ratio of the unit cell and, therefore, varies between the calculations.

For small core thickness values ($h_c < 17$ mm), shear crimping is the critical mode. As the core thickness reaches about 17 mm, the failure mode switches to snake mode wrinkling. This wrinkling mode leads to a double eigenvalue of φ^{cr} with different values that are symmetric with respect to the loading direction. At $h_c = 22$ mm the mode switches to ideal symmetric hourglass wrinkling in loading direction, and the critical load level decreases. As h_c reaches 164 mm the mode type remains an hourglass mode, but a mode jump leads to a double eigenvalue, and the mode is not symmetric anymore.

The FE models were evaluated for different core thickness values. For the special case of $h_c = 175$ mm the buckling modes as obtained by the small and the large unit cell models, respectively, are shown in Fig. 13.17. The corresponding result is close to a bifurcation point and, therefore, of special interest. For further details see.[46]

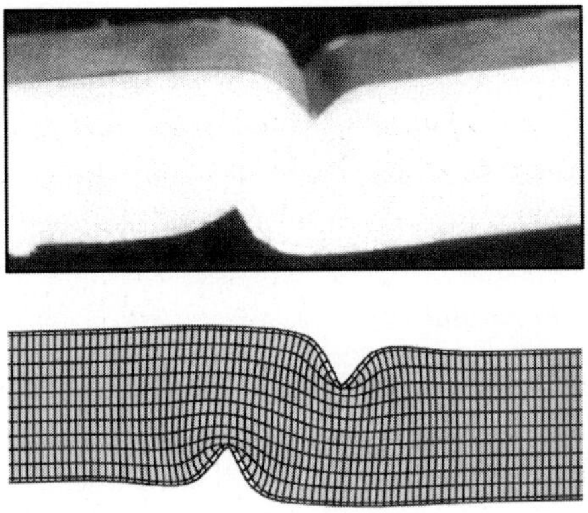

Fig. 13.18. Localisation — formation of discrete folds.

Post-Critical Behaviour — Localised Folds

In the post-bifurcation regime (which is typically an unstable combination of local and global instability modes) starting from the periodic buckling mode a strong localisation in the deformation pattern leading to single folds can be observed experimentally as well as numerically, see Fig. 13.18, which shows experimentally observed and numerically predicted results for axially compressed sandwich beams. Simulations of such localised deformation mechanisms require a full discretisation of the sandwich structure, i.e., in corresponding finite element models the faces have to be modelled by shell elements and the core has to be represented by continuum elements.

Sandwiches with Homogeneous Cores — Summary

A semi-analytical framework for predicting wrinkling of face sheets for symmetric and unsymmetric sandwiches with orthotropic face sheets and orthotropic, homogeneous core materials was presented along with corresponding unit cell models.

All finite element verification models agree with the analytical model in terms of critical loads and mode shapes, despite the complexity of the example problem which was chosen to trigger effects that are not *a priori* covered by the analytical model. The presented methods appear to be well-suited for assessing the onset of buckling instabilities in the face sheets.

13.3.2 *Sandwiches with Honeycomb Cores*

As mentioned above, not only wrinkling has to be considered as soon as sandwiches with thin faces and structural cores (e.g., honeycomb cores) are treated, but also other forms of stability loss like dimpling and/or core wall buckling should be taken into account as possible buckling modes.

Analytical Buckling Formulae

Analytical buckling formulae for different sandwich failure modes — global buckling, face sheet wrinkling of sandwiches with thin or thick cores, face sheet dimpling, face/core yielding, and cell wall buckling under multi-axial states of in-plane compression — are summarised in Tab. 13.1 along with references to the respective source literature.

Table 13.1. Overview of analytical failure models for various modes of instability in sandwich plates.

Buckling Mode	Formula	Literature
global beam buckling	$\frac{1}{P^{\mathrm{G}}} = \frac{1}{P^{\mathrm{E}}} + \frac{1}{P^{\mathrm{S}}}$ P^{E} ... Euler buckling load P^{S} ... shear buckling load P^{G} ... effective buckling load	43
wrinkling thin cores	$F^{\mathrm{W,thin}} = 2\, t_{\mathrm{f}} \sqrt{\frac{E_{\mathrm{f}}\, \bar{E}_z\, t_{\mathrm{f}}}{6\,(1-\nu_{\mathrm{f}}^2)\, h_{\mathrm{c}}}}$	50
wrinkling thick cores	$F^{\mathrm{W,thick}} = 0.85\, t_{\mathrm{f}} \sqrt[3]{E_{\mathrm{f}}\, (k^{\mathrm{thick}})^2}$	45
face sheet dimpling	$F^{\mathrm{D}} = \frac{2\, E_{\mathrm{f}}\, t_{\mathrm{f}}}{1-\nu_{\mathrm{f}}^2} \left(\frac{t_{\mathrm{f}}}{\frac{SW}{2}} \right)^2$	51
cell wall buckling	Classical Lamination Theory (CLT) + kinematic relation + plate buckling	52
face sheet yielding	CLT + von Mises plasticity	
multi-axial loading	$\frac{1}{\lambda} = \sqrt[p]{\left(\frac{F_{xx}^{\mathrm{N}} + F_{xx}^{\mathrm{M}}}{F_{xx}^{\mathrm{W,D}}} \right)^p + \left(\frac{F_{yy}^{\mathrm{N}} + F_{yy}^{\mathrm{M}}}{F_{yy}^{\mathrm{W,D}}} \right)^p}$ with $p = 3$, $F_{ij}^{\mathrm{N}} = \frac{N_{ij}}{2}$, $F_{ij}^{\mathrm{M}} = \frac{M_{ij}}{h_{\mathrm{c}}}$, $()^{\mathrm{W,D}}$ ... 'wrinkling or dimpling'	53

FEM Modeling

Numerical predictions for the buckling loads are given by linear finite element buckling analyses of plane periodic finite element unit cell models

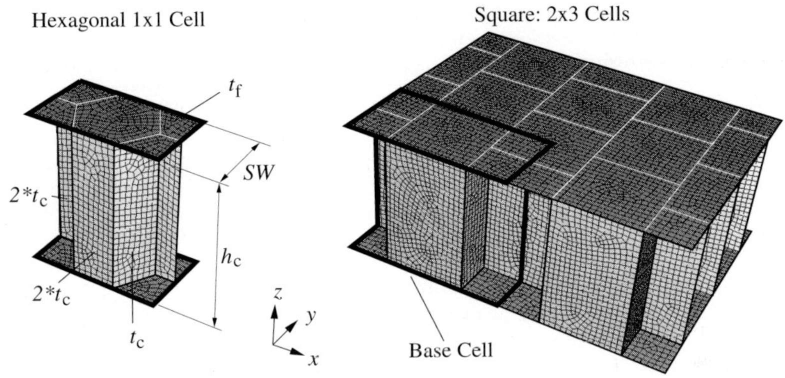

Fig. 13.19. Finite element unit cell models: Hexagonal 1×1 cell model (left, h_c=10 mm, SW=4.7526 mm) and rectangular (square) 2×3 cell model (right). 8-noded, quadratic shell elements are used.

allowing for shell-like macroscopic deformation modes. Figure 13.19 shows possible setups of the unit cell models which are products of the general pre-processing scheme used for the present study.

Selected Results

In this section, the critical mode of buckling will be discussed for sandwiches with honeycomb cores made from aluminium or Nomex® material as well as face sheets made from aluminium or steel. All relevant core and face sheet geometry and material data for the presented examples are placed in Tab. 13.2. The effects of various material combinations, geometrical parameters and loading conditions on the buckling behaviour of the sandwich will be discussed in the following. In the respective diagrams, the reference membrane load is 1 N/mm, and the reference bending moment is defined as 1 Nmm/mm.

Figure 13.20 shows the failure behaviour of a sandwich structure under uniaxial compressive loading as a function of the face sheet thickness. Possible buckling modes comprise dimpling, face sheet yielding, and cell wall buckling. The aluminium honeycomb core of the considered sandwich has a constant thickness of h_c = 10 mm. The aluminium face sheets have thicknesses ranging from 0.01 to 0.5 mm.

For all cases a very good agreement between the analytical and the finite element results is found. A 1×2 unit cell is found to be sufficient in the considered cases, as long as global buckling is not relevant. Under uniaxial

Table 13.2. Honeycomb core/face sheet dimensions and material parameters.

Param.	Description	Al	Nomex	Unit
SW	cell size	4.7625	4.7625	mm
h_c	core height	10...50	10...50	mm
t_c	cell wall thickness	0.05	0.05	mm
E_c	Young's modulus of core	60000.	1600.	MPa
ν_c	Poisson ratio of core	0.3	0.3	-
σ_Y	wall yield stress of core	220.	220.	MPa
$\bar{E}_x$	eff. Young's mod. of core*	0.94	0.0251	MPa
$\bar{E}_y$	eff. Young's mod. of core*	0.94	0.0251	MPa
$\bar{E}_z$	eff. Young's mod. of core*	1679.	44.794	MPa
$\bar{G}_{yz}$	eff. shear mod. of core*	242.47	6.4658	MPa
$\bar{G}_{xz}$	eff. shear mod. of core*	363.83	9.7021	MPa
$\bar{G}_{xy}$	eff. shear mod. of core*	0.562	0.01499	MPa
$\bar{\nu}_{yz}$	eff. Poisson ratio of core*	1.7E-4	1.67E-4	-
$\bar{\nu}_{xz}$	eff. Poisson ratio of core*	1.7E-4	1.67E-4	-
$\bar{\nu}_{xy}$	eff. Poisson ratio of core*	0.999	0.999	-
k_x^{thick}	foundation stiffness*,†	364.569	9.728	MPa
k_y^{thick}	foundation stiffness*,†	364.217	9.719	MPa
Face sheet material:		Al	Steel	Unit
t_f	face sheet thickness	0.01...0.5	0.01...0.5	mm
E_f	face sheet Young's modulus	60000.	210000.	MPa
ν_f	face sheet Poisson's ratio	0.3	0.3	-
σ_Y	face sheet yield stress	220.	400.	MPa

* computed with spatially periodic finite element models
† based on[45]

compression of the sandwich (Fig. 13.20) dimpling and cell wall buckling are the critical failure modes, and wrinkling or face sheet yielding does not occur. Dimpling is the relevant mode for face sheet thicknesses lower than 0.1 mm, where a full dimpling wave develops over one honeycomb cell (point 'A' in Fig. 13.20). For cell wall buckling predictions a constant plate buckling factor of $k = 2.9$ has been found to be appropriate. One buckle half wave is visible within the buckled cell walls (point 'B' in Fig. 13.20).

A transition from dimpling to wrinkling (see Fig. 13.21) is observed in the case of Nomex cores. If these sandwiches are loaded by a bending moment M_{xx} the critical failure mode for very thin face sheets is dimpling (point 'A' in Fig. 13.21), whereas a combination of dimpling and wrinkling (point 'B' in Fig. 13.21) becomes relevant for moderately thick faces. The wrinkling wave length is periodic with respect to the basic cell (1×1 cell corresponding to Fig. 13.19, left). Thicker face sheets (point 'C' in Fig. 13.21)

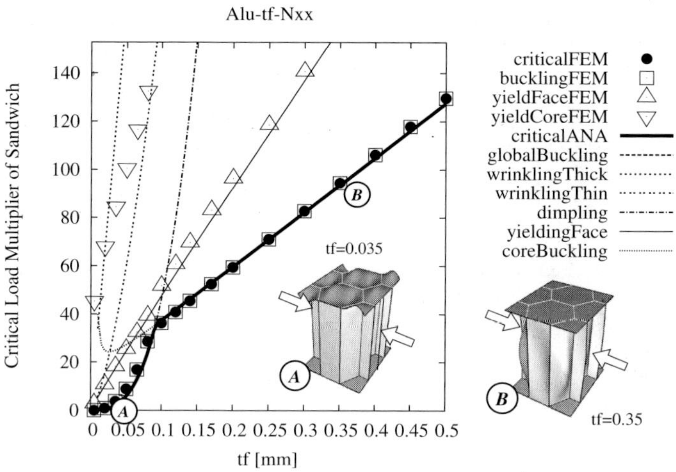

Fig. 13.20. Critical load multipliers λ over face sheet thickness of an aluminium (face sheet material)/aluminium (hexagonal core material) sandwich ($h_c = 10$ mm) under a uniaxial compressive load N_{xx}. Lines are analytical results, symbols are FE results. The thick line and the solid circles represent the relevant critical load multipliers.

show a typical wrinkling pattern with increasing wave lengths, where a deviation of the analytical (homogeneous core assumption) and finite element computations (cellular core) is visible. However, at these load levels face sheet yielding appears prior to local buckling.

Dimpling or wrinkling, as well as global buckling of thick and thin sandwiches is observed when steel face sheets are attached to a Nomex core (Fig. 13.22). The finite element models for these cases consist of 35 base cells in x-direction. For a constant core height ($h_c = 25$ mm) increasing face sheet thicknesses lead to dimpling for $t_f < 0.06$ mm. In the range of $0.06 < t_f < 0.2$ mm, wrinkling calculated with thick core assumptions (see Tab. 13.1) is the relevant predicted mode (point 'A' in Fig. 13.22), which is followed by wrinkling calculated with thin core assumptions (see Tab. 13.1). In the second case an antisymmetric wrinkling pattern (influence of opposite face sheet) is visible. Finally, global, transversal shear dominated buckling occurs for thicker faces and global dimensions as chosen in the considered examples (point 'C' in Fig. 13.22). The analytical predictions of the wrinkling stresses are rather poor (points 'A' and 'B' in Fig. 13.22). This is due to the fact that in these cases one half wave (HW) is running over only 0.73 (point 'A' in Fig. 13.22) and 1.25 (point 'B' in Fig. 13.22) honeycomb cells, respectively, which is less than the suggested

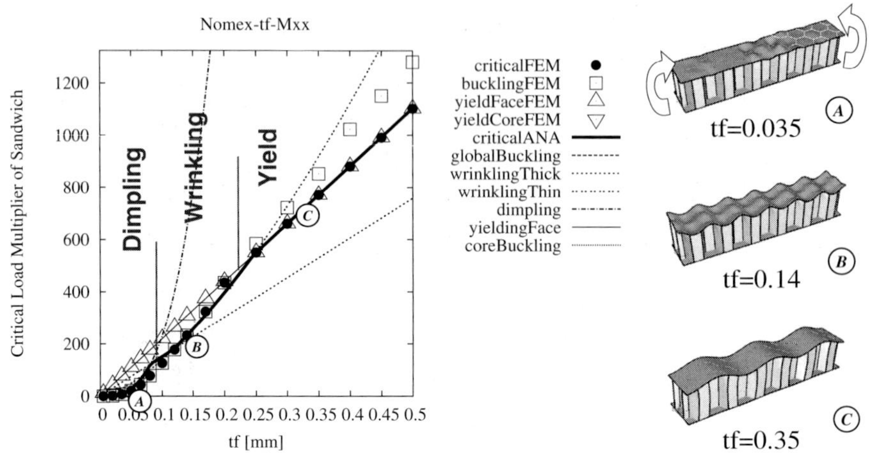

Fig. 13.21. Critical load multiplier λ over face sheet thickness of an aluminium (face sheet) - Nomex (hexagonal core) sandwich (h_c=10 mm) under uni-axial bending M_{xx} (left) as well as the corresponding buckling pattern (right).

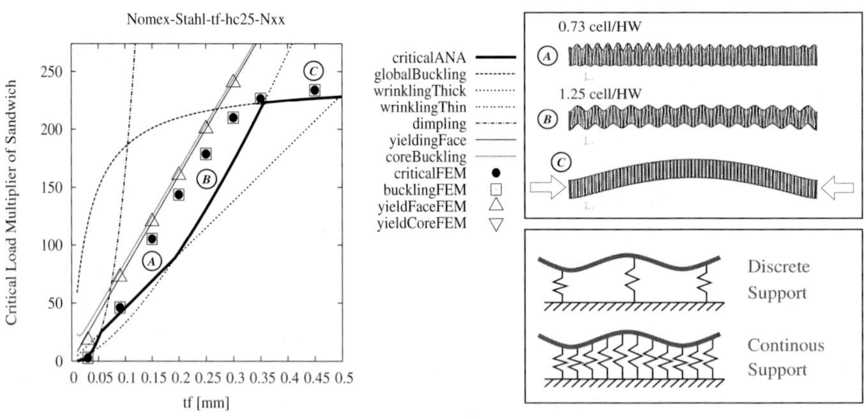

Fig. 13.22. Critical load multiplier λ over face sheet thickness of a steel (face sheet) - Nomex (hexagonal core) sandwich (h_c=25 mm) under uni-axial compression load N_{xx} (left), corresponding buckling pattern (right, top) and sketch of discrete/continuous support (right, bottom). Cells/HW ... cells per half wave.

value (see[54]) of at least 2 cells/HW for adequate analytical wrinkling predictions, i.e., in these cases the continuous support assumption (see Fig. 13.22, right/bottom) is not matched. However, a very good analytical prediction

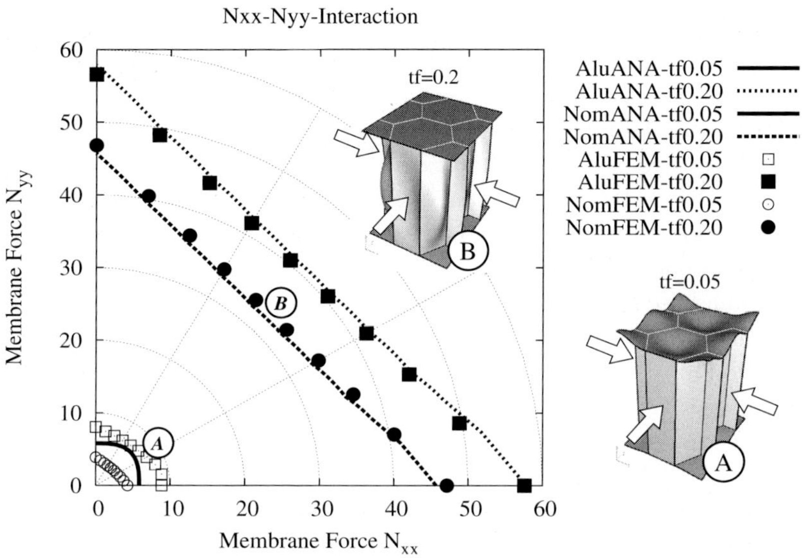

Fig. 13.23. Critical load combinations of sandwiches with hexagonal cores (h_c=10 mm) under bi-axial compression N_{xx} - N_{yy}. Dimpling loads are independent of core material.

of global buckling is visible. The buckling length for the analytical prediction of this case is chosen as the overall length of the finite element model in x-direction.

Results for local buckling loads of sandwiches under biaxial loading are shown in Fig. 13.23. Two core materials (aluminium and Nomex) as well as two different aluminium face sheet thicknesses are considered. Thicker face sheets (t_f = 0.2 mm) lead to cell wall buckling, thin face sheets (t_f = 0.05 mm) lead to dimpling. Problems with the suggested dimpling interaction formula become obvious. For example, the rather "soft" Nomex core leads to an overestimation of the dimpling loads (Fig. 13.23). In the case of an aluminium core the obtained agreement is better. The cell wall buckling predictions in Fig. 13.23, which were obtained with the analytical model, agree well with the finite element results.

Honeycomb Sandwiches — Summary

Dimpling can only be expected for very thin face sheets ($t_f < 0.1$ mm) in the presented settings. Generally, the agreement between analytical and numerical dimpling predictions is good. However, there are parameter regimes for which high deviations of the analytical dimpling predictions from FE results are visible (Fig. 13.23). This is because the dimpling formula does not take the stiffness of the different core materials into account. Cell wall buckling is only found in hexagonal cores with relatively thick face sheets, i.e., in the considered cases dimpling can only be expected for very thin face sheets ($t_{\mathrm{f}} > 0.05 \ldots 0.1$ mm). The stiffness loss of the core due to cell wall buckling can influence the postbuckling behaviour of the sandwich (next visible buckling mode, e.g., wrinkling) significantly. A cell wall buckling formula based on CLT stress computations, which was introduced in,[52] is well suited for in-plane loading of the sandwich. Plastic yielding of face sheets is found to be the relevant failure mode for sandwiches under bending and face sheet thicknesses being greater than 0.1 mm (aluminium) ... 0.25 mm (Nomex). Accurate membrane stresses in the face sheets can only be calculated if a CLT approach is used for the analytical computations, i.e., if the core stiffness is taken into account. Wrinkling and global transverse shear dominated buckling is observed for very stiff face sheets (steel) and very soft honeycomb cores (Nomex). The finite element models for these analyses are larger than the ones for dimpling predictions. If at least 10 wrinkling waves are realized within the FE model, the error which comes from the periodicity constraint (only discrete wavelength numbers possible within the FE model) becomes negligible. Poor correlation between analytical and finite element results (Fig. 13.22) can be explained by the over-simplified assumption of a continuous core support (homogenised core material) in the analytical model. For this assumption to be meaningful one wrinkling half wave has to go over more than two honeycomb cells in order to get reliable analytical wrinkling predictions. Otherwise, the core has to be considered as a discrete structure. The error which comes from the continuous core assumptions can be considerable. In the considered cases this error is up to 35%.

Summarising, it can be said that for a wide range of parameters good agreement between analytical and finite element results is found. Furthermore, finite element unit cell models appear to be well suited for accurate predictions of the critical sandwich failure loads and represent a cheap and reliable method (compared to experiments) for validation of new analytical design formulae as well as for investigations of new core geometries.

13.3.3 *Corrugated Board*

Introduction

In the previous section the focus of investigation was placed on structured honeycomb cores, the axis of extrusion of which is perpendicular to the mid-plane of the sandwich. In this section we present an example of a structured sandwich core with a geometry which can be generated by sweeping a straight line along a sinus curve in a plane that is perpendicular to the sandwich mid-plane. This kind of structured core is very well known from everyday experience, since it is heavily used in corrugated board for packaging purposes.

Corrugated paper or board is produced in large volumes, and, consequently, attracts considerable commercial and scientific interest. If corrugated board is used in packaging it is required to remain a stable container even under rough environmental conditions and to protect the enclosed goods from being crushed or being subjected to impact loads. Naturally, these requirements are closely tied to avoiding structural instabilities at all structural levels, from the sinusoidal flute and the face sheets or liner, to whole corrugated boards.

As far as local instabilities under in-plane loading are concerned, buckling of the face sheets, which are called 'liners' in the context of corrugated paper board, and the wavy core, which is called 'flute', are the principal modes of buckling and can be treated by unit cells as shown in Fig. 13.24. The same minimisation of the critical load intensity by variation of the size of the unit cell as described in Sec. 13.3.1 is required. Buckling of the liners is shown in Fig. 13.24(a), while a typical eigenmode of flute buckling is shown in Fig. 13.24(b). In both cases, the applied load is a compressive membrane force N_{11} aligned with the generatrix of the flute.[55]

Optimisation of Corrugated Board Geometry

If the geometric parameters of corrugated board are chosen such that liner and flute buckle independently as shown in Fig. 13.24 (a) and (b), the board may not be optimal with respect to its buckling resistance under a compressive membrane force N_{11}. Indeed, an attempt to design corrugated boards with improved buckling resistance as documented by Flatscher[55] produces a configuration that shows simultaneous buckling of liner and flute, compare Fig. 13.24 (c). The subject of the numerical optimisation scheme that was applied for obtaining this design was to minimise the

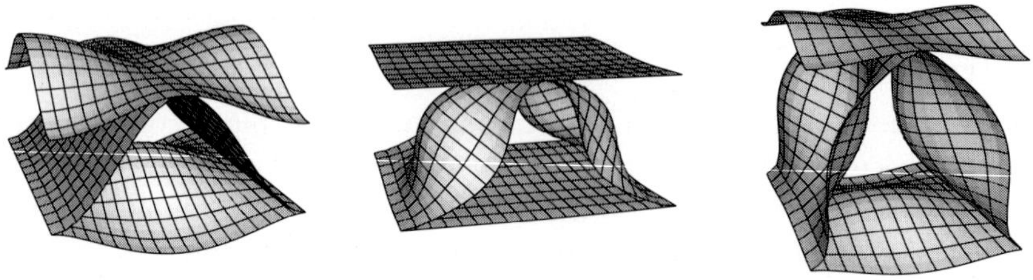

(a) (b) (c)

Fig. 13.24. Predicted local buckling modes of corrugated boards: buckling of face sheets (a), core/flute buckling (b) and simultaneous buckling.[55]

weight per plate area of a corrugated board under the constraints of a given initial elastic buckling load $N_{11}^* = 26.21$ N/mm and a required bending stiffness of $B = 17277$ Nmm with regard to bending about the 2-axis. For an isotropic linear elastic bulk material ($E = 8000$ MPa, $\nu = 0.3$) the resulting configuration can be described by the height of the flute, $h = l_3 = 4.2$ mm, the wavelength of the flute, $b = l_2 = 4.83$ mm, and the thicknesses of flute and liner, $t_F = 0.136$ mm and $t_L = 0.152$ mm, respectively.

Optimal or close-to optimal structures generally show little load-carrying capacity beyond their design loads and a high sensitivity to imperfections. It is, therefore, good practice to investigate the non-linear behaviour of such structures (compare, e.g.,[56]) in order to assess their robustness. For the corrugated board described above this was done in[55] for a plate with a length of $l_1 = 60$ mm and infinite width $l_2 = \infty$, which was accounted for by periodic boundary conditions in the 2-direction. This plate was constrained between rigid, freely rotating end plates and subjected to a compressive membrane force in the 1-direction. Small geometric imperfections affine to a superposition of the first three buckling eigenmodes were imposed to trigger elastic buckling. The corresponding predicted force-versus-displacement diagram can be found in Fig. 13.25. The critical compressive membrane force $N_{11}^* = 23.93$ N/mm as predicted by a linear buckling analysis is exceeded by the limit load by approximately 20%. Shortly before the limit load is reached, a periodic buckling pattern appears over the whole length of the model, see point ① in the diagram 13.25.

For a force-controlled process, reaching the limit load instantly triggers dynamic collapse, as the dashed line in Fig. 13.25 is meant to indi-

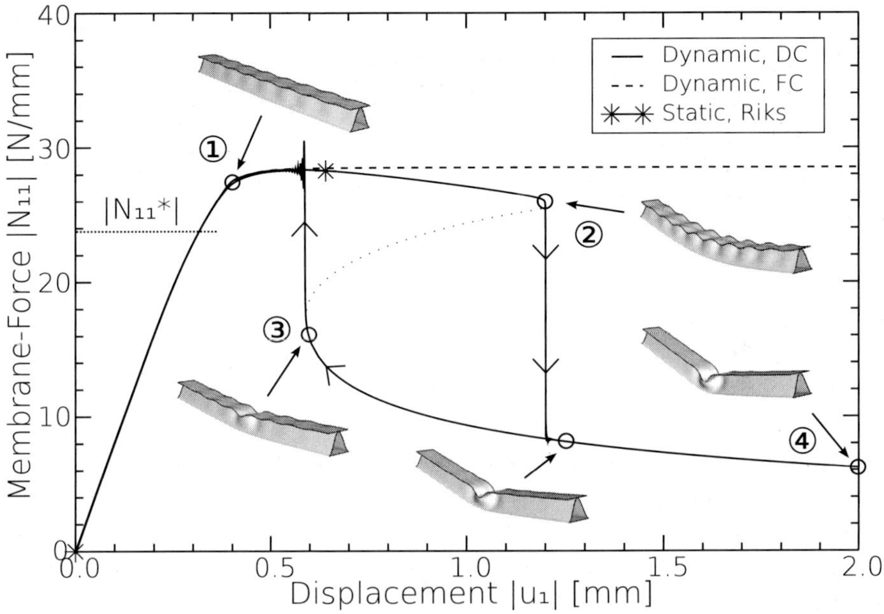

Fig. 13.25. Predicted membrane force versus end displacement diagram for the uniaxial in-plane compression of a corrugated board of infinite width.[55]

cate. Application of displacement controls or of an numerical arc-length incrementation method allows for the detection of the turning point ② in Fig. 13.25, which represents the occurrence of global buckling along with a large increase of the buckle wave amplitude. Upon passing the turning point, the plate exhibits snap-back behaviour that coincides with the localisation of the deformation in a single, central fold. Since the unstable snap-back equilibrium path implies a decrease of the applied load, the elastic buckling waves vanish but for the central one, see point ③ in Fig. 13.25.

As the deformation progresses the dominant fold on the upper liner of the corrugated board acts as a joint, whereas the elastically unloaded sections of the board straighten out, compare point ④ in Fig. 13.25. This unstable behaviour could only be predicted by means of a dynamic, displacement-controlled (DC) simulation.

The load-carrying capacity of the board is clearly exhausted as soon as the limit load is reached and global buckling sets in. The reserve against failure after the onset of buckling is smaller for an 'optimised' configuration than for a configuration that does not show simultaneous buckling of liner and flute. This behaviour certainly has to be considered whenever attempts to improve corrugated board structures are undertaken.

Corrugated Board — Summary

In this section a brief overview over the mechanics of corrugated board with respect to the occurrence of instabilities on the global (board) and local (liner and flute) level was given. The detailed representation of the periodic wavy core is still not feasible within simulations of whole corrugated containers. Appropriate homogenisation procedures allow for an accurate representation of the effective stiffness of the considered sandwich structure within the capabilities of shell theory; the prediction of global, elastic buckling loads is certainly possible with this information. As long as the mechanical postbuckling response does not include material nonlinearities, local buckling or other local failure mechanisms, it is furthermore possible to predict the geometrically nonlinear behaviour of corrugated boxes fairly accurately, compare.[57]

With regard to the prediction of the onset of local buckling from homogenised stress data, failure criteria operating on homogenised stress components were proposed, see, e.g.,.[58] A representation of the effects of local buckling on the effective properties of the board, presumably by means of stiffness degradation, certainly leaves room for further research.

Due to the wide variation in environmental and operating conditions of corrugated board applications, there are certainly more aspects to consider when the stability of corrugated board is to be assessed. The review of literature on the mechanical and hygroscopic performance and analysis of corrugated board presented in[59] may be found helpful in this respect.

13.4 Conclusions and Summary

Instabilities on the micro- and meso-structural levels of materials and sandwich structures may limit the load-carrying capacity of components in a significant way. This chapter made an attempt at presenting various relevant instability modes and analytical as well as numerical means for predicting them.

The relevant mechanisms for the onset and the propagation of instabilities in cellular materials were discussed. These mechanisms include buckling of slender struts in the case of open-cell sponges and buckling of thin cell-walls in the case of closed-cell foam. With respect to sandwich structures, a presentation of a multitude of possible modes of instability in sandwiches with structured or unstructured cores was given, among them buckling of face sheets including dimpling, as well as buckling of the walls of structured cores in honeycomb sandwiches and corrugated board.

13.5 References

1. A. G. Evans, J. W. Hutchinson, and M. F. Ashby, *Prog. Mat. Sci.* **43**, 171–221, (1999).
2. A. G. Evans, J. W. Hutchinson, N. A. Fleck, M. F. Ashby, and H. N. G. Wadley, *Prog. Mat. Sci.* **46**, 309–327, (2001).
3. L. J. Gibson, *J. Biomech.* **38**, 377–399, (2005).
4. U. Müller, W. Gindl, and A. Teischinger, *IAWA J.* **24**(2), 117–128, (2003).
5. L. J. Gibson, *Mater. Sci. Engng.* **A110**, 1–36, (1989).
6. L. J. Gibson and M. F. Ashby, *Cellular Solids: Structure and Properties.* (Cambridge University Press, Cambridge, New York, Melbourne, 1997), 2nd edition.
7. T. Daxner, *Multi-Scale Modeling and Simulation of Metallic Foams.* Fortschritt-Berichte VDI, Reihe 18, No. 285, (VDI Verlag, Düsseldorf, 2003).
8. L. Gong and S. Kyriakides, *Int. J. Sol. Struct.* **42**(5–6), 1381–1399, (2005).
9. H. X. Zhu, N. J. Mills, and J. F. Knott, *J. Mech. Phys. Solids.* **45**(11/12), 1875–1904, (1997).
10. S. Youssef, E. Maire, and R. Gaertner, *Acta Mater.* **53**(3), 719–730, (2005).
11. M. Laroussi, K. Sab, and A. Alaoui, *Int. J. Sol. Struct.* **39**(13–14), 3599–3623, (2002).
12. L. Gong, S. Kyriakides, and N. Triantafyllidis, *J. Mech. Phys. Solids.* **53**, 771–794, (2005).
13. J. Stampfl, A. Wöß, S. Seidler, H. Fouad, A. Pisaipan, F. Schwager, and R. Liska, *Macromol. Symp.* **217**, 99–107, (2004).
14. A. Wöß, J. Stampfl, and P. Fratzl, *Adv. Eng. Mat.* **6**(3), 134–138, (2004).
15. W. C. Hayes and D. R. Carter, *J. Biomed. Mater. Res.* **10**(4), 537–544, (1976).
16. P. R. Townsend, R. M. Rose, and E. L. Radin, *J. Biomech.* **8**(3–4), 199–201, (1975).
17. J. C. Runkle and J. Pugh, *Bull. Hosp. Joint. Dis.* **36**(1), 2–10, (1975).
18. N. Chan and K. E. Evans, *J. Mat. Sci.* **32**, 5725–5736, (1997).
19. Y. Wang and A. M. Cuitiño, *J. Mech. Phys. Solids.* **48**, 961–988, (2000).
20. V. Shulmeister, M. W. D. Van der Burg, E. Van der Giessen, and R. Marissen, *Mech. Mat.* **30**, 125–140, (1998).
21. T. C. Triantafillou, J. Zhang, T. L. Shercliff, L. J. Gibson, and M. F. Ashby, *Int. J. Mech. Sci.* **31**(9), 665–678, (1989).
22. M. H. Luxner, J. Stampfl, and H. E. Pettermann, *J. Mat. Sci.* **44**(20), 5859–5866, (2005).

23. M. Luxner, J. Stampfl, A. Woesz, P. Fratzl, and H. E. Pettermann. In *Proceedings of the III European Conference on Computational Mechanics (ECCM-2006)*, Lisbon, (2006).
24. M. H. Luxner. *Modeling and Simulation of Highly Porous Open Cell Structures—Elasto-Plasticity and Localization versus Disorder and Defects.* PhD thesis, Vienna University of Technology, Vienna (June, 2006).
25. M. H. Luxner, J. Stampfl, and H. E. Pettermann, *Int. J. Sol. Struct.* **44**, 2990–3003, (2007).
26. J. Banhart, *Prog. Mat. Sci.* **46**, 559–632, (2001).
27. B. Kriszt, B. Foroughi, K. Faure, and H. P. Degischer. In eds. J. Banhart, M. F. Ashby, and N. A. Fleck, *Metal Foams and Porous Metal Structures*, pp. 241–246, Bremen, (1999). Verlag MIT.
28. K. Y. G. McCullough, N. A. Fleck, and M. F. Ashby, *Acta Mater.* **47**(8), 2323–2330, (1999).
29. P. H. Thornton and C. L. Magee, *Metall. Trans. A.* **6A**, 1253–1263, (1975).
30. A.-F. Bastawros, H. Bart-Smith, and A. G. Evans, *J. Mech. Phys. Solids.* **48**, 301–322, (2000).
31. N. J. Mills and H. X. Zhu, *J. Mech. Phys. Solids.* **47**, 669–695, (1999).
32. S. P. Santosa and T. Wierzbicki, *J. Mech. Phys. Solids.* **46**(4), 645–669, (1998).
33. A. E. Simone and L. J. Gibson, *Acta Mater.* **46**(6), 3929–3935, (1998).
34. M. Reinfried, U. Waag, G. Stephani, and B. Kieback. In eds. R. Singer, C. Körner, and V. Altstädt, *Cellular Metals and Polymers 2004*, pp. 93–98, Zürich, (2004). Trans Tech Publications.
35. O. Friedl, C. Motz, J. Färber, M. Stoiber, and R. Pippan. In eds. H. Nakajima and N. Kanetake, *Porous Metals and Metal Foaming Technology*, pp. 479–482, Sendai, (2006). The Japan Institute of Metals.
36. T. Dillard, F. N'Guyen, E. Maire, L. Salvo, S. Forest, Y. Bienvenu, J.-D. Bartout, M. Croset, R. Dendievel, and P. Cloetens, *Phil. Mag.* **85**(19), 2147–2175, (2005).
37. V. Paserin, S. Marcunson, J. Shu, and D. S. Wilkinson, *Adv. Eng. Mat.* **6** (6), 454–459, (2004).
38. T. Dillard. *Caractérisation et simulation numérique du comportement mécanique des mousses de nickel: morphologie tridimensionnelle, réponse élastoplastique et rupture.* PhD thesis, Ecole des Mines de Paris, Paris (March, 2004).
39. T. Daxner, R. D. Bitsche, and H. J. Böhm. In eds. T. Chandra, K. Tsuzaki, M. Militzer, and C. Ravindran, *THERMEC 2006*, pp. 1857–1862, Stafa-Zurich, (2006). Trans Tech Publications Ltd.
40. D. Weaire and R. Phelan, *Phil. Mag. Letter.* **69**, 107–110, (1994).
41. D. Weaire, Ed., *The Kelvin Problem.* (Taylor & Francis, London, 1996).
42. K. Brakke, *Exp. Math.* **1**(2), 141–165, (1992).
43. F. Plantema, *Sandwich Construction.* (John Wiley and Sons, Inc., New York, USA, 1966).
44. D. Zenkert, *An Introduction to Sandwich Construction.* (EMAS Ltd., Solihull, UK, 1995).

45. W. K. Vonach and F. Rammerstorfer, *Struct. Eng. Mech.* **12**, 363–376, (2001).
46. W. K. Vonach, *A General Solution to the Wrinkling Problem of Sandwiches.* Fortschritt-Berichte VDI Reihe 18 Nr. 268, (VDI-Verlag, Düsseldorf, Germany, 2001).
47. F. G. Rammerstorfer and W. K. Vonach. In ed. K. Bathe, *Computational Fluid and Solid Mechanics 2003*, pp. 727–731, Amsterdam, (2003). Elsevier.
48. T. Suzuki and P. K. L. Yu, *J. Mech. Phys. Solids.* **46**(1), 115–138, (1998).
49. B. K. Hadi, *Comput. Struct.* **53**(4), 477–482, (2001).
50. S. Yusuff, *J. R. Aeronaut. Soc.* **59**, 30–36, (1955).
51. C. Norris and W. Kommers, *FPL Report.* **1817**, (1950).
52. D. H. Pahr and F. G. Rammerstorfer, *CMES - Comp. Model. Eng.* **12**, 229–242, (2006).
53. R. Ley, W. Lin, and M. Uy. Facesheet wrinkling in sandwich structures (Contractor Report CR-1999-208994). Technical report, NASA, (1999).
54. L. Lamberti, S. Venkataraman, R. Haftka, and T. Johnson, *Int. J. Numer. Meth. Eng.* **57**, 1351–1380, (2003).
55. T. Flatscher. Modellierung der Steifigkeit und Stabilität von Wellpappe. Diploma thesis, Vienna University of Technology, Vienna, (2006).
56. T. Nordstrand, *Compos. Struct.* **30**, 441–451, (1995).
57. M. E. Biancolini and C. Brutti, *Pack. Tech. Sci.* **16**, 47–60, (2003).
58. P. Patel, T. Nordstrand, and L. A. Carlsson, *Compos. Struct.* **39**(1–2), 93–110, (1997).
59. S.-L. Gelin. Mechanical and hygroscopic performance and analysis of corrugated board - a literature review. Report 384, KTH Solid Mechanics, Stockholm, Sweden, (2004).

INDEX

Computational and Experimental Methods in Structures